3rd International Conference on Internet of Things, Artificial Intelligence and Mechanical Automation (IoTAIMA 2022)

Journal of Physics: Conference Series Volume 2365

Shanghai, China
22-24 July 2022

ISBN: 978-1-7138-6473-8
ISSN: 1742-6588

Printed from e-media with permission by:

Curran Associates, Inc.
57 Morehouse Lane
Red Hook, NY 12571

Some format issues inherent in the e-media version may also appear in this print version.

Institute of Physics
Dirac House, Temple Back
Bristol BS1 6BE UK

Phone: 44 1 17 929 7481
Fax: 44 1 17 920 0979

techtracking@iop.org

Additional copies of this publication are available from:

Curran Associates, Inc.
57 Morehouse Lane
Red Hook, NY 12571 USA
Phone: 845-758-0400
Fax: 845-758-2633
Email: curran@proceedings.com
Web: www.proceedings.com

TABLE OF CONTENTS

Journal of Physics: Conference Series **2365** (2022) 011001 doi:10.1088/1742-6596/2365/1/011001

Preface

The 2022 3rd International Conference on Internet of Things, Artificial Intelligence and Mechanical Automation (IoTAIMA 2022) was held virtually on July 22-24, 2022 in Shanghai, China. The conference provided a communication forum for the international experts and researchers in the fields of Internet of Things, Artificial Intelligence and Mechanical Automation to share research achievements, explore solutions to research challenges, encourage international collaborations, and boost development of science and technology.

In the conference, we were greatly honored to have Prof. Ke Zhang from Shanghai Institute of Technology, China to serve as our Conference Chair. 120 individuals around the world took part in the conference. Divided into three parts, the conference agenda included keynote speeches, oral presentations and online Q&A discussion. Primarily, the keynote speakers were each allocated 30-45 minutes to address their speeches. Then in the next part, oral presentations, the excellent papers we had selected were presented by their authors one by one.

Four distinguished professors were invited to hold keynote speeches during the conference. Among them, Prof. Xiaodong Liu from Dalian University of Technology performed a thought-provoking speech on Knowledge Discovery and Semantic Learning in the Framework of AFS Theory. Under the background that AFS theory gains increasing interest from various research fields, this survey aimed to report the current developments of theoretical research and practical advances on AFS theory. Firstly some notions and foundations of AFS theory with an illustrative example were reviewed and then various extensions of AFS theory for knowledge discovery were analyzed. And then Prof. Yudong Zhang from University of Leicester addressed a speech: Artificial Intelligence Theories and Methods for COVID-19 Recognition. In this research, CT scan, a medical imaging technique used in radiology to get detailed images of the body noninvasively for diagnostic purposes, was applied to discuss the recent AI theories and methods in chest CT-based COVID-19 diagnosis. Two other chest-related diseases: secondary pulmonary tuberculosis and community-acquired pneumonia, were covered in the study. Their insightful speeches had triggered heated discussion. And every participant praised this conference for disseminating useful and insightful knowledge.

We are glad to share with you that we received lots of submissions from the conference and we selected a bunch of high-quality papers and compiled them into the proceedings after rigorously reviewing them. These papers feature but are not limited to the following topics: Radio Frequency Identification, New Sensing Technology, Intelligent Perception Systems, Unmanned System Control Technology, Automation etc. All the papers have been checked through rigorous review and processes to meet the requirements of publication.

We would like to acknowledge all of those who supported IoTAIMA 2022 and made it a great success. Particularly, we would like to thank the Journal of Physics: Conference Series, for all the efforts of its colleagues in publishing this paper volume. We sincerely hope that the IoTAIMA 2022 turned out to be a forum for excellent discussions that enable new ideas to come about, promoting collaborative research.

Committee of IoTAIMA 2022

Committee member

Conference Chair
Prof. Ke Zhang, Vice dean of the School of Mechanical Engineering, Shanghai Institute of Technology, China

Organizing Committee Chair
Prof. Y.Q. He, Tongji University, China

Organizing Committee
Prof. Jixiang Lu, Northwestern Polytechical University, China
Prof. Enchang Sun, Beijing University of Technology, China
Prof. Zhe Chen, Dalian University of Technology, China
Prof. Xiaolin Jia, Southwest University of Science and Technology, China
Prof. Hailang Pan, Nanjing University of Science and Technology, China

International Technical Committee
Assoc. Prof. Jing Zhao, Xi'an Mingde Institute of Technology, China
Prof. Jinping Liu, Hunan Normal University,China
Prof. Heming Jia, Sanming University, China
Prof. Ying Chen, Jiangnan University, China
Prof. Tianmin Guan, Dalian Jiaotong University, China
Prof. Shuihua Han, Xiamen University, China
Prof. Xiaodong Liu, Dalian University of Technology, China
Prof. Xueliang Ping, Jiangnan University, China
Prof. Zhiying Yang, Information Engineering College of Shanghai Maritime University, China
Prof. Li Guo, Hunan University, China
Prof. Qixin Cao, Shanghai Jiao Tong University, China
Prof. Yajun Liu, South China University of Technology, China
Prof. Hongshan Yang, Transwarp Technology (Shanghai) Co., Ltd, China
Prof. Ghasem Mirjalily, Yazd University, IRAN
Prof. Rza Bashirov, Eastern Mediterranean University, North Cyprus
Prof. Emin BAYRAKTAR, Supmeca-PARIS, France
Prof. Jianmei He, Kogakuin University, Japan
Prof. Wei Jiang, Dalian University of Technology, China
Prof. Xueye Chen, Liaoning University of Technology, China
Prof./Dr. Weidong Zhu, University of Maryland, USA
A. Prof. Haoran Feng, Emergen Co., Ltd., China
A. Prof. Bo Wang, Shanghai Institute of Technology, China
A. Prof. Hongfang Qi, Wuhan HuaXia University of Technology School, China
A. Prof. Guanglei Wu, Dalian University of Technology, China
A. Prof. Gang Chen, Nanjing University of Science and Technology, China
A. Prof. Wenzhong Nie, Shanghai Institute of Technology, China
A. Prof. Zhao Zhang, University of Science and Technology Liaoning, China
A. Prof. Canguo Zhang, Hebei University of Architecture, China
A. Prof. Chandra Shekhar, Birla Institute of Technology & Science Pilani, India
A. Prof. Hongfang Qi, Wuhan HuaXia University of Technology, China
A. Prof. Şuayip Yüzbaşı, Akdeniz University, Turkey
Dr. Zhengming Gao, Jingchu university of technology, China
Dr. Shweta Upadhyaya, Amity University, India
SN ENGR Tian Cao, Nanjing Xiaozhuang University, China
ENGR Jianyin Tang, Hubei Business College, China

Journal of Physics: Conference Series **2365** (2022) 011002 doi:10.1088/1742-6596/2365/1/011002

Peer Review Statement

All papers published in this volume have been reviewed through processes administered by the Editors. Reviews were conducted by expert referees to the professional and scientific standards expected of a proceedings journal published by IOP Publishing Publishing.

- **Type of peer review:** Single Anonymous
- **Conference submission management system:** Morressier
- **Number of submissions received:** 102
- **Number of submissions sent for review:** 98
- **Number of submissions accepted:** 55
- **Acceptance Rate (Submissions Accepted / Submissions Received × 100):** 53.9
- **Average number of reviews per paper:** 2
- **Total number of reviewers involved:** 22
- **Contact person for queries:**
  Name: Siting(Cindy) Chen
  Email: cindy.chen@gsrassn.org
  Affiliation: Global Scientific Research Association

Journal of Physics: Conference Series **2365** (2022) 012001 doi:10.1088/1742-6596/2365/1/012001

Dynamic stability study of velocity time lag feedback control of axially moving viscoelastic beams

ChangHao Yan, YouQi Tang, Bo Wang

Shanghai Institute of Technology, College of Mechanical Engineering, Shanghai 201418, China

ych2021jy@163.com

Abstract: In engineering fields, such as in deep space processing, paper industry, steel making industry and aerospace field etc., transverse vibration caused by axial motion not only affects the service life of the structure, but also may cause major accidents. Therefore, it is necessary to control the vibration. However, time lag is unavoidable in the process of vibration control because of the time lag in the signal transmission. On the one hand, it not only affects the robustness of the system and produces a very complex dynamic response; on the other hand, the amount of time lag can be used as a parameter to control the vibration. In this paper, a new model of time lag feedback is established, considering the effect of mechanical properties of the linear vibration model of an axially moving viscoelastic beam. The direct multi-scale analysis is applied to analyze the parameter stability of the beam with and without the addition of time lag feedback control model. The subharmonic parameters are obtained by solvability conditions and the Routh-Hurwitz criterion.

1. Introduction

Active control of time-lagged nonlinear systems has become a hot research topic in recent years and has become a key technology for solving large beam vibrations. With the improvement of the speed and control stability aspects of the control, the time lag phenomenon in the controller has become a non-negligible problem. For example, Zhao Tong et al[1] analyzed the two problems of the maximum amount of time lag stability and the optimal ratio of control rate by the control variable method. In addition Hu[2,3] conducted an in-depth study with the time lag problem in vibration control, they analyzed for the damping in active control and analyzed the critical conditions for time lag instability by classifying the damping into three cases according to a certain condition. The asymptotic jump of the joint time-lag-free feedback system is analyzed to make the stability region in the feedback system. Udwadia et al[4] found that if the time lag is neglected it causes a decrease in the stability of the system and if it is introduced it improves the robustness of the system. In 1997, Hassard[5] extended Routh-Hurwitz to time-lag systems, and the corresponding methods to study the instability of time-lag differential bins are: Rouche's theorem; Lyapounov's generalized function[6], etc. There has been a great progress in the research of active control methods for linear systems with time lag, such as Cai[7] from Shanghai Jiao Tong University, who used modal space control in independent space and the modal control law was designed using discrete variable structure control.

2. Mathematical model for velocity time lag feedback control of axially moving viscoelastic beams

The physical model of an axial motion time-lag feedback controlled Euler beam with viscoelastic coefficient α, density $\rho(z)$, cross-sectional area A (height h and width b), Young's modulus E, shear modulus G, moment of inertia I, time lag τ, initial axial tension P, and time-dependent velocity $\Gamma(t)$ between two simple supports separated by length l for axial motion is shown in Figure 1, assuming that the cross-section of the Euler is in the horizontal plane when it is deformed and bent.

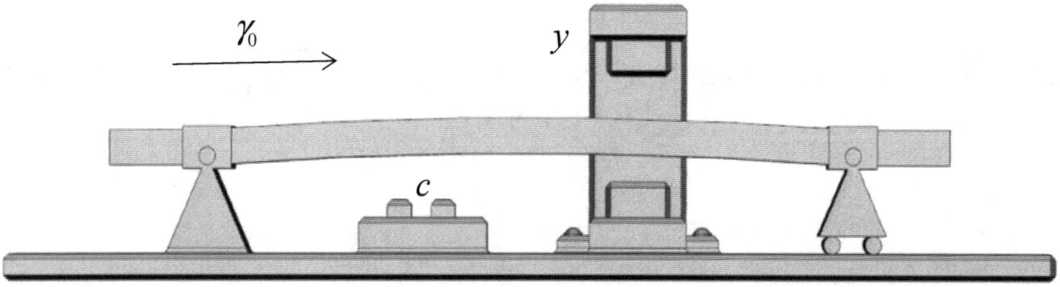

Figure 1. The physical model of an axially moving viscoelastic beam with velocity time delay feedback

The dimensionless governing equation is

$$\begin{aligned} & v_{,tt} + 2\gamma_0 v_{,xt} + \left[(1-\eta)\gamma_0^2 - 1 \right] v_{,xx} + k_f^2 v_{,xxxx} \\ & = -\varepsilon \left\{ \alpha \left(v_{,xxxxt} + \gamma_0 v_{,xxxxx} \right) + 2\gamma_1 \sin(\omega t) \left[v_{,xt} + (1-\eta)\gamma_0 v_{,xx} \right] \right. \\ & \quad \left. + (1-x)\omega\gamma_1 \cos(\omega t) v_{,xx} + g(v_{,t} - o v_{\tau,t}) \right\} - O\left(\varepsilon^2\right) \end{aligned} \tag{1}$$

This equation is a dimensionless equation, where "," denotes partial differential. t and x are the transverse displacements at time t and x, respectively, where α is the viscoelastic coefficient, k_f is the stiffness, τ is the time lag, η is the support stiffness, where ε is also a dimensionless parameter, characterizing the viscoelastic coefficient of the Euler beam as a small quantity, g is the feedback gain coefficient, o is the weighting factor; γ_0 , γ_1 and ω are dimensionless parameters, which are expressed as the average velocity, amplitude and frequency of axial velocity pulsation, respectively.

3. Multi-scale analysis

The multiscale method is applied to find a consistent and valid solution of the form of equation (1) as

$$v(x,t;\varepsilon) = v_0(x,T_0,T_1) + \varepsilon v_1(x,T_0,T_1) + O\left(\varepsilon^2\right) \tag{2}$$

$$v_\tau(x,\tau;\varepsilon) = v_{\tau 0}(x,T_0-\tau,T_1) + \varepsilon v_{\tau 1}(x,T_0-\tau,T_1) + O\left(\varepsilon^2\right) \tag{3}$$

$T_0=t$ is the fast time scale due to some inherent frequency of the moving beam; $T_1=\varepsilon t$ is the slow time scale of the amplitude and phase slow change induced by the small perturbation of the velocity of the viscoelastic coefficient.

Bringing equations (2) and (3) into equation (1), then separating ε^0 and ε^1 of different orders and making their coefficients equal gives

$$\varepsilon^0 : \; v_{0,T_0T_0} + 2\gamma_0 v_{0,xT_0} + \left[(1-\eta)\gamma_0^2 - 1 \right] v_{0,xx} + k_f^2 v_{0,xxxx} + v_{\tau 0} = 0 \tag{4}$$

$$\begin{aligned}\varepsilon^1:\ & v_{1,T_0T_0}+2\gamma_0 v_{1,xT_0}+\left[(1-\eta)\gamma_0^2-1\right]v_{1,xx}+k_f^2 v_{1,xxxx}+v_{1\tau,T_0}= \\ & -\left\{2\left(v_{0,T_0T_1}+\gamma_0 v_{0,xT_1}\right)+\alpha\left(v_{0,xxxxT_0}+\gamma_0 v_{0,xxxxx}\right)+\right. \\ & \left.(1-x)\omega\gamma_1\cos(\omega t)v_{0,xx}+v_{0\tau,T_1}+2\gamma_1\sin(\omega t)\left[v_{0,xT_0}+(1-\eta)\gamma_0 v_{0,xx}\right]\right\}\end{aligned} \tag{5}$$

Assuming that its solution has the form of separated variables

$$v_0(x,T_0,T_1)=\sum_{n=1}^{\infty}\left[A_n(T_1)\varphi_n(x)\mathrm{e}^{\mathrm{i}\omega_n T_0}+cc\right] \tag{6}$$

$$v_\tau(x,T_0-\tau;\varepsilon)=\sum_{n=1}^{\infty}\left[A_n(\tau_1)\varphi_n\mathrm{e}^{\mathrm{i}\omega_n T_0-\tau_0}+cc\right] \tag{7}$$

A_n is the complex function to be determined; φ_n is the nth-order modal function of the linearly derived system transverse displacement, ω_n is the nth-order intrinsic frequency. Substituting equation (6) into (4) and eliminating the non-zero exponential term, derive

$$-\omega_n^2\varphi_n+2\mathrm{i}\,\omega_n\gamma_0\varphi_{n,x}+\left[(1-\eta)\gamma_0^2-1\right]\varphi_{n,xx}+k_f^2\varphi_{n,xxxx}=0 \tag{8}$$

The form of the solution of equation (8) can be expressed as

$$\varphi_n(x)=C_{1n}\left(e^{i\beta_{1n}x}+C_{2n}e^{i\beta_{2n}x}+C_{3n}e^{i\beta_{3n}x}+C_{4n}e^{i\beta_{4n}x}\right) \tag{9}$$

The solution of the system (9) of fourth-order linear linear complex coefficient ordinary differential equations can be written as

$$\begin{aligned}&\left[\mathrm{e}^{\mathrm{i}(\beta_{n1}+\beta_{n2})}+\mathrm{e}^{\mathrm{i}(\beta_{n3}+\beta_{n4})}\right]\left(\beta_{n1}^2-\beta_{n2}^2\right)\left(\beta_{n3}^2-\beta_{n4}^2\right)+\left[\mathrm{e}^{\mathrm{i}(\beta_{n1}+\beta_{n3})}+\mathrm{e}^{\mathrm{i}(\beta_{n2}+\beta_{n4})}\right]\cdot \\ &\left(\beta_{n3}^2-\beta_{n1}^2\right)\left(\beta_{n2}^2-\beta_{n4}^2\right)+\left[\mathrm{e}^{\mathrm{i}(\beta_{n2}+\beta_{n3})}+\mathrm{e}^{\mathrm{i}(\beta_{n1}+\beta_{n4})}\right]\left(\beta_{n2}^2-\beta_{n3}^2\right)\left(\beta_{n1}^2-\beta_{n4}^2\right)=0\end{aligned} \tag{10}$$

To study the parametric resonance, the solution of equation (9) can be written as

$$v_0(x,T_0,T_1)=\varphi_m(x)A_m(T_1)\mathrm{e}^{\mathrm{i}\omega_m T_0}+\varphi_n(x)A_n(T_1)\mathrm{e}^{\mathrm{i}\omega_n T_0}+bb \tag{11}$$

$$v_{\tau 0}(x,T_0,T_1)=\varphi_m(x)A_m(T_1)\mathrm{e}^{\mathrm{i}\omega_m(T_0-\tau)}+\varphi_n(x)A_n(T_1)\mathrm{e}^{\mathrm{i}\omega_n(T_0-\tau)}+bb \tag{12}$$

where the conjugate complex of each term is denoted by bb, ω_m and ω_n denote the intrinsic frequencies of the mth and nth order linearly derived systems, respectively.

4. Second harmonic parameter resonance

Resonance also occurs when the pulsation frequency ω of the velocity approaches twice the intrinsic frequency of some order of the linearly derived system. Now we introduce the detuning parameter σ, which represents the pulsation frequency ω varying around $2\omega_m$.

$$\omega=2\omega_m+\varepsilon\sigma \tag{13}$$

Substituting equation (13) into (12), according to the nature of the inner product and applying the solvability condition, the collation is obtained

$$\dot{A}_m+\gamma_1 c_m\overline{A}_m\,\mathrm{e}^{\mathrm{i}\sigma T_1}+(\alpha d_m+g e_m)A_m=0 \tag{14}$$

Introducing transformation

$$A_m = \left[p(T_1) + \mathrm{i} q(T_1) \right] \mathrm{e}^{0.5\mathrm{i}\sigma T_1} \tag{15}$$

where p and q are real functions of T_1. Substituting equation (15) into (14) and separating the real and imaginary parts of the result, derive

$$\begin{aligned} \dot{p} &= -\left[\alpha d_m + \gamma_1 \mathrm{Re}(c_m) \right] p + \left[0.5\sigma - \gamma_1 \mathrm{Im}(c_m) \right] q \\ \dot{q} &= -\left[0.5\sigma + \gamma_1 \mathrm{Im}(c_m) \right] p - \left[\alpha d_m - \gamma_1 \mathrm{Re}(c_m) \right] q \end{aligned} \tag{16}$$

Directly calculate the determinant of the characteristic equation of its coefficient matrix

$$\lambda^2 + (\alpha d_m + g e_m)\lambda + \left(0.25\sigma^2 + \alpha^2 d_m^2 - \gamma_1^2 |c_m|^2\right) = 0 \tag{17}$$

According to the Rauschlewitz stability criterion, the conditions for the stability of the zero solution of equation (17)

$$|\gamma_1| < \sqrt{\frac{0.25\sigma^2 + \alpha^2 d_m^2}{|c_m|^2}} \tag{18}$$

When time lag feedback control is not introduced, the parameters η=0.5, k_f=0.8, g=0, o=0, time lag τ=0 and γ_0=2. The effect of different viscoelastic coefficients on the resonant instability region of the 2nd order harmonic parameter without the introduction of time-lag feedback control can be seen in Figure 2 (a), which increases with the increase of the viscoelastic coefficient. In the case of a given γ_1, the instability region becomes smaller. In the subharmonic resonance as well as the combined resonance, the law of the instability region affected by viscoelasticity is the same, with the increase of the subharmonic resonance parameters, the more obvious the effect of the viscoelastic coefficient. As shown in Figure 2 (b), the effect of different stiffnesses on the resonant instability region of the 2nd order harmonic parameter without the introduction of time lag feedback control can be seen in the image. When σ is given, it can be seen that the stability region becomes smaller when larger, while when γ_1 is fixed it can be seen that the instability range gradually increases. The subharmonic resonance as well as the combined common, the stiffness k_f shows a negative correlation with the instability region.

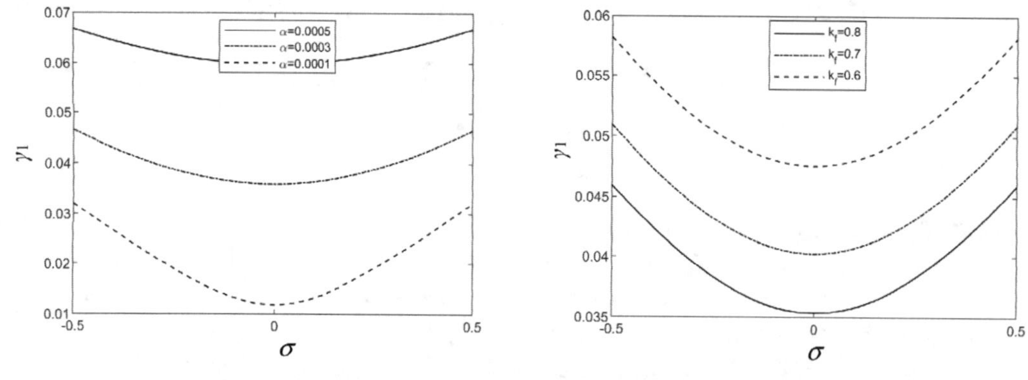

(a) Different viscoelastic coefficients (b) Different stiffness coefficients

Figure 2. Effect of different parameters on the resonant instability region of the 2nd order harmonic parameter without the introduction of time lag feedback control

Introducing time lag feedback control, given initial parameters η=0.5, k_f =0.8, γ_0=2, g=0.2, o=1.5, time lag τ=0.2 and γ_0=2, and η=0.5, k_f =0.8, g=0, o=0, time lag τ=0 and γ_0=2， It can be seen that the stability interval increases in comparison to the case without adding time lag, and the same conclusion is reached for the second order. The stability interval becomes significantly larger than the stability

Journal of Physics: Conference Series **2365** (2022) 012001 doi:10.1088/1742-6596/2365/1/012001

interval without increasing time lag.

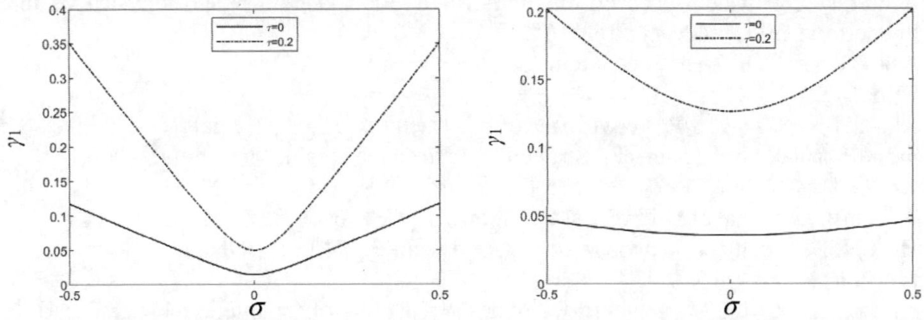

(a) 1st order harmonic parameter resonance (b) 2nd order harmonic parameter resonance

Figure 3. Comparison of the changes in the stability domain with and without the introduction of time lag feedback control

When time lag feedback control is not introduced, set the parameter η=0.5, α=0.0003, γ_0=1.5, g=0, o=1, time lag τ=0, with the introduction of time lag feedback control, set the parameter η=0.5, α=0.0003, γ_0=1.5, g=0.2, o =1.5, time lag τ=0.2. The solid line indicates that the time lag feedback control is not introduced, and the dashed line indicates that the time lag feedback control is introduced. It can be seen that the stability region threshold of the first-order system is increased when the time lag feedback control is introduced, and the second-order system can also improve the stability interval to a certain extent although there is an intersection point.

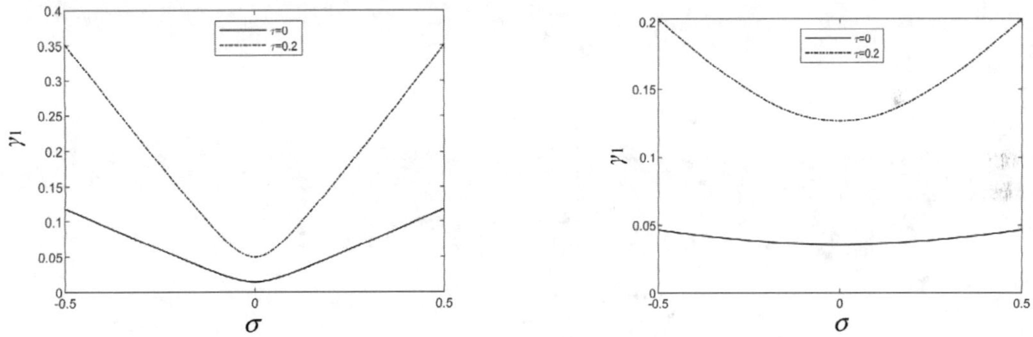

(a) 1st order harmonic parameter resonance (b) 2nd order harmonic parameter resonance

Figure 4. Comparison of the changes in the stability domain after the introduction of time lag feedback control

The above parameters are given by the analysis based on the Galerkin method and the modal function method.

5. Conclusion

In this paper, the modal functions of a viscoelastic Euler beam with axial motion functionalities under time-lag feedback control are studied, focusing on the subharmonic parameter resonance when the pulsation frequency of the velocity approaches two times the intrinsic frequency of a certain order of the linearly derived system. In subharmonic resonance, for a given detuning parameter σ, larger viscoelastic coefficients α, both without and with increased time-lag feedback control, increase the stability threshold, and the viscoelastic coefficient has a greater effect on higher order frequencies. Keeping other parameters constant and changing the support stiffness, the stiffness shows a negative correlation with the instability region.

References

[1] Just W. On the eigenvalue spectrum for time-delayed Floquet problems[J]. Physica D Nonlinear Phenomena, 2000, 142(1-2): 153-165.
[2] Kopfova J, Kopf T Differential equations, hysteresis, and time delay. Z Angew Math Phys, 2002, 53:676–691.
[3] Udwadia F E, Phohomsiri P. Active control of structures using time delayed positive feedback proportional control designs[J]. Structural Control and Health Monitoring, 2006.
[4] Hassard B D. Counting Roots of the Characteristic Equation for Linear Delay-Differential Systems[J]. Journal of Differential Equations, 1997, 136(2):222-235.
[5] Nayfeh A H. Perturbation methods in nonlinear dynamics[M] Lecture Notes in Physics. Springer Berlin Heidelberg, 1986.
[6] Masound Z, Daqaq M F, Nayeeh N H. Pendulation reduction of small telescopic cranes[J]. Journal of Vibration and Control, 2004,10;1167.
[7] Cai GP and Huang JZ, Yang SX. An optimal control method for linear systems with time delay. Journal of Computers and Structures, 2003, 81(15): 1539-1546.

Inverse kinematics solution of CMOR short carrier manipulator

Heng Zhang[1], Xiang Dong[1*], Ning Fang[1], Jun Zhang[2], Yong Cheng[2]

[1]School of Electrical Engineering and Automation, Anhui University, Hefei, 230000, Anhui Province, China

[2]Institute of Plasma Physics, Hefei Institute of Physical Science, Chinese Academy of Sciences, Hefei, 230000, Anhui Province, China

*xdong@ahu.edu.cn

Abstract: CMOR is a heavy-duty manipulator system applied to the maintenance of fusion reactor components. On the basis of fully considering the configuration characteristics of the manipulator, the geometric solution based on coordinate transformation combined with the analytical method is used to solve the inverse kinematics of the manipulator. First, the modified D-H method is used to establish the link coordinate system model of the CMOR carrier manipulator, and the forward kinematics is deduced. Then, based on the end pose ${}^{0}T_{7}$ and the geometric structure of the manipulator, the movement range of the redundant degree of freedom d1 is deduced, and the remaining joint angles are obtained through the mapping solution and verified in MATLAB. The simulation results show that the method has high precision, reduces the overall calculation amount, and is conducive to the real-time and accurate maintenance movement of the carrier manipulator, which is of great significance for the maintenance of fusion reactors.

1. Introduction

The Chinese Fusion Engineering Testing Reactor (CFETR) is a major scientific project independently designed and developed by China in parallel with international cooperation. The internal components of the fusion experimental reactor will be damaged when operating in the extreme environment of strong radiation, strong magnetic field, high temperature and vacuum, so it is necessary to use teleoperated robots to maintain the components. The CMOR (CFETR Multipurpose Overload Robot) system is a multi-degree-of-freedom robot system used for the maintenance of components in the CFETR vacuum chamber. It can replace humans to complete complex and special work tasks under high temperature, radiation and vacuum conditions[1].

The same end pose of the redundant degree-of-freedom manipulator will correspond to the inverse kinematics solution of an infinite number of groups[2-3], which has good obstacle avoidance ability. and better flexibility. The existing research mainly includes algebraic numerical methods such as Jacobian matrix pseudo-inverse method, quadratic programming inverse algorithm[4]; iterative algorithms such as neural network algorithm[5], particle swarm algorithm[6], weighted minimum algorithm Norm algorithm, etc. Degenerate degrees of freedom method; Analytical method.

The algebraic numerical method generally has the problem of low precision; the iterative method has a large amount of computation and a slow solution speed, which cannot meet the requirements of real-time servo control of the robot; the degenerate degree of freedom method reduces the flexibility of the manipulator; the analytical method uses the seat separation method to solve the problem. But the redundant degrees of freedom decoupling is not achieved. This paper adopts the geometric solution method based on coordinate transformation combined with analytical method to avoid reducing the degrees of freedom, realize the decoupling of redundant degrees of freedom, reduce the overall calculation amount, and improve the calculation speed.

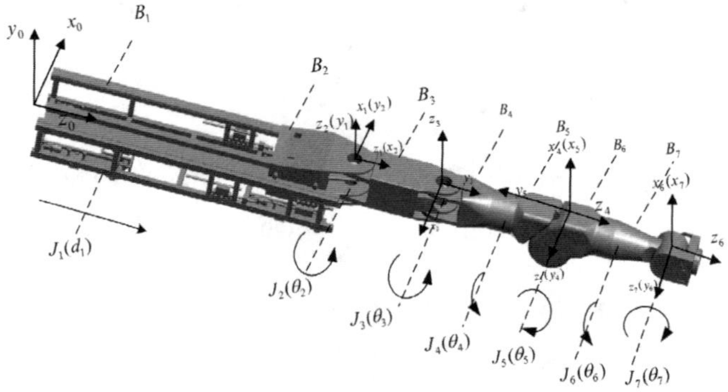

Figure 1. Structure and size of the delivery manipulator and DH model.

Table 1 Modified DH parameters

i	a_{i-1}	α_{i-1}	d_i	θ_i
1	0	0	d_1	0
2	0	π/2	0	π/2
3	B_3	0	0	-π/2
4	0	-π/2	B_4+B_5	0
5	0	-π/2	0	0
6	0	π/2	B_6+B_7	0
7	0	-π/2	0	0

2. Forward kinematics analysis

2.1 DH parameters of CMOR short manipulator

The structure, size and modified DH model of the CMOR short transport manipulator studied in this paper are shown in Figure 1. where the coordinate system $(\overrightarrow{x_0}, \overrightarrow{y_0}, \overrightarrow{z_0})$is the base coordinate system of the carrier manipulator, the coordinate system $(\overrightarrow{x_1}, \overrightarrow{y_1}, \overrightarrow{z_1})$coincides with the origin of the joint 2 coordinate system. According to the established coordinate system, the modified DH parameter table of the carrier manipulator can be obtained as shown in Table 1.

2.2 Forward kinematics solution

The formula of the homogeneous transformation matrix of adjacent links established by the modified DH method is:${}^{i-1}_{i}T$

$$ {}^{i-1}_{i}T=\begin{bmatrix} c\theta_i & -s\theta_i & 0 & \alpha_{i-1} \\ s\theta_i c\alpha_{i-1} & c\theta_i c\alpha_{i-1} & -s\alpha_{i-1} & -s\alpha_{i-1}d_i \\ s\theta_i s\alpha_{i-1} & c\theta_i s\alpha_{i-1} & c\alpha_{i-1} & c\alpha_{i-1}d_i \\ 0 & 0 & 0 & 1 \end{bmatrix} \quad (1) $$

Among them, ${}^{i-1}_{i}T$ is the homogeneous transformation matrix of the connecting rod coordinate system i relative to i-1; θ_i is the rotation angle in the DH parameter; c is the cosine cos; s is the sine sin; a_{i-1} is the DH parameter d_i is the intercept in the DH parameter; α_{i-1} is the torsion angle in the DH parameter.

Multiply the transformation matrix of each link to obtain the transformation equation of the manipulator as:

$$
{}_7^0P=\left[{}_7^0p_x \quad {}_7^0p_y \quad {}_7^0p_z\right]^T=
\begin{bmatrix}
L_1c\theta_2-L_2(c\theta_2s\theta_3+c\theta_3s\theta_2)+L_3\left[c\theta_4s\theta_5(c\theta_2c\theta_3-s\theta_2s\theta_3)-c\theta_5(c\theta_2s\theta_3+c\theta_3s\theta_2)\right]\\
L_3s\theta_4s\theta_5\\
d_1+L_1s\theta_2+L_2(c\theta_2c\theta_3-s\theta_2s\theta_3)+L_3\left[c\theta_4s\theta_5(c\theta_2s\theta_3+s\theta_2c\theta_3)+c\theta_5(c\theta_2c\theta_3-s\theta_3s\theta_2)\right]\\
\\
\end{bmatrix}
\tag{2}
$$

$$
{}_7^0R=\begin{bmatrix}
{}_7^0n_x & {}_7^0o_x & {}_7^0a_x\\
{}_7^0n_y & {}_7^0o_y & {}_7^0a_y\\
{}_7^0n_z & {}_7^0o_z & {}_7^0a_z
\end{bmatrix}
\tag{3}
$$

The position and pose of the end coordinate system {7} relative to the base coordinate system {0} are described in the ${}_7^0R$ formula, where ${}_7^0T$ is the end pose matrix and ${}_7^0P$ is the end position matrix.

3. Inverse kinematics analysis

The method to obtain the angle value of each joint under the condition of a given end pose is the inverse kinematics of the manipulator.

3.1 Solving the joint angle θ_7

$$
\begin{cases}
{}_6^0n_yc\theta_7-{}_6^0a_ys\theta_7={}_7^0n_y\\
{}_6^0a_yc\theta_7+{}_6^0n_ys\theta_7=-{}_7^0o_y\\
L_3\,{}_6^0a_y={}_7^0P_y
\end{cases}
\tag{4}
$$

According to the structure of the end of the manipulator, the movement of joint 7 does not affect the position of joint 6. From ${}_6^0T={}_1^0T{}_2^1T{}_3^2T{}_4^3T{}_5^4T{}_6^5T$, knowing ${}_7^0T={}_6^0T{}_7^6T$, it can be solved by formula (4) θ_7.

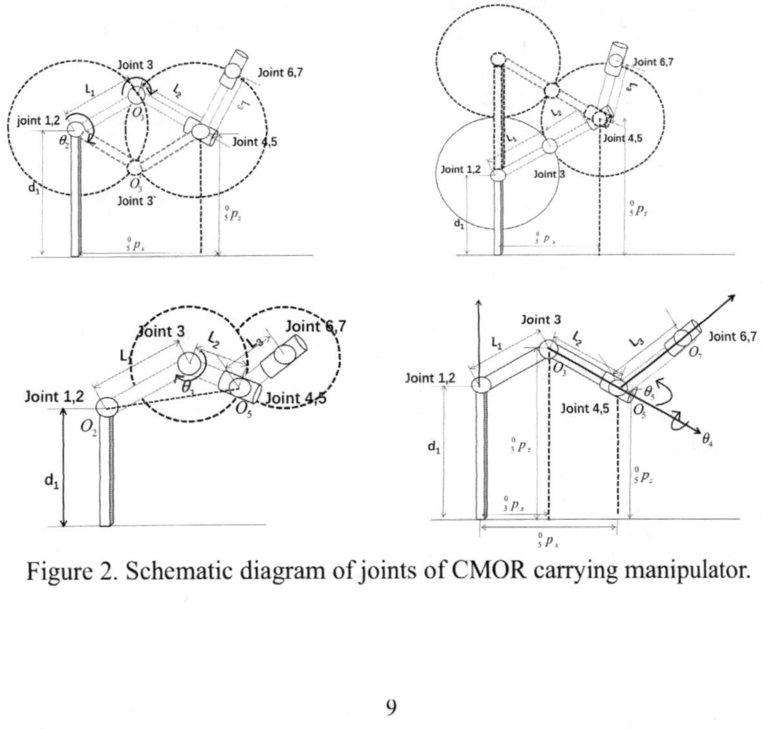

Figure 2. Schematic diagram of joints of CMOR carrying manipulator.

3.2 value range of the moving joint d_1

From equation (2), it can be seen that the change of the moving joint d_1 only affects its position component ${}_7^0P_z$ in the Z-axis direction, so d_1 can be defined as a redundant variable, and the other joint variables $\theta_2 \sim \theta_6$ of the manipulator can be obtained when the redundant variable d_1 is properly valued.

The movement of joint 6 has no effect on the position of joint 5 and joint 5 moves in the XOZ plane. Therefore ${}_5^0P = 0$, the pose matrix of joint 6 relative to the base coordinate {0} can be converted into the position matrix of joint 5 relative to the base coordinate, that is

$$ {}_5^0P = \begin{bmatrix} {}_5^0p_x & {}_5^0p_y & {}_5^0p_z & 1 \end{bmatrix}^T = {}_6^0T \times \begin{bmatrix} 0 & 0 & -L_3 & 1 \end{bmatrix}^T \tag{5} $$

Available

$$ \begin{cases} {}_5^0p_x = {}_7^0p_x - L_3\, {}_6^0a_x \\ {}_5^0p_y = {}_7^0p_y - L_3\, {}_6^0a_y = 0 \\ {}_5^0p_z = {}_7^0p_z - L_3\, {}_6^0a_z \end{cases} \tag{6} $$

Figure 2 first row shows the normal working state of the robot manipulator. The origin of joint point 5 and joint point 2 are used as the center of the circle, and L_2 and L_1 are the radii to make a circle. The intersection of the two circles is the position of the origin of joint point 3. The relationship determines the possible value range of d_1. As shown in the upper right of Figure 2, when the two circles are tangent, the value range of d_1 can be obtained.

$$ d_1 = {}_5^0p_z \pm (L_1 + L_2)\sin\left[\arccos\left(\frac{\left|{}_5^0p_x\right|}{L_1 + L_2}\right)\right] \tag{7} $$

3.3 Solve the joint rotation angle θ_2,θ_3

According to the upper left of Figure 2, the origin of joint point 3 ${}_3^0P$ is $({}_3^0P_x, {}_3^0P_z)$, $({}_3^0P'_x, {}_3^0P'_z)$.

$$ \theta_2 = \begin{cases} \pm\arctan\dfrac{\left|{}_3^0p_x\right|}{\left|{}_3^0p_z - d_1\right|} & {}_3^0p_z - d_1 > 0; ({}_3^0p_x < 0, {}_3^0p_x > 0) \\ \pm\left(\pi - \arctan\dfrac{\left|{}_3^0p_x\right|}{\left|{}_3^0p_z - d_1\right|}\right) & {}_3^0p_z - d_1 < 0; ({}_3^0p_x < 0, {}_3^0p_x > 0) \\ \pm\dfrac{\pi}{2} & {}_3^0p_z - d_1 = 0; ({}_3^0p_x < 0, {}_3^0p_x > 0) \end{cases} \tag{8} $$

According to the bottom left of Figure 2, the joint point 5, the joint point 3, and the joint point 2 form a triangle. $P_{23} = O_3 - O_2 = ({}_3^0P_x, 0, {}_3^0P_z - d_1)$, $P_{35} = O_5 - O_3 = ({}_5^0P_x - {}_3^0P_x, 0, {}_5^0P_z - {}_3^0P_z)$, define $n = P_{23} - P_{35}$.

$$ \theta_3 = \begin{cases} \pi - \arccos\dfrac{L_1^2 + L_2^2 - \left[{}_5^0P_x^2 + ({}_5^0P_z - d_1)^2\right]}{2L_1L_2} & n(1,2) < 0 \\ \arccos\dfrac{L_1^2 + L_2^2 - \left[{}_5^0P_x^2 + ({}_5^0P_z - d_1)^2\right]}{2L_1L_2} - \pi & n(1,2) > 0 \end{cases} \tag{9} $$

3.4 Solve the joint angle $\theta_4 \sim \theta_6$

As shown in the lower right of Figure2, $P_{57} = ({}_7^0P_x - {}_5^0P_x, {}_7^0P_y, {}_7^0P_z - {}_5^0P_z)$, is defined $n_1 = P_{35} * P_{57}$, then

Journal of Physics: Conference Series **2365** (2022) 012002 doi:10.1088/1742-6596/2365/1/012002

$$\theta_5 = \pm\left(\pi - \arccos\frac{n_1}{|n_1|}\right) \tag{10}$$

Knowing from formula (5) $L_3 S\theta_4 S\theta_5 = {}^0_7P_y$, then

$$\theta_4 = \pm\arcsin\frac{{}^0_7P_y}{L_3 S\theta_5} \tag{11}$$

In the case of obtaining d_1,$\theta_2 \sim \theta_5$,θ_7,can be obtained ${}^5_6T = {}^0_5T^{-1}\,{}^0_6T$, then ${}^5_6T(1,1) = c\theta_6$, obtained

$$\theta_6 = \arccos\left({}^5_6T(1,1)\right) \tag{12}$$

4. Validity verification

The inverse kinematics solution process of the manipulator is shown on the left side of Figure 3.

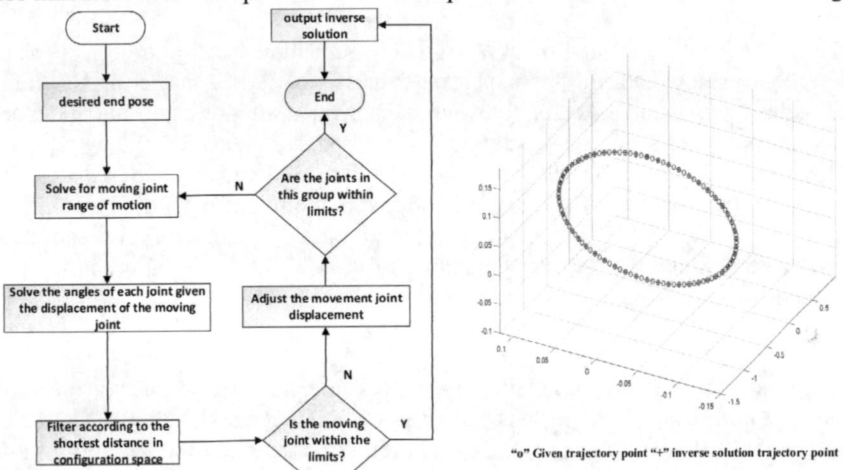

Figure 3. CMOR inverse kinematics solution process chart and circular trajectory tracking.

The initial value of the ${}^0_5P_z$ mobile joint d_1 can be set to optimize the inverse solution according to the shortest distance in the configuration space, and finally judge whether the joint angle is within the maximum range, and output the joint angle if it meets the requirements. Otherwise, readjust the mobile joint d_1 and perform the inverse solution calculation and Preferred.

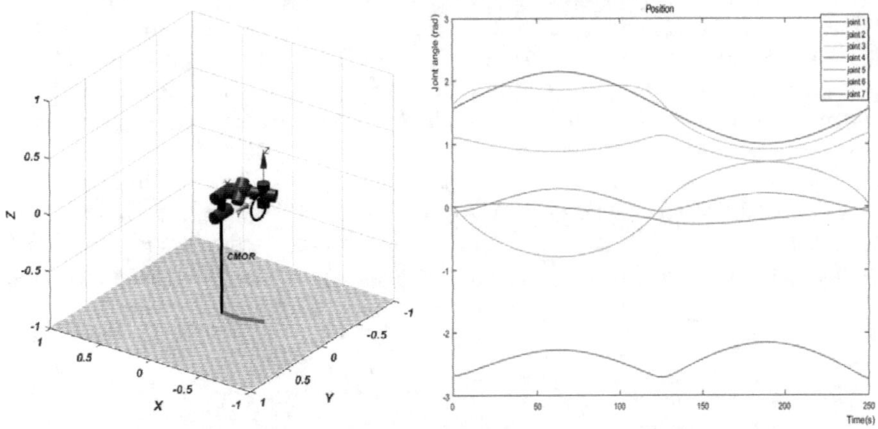

Figure 4. Verify the inverse kinematics of the robotic manipulator.

In the case of fixed mobile joints, the inverse kinematics solution method studied above is used to control the manipulator to move along a given circular trajectory to verify the correctness and accuracy of the inverse solution. Sampling 50 points on a given circular trajectory, solve and optimize the inverse solution of the manipulator according to the above research method, and obtain the comparison between the pose of the end of the manipulator and the position of the trajectory point of the sampling point through the selected inverse solution, as shown on the right side of Figure 3. The inverse solution trajectory point coincides with the given trajectory point, and the position error of the trajectory point is in the order of 10^{-14}, that is, the inverse solution of the manipulator according to the above research method is correct and has high precision. Robotic Toolbox is used in MATLAB to build a 3D visual kinematics model of the CMOR carrier manipulator. According to the optimal inverse solution, after the manipulator completes the circular trajectory motion, its end trajectory can be displayed in the 3D visualization module, as shown on the left side of Figure 4. The angles of each joint during the movement of the robotic manipulator are shown on the right side of Figure 4, and the changes of the joint angles are smooth and stable.

5. Conclusion

Solving the kinematics accurately and quickly lays the foundation for fine trajectory planning and control of the robotic manipulator. In this paper, on the basis of fully considering the configuration characteristics of the manipulator, a method for solving the inverse solution of the manipulator based on the redundant degrees of freedom pose separation is proposed, which avoids reducing the degrees of freedom, realizes the decoupling of the redundant degrees of freedom, and reduces the overall Computation. Finally, the accuracy of the forward and inverse solutions is verified by simulation. The safe motion planning of the carrier manipulator is very important for teleoperation and maintenance. Future work will further study the trajectory planning of the CMOR carrier arm and the cooperative control of the manipulator.

Acknowledgment

This work was partially supported by the National Key Research and Development Project Monitoring and Prevention of Major Natural Disasters Special Program under Grant No. 2020YFC1512202; partly supported by the Anhui University Cooperative Innovation Project under Grant GXXT-2019-003.

References

[1] Yao, Z., Cheng, Y., Pan, H., Yang, Y., & Wu, H. (2022). Optimal design of cfetr multipurpose overload robot based on advantage posture. *Journal of Fusion Energy*, 41(1), 1-10.

[2] Zaplana Isiah, Hadfield Hugo & Lasenby Joan. (2022). Closed-form solutions for the inverse kinematics of serial robots using conformal geometric algebra. *Mechanism and Machine Theory*. doi: 10.1016/J.MECHMACHTHEORY.2022.104835.

[3] Song, G., Su, S., Li, Y., Zhao, X., & Zhao, Y.. (2020). A closed-loop framework for the inverse kinematics of the 7 degrees of freedom manipulator. *Robotica*, 1-10.

[4] Khudher, D., & Powell, R.. (2017). Quadratic programming for inverse kinematics control of a hexapod robot with inequality constraints. *International Conference on Robotics: Current Trends & Future Challenges*. IEEE.

[5] Shah, S. K., Mishra, R., & Ray, L. S.. (2020). Solution and validation of inverse kinematics using Deep Artificial neural network. *Materials Today: Proceedings* (Pt 2).

[6] Deng, H., & Xie, C.. (2021). An improved particle swarm optimization algorithm for inverse kinematics solution of multi-dof serial robotic manipulators. *Soft Computing*, 1-14.

Research and application of generator excitation overvoltage protection device tester

Dongliang Su[1]*, Tao Qiu[1] and Qihui Yin[2]

[1] Shandong Zhongshi Yitong Group Co., Ltd, Shandong, China

[2] Shandong Minker Electric Technology Co., Ltd, Shandong, China

Corresponding author's e-mail address: 2857139947@qq.com

Abstract. In order to solve the problems of complex operation steps, insufficient measurement accuracy and poor safety of the current generator excitation overvoltage protection device tester, the testing technology of generator excitation overvoltage protection device was deeply studied in this paper. Firstly, the overall architecture of the excitation overvoltage protection device tester was constructed. Secondly, it was designed from two aspects of software and hardware. Finally, combined with the actual application scenario, the developed tester was used to measure the action voltage and working voltage fluctuation range of the excitation overvoltage protection device. It was verified that the tester developed in this paper is simple to operate and the measurement results have high accuracy.

1. Introduction

Excitation system is the equipment that provides excitation current to the generator. It is an important part of the generator set, the stability of which directly affects the stability of the unit operation. However, when the generator is running, strong overvoltage often occurs in the excitation circuit due to over excitation, fault, asynchronous generator, asynchronous closing, non full phase operation, rectifier bridge commutation and other reasons. These overvoltages will damage the insulation of the equipment, affect the service life of the generator, and even damage the rotor or stator in serious cases, endangering the unit and personal safety [1]. Therefore, the generator excitation device must be equipped with reliable overvoltage protection device to quickly cut off the excitation power supply and de-excitation in case of overvoltage of the motor, so as to protect the excitation winding [2,3,4]. In order to ensure the reliability and safety of generator excitation overvoltage protection device, it is necessary to conduct overvoltage test regularly.

At present, there are many problems in the test methods of generator excitation overvoltage protection device in China: first, the test methods are not unified, and there are great differences in the content of data acquisition [5,6]. Second, the operation is complex and the measurement accuracy is low. At present, megger, voltage withstand device, oscilloscope and other equipment are often used for testing. Moreover, the test voltages required for multiple tests are different, and the test data often need to be measured for many times. The operation steps and wiring process are complex, and the weight and volume of the test equipment are large, which is not only not conducive to the field test, but also difficult to obtain accurate measurement results [7]. Third, poor safety. Complex wiring methods are more likely to cause faults, and some test methods may also damage finished equipment, resulting in high potential safety hazards [8,9,10]. Therefore, it is necessary to study the tester of excitation overvoltage protection device which is safe and easy to operate.

To solve the above problems, this paper first constructed the overall architecture of the generator excitation overvoltage protection device tester. Then, considering the test function requirements and safety performance requirements of the tester, the hardware design and software design were carried out respectively. Finally, the tester developed in this paper was used to test the MB32 generator overvoltage protection device in the Dongjiakou 2×35W kW thermal power project of Huaneng Qingdao Thermal Power Co., Ltd., which verified the effectiveness and accuracy of the tester applied to the field maintenance of excitation overvoltage protection device.

2. Overall structure of generator excitation overvoltage protection device tester

According to the test requirements, the test of excitation overvoltage protection device was divided into two parts: performance test and voltage fluctuation test. The performance test mainly tests whether the device can make protection action and action voltage value in time in case of overvoltage. The voltage fluctuation test is the working condition of the test device under the condition of voltage fluctuation [11,12]. Therefore, the tester needs to have the following functions: output adjustable DC voltage; Provide interfaces for testing voltages in different ranges; Acquisition and processing of voltage, current, temperature and other signals; Over current and over temperature protection function. The overall architecture of the generator excitation overvoltage protection device tester constructed in this paper is shown in Figure 1.

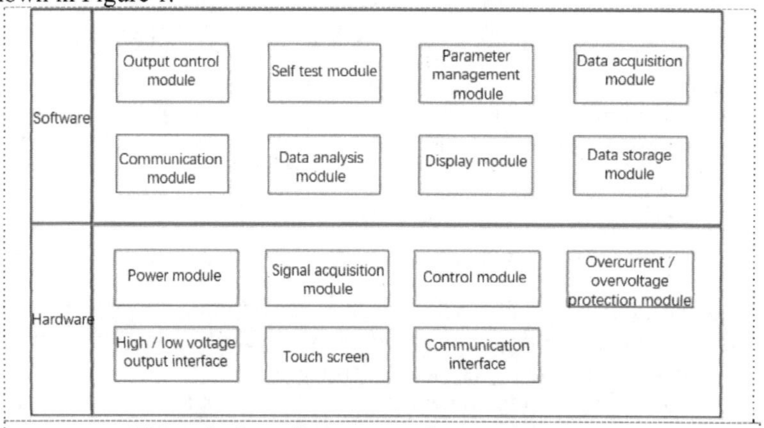

Figure 1. Overall structure of overvoltage tester.

The overall architecture of generator excitation device overvoltage tester integrates hardware architecture and software architecture. The hardware architecture can realize many functions, such as test voltage output, signal acquisition, output voltage control, data processing, overvoltage / overtemperature protection, human-computer interaction and so on. The software architecture can realize the functions of instrument self-test, test mode selection, test voltage control, equipment communication, data analysis, test result display and storage. Using RS232 communication, the tester interacts with the equipment through the touch screen, and the control equipment converts 220V AC into test voltage to test the tested device.

3. Hardware design

The hardware design of the tester includes power output module, control module, signal acquisition module, touch screen, over temperature/over voltage protection module, output port and so on. The schematic diagram of hardware design is shown in Figure 2.

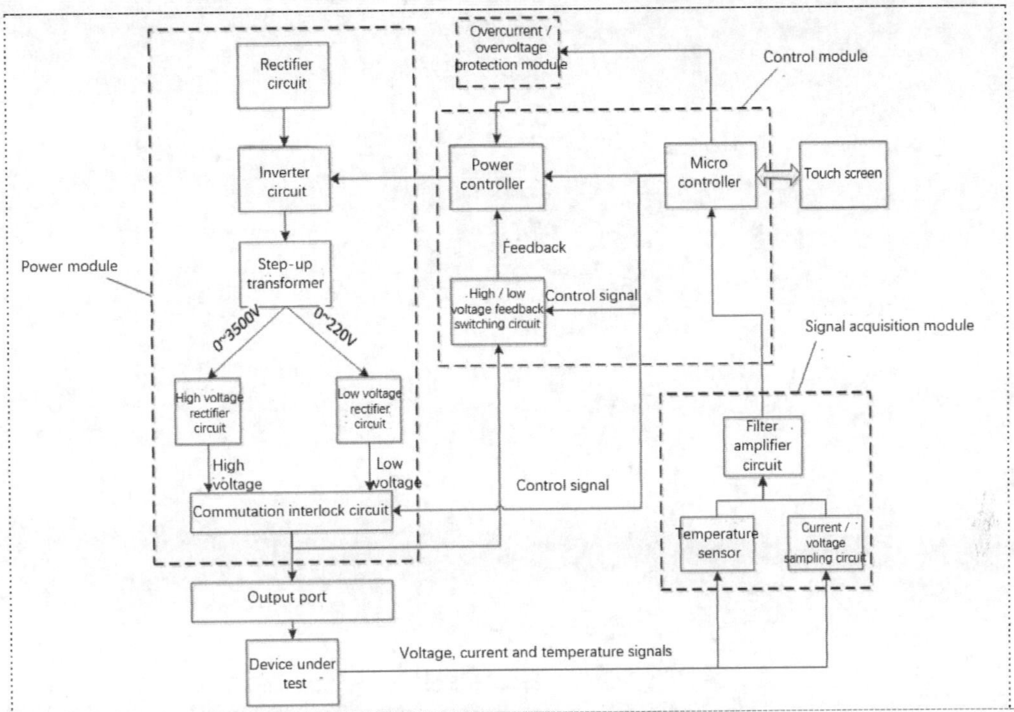

Figure 2. Hardware design schematic diagram.

The power module includes rectifier circuit, inverter, step-up transformer, high/low voltage rectifier circuit and commutation interlock circuit. 220V AC voltage was converted into DC voltage through rectifier circuit and into high-voltage AC voltage through inverter and step-up transformer. Through the voltage division at the secondary side of the step-up transformer, an AC voltage with a range of 0 ~ 3500V and a range of 0 ~ 220V were obtained. It was changed into DC voltage through the rectifier circuit, and the direction was switched through the commutation interlock circuit. Finally, the test voltage was output from the output port.

The signal acquisition module includes temperature sensor, voltage / current sampling circuit, filter amplification circuit, etc. It can collect analog signals such as voltage, current and temperature in real time, filter and amplify these information and upload them to the microcontroller.

The control module includes microcontroller, PWM controller and high / low voltage feedback switching circuit. The microcontroller sent control signals to PWM controller, commutation interlock circuit, high / low voltage feedback switching circuit and overcurrent / overtemperature protection module. The PWM controller generates a PWM signal with adjustable pulse width and sends it to the gate of the high-frequency switching device of the inverter circuit, so as to control the output voltage of the power module. The high / low voltage feedback switching circuit switches the feedback channel of the output voltage according to the control signal. The microcontroller processes the collected signal and outputs it to the display for display.

4. Software design

The software design includes self-inspection module, data analysis module, output control module, communication module, parameter management module, data acquisition module, display module and data storage module. The software control flow chart is shown in Figure 3.

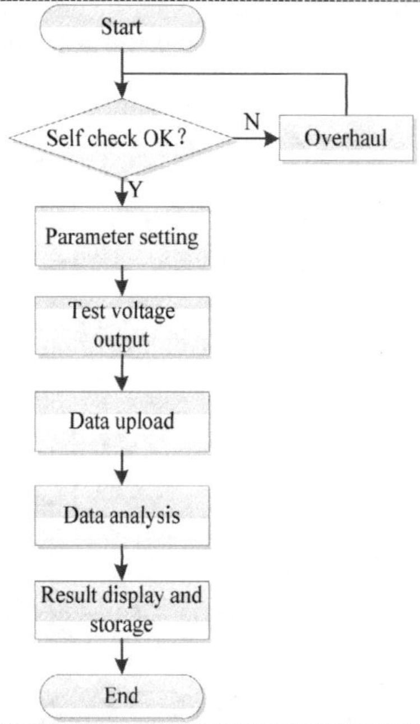

Figure 3. Software control flow chart.

The self-inspection module monitored the voltage, current and temperature of the tester, and transmitted the self-inspection results (Pass or Fail) to the output control module. In case of overvoltage, overtemperature and other conditions, the voltage output was cut off in time to ensure the safety of the test equipment.

The parameter management module includes test mode selection interface and parameter setting interface. Firstly, through the test mode selection interface, four test modes can be selected: automatic high-voltage output, manual high-voltage output, automatic low-voltage output and manual low-voltage output. After that, preset the change parameters of output voltage in the parameter setting interface according to the test conditions. The parameters set under different test modes are different: In the high voltage output mode, the setting parameters include test direction, initial voltage, initial terminal rate, turning voltage, terminal rate and final voltage; In the low-voltage output mode, the setting parameters include voltage range (110V or 220V), initial voltage, maximum current, step voltage, step time and target voltage.

The communication module adopted RS232 communication, which can realize the functions of test control, communication requirements between hardware equipment and control module, human-computer interaction and so on.

After the parameters were set, the output control module sent control instructions to the hardware equipment through the communication module according to the preset parameters to control the output voltage of the test device. The test data was then uploaded to the data analysis module through the data acquisition module for data conversion and processing, and the analysis and calculation of various test indicators were completed. The final results were displayed by the display module. At the same time, the test data was stored in the history record according to the test time for query and call.

5. Field test verification and data analysis

The tester developed in this paper has been tested and verified in the Dongjiakou 2×35W kW thermal power project of Huaneng Qingdao Thermal Power Co., Ltd. The generator excitation system of the project adopts GEX-6301 excitation device produced by Kedian Yiheng, with an output voltage of 196V and an output current of 384A. Its overvoltage protection device adopts MB32 generator overvoltage protection device produced by Anhui Huidian technology, with energy capacity of 0.24MJ and rated protection voltage of 1500V. The tester developed in this paper was used to test the overvoltage protection device. The test site is shown in Figure 4.

Figure 4. Diagram of test site .

5.1. *Performance test of excitation overvoltage protection device*

In order to test the protection action and action voltage value of the overvoltage protection device, first connected the tested device to the high-voltage output port of the tester, selected the test mode as the high-voltage output mode, and set the test parameters. After the test, the output voltage of the tester started from 1000V and gradually increases at a preset rate. Monitored the output voltage and current in real time until the overvoltage protection device act, sent out the "Rotor Overvoltage fault" signal, the output stopped automatically, and read the voltage and current curve and action voltage value. The test results are shown in Figure 5.

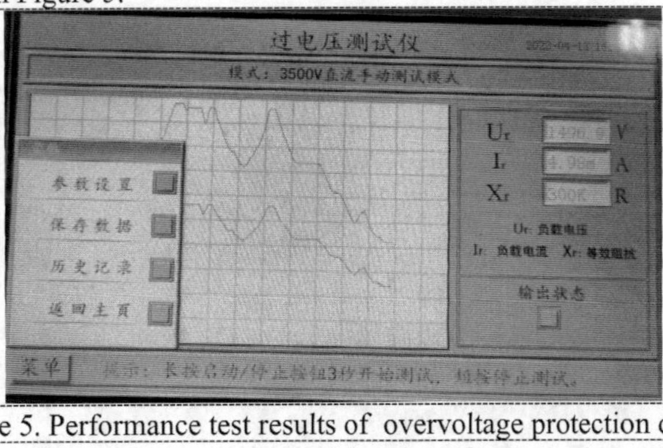

Figure 5. Performance test results of overvoltage protection device.

In order to verify the measurement accuracy of the tester for the action voltage value, repeated the above test for 16 times, recorded the action voltage value, and compared it with the action voltage value measured by the previous test method. The comparison of the test results is shown in Figure 6.

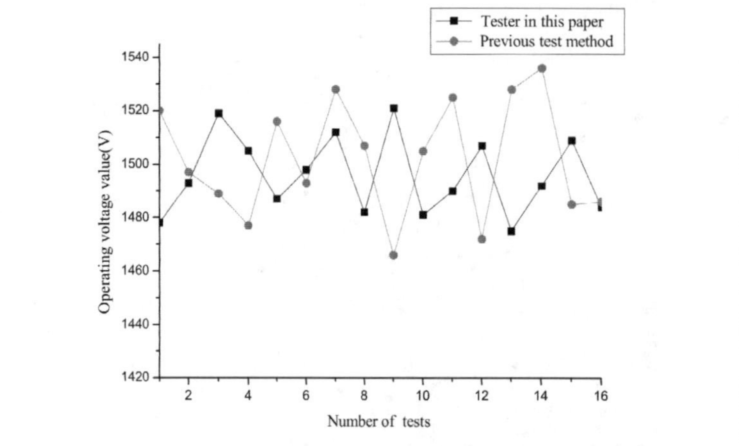

Figure 6. Comparison diagram of measurement results.

The operating voltage of the device measured by the tester developed in this paper was 1475 ~ 1521V, and the operating voltage measured by the previous test method was 1466 ~ 1536V. It can be concluded that the tester developed in this paper reduced the error range of the action voltage value of the overvoltage protection device, and the measurement accuracy was higher.

5.2. Voltage fluctuation test of excitation overvoltage protection device

Since the power supply voltage of the excitation overvoltage protection device will fluctuate in the actual working process, the voltage fluctuation test is required to test the working condition of the device under the condition of voltage fluctuation. Connect the tested device to the low-voltage output port of the tester, adjust the test mode to the low-voltage output mode, and set the test parameters. After the test, the output voltage of the tester started from 220V and gradually decreased in a step-by-step manner. Monitored the output voltage and the working state of the excitation overvoltage protection devices, and stopped the voltage output when the excitation overvoltage protection device stopped working or the current value reached the maximum current. As show in Figure 7 for the test results.

Figure 7. Voltage fluctuation test results of overvoltage protection device.

It can be concluded that when the output voltage was lower than 212.8V, the excitation overvoltage protection device stopped working. At this time, the output current was 209.5mA and the output power was 44.58W. Therefore, it can be concluded that the voltage range that the excitation overvoltage protection device can work normally was 212.8V ~ 220V.

6. Conclusions

The reliability of excitation overvoltage protection device is directly related to the safety of generator set and power grid, and it is an important protection device for generator set. The tester developed in this paper tests the generator excitation overvoltage protection device, which is of great significance for the on-site electrical maintenance and commissioning, as well as the safe and reliable operation of the generator set. The tester of excitation overvoltage protection device developed in this paper has the following advantages:

1) It has two output modes of high voltage output and low voltage output, and can realize the functions of output voltage commutation, high and low voltage output interlock, high and low voltage feedback switching, etc. The test of excitation overvoltage protection device and its controller can be completed through a set of devices.

2) After the test data was measured at one time under fixed working conditions, the test data was directly read, which simplified the test steps and wiring process, facilitated the field test and improves the test efficiency.

3) It is beneficial to standardize the test process of generator excitation overvoltage protection device, reduce the faults caused by insufficient communication under complex working conditions, and improve the safety of rotor overvoltage test process.

References

[1] Liu Yulin, Zhu Hongwei, Gu Jianrong. A detection method of Generator Rotor Overvoltage protection device [J]. Huadian Technology. 2018, Vol. 40, No. 2: 16-18.Another reference

[2] MA Zuo-fu, CHEN Xiao-ming, YAN Feng. A test method for BOD action value of synchronous generator excitation system[J]. Mechanical & Electrical Technique of Hydropower Station, 2021, 11:49-51.

[3] Technical specification for generator's demagnetization and Rotor overvoltage's protection device:DL/T 294.3-2019[S]. Beijing: China Electric Power Press, 2019.

[4] Li Yi. Development and application of test equipment for Rotor Overvoltage protection device[J]. Sichuan Water Power, 2013, 32(3):184-186.

[5] Zhu Shizhang. Generator excitation system test[M]. Beijing: China Electric Power Press， 2005.

[6] Gong Jian. Performance test of overvoltage protection device for excitation system of a Hydropower Station[J]. Sichuan Water Power, 2018, 37(6):157-159.

[7] Du Jiuqiang, Duan Zhenguo. Improvement of generator's rotor demagnetization and overvoltage protection device[J]. Electric Power Automation Equipment, 2009(6).

[8] K Ramesh and Y Mallikharjuna Rao. POWER TRANSFORMER PROTECTION USING WAVELETS[J]. International Journal of Electrical and Electronic Engineering & Telecommunications, Vol. 3, No. 4, pp. 14-21, October 2014.

[9] Chen Zhijie.Research and Analysis on parameter setting of generator excitation system[J]. 2021. 12:6-9. doi:10. 3969/j. issn. 1009-3230. 2021. 12. 003.

[10] Technical specification for static rectifier excitation system of large and medium-sized hydro generators:DL/T 583—2018[S]. Beijing: China Electric Power Press, 2018.

[11] Guide for field test of excitation system of large generator: DL 1166—2012[S]. Beijing: China Electric Power Press, 2012.

[12] ZHANG QixueNIU HonghaiBAI ChuanjunXIONG LianggenWANG Kai. Subsynchronous serf-excitation overcurrent protection for turbine generator[J]. Electric Power Automation Equipment, 2013, 33(2):166-170.

Early Weak Fault Diagnosis of Rolling Bearing Based on Duffing System and LMCD

Chenyang Hu[1,a], **Shijie Shan**[2,b], **Ning Li,**[1,c], **Jizhen Li**[1,d], **Zhao Li**[1,e]

[1]Xi'an North Qinghua Electromechanical Co., Ltd., Xi'an 710025.

[2]Faculty of Mechanical and Precision Instrument Engineering, Xi'an University of Technology, Xi'an 710048.

[a]372524820@qq.com,[b]284534560@qq.com,[c]395574220@qq.com,[d]378096963@qq.com,[e]songshu130@163.com

Abstract. The early weak fault signals of bearings are usually hidden in a large amount of noise, and the components of the noise are more complicated. Therefore, it is difficult to extract the early weak fault features of bearings using conventional time-frequency methods. In this paper, the traditional Holms-Duffing system has been improved to enhance the anti-noise performance of the system, and then the Duffing system can be used to detect the signal in a broadband frequency through time transformation. At the same time, the four Duffing systems array model has been used to solve the misjudgment problem of the Duffing system caused by the phase difference. When the Duffing system is in a large period state, the complexity of the displacement sequence is smaller than that in the chaotic state. According to this characteristic, a method for diagnosing weak bearing faults based on the Duffing system and LMC difference entropy(LMCD) is proposed. In this paper, simulation signal with strong noise components and the real fault signals have been used to verify the suggested method. The proposed methods in this paper have certain practical value in the field of fault diagnosis.

1. Introduction

Rolling bearings are widely used in the field of mechanical equipment, and their health status will directly affects the working performance of the equipment[1-2]. Due to the complex working conditions, the bearing is prone to failure, so it is of great significance to detect and diagnose the bearing fault[3].

How to effectively extract the fault characteristic frequency from the bearing vibration signal is an important subject in the field of bearing fault diagnosis. The bearing vibration signals of rotating machinery are often doped with noises from a variety of different vibration sources, it is difficult to extract and identify the characteristics of weak fault under strong noise background[4]. When the built-in driving power frequency of the Duffing system is the same as the frequency of the external excitation signal, the phase plane trajectory of the Duffing system will change. In recent years, this rule has also been applied to diagnose bearing faults[5]. However, the detection accuracy and reliability of the Duffing system still have shortcomings in specific applications, which is worthy of further study.

Firstly, the mathematical model of Duffing system has been improved to enhance the detection accuracy of weak periodic signals under strong noise interference. Then the complexity of displacement sequence of Duffing system has been extracted by LMCD algorithm. It can be found the LMCD value of displacement sequence of Duffing system in the periodic state is smaller than that of Duffing system in the chaotic state. Finally, this rule has been used in bearing fault detection successfully.

2. Duffing chaotic system

2.1. Phase jump principle of duffing system

The equation of state of the traditional Holms-Duffing system is:

$$\begin{cases} \dot{x}_1 = x_2 \\ \dot{x}_2 = -kx_2 + x_1 - x_1^3 + r\cos(\omega t) \end{cases} \tag{1}$$

Here k is the damping ratio; $-x_1+x_1^3$ is a nonlinear force; r is the amplitude of the built-in periodic driving power, and ω is the angular frequency of the built-in periodic driving power. The research shows that when ω=1rad/s and k=0.5, the chaotic threshold of the system is 0.826[6].

According to the research,when the parameter r exceeds the chaotic threshold of the system, the phase state of the system jumps from chaos state to large periodic state. When a periodic signal has been added to Duffing system which was in critical state. If this signal contains a component whose frequency is same as the frequency of periodic driving power, the amplitude of the total periodic driving power of the system will be larger than the chaotic threshold of the system, so the phase state of system will jump. Therefore, according to the state jump of the system, it can be judged there is a weak periodic signal with the same frequency as the built-in periodic driving power in the external excitation signal.

In order to ensure Duffing system can be used to detect weak signals in a broadband frequency, generalized time-scale transformation has been introduced. In equation (1), let t=pτ, τ represents a generalized time variable, then:

$$\begin{cases} \dot{x}_1 = px_2 \\ \dot{x}_2 = p[-kx_2 + x_1 - x_1^3 + r\cos(pt)] \end{cases} \tag{2}$$

Through the above transformation, the original dynamic characteristics of the system have not been changed, so the two systems before and after the transformation have the same chaotic threshold.

2.2. Improvement of fault detection ability of duffing system

Although Duffing system has strong noise immunity, in order to achieve state jump of the system, the amplitude of the periodic excitation signal is required to be larger when the noise is larger. In this case the ability to detect weak signals will be reduced. Therefore, the Duffing system has been improved to enhance the anti-noise performance of the system. The improved mathematical model of the Duffing system is as follows:

$$\begin{cases} \dot{x}_1 = px_2 \\ \dot{x}_2 = p(-0.75(x_1^4 - x_1^2)x_2 + 0.5x_1^3 \\ \quad -0.5x_1^5 + r\cos(pt)) \end{cases} \tag{3}$$

Before the Duffing system is used to detect weak signals, it is necessary to determine the chaotic threshold of the system, r is set as the chaotic threshold so that the system is in critical large period state. In this paper, Jaccbo method has been used to calculate the maximum Lyapunov exponent of the system, and the chaotic threshold of the system is determined by the maximum Lyapunov exponent. It can be determined that the chaotic threshold of the Duffing system is 0.382.

The anti-noise performance of the improved Duffing system is studied by experiments. Firstly, the periodic signal 0.01cos(400t) has been added to the two systems before and after the improvement. Then Gaussian white noise also has been added to the two systems. SNR of the signal is changed by adjusting the intensity of noise, the range of SNR is from -30dB to -20dB. Two kinds of detection systems has been used to diagnose 100 kinds of simulate signals with different SNR.

When the SNR is -30dB, the detection accuracy of the original system is 45%, while the accuracy of the improved system is 49%. With the increasing noise intensity, the detection accuracy of the improved system is always greater than that of the original system. It can be found the anti-noise performance of the improved Duffing system proposed in this paper is better than that of the original Duffing system.

2.3. Application of four duffing systems array model

In equation (3) the *r* is replaced by chaotic threshold, then the equation (4) is obtained:

$$\begin{cases} \dot{x}_1 = px_2 \\ \dot{x}_2 = p(-0.75(x_1^4 - x_1^2)x_2 + 0.5x_1^3 \\ \quad -0.5x_1^5 + 0.382\cos(pt)) \end{cases} \tag{4}$$

The periodic weak signal acos(ωt+φ) is added to the improved Duffing system, then the total periodic driving power of the system is:

$$\begin{aligned} F(t) &= 0.382\cos(pt) + a\cos(pt+\phi) \\ &= f\cos(\omega t + \theta) \end{aligned} \tag{5}$$

In equation (5):

$$f = \sqrt{f^2 + 2fa\cos\varphi + a^2} \tag{6}$$

$$\theta = \arctan[a\sin\varphi/(0.382 + a\cos(\varphi))] \tag{7}$$

When the amplitude f of total periodic driving power of the system is larger than the chaotic threshold, the state of Duffing system will change. Therefore, if f is larger than chaotic threshold the equation (8) can be obtained.

$$-\arccos[-a/(2Fd)] < \theta < \arccos[-a/(2Fd)] \tag{8}$$

In equation (8): Fd is chaotic threshold.

Because a<<Fd, the range of θ is (-π/2,π/2). When θ is in this range, the phase state of the system will change. On the contrary, the system will be misjudged. In order to solve this problem, some researchers add π (phase angle) to the built-in periodic driving power of Duffing system. Combine this system with the system without adding phase angles to form a double Duffing system array detection model[7]. However, many experiments proved that the double Duffing system array detection model can't avoid the misjudgment completely. Therefore this paper proposes a four Duffing systems array model to solve the misjudgment phenomenon. The phase angles of each Duffing system are 0, π/2, π and 3π/2 respectively. The detection model is shown in equation (9):

$$\begin{cases} \dot{x}_1 = px_2 \\ \dot{x}_2 = p(-0.75(x_1^4 - x_1^2)x_2 + 0.5x_1^3 \\ \quad -0.5x_1^5 + 0.382\cos(pt+\mu)) \end{cases} \tag{9}$$

In formula (9), μ is 0, π/2, 3π/2 and π in turn.

When the improved Duffing system is used to detect weak periodic signals, the signals should be added to four detection systems with phase angles of 0, π/2, π and 3π/2 respectively. If the phase plane states of the all four Duffing systems have not changed, it is judged that the frequency of the weak signal is different from the built-in periodic driving power of the system. If one of the phase planes state of the four Duffing systems changes, it can be judged that the same frequency signal is detected.

3. Nonlinear complexity analysis of lmcd

3.1. Lmcd algorithm for symbol sequence

Entropy as an important metric is used to describe the disordered state of matter in the microscopic environment, and is often used to indicate the degree of disorder in some systems. In 1995, LMCD were proposed by Lopez-Ruiz and it is used as a measure of the complexity of nonlinear systems.

The LMCD algorithm of Duffing system displacement time series is as follows:

Step1: If $X=\{x_1,x_2,\ldots,x_N\}$ is the output displacement time series of the Duffing system.

Step2: The extreme points in the sequence X are formed a new discrete sequence $Y=\{y_1,y_2,\ldots,y_{N1}\}$.

Step3: The sequence Y is symbolized by equation (10) , then the symbol sequence $L=\{L_1,L_2,\ldots,L_{N2}\}$ can be obtained, N2=N1, the $L_i \in B(B=0,1)$.

$$L_i(x_i)=\begin{cases}0: x_i > 1.5 or x_i < -1.5 \\ 1: -1.5 \le x_i \le 1.5\end{cases} \tag{10}$$

Step4: The symbol sequence L is encoded by equation (11), then the coding segment sequence $\overline{L}$ is obtained:

$$\overline{L}_{(i)}=\begin{bmatrix} L_1 & L_2 & \cdots & L_m \\ L_{(1+t)} & L_{(2+t)} & \cdots & L_{(m+t)} \\ \vdots & \vdots & \vdots & \vdots \\ L_{(1+nt)} & L_{(2+nt)} & \cdots & L_{(m+nt)} \end{bmatrix}=\begin{bmatrix} \overline{L}_1 \\ \overline{L}_2 \\ \vdots \\ \overline{L}_{(n+1)} \end{bmatrix} \tag{11}$$

In equation (11), m is the length of the coding segment sequence, and t is the coding delay factor.

Step5: The probability of each mode in the coding segment sequence is recorded as $p_i=\{p_1,p_2,\ldots,p_n\}$, here n is the total number of possible modes generated by the coding segment sequence.

Step6: Calculating the normalized information entropy Hs;

$$H_S=-\frac{1}{\ln n}\sum_{j=1}^{n1} p_j \ln p_j \tag{12}$$

In equation (12), n1 is the total number of modes with a probability greater than 0.

Step7: Calculating the non-equilibrium state D of coding segment sequence:

$$D_{LMC}=\sum_{j=1}^{n1}(p_j-1/n)^2 \tag{13}$$

Step8: Equation (14) is used to calculate the LMCD value of the sequence:

$$\begin{aligned} C_{LMCD} &= D_{LMC}\cdot H_S \\ &= -\frac{1}{\ln n}\sum_{j=1}^{n1}(p_j-1/n)^2\cdot\sum_{j=1}^{n1}p_j\ln p_j \end{aligned} \tag{14}$$

3.2. Lmcd analysis of displacement sequence of duffing system

The complexity of the output displacement sequence of the improved Duffing system has been analyzed by LMCD. The variation range of system's parameter *r* is set from 0.3 to 0.45 and the parameter *p* of system is set to 1. In addition, the length *m* of the coding segment sequence is set to 3 and 4, and the delay factor *t* is set to 1.

According to the research on the improved Duffing system in Section 2.2, when the parameter r is less than 0.382, the system is in chaotic state, and the displacement sequence of the system has strong disorder, so the sequence has higher complexity. When the parameter r is greater than 0.382, the system is in a lager periodic state, and the displacement sequence fluctuates regularly, and the sequence complexity is low. When parameter r is greater than the chaotic threshold (0.382), the LMCD of the system displacement sequence tends to 0. The LMCD value decreases with the decrease of system complexity, and the phase jump would be occured at the initial stage when Duffing system enters the big periodic state.

It is concluded when Duffing system is in large periodic state, the LMCD value is 0. Otherwise, the LMCD value is greater than 0. Therefore, the state type of the system can be judged by the LMCD value, so the weak fault signal can be detected.

4. Diagnosing of bearing weak fault simulation signals

The steps of diagnosing bearing weak fault using output displacement symbol sequence LMCD value of Duffing system are as follows:

1) Calculating the theoretical characteristic angular frequency of the bearing fault according to the traditional bearing failure frequency calculation equation.
2) Selecting a range of variation of the parameter p.
3) The signal is input the four Duffing systems array detection model, and LMCD of the displacement sequence is calculated, and the graph of LMCD varies with parameter P is drawn.
4) Searching minimum LMCD value in step 3, the corresponding p is used to compare with theoretical fault frequency of the bearing, and then judging whether there is a bearing fault occured .

In order to verify the applicability of the above models and methods in the detection of bearing weak faults, the simulation signal has been added to the detection model. The simulation signal is shown as follows:

$$\begin{cases} s1(t) = 0.2\cos(300t + \varphi) \\ s2(t) = 0.1\cos(\pi t + 0.01\sin(150t)) \\ s3(t) = sqrt(1)randn(1,n) \\ s(t) = s1(t) + s2(t) + s3(t) \end{cases} \tag{15}$$

In equation (15), $s1(t)$ is the weak signal to be measured, φ is the phase angle randomly generated, $s2(t)$ is the amplitude and frequency modulation signal, and $s3(t)$ is the noise.

From equation (15) we can see that the characteristic frequency of the simulated signal fault is 300rad/s, so 250 ~350 can be set as the variation range of parameter p. The length m of the coding segment sequence is set to 3, and the delay factor t is set to 1. The simulation signal $s(t)$ has been respectively added to the four Duffing systems array detection model, and the graph of the displacement output sequence LMCD has been obtained as a function of the parameter p. As shown in Figure 1.

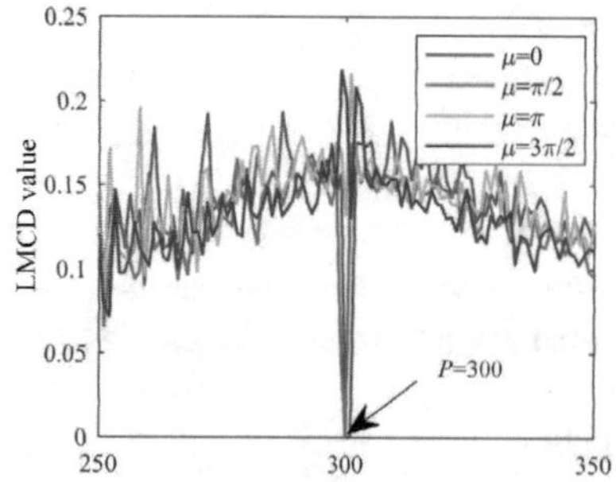

Figure 1. Displacement sequence LMCD with parameter p

It can be seen from Figure 1 that when the parameter p is 300, the value of LMCD is the minimum. The minimum point is equal to the fault frequency of the simulation signal. Several simulation experiments verify the effectiveness of the proposed method and it can be used for bearing fault diagnosis.

5. Diagnosing of actual bearing weak fault signal

The actual weak fault data are provided by Case Western Reserve university Bearing Experimental Center in USA. The data sampling frequency is 12 kHz, and the rotation speed of the spindle R is 1797r/min. The bearing parameters are shown in Table 1.

Table 1. JEM SKF 6205-2RS Bearing Structural Parameters

Number of Rolling Bodies N	9
Ball Diameter d	7.94mm
Contact Angle α	0^0
Section Diameter D	39.04mm

By calculation it can be deduced that the theoretical frequencies of the inner ring and outer ring are 1018rad/s and 674rad/s. Therefore, the variation range of weak signal detection model parameter p is set to 900~1100 and 600~800 respectively. The fault signals of bearing inner and outer ring are added to the four Duffing systems array detection model respectively. After calculation, the graph of the displacement output sequence LMCD has been obtained, it can be viewed as a function of the parameter p. As shown in Figure 2.

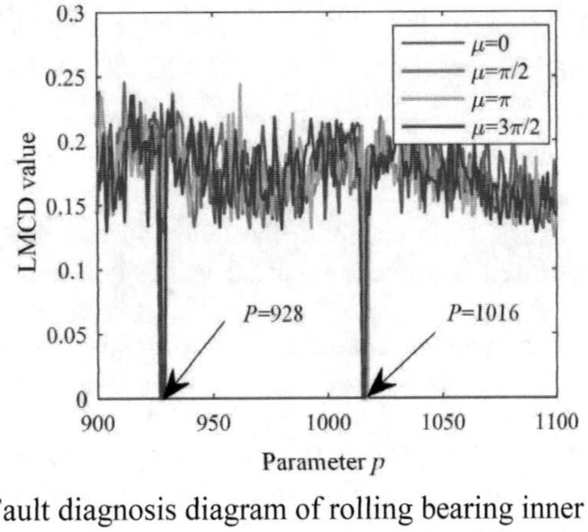

(a)Fault diagnosis diagram of rolling bearing inner ring

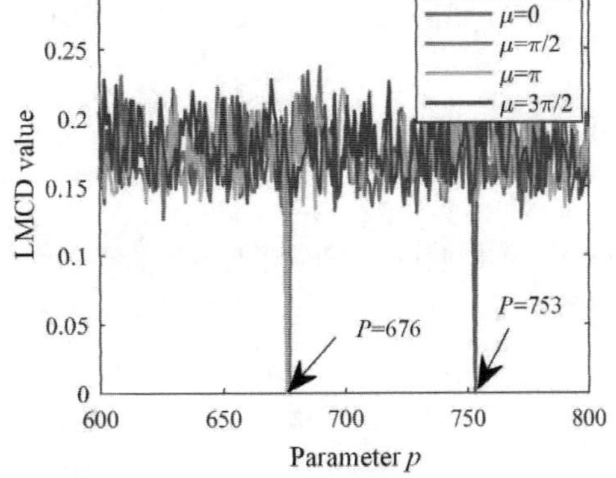

(b)Fault diagnosis diagram of rolling bearing outer ring

Figure 2. Displacement sequence LMCD with parameter p

It can be found from Figure 2(a) that when the p is 928 and 1016, the LMCD takes minimum value. Among them, 1016 is about equal to the fault frequency of bearing inner ring, and 928 is about 5 times the rotation frequency. It also can be seen from Figure 2(b) that when p is 676 and 753, the LMCD takes minimum value. Among them, 676 is the fault frequency of bearing outer ring, and 753 is about equal to 4 times the rotation frequency.

The fault frequency detected by the method proposed in this paper has a small difference with the theoretical value, it is a reasonable phenomenon and within the allowable range. The LMCD value at fault frequency is obviously different from other non-fault frequencies.

6. Conclusion

In this paper, the LMCD value of displacement sequence of the Duffing system is calculated, and the characteristic of this value is used to detect the early weak fault of the bearing. The analysis results show that:

1) After improving the Duffing system, the anti-noise performance is improved, and it can be used to detect signals in a broadband frequency.

2) Based on the four Duffing systems array model, the misjudgment phenomenon caused by the phase difference of the built-in driving power of the system can be effectively avoided.,the detection accuracy of the system can be improved.

3) Based on variation rule of the displacement output sequence LMCD value, by continuously changing the parameter p of the detection model， the fault signal frequency can be detected precisely.

This investigation belongs to the field of bearing weak fault diagnosis. The proposed methods in this paper have certain theoretical and practical reference value.

References

[1] Liu Wenpeng, Yang Shaopu, Li Qiang, Liu Yongqiang, Gu Xiaohui. An adaptive resonance demodulation method and its application in fault diagnosis of railway bearings [J]. Vibration and shock,2021,40(17):290-297.

[2] Zhao Kaihui,Wu Sicheng,Li Tao,He Caichun,Cha Guotao.Research on Fault Diagnosis Method of Rolling Bearing Based on Inception-BLSTM[J].Vibration and Shock,2021,40(17):290-297.

[3] Tiwari P, Upadhyay S H. Novel self-adaptive vibration signal analysis: Concealed component decomposition and its application in bearing fault diagnosis[J]. Journal of Sound and Vibration, 2021, 502: 116079.

[4] Shan Shijie,Wang Kai,Qie Xuliang,Zheng Dan,Dai Xueqing,Shi Jiale.Investigation on Bearing Weak Fault Diagnosis under Colored Noise[J], Proceedings of the 32th Chinese Control and Decision Conference, CCDC 2020 5097-5101.

[5] Zhang Gang, Wu Xia. Stochastic resonance research and bearing fault diagnosis based on the strong coupling system of Duffing and Van der Pol[J]. Vibration and Shock, 2020, 39(19): 266-276.

[6] Huang Zehui, Li Ya'an, Chen Zhe, Lian Liu. Threshold determination method of Duffing chaotic system based on multi-scale entropy [J]. Acta Physica Sinica, 2020,69(16):56-64.

[7] Hong-Wei Li,Ji-Wei Li,Quan Su,Jun-Peng Liu,Bin Sun. Analysis of the characteristics of two-phase flow pattern in Y-shape mixed microchannel based on double coupled Duffing vibration simulation detection system[J]. Experimental Thermal and Fluid Science,2018,91:432-442.

Research on foreign object detection method in track area based on Mask-RCNN

Guoyi Ji[1], Wen Chen[1], Jinbai Zou[1], Shiyan Wei[1]

[1]Shanghai Institute of Technology, China

Corresponding author: 1245551436@qq.com

Abstract. As the train speeds up, the damage caused by the collisions between trains and foreign objects are becoming more and more severe. Therefore, it is of great significance to monitor the intrusion of foreign objects in the track environment. In this paper, transfer learning is introduced into Mask-RCNN deep learning model. And the data set of rail image is used to train the model, which improves the effect of rail segmentation. The trained model is used to segment the rail in the picture, and the rail vulnerable area is divided based on the segmentation results. The sliding window ORB feature matching algorithm is used to calculate the similarity of vulnerable area. The detection of foreign objects in the area where is easy to invade is realized, and the detection reliability is improved. Experiments represent that this method has high accuracy, strong practicability, good robustness and universality.

1. Introduction

Foreign objects in the railway area will pose a severe threat to the train operation, so monitoring the intrusion status of foreign objects becomes the top priority[1]. When it comes to the detection of track line, the feature-based track detection algorithm[2] has impoverished applicability in complex environments, and the model-based track detection algorithm[3] has complex calculations. In recent years, machine vision has become a new direction for the detection of track line. With the wide application of video surveillance systems, a large number of images have laid the data foundation for machine vision methods. In the aspect of foreign object detection, the traditional method for the frame difference can merely detect the contour information of the foreground target, the extraction of the target is incomplete, and the optical flow method is slow and the real-time performance is not strong. The ORB[4] (Oriented FAST and Rotated BRIEF)feature matching algorithm can quickly detect feature points in the picture, with a minor amount of calculation and high accuracy.

In the pictures taken by the camera, the image range is wide, and the information of objects without threat in the distance will also be recorded. It will waste an army of computing power and time if the detection of foreign object is directly carried out on the pictures collected by the camera. In this paper, Mask-RCNN (Mask-Region CNN) neural network model is used to detect track lines. The detected track lines are extended to the left, right and far to obtain the vulnerable area. The sliding window ORB feature matching algorithm is used to detect the similarity between the vulnerable area and the background picture, and to judge the foreign objects intrusion status in the vulnerable area.

2. Algorithm research

2.1. Mask-Rcnn

In 2018, He et al. Proposed the Mask-RCNN[5] algorithm, which is an instance segmentation algorithm with two branches. It can achieve classification, regression and segmentation at the same time. The architecture is flexible and has good performance in instance segmentation. The surface of the frequently used rail will be bright due to friction[6], which is fairly different from the surrounding environment. This feature facilitates the use of neural network for instance segmentation. Therefore, this paper uses Mask-RCNN to segment the track line and obtain the rail position. Nevertheless, the convergence speed of the traditional Mask-RCNN network is slow, so this paper introduces the pre-training model of Mask-RCNN on coco data set, so that the network training has a high starting point. Based on this pre-training model, the data set of track is used to train the network. The Mask-RCNN neural network architecture adopted in this paper is shown in figure 1.

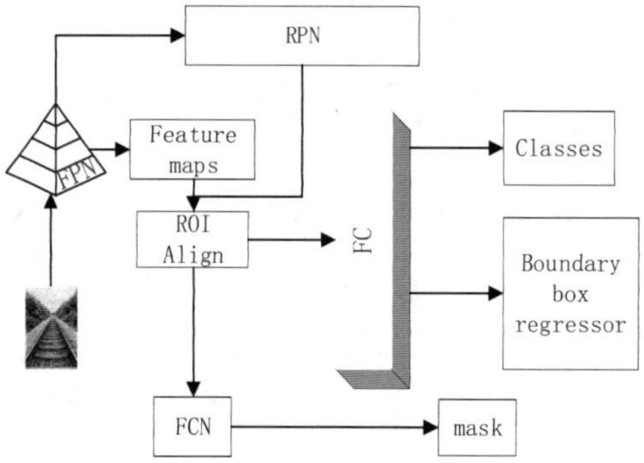

Figure 1. Network architecture of Mask-RCNN

Resnet101 residual neural network and the feature pyramid are used to extract the features of the input pictures, and five size feature maps are generated. Three candidate boxes of different sizes are generated at each central point of each feature map. The region proposal network (RPN) is used to perform operations such as non maximum suppression(NMS), filter the candidate boxes, and finally obtain regions of interest (ROI). The traditional pooling operation will generate offset when the coordinates are floating-point numbers after two quantization. Mask-RCNN creatively uses the RoI Align(Region of Interest Align) to avoid candidate frame offset and greatly improve the pixel level segmentation accuracy[7]. Use equation (1) to calculate which feature layer the ROI belongs to, so as to accurately map the ROI to each feature layer.

$$k = k_0 + \log_2(\sqrt{wh} / 244) \quad (1)$$

Where w and h respectively represent the width and height of ROI, k_0 is 4, which is the mapped feature layer when w and h are 244, and k is the mapped feature layer after calculation.

After RoI Align processing, the results will be sent to convolution layer and full connection layer. The steps of the mask prediction, category prediction and position regression are carried out to obtain the results of instance segmentation.

2.2. Sliding window ORB feature matching method

ORB[8] is a fast feature extraction and description algorithm, which is mainly divided into two steps. The first step is to use FAST (Feature from Accelerated Segment Test) algorithm to find feature points in the image. The second step is to use BRIEF (Binary Robust Independent Elementary Features) algorithm to construct binary descriptors for feature points, and compare the similarity of feature points by calculating the distance between descriptors.

In the FAST algorithm, there is a step to find features. In this step, a pixel in the image is in comparison with its surrounding pixels. When the gray value difference between the pixel and the surrounding pixels exceeds the threshold, the two pixels are considered to be different pixels. In this paper, the pixel is compared with the surrounding 16 pixels. If more than 12 pixels are different from the pixel, the pixel is determined as a feature point. In the process of similarity comparison using BRIEF algorithm, binary feature vector descriptors are created based on feature points. Calculate the Hamming distance between the descriptors of the two images, and reserve two optimal values for each feature point. If the ratio of the two optimal values exceeds 0.7, the number of the same points will increase by 1. Calculate the proportion of the same points in the total detection points, and the consequence is the similarity of the two images, and the calculation speed is fast.

The traditional ORB feature matching algorithm compares the entire image with the background image, which will lead to the decrease of the overall influence of small foreign objects and foreign objects far from the camera in the actual scene, and the overall result of similarity is high, thus increasing the possibility of misjudgment. Therefore, this paper improves the algorithm and uses the sliding window ORB feature matching algorithm to detect the detected images, which effectively enhances the influence of small and far foreign objects in the determination of foreign objects. The specific steps are as follows:

(1) Unify the size of the image to be detected and the background image to 512 * 512.

(2) The image to be detected and the background image are divided into 16 regions. The same size window is used to slide on the two images to make the detection area one to one correspondence. The ORB algorithm is used to detect the image in the sliding window, and the similarity is obtained.

(3) Record the similarity detection results of each window, and take the minimum value as the similarity of the whole picture, as shown in equation (2).

$$M = MIN_{i=1}^{N}(M, f(i)) \tag{2}$$

Where M is the similarity detection result of ORB algorithm for the whole picture, and $f(i)$ is the similarity detection result of the ith window.

3. Experimental simulation

3.1. Track line detection

In this paper, we collected a large number of distant images of track lines, and randomly changed the brightness and contrast of the images to simulate different environmental conditions. Finally, 726 original images were obtained. The image size is unified to 512 * 512, and the track line is marked by the labeling software called labelme . The label is 'CL', 'CR', 'SL', 'SR' which corresponds to the left and right tracks in the curved track, and the left and right tracks in the straight track. Another 45 track line images, 30 images with foreign objects in the track environment and 2 background images were collected to verify the reliability of the method. Enter the dataset into the Mask-RCNN network model for training, and iterate 100 times. Comparing the loss between the true value and the predicted value of the network model and the original network model in the first 10 rounds of training, the data are shown in table 1.

Table 1. Loss comparison of different network models.

	1	2	3	4	5	6	7	8	9	10
Original network model	2.852	1.530	1.210	0.985	0.855	0.720	0.639	0.558	0.510	0.473
Network model in this paper	1.009	0.477	0.333	0.276	0.235	0.195	0.175	0.158	0.141	0.130

It can be seen from the data in table 1 that with the increase of training times, the loss value decreases continuously. In the first round of training, the loss value of the original network model is high, while the loss value of the network model in this paper has reached a low level and has been maintained at a low level, and the convergence rate of the model has been greatly improved. The performance of the original network model and the network model in this paper is evaluated using the coco dataset evaluation index in the 45 additional images collected in this paper. When IOU is 0.5, it is concluded that the accuracy of the latter in judging the type of track line is 39.7 % higher than that of the former. The detection results of the network model in this paper are shown in figure 2, which demonstrates the network model can segment the track lines well.

Figure 2. Segmentation results of Mask-RCNN model in this paper

3.2. Division of vulnerable zone

To determine the vulnerable area, it is necessary to determine the location of the track line. Input the images to be detected into the Mask-RCNN network, and output the types of track lines and mask information in the detected images. Get the position of the track line from the mask. The pixels in the detection results classified as the left orbit extend to the left, and the pixels in the right orbit extend to the right. The maximum and minimum values of the extended pixels in the image coordinate system are calculated respectively, and the quadrilateral vertex coordinates of the vulnerable area are obtained. The distant vertex coordinates are extended to 15 pixels, and the final area is the vulnerable area explored in this paper. Taking the images of straight rail and curved rail in the test set as an example, the segmentation results, the mask and the vulnerable areas are shown in figure 3. It can be seen from figure 3 that this method has good effect on dividing the vulnerable areas of curved rail and straight rail.

a)

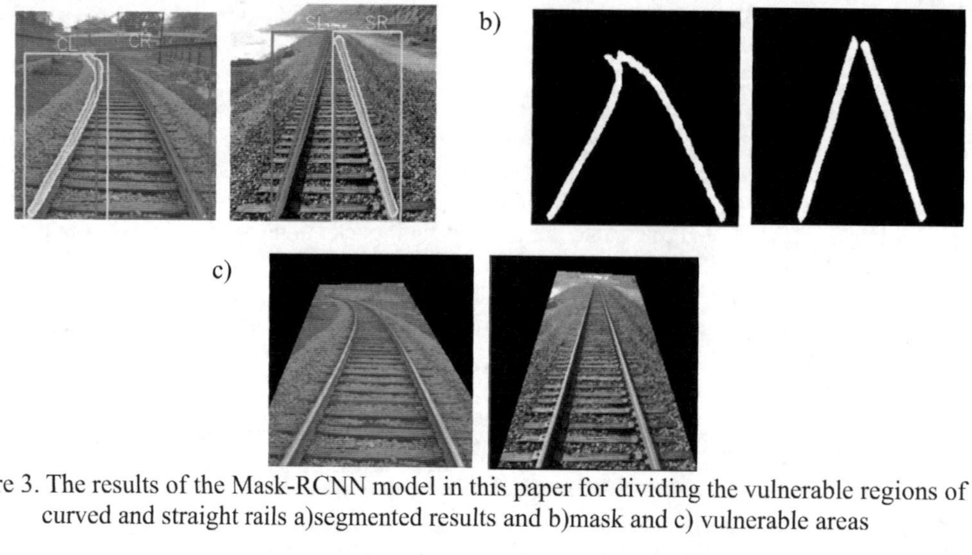

Figure 3. The results of the Mask-RCNN model in this paper for dividing the vulnerable regions of curved and straight rails a)segmented results and b)mask and c) vulnerable areas

3.3. Sliding window ORB feature matching

In this paper, the picture of the same orbital environment without foreign matters is taken as the background picture. The moving window ORB feature matching is used to detect the detected image and the background image. Set the similarity threshold to 0.9. When the final similarity consequence is greater than the threshold, it is judged that there is no foreign matter, and the picture is taken as the background picture. Otherwise, there is foreign matter and an alarm is given. Due to the continuous updating of background images, the influence of light, weather and other factors on image feature extraction can be ignored. In the test set, the sliding window ORB feature matching and the traditional ORB feature matching were used to detect 30 pictures with foreign objects in the vulnerable area. The detection results of the first 10 pictures were displayed, as shown in table 2.

Table 2. Comparison of similarity calculation results between sliding window ORB algorithm and traditional ORB algorithm.

	1	2	3	4	5	6	7	8	9	10
Sliding window ORB algorithm	0.005	0.750	0.652	0.704	0.550	0.500	0.777	0.777	0.700	0.118
Traditional ORB algorithm	0.853	0.923	0.887	0.880	0.677	0.864	0.810	0.771	0.808	0.759

It can be seen from the table 2 that under the same experimental conditions, the similarity obtained by the sliding window ORB feature matching algorithm is generally lower than that obtained by the traditional ORB feature matching algorithm. Moreover, in some pictures with foreign objects, the similarity detected by the traditional ORB feature matching algorithm has exceeded the threshold value of 0.9, and it is judged that there is no foreign object, resulting in misjudgment. In the test set of 30 track images with foreign objects, there is no misjudgment in the detection result of sliding window ORB feature matching algorithm, while the accuracy of the traditional ORB feature matching algorithm is 80%. The experimental results represent that the sliding window ORB feature matching algorithm has excellent performance in foreign objects.

4. conclusion

Aiming at the problem of foreign objects intrusion in the track line, which threatens traffic safety and personnel safety, this paper proposes a method for foreign objects detection in track area based on Mask-RCNN. The transfer learning mechanism is introduced into the traditional Mask-RCNN network to improve the convergence speed and segmentation accuracy of the network model. The Mask-RCNN network in this paper is used to detect track lines, and the mask data and category data in the detection results are used to obtain the vulnerable area. The sliding window ORB feature matching algorithm is used to detect the similarity between the detection image and the background image, and the presence of foreign objects in the detection image is judged. The accuracy of determination for foreign object is improved. Experiments represent that this method improves the detection accuracy and has a good performance on detection.

Reference

[1] Xiao T, Xu Y and Xu X. (2019)Methods for detecting and controlling foreign body invasion in urban rail transit tunnels. Electrical Engineering,20(S1):48-52+56.

[2] Song Y, Pan D and Han Kn. (2018)Track detection approach in complex environment based on saliency features.Journal of Railway Science and Engineering,15(04):871-879.

[3] Guo B, Dong Y. (2017)Railway track detection algorithm based on piecewise curve model. Journal of Railway Science and Engineering,14(02):355-363.

[4] Zhao Y, Xiong Z , Duan S,Zhou S AND Cui Y. (2020).Improved ORB Based Image Registration Acceleration Algorithm in Visual-Inertial Navigation System. In:2020 Chinese Automation Congress (CAC).Shanghai. 5714-5718.

[5] He K , Gkioxari G , Dollár P and Girshick R. (2017). Mask R-CNN. In:2017 IEEE International Conference on Computer Vision (ICCV). Venice, Italy. pp. 2980-2988.

[6] Huang Z, Chang N, Zhen S, Liang M, Wu Z, Liang C and Li T. (2018)Research on obstacle distance detection method of railway rail. Journal of Railway Science and Engineering, 15(3):763-769.

[7] Wang J, Sun R and Wang Y. (2021)Gesture Recognition Based on Dual Channel Mask RCNN. Journal of Changchun University of Science and Technology(Natural Science Edition),44(03):109-117.
[8] Jiao Y, Wang Z, (2019).Stereo vision SLAM system based on ORB. In:2019 IEEE 4th Advanced Information Technology, Electronic and Automation Control Conference (IAEAC). Chengdu, China. 2019: 418-421.

Optimization of cavitation characteristics of axial flow submersible pump

Hansheng Pei, Xuemin Wang

Shanghai Institute of Technology, Shanghai 201418, China

13162913202@163.com

Abstract. In view of the problem that the actual performance of axial-flow pump is less than the theoretical design value, Taking the axial-flow pump with a designed flow rate of 350L/s as the research object. The theoretical relationship between impeller head and outlet installation angle was studied. The influence of different blade angles on the external characteristics of the impeller was analyzed. Through the analysis of different external characteristics corresponding to different blade outlet installation angles, the influence of blade installation angles on shaft oil pump performance is verified, and the theoretical basis for the structural optimization of axial flow pump is provided.

1. Introduction

Axial flow pump has the characteristics of large flow, low head, and large specific speed. It is widely used in flood drainage and flood control, agricultural irrigation, and water conservancy projects. The blade outlet installation angle is the main reason that affects the performance of the axial flow pump. In the design of an axial flow pump, it is great significance to optimize the blade outlet installation angle.

At present, scholars have conducted a lot of research on how to improve the performance of the axial-flow pump in the design process. Hou[1]taking a horizontal axial flow pump as the research object, analyze the influence of different blade installation angles on cavitation. It was found that the cavitation of the axial flow pump first occurred at the tip of the inlet blade side, and the critical cavitation allowance decreased with the decrease of the blade installation angle. Li[2] Take a three-dimensional Reynolds average N-S equation to simulate the axial flow pump blades with different inlet blade mounting angles. It was found that under design conditions, the hydraulic performance of the blades could be optimized by increasing the inlet blade installation angle by 2°. Han[3] analyzed the flow field of the axial-flow pump with different blade installation angles. It was found that the optimal hydraulic performance could be obtained when the inlet installation angle of the blade increased by 2° and the outlet installation angle increased by 6° under design conditions. Increasing blade outlet installation angle in off-design conditions can increase efficient working range in large flow conditions. Li[4] Take simple algorithm to simulate the axial flow pump. Found that too small or too large blade inlet angle will affect the efficiency and cavitation performance of axial flow pump, Liu [5-6] by changing the blade installation angle to improve the guide vane rectification effect, the efficiency of axial flow pump has been improved, Shi[7] take whole flow channel of the axial flow pump with a different blade mounting angles was numerically simulated. The appropriate blade mounting angle of the axial flow pump was given by predicting the static pressure distribution, velocity distribution, and head of the axial flow pump under different blade mounting angles. Yu[8] by adjusting inlet angle and moving the inlet edge forward, found that the external characteristics and cavitation performance of the submersible sewage pump at a high specific speed can be improved without changing the impeller size. Wang [9] analyzed the influence of blade installation angle on the performance of axial flow pump, and it was found that the influence of blade installation angle on the efficiency and pressure distribution of impeller had little difference in the optimal working condition. Wu[10] found that the head flow rate and efficiency of the axial flow pump increased with the increase of blade mounting angle. Li[11] researched aviation characteristics of axial flow pumps at different blade mounting angles. It was found that cavitation resistance of axial flow pump decreased with the increase of blade mounting angles.

So far, there are few studies on blade outlet placement angle in axial flow pump design based on the calculation theory of blade outlet placement angle of axial flow pump, this paper puts forward the correction coefficient of blade outlet placement angle, and finds the best design method in the process of impeller design of axial flow pump.

2. Blade outlet placement Angle theory

2.1. Infinite blade design theory and cylinder independence hypothesis

The liquid flow in the axial flow pump is cylindric flow with the pump axis as the centerline, there is no radial velocity, and the cylindric flow on different radii is unrelated to each other. Therefore, in the design of axial flow pump impeller, the cylinder independence assumption is often used to simplify the complex flow in the axial flow pump. After cutting the cylinder surface along the bus, the cylinder surface intersects with each blade, and the intersecting section can be expanded on a plane. The line out of each cascade airfoil on the expanded plane is straight, and the spacing between each cascade is the same, and the distance is $\frac{2\pi R}{Z}$, where Z is the number of blades and R is the radius of the cylinder surface. It can be seen that the spacing between cascades varies with the radius of the cylinder. Therefore, after cutting the cylinder with a different radius along the bus line, an infinite number of cascades can be developed on the plane. In the actual design process, only one airfoil in the flow around the cascade is studied to represent the flow of the whole cascade. This theoretical design method is the infinite blade theory in impeller design process, which is a method to simplify the design of blade airfoil based on the liquid flow characteristics in axial flow pump and the assumption of cylinder independence.

2.2. Determination of blade outlet placement Angle

In the design process of the axial flow pump, the blade outlet placement angle depends on the design parameters, according to the axial flow pump speed triangle theory, the value of the blade outlet placement angle is mainly determined by the blade outlet circumferential velocity. Because the actual axial-flow pump is not infinite blade, but finite blade, the circumferential velocity at the outlet can not reach the design value when the impeller with finite blade works under the condition that the circumferential velocity is calculated by using the theory of infinite blade. Axial flow pump head size mainly depends on the outlet circumferential speed v_{u_2}. Therefore when the blade shape designed by infinite blades is limited with multiple blades, the actual working outlet circumferential velocity is less than the design value, which is the main reason why the axial flow pump is less than the design value in actual working. Therefore, according to the velocity triangle theory, in the design process of axial flow pump, the outlet circumferential velocity should be corrected by increasing the blade outlet placement angle.

3. Computation model and numerical method

3.1. Calculation model

The axial flow pump calculation model adopted in this paper designed flow is 350*L/s*, head 4.58*m*, specific speed 1000, impeller blade diameter 300*mm*, hub diameter 120*mm*, and the number of blade is 4, speed *n*=1450*r/min*. The blade is designed by streamline method.,and divided into five radial surfaces. Firstly, according to the infinite blade theory and design parameters, the impeller outlet placement angle at five design lines was calculated, and then the impeller model was established in 3d modeling software. The calculation model is shown in figure 1.

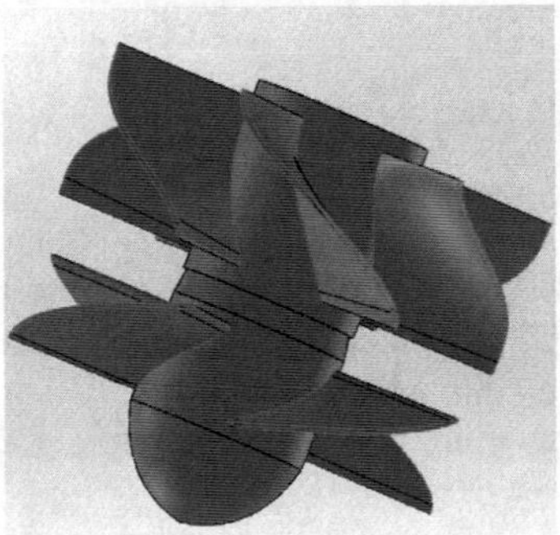

Figure 1. Calculation model of axial flow pump

3.2. Meshing and independent verification

In computational fluid mechanics, the mesh density has a certain influence on the accuracy of simulation results. In general, the much the mesh number is, the more accurate the calculated parameters such as head and efficiency are. However, too many grids will increase the time cost of calculation. Choosing an appropriate grid number can not only get excellent numerical results but also reduce the calculation time. So the influence of different meshes on the calculation results is considered. In this simulation, the polyhedral mesh was selected, the boundary layer was set as 10 layers, and the growth rate was 1.2. Due to the complexity of flow channels in the impeller and guide vane area, grid encryption was carried out in the area where the curvature radius of the impeller and guide vane was smaller. The relationship between different mesh numbers and the head is shown in figure. 2.

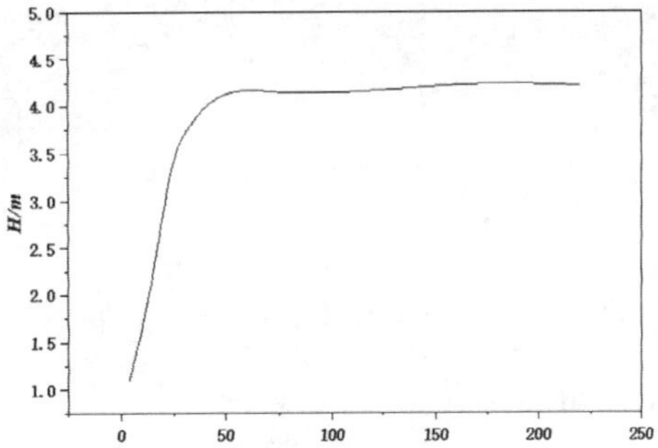

Figure 2. Relationship between grid number and head

It can be seen from the figure that when the number of grids is about 700,000, the increase of grids has little influence on the head of the axial-flow pump. So the grid model for calculating the axial-flow pump is about 800,000.

3.3. Governing equations and boundary conditions

In this numerical simulation, the mass flow inlet boundary condition and pressure outlet boundary condition are adopted, the reference pressure is 1bar, and the wall boundary condition is non-slip boundary condition. The impeller runner region is the rotating region, the guide vane, inlet and outlet are the static fluid region, and the rotation speed is the blade design rotation speed of 1450*r/min*. A three-dimensional incompressible N-S equation was used for the calculation model, with high resule as the convective discretization method. The convergence criterion was selected based on the average residual RSM of all control bodies, and the acceptance criterion was 10-5.

3.4. Analysis of numerical simulation results

According to the numerical simulation results, under the initial design of the blade shape based on the infinite blade theory, the axial flow pump head is 4.19*m*, less than the design head of 4.58m, the actual torque is 116.76*nm*, the shaft power is 17.72kW, the actual effective power is 14.35kW, and the efficiency is 80.98%.

According to the axial flow pump head calculation equation(1):

$$H = \frac{u_2 v_{u_2} - u_1 v_{u_1}}{g} \tag{1}$$

Where H is the head of the axial-flow pump, u_1 is the circumferential velocity at the inlet, u_2 is the circumferential velocity at the outlet, and v_{u1} is the circumferential velocity at the inlet. Since water enters the axial-flow pump vertically, the value of v_{u1} is usually 0 and v_{uz} is the circumferential velocity at the outlet, g is the gravitational acceleration. The circumferential velocity at the outlet u_2 is related to the speed of the driving shaft, and its expression is(2):

$$u_2 = \frac{nD\pi}{60} \tag{2}$$

Where D is the outer diameter of the impeller, and n is the actual speed of the impeller. Under normal circumstances, the impeller diameter D and speed are determined values is a certain value because the axial flow pump v_{u1}=0.The gravity g is a certain value. According to the calculation equation of the head, the head of the axial flow pump mainly depends on the circumferential velocity of the blade outlet. Therefore, to increase the circumferential velocity of blade outlet, the blade outlet placement angle should be modified during design. In the design of an axial flow pump to achieve the actual head requirements of the outlet circumferential velocity value should be 1.97m/s.The actual simulation circumferential velocity is 1.8m/s.Axial flow pump blade outlet placement angle and outlet circumferential velocity are shown in figure 3.

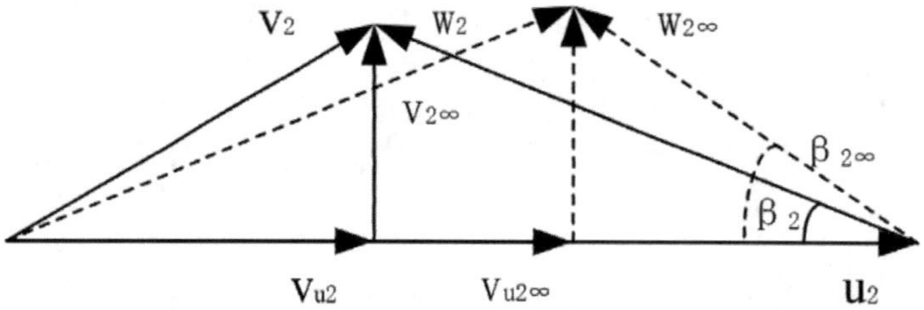

Figure. 3. Axial flow pump velocity triangle and outlet installation Angle

Figure 4 is the pressure distribution cloud diagram of impeller pressure surface and suction surface without correction coefficient. It can be seen that at the part of the blade inlet, Some areas of low pressure are generated. The suction surface of the impeller rises continuously from the blade inlet to the blade outlet. The pressure value in the low-pressure area is lower than the critical value of water gasification under atmospheric pressure, so the cavitation phenomenon will occur in the low-pressure area at the blade inlet, leading to the reduction of performance. The cavitation region is shown in figure 5

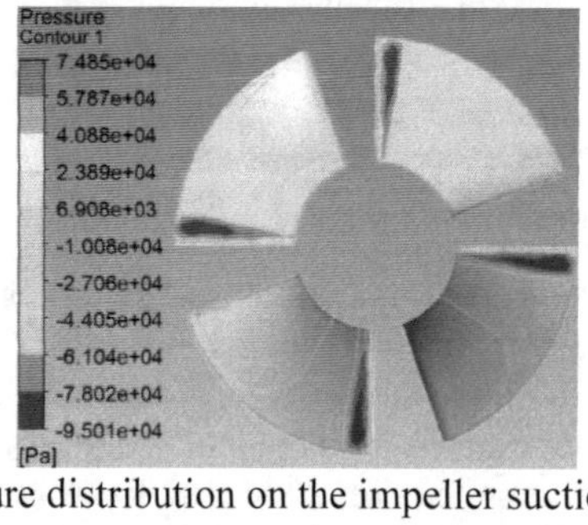

(a) Pressure distribution on the impeller suction surface

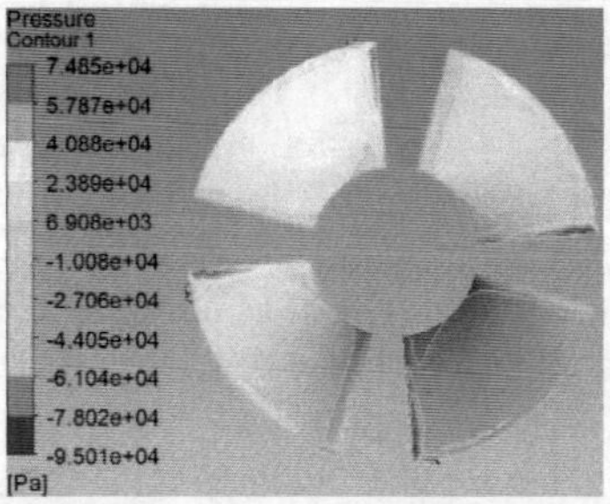

(b) Pressure distribution on the impeller pressure surface

Figure. 4. Cloud diagram of initial design blade pressure distribution

Figure. 5. Area of cavitation of initial design value

4. Optimization Design

4.1. Selection of adjustment coefficient of blade outlet placement Angle of axial flow pump

To meet the design requirements of axial flow pump head under the actual working environment, four optimization schemes were designed, and the correction coefficients of the blade outlet placement angle were selected as 1.1, 1.2, 1.3 and 1.4 times respectively. Three-dimensional modeling and numerical simulation were conducted for the impeller, and the variation rules of the impeller head under different correction coefficients were statistically analyzed. The head comparison between the four models and the original model is shown in figure 6. Table 1 shows the simulation value of blade outlet circumferential velocity under different outlet placement angle correction coefficients.

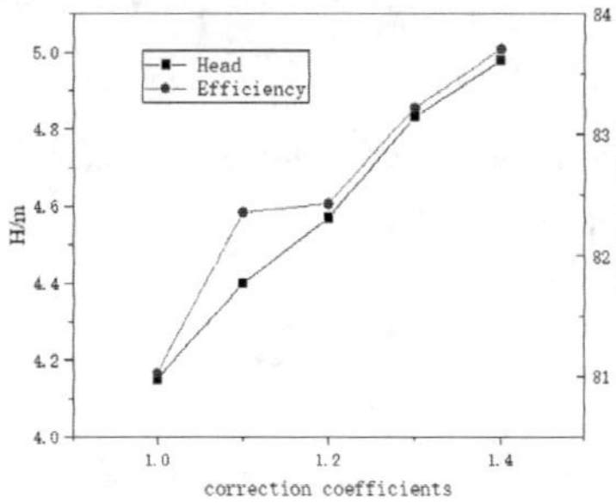

Figure. 6. Performance comparison results of different correction coefficients

Table 1. Outlet circumferential velocity with different correction coefficients.

Parameter	1.0	1.1	1.2	1.3	1.4
V_{u2}	1.8	1.89	1.96	2.07	2.14

It can be seen that with the increase of the correction coefficient of the blade outlet position angle, the circumferential velocity of the blade outlet of the axial flow pump is also increasing, and the actual head of the axial flow pump is also increasing, but in the design process of the axial flow pump, in addition to the head to achieve the actual work needs, also need to take into account the influence of other characteristics of the axial flow pump.

Table 2 shows the statistical results of external characteristics of the axial-flow pump, where H is head, T is torque and η is efficiency. According to Table 2, the torque and head of the axial-flow pump increase with the increasing correction coefficient of outlet placement angle, and the efficiency also increases. When the correction coefficient reaches 1.3, the head is greater than the design head value.

Table 2. Comparison table of external characteristics of different correction coefficients

Parameter	1.0	1.1	1.2	1.3	1.4
H/m	4.15	4.4	4.57	4.83	4.98
T/nm	115.55	120.52	125.07	130.93	134.21
η/%	81.2	82.36	82.43	83.22	83.71

Figure 7 shows four kinds of optimization schemes under the suction side of the impeller pressure distribution chart, it can be seen that, with the increase of outlet correction coefficient, pressure on the suction side of the blade outlet is also increased, this due to the increase of blade outlet angle, make blade outlet axial velocity increases, more fluid flow to the blade tip extruding, makes the blade outlet pressure increased. The maximum pressure on the suction surface of the impeller decreases, and the pressure distribution on the suction surface of the impeller becomes more uniform with the increase of the correction coefficient of the blade outlet placement angle.

(a)1.1 times correction factor

(b)1.2 times correction factor

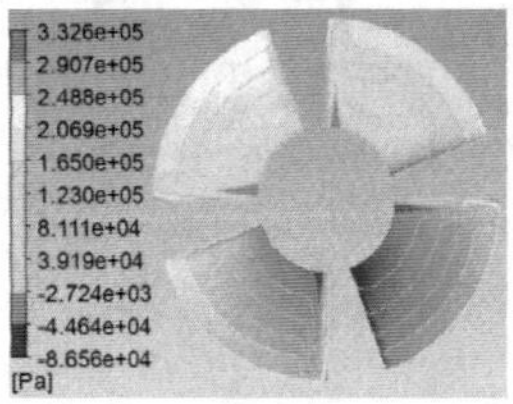

(c) 1.3-fold correction factor

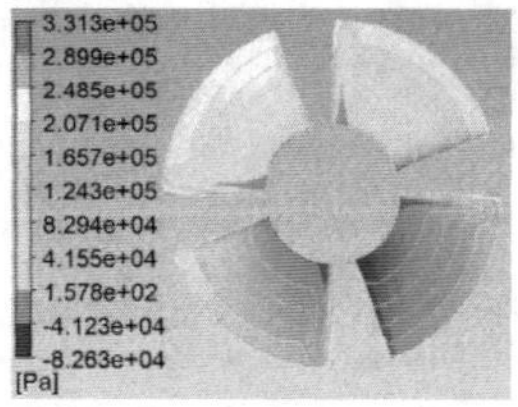

(d) 1.4-fold correction factor

Figure. 7. Cloud diagram of pressure distribution under different correction coefficients

Figure. 8 shows the gas volume distribution under different correction coefficients. It can be seen that under the 1.1 times correction coefficient, the highest part of the gas volume fraction reaches 0.4. With the increase of the correction coefficient of outlet placement angle, the maximum gas volume fraction decreases; under the 1.3 times correction coefficient, the gas volume occupies the least area. Cavitation occurs mainly at the outside edge of the blade inlet suction surface. The relative velocity of this part is the maximum value of the impeller area, and the pressure generated is lower than the critical value of liquid water gasification pressure at room temperature, which is relatively consistent with the actual situation.

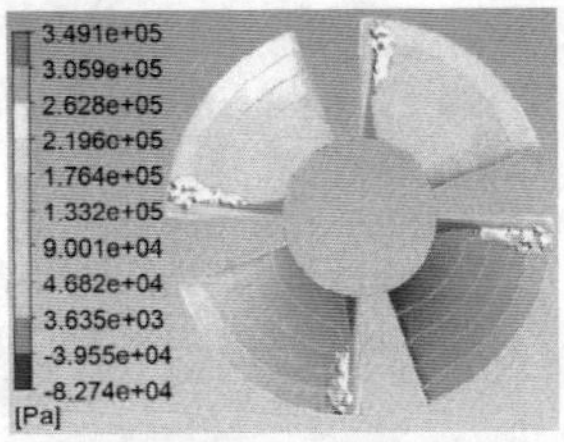

(a)1.1 correction factor

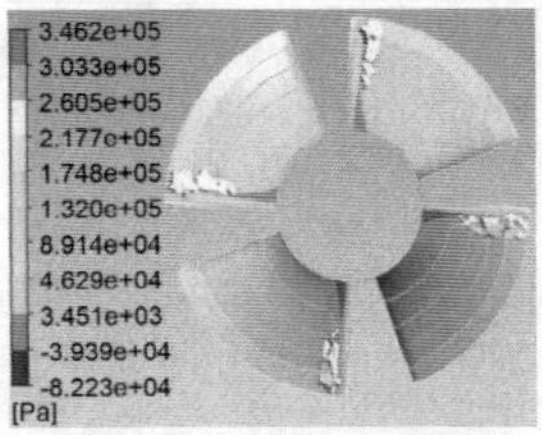

(b)1.2 correction factor

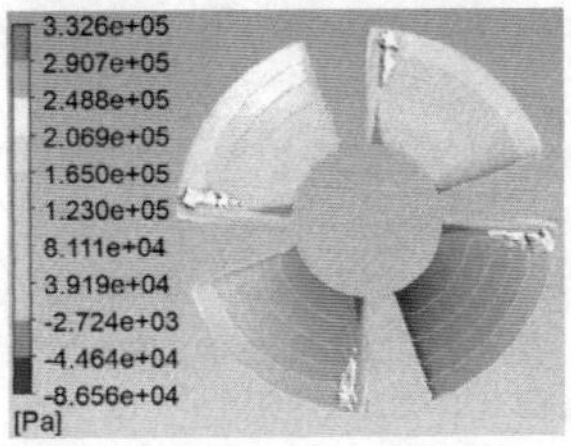

(c) 1.3 correction factor

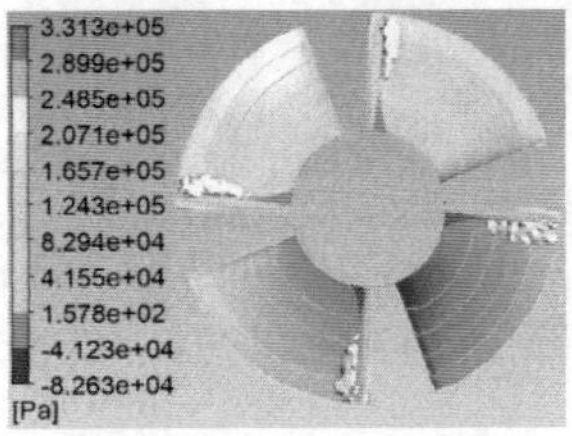

(d) 1.4 correction factor

Figure. 8. Cloud diagram of water vapor volume fraction distribution with different correction coefficients

5. Result analyze

Through introduction blade outlet angle correction coefficient on the initial design is optimized. The blade outlet placement angle of the impeller has changed, and the increase of blade outlet placement angle is the main reason to increase outlet circumferential velocity, from 1.3 times correction coefficient, The actual head of the impeller was 4.83m, which began to meet the actual design requirements, and the working efficiency reached 83.22%. The actual working head optimized by 1.4 times correction coefficient was 4.98m, and the working efficiency was 83.71%. With the increase of blade outlet placement angle, the hydraulic performance of axial flow pump is improved. However, the blade gasification area under the 1.3 times correction coefficient is significantly smaller than the 1.4 times correction coefficient, and the maximum volume fraction under the 1.3 times correction coefficient is smaller than the 1.4 times correction coefficient. Therefore, considering the head, efficiency, and blade cavitation performance, In this axial flow pump design process using infinite blade theory design because of the choice of 1.3 times the correction coefficient of blade outlet placement angle optimization design, to make it can achieve the best comprehensive performance effect.

6. Conclusion

By comparing the actual low working head and cavitation problems of axial-flow pump with the rotational speed of 1000, the outlet placement angle correction coefficient was proposed. CFD technology was used to simulate and analyze the internal flow field of axial-flow pump, and the following conclusions were obtained:

(1) Through the theoretical analysis of axial flow pump outlet installation angle optimization design, the feasibility of the optimized design was verified by CFD simulation, providing a strong basis for the performance optimization of the axial flow pump.

(2) Cavitation initiation of axial-flow pump mainly occurs at the inlet of blade suction surface. CFD calculation can accurately predict hydraulic characteristics such as pump head and the occurrence of cavitation. The occurrence of cavitation will affect blade pressure distribution, and then affect external characteristic parameters of the axial-flow pump, such as torque, efficiency, head and so on.

(3) The outlet installation angle of axial flow pump has influence on the performance of axial flow pump. In the axial flow pump structure of the theoretical design, outlet mounting angle should be corrected to ensure that the actual operation meets the design requirements.

References

[1] Hou C. Numerical calculation of cavitation performance of Axial flow Pump at different Blade Mounting Angles [D]. Xi 'an University of Technology, 2019.
[2] Li C, DING Hong. Analysis of Influence of Inlet Installation Angle on Performance of Axial Flow Pump Blade [J]. Heilongjiang Science and Technology Information, 2017.135-136.
[3] Han J, LI Chen, ZHONG Junjun, et al. Journal of Dalian Maritime University, 2014(3):5.89-93.
[4] Li T Q, SONG Shaolei, SHU Chunying. Water Pump Technology, 2013(3):3.27-29.
[5] Liu C J,Wang Y S,Wang L,Research on commutating effect optimization of waterjet axial flow pump Stator stator based on CFD method[J]. Journal of Ship Mechanics, 2010,14(5) : 466-470.
[6] Tang F P , Wang G Q. Influence of outlet guide vanes upon performances of water jet axial-flow pump[J]. Chuan Bo Li Xue Journal of Ship Mechanics, 2006, 10(6):19-26.
[7] Shi G P Guo X M, Chen H L. Analysis of internal flow characteristics of axial flow pump with different blade mounting angles [J]. Hydroelectric power,2012,38(1):84-87.
[8] Yu Z L, You F S, Lu H Z, et al. Cavitation Performance Analysis and Optimization of High Specific Speed Submersible Sewage Pump under Large Flow Condition [J]. General Machinery, 2019, 10:19-25.
[9] Wang T, Kong F Y,Yang S S,et al.Effects of blade angle distributions on performance of pump as turbine[J]. Transactions of the Chinese Society for Agricultural Machinery, 2015, 46(10): 80-85.
[10] Wu X F, Lu Y D, Tan M G et al. Transactions of the Chinese Society of Agricultural Engineering, 2018, 34(17): 54-61.
[11] Li P Z. Numerical Study on Cavitation Characteristics of Axial-flow Pump with Water Jet Propulsion [D]. Dalian Maritime University, Dalian, China, 2017.

Motion analysis of armored vehicle suspension based on ADAMS

***XinjianYuan**

Engineering Training Center, Natong Institute OF Technology, Nantong Jiangsu, China

*Email: 123562887@qq.com.

ABSTRACT: This paper takes the suspension of a 8×8 wheeled armored vehicle as the research object, uses Pro-E to model the armored vehicle, uses ADAMS simulation software to simulate the kinematics of the established suspension model, and analyzes the motion characteristics of the armored vehicle suspension from three aspects: positioning parameters, track width and suspension stiffness. It provides a reference for the optimal design of operational stability and ride comfort of wheeled armored vehicles.

1. introduction

As a relatively new type of armored vehicle equipment, wheeled armored vehicles have unique advantages in mobility, comfort, fast action, safety and trafficability. The design and research and development of wheeled armored vehicles has become one of the mainstream trends in the world armored vehicle market, which is of great significance to the improvement of military combat capability. There are many factors that affect the operational stability and ride comfort of wheeled armored vehicles, and the suspension system is an important one among many factors. Because wheeled armored vehicles have special requirements for operational stability and ride comfort, the research on the suspension system of wheeled armored vehicles is more important. The research on the wheeled suspension system is relatively few at home and abroad, so it is necessary to study it.

2. structural analysis of some 8 × 8 wheeled army armored vehicle

This paper Taking the 8×8 armored vehicle suspension as the research object, which is mainly composed of chassis and turret. The ISO coordinate system is used to establish the model[1].

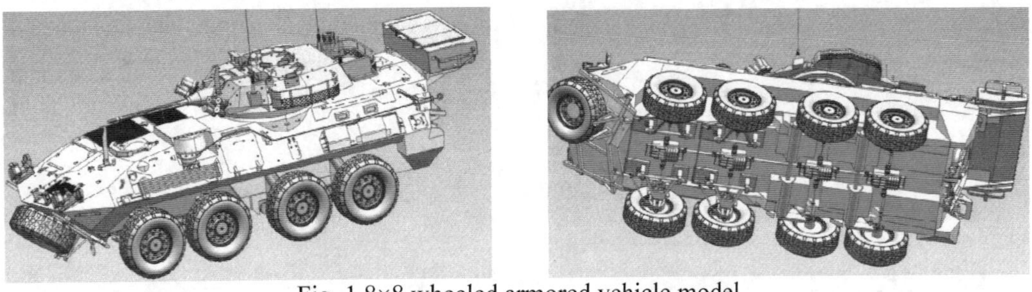

Fig. 1 8×8 wheeled armored vehicle model

Table 1 Shows the main parameters of the whole vehicle model of the armored vehicle.

Table 1 Vehicle model parameters

Parameter name and unit		numerical value
Front axle load	Sprung mass（Kg）	4766
	Unsprung mass（Kg）	500
Rear axle load	Sprung mass（Kg）	5233
	Unsprung mass（Kg）	500
Overall dimensions（mm）		8000×3000×2500
barycentric coordinate（mm）		2500×0×950
maximum speed（Km/h）		120
front-wheel camber（°）		1
toe-in gauguste（°）		0.2
caster angle（°）		1
Kingpin Inclination Angle（°）		8
Wheel Base（mm）		1420-2000-1420
Tread（mm）		2600
Tire type		16.00R20-18RP

The armored vehicle has four groups of axles. The first two groups of axles adopt McPherson independent suspension, the shock absorber adopts hydraulic two-way barrel shock absorber, and the two wheels are connected by transverse stabilizer bar. The up jump stroke and down jump stroke are 126.5mm and 83.5mm respectively. The suspension of the first two groups of axles is referred to as "front suspension"; The rear two groups of axle suspensions are referred to as "rear suspension" for short, and simple torsion beam suspension is adopted. The shock absorber still adopts two-way cylindrical hydraulic shock absorber, and the up jump stroke and down jump stroke are 143.5mm and 95.5mm respectively. Table 2 shows the suspension parameters of front and rear suspension.

Table 2 Suspension parameters

	Spring stiffness	Damping coefficient of shock absorber	Suspension space
Front suspension	20000 N/m	1.2 N.s/m	0.144 m
After the suspension	35000 N/m	0.9 N.s/m	0.12 m

3. establishment of vehicle model

Pro-E 3D drawing software is used to model the armored vehicle, as shown in Figure 2. Figure 3 shows the front suspension model and Figure 4 shows the rear suspension model.

Fig. 2 Assembly model of integral suspension of a 8×8 wheeled armored vehicle

Fig. 3 Front suspension model

Fig. 4 Rear suspension model

Fiala tire model is selected, which weakens various tire sideslip phenomena during armored vehicle driving, strengthens the calculation of lateral force and ensures the reliability of lateral force calculation.

4. calculation method of adams simulation software

4.1 Establishment and solution of dynamic equation

The motion equation established in ADAMS software is Lagrange multiplication [2], as shown in formula (1):

$$\frac{d}{dt}\left(\frac{\partial T}{\partial \dot{q}}\right)^T - \left(\frac{\partial T}{\partial q}\right) + \phi_q^T \rho + \theta_{\dot{q}}^T \mu = Q \qquad (1)$$

Transform the Lagrange multiplication equation of motion formula (1) [3], and obtain formula (2):

$$\begin{cases} F(q,u,\dot{u},\lambda,t) = 0 \\ G(u,\dot{q}) = u - \dot{q} = 0 \\ \Phi(q,t) = 0 \end{cases} \qquad (2)$$

Where: $\dot{q}, u$ represents the generalized velocity array; Φ represents the array of algebraic equations describing constraints; λ represents the force and reaction. Define the corresponding position objective function L_0, see formula (3)

$$L_0 = \frac{1}{2}\sum_{i=1}^{n} W_i (q_i - q_{0i})^2 + \sum_{j=1}^{m} \lambda_j^0 \Phi_j \qquad (3)$$

Where, n is the total number of generalized coordinates of the system; m represents the number of system constraint equations; Φ_j and λ_j^0 respectively represent the constraint equation and the Laplace multiplier corresponding to the constraint equation; q_{0i} represents the initial coordinate value set by the user or the default coordinate value set by the program; W_i represents the weighting coefficient corresponding to q_{0i}. If L_0 is the minimum value, then $\frac{\partial L_0}{\partial q_i} = 0$ and $\frac{L_0}{\partial \lambda_j^0} = 0$ can obtain:

$$W_i (q_i - q_{0i}) + \sum_{j=1}^{m} \lambda_j^0 \frac{\partial \Phi_j}{\partial q_i} = 0 \ , \ i = 1,2,\ldots,n; \ \Phi_j = 0, \ j = 1,2,\ldots,m.$$

The corresponding function form is: $\begin{cases} f_i(q_k, \lambda_{0i}) = 0 & k = 1,2,\ldots,n \\ g_j(q_k) = 0 & l = 1,2,\ldots,m \end{cases}$

Journal of Physics: Conference Series **2365** (2022) 012007 doi:10.1088/1742-6596/2365/1/012007

Then define the corresponding speed objective function L_1, as shown in formula (4):

$$L_1 = \frac{1}{2}\sum_{i=1}^{n} W_i'(\dot{q}_0 - \dot{q}_{0i})^2 + \sum_{j=1}^{m} \lambda_j' \frac{d\Phi_j}{dt} \tag{4}$$

Where, $\dot{q}_{0i}$ represents the initial speed value set by the user or the default speed value set by the program; W_i' represents the weighting coefficient corresponding to $\dot{q}_{0i}$; $\frac{d\Phi_j}{dt}$ represents the velocity constraint equation; λ_j' represents the Laplace multiplier corresponding to the velocity constraint equation. Take L_1 as the minimum value and get formula (5) from $\frac{\partial L_1}{\partial q_i} = 0$ and $\frac{L_1}{\partial \lambda_j^0} = 0$:

$$\begin{cases} \frac{\partial L_1}{\partial \dot{q}_i} = W_i'(\dot{q}_i - \dot{q}_{0i}) + \sum_{j}^{n} \lambda_j' \left(\frac{\partial \Phi_j}{\partial q_i} \right) = 0 & i = 1,2,\ldots,n \\ \frac{\partial L_1}{\partial \lambda_j'} = \sum_{k}^{n} \left(\frac{\partial \Phi_j}{\partial q_k} \right) \dot{q}_k + \frac{\partial \Phi_j}{\partial t} = 0 & j = 1,2,\ldots,m \end{cases} \tag{5}$$

The above formula is a linear equation about $\dot{q}_k$ and λ_j', which can be solved directly in formula (5).The analysis of the initial acceleration and the initial Laplace multiplier can be directly determined by the second derivative of the system dynamics equation and the system constraint equation. Write the system dynamics equation in matrix form into component form [4]:

$$\begin{cases} \sum_{k=1}^{n} (m_{ik} q_k) \ddot{q}_k + \sum_{j=1}^{m} \lambda_j \frac{\partial \Phi_j}{\partial q_i} = Q(q_k, \dot{q}_k, t) & i = 1,2,\ldots,n \\ \frac{d^2 \Phi_j}{dt^2} = \sum_{i=1}^{n} \left(\frac{\partial \Phi_j}{\partial q_i} \right) \ddot{q}_i - h_j(q_k, \dot{q}_k, t) = 0 & j = 1,2,\ldots,m \end{cases} \tag{6}$$

$$h_j = -\left\{ \frac{\partial^2 \Phi_j}{\partial^2 t} + \sum_{i=1}^{n} \frac{\partial}{\partial t}\left(\frac{\partial \Phi_j}{\partial q_i} \right) \dot{q}_i + \sum_{i=1}^{n} \frac{\partial}{\partial q_i}\left(\frac{\partial \Phi_j}{\partial t} \right) \dot{q}_i + \sum_{i=1}^{n}\sum_{k=1}^{n} \left(\frac{\partial^2 \Phi_j}{\partial q_k \partial q_i} \right) \dot{q}_k \dot{q}_i \right\} \tag{7}$$

$\ddot{q}_k$ and λ_j' can be solved directly in formula (6) and formula (7).

4.2 Determination of objective function and optimization design variables

Through single-factor orthogonal test of design variables, the main design variables sensitive to the influence of wheel tread can be found out [5]. In the orthogonal test, the suspension structure size parameters are taken as variables, and the goal is to minimize the change of wheel base. According to the single-factor orthogonal test, parameters of design variables that have a great influence on the objective function can be obtained [6].

5. suspension motion analysis

5.1 nfluence analysis of front suspension wheel alignment parameters

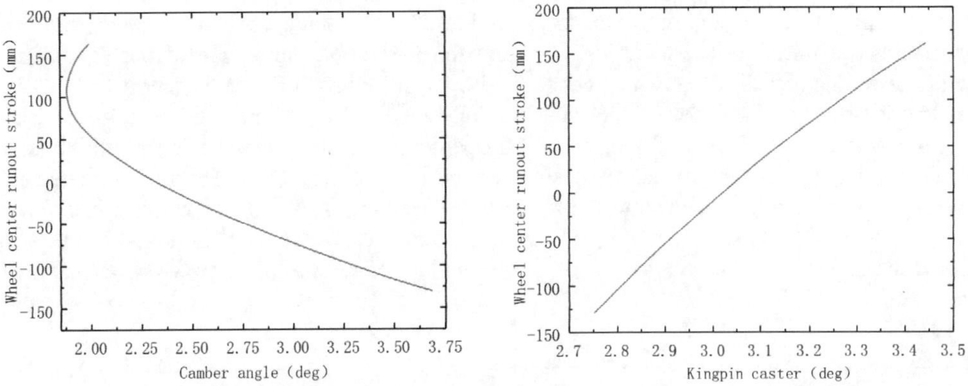

Fig. 5 Variation curve of camber and caster angle with wheel runout displacement

When the established suspension virtual prototype model is in the static equilibrium state, the initial value of wheel camber remains , and the variation relationship of wheel camber value with the runout stroke of wheel center is shown in Figure 5. The runout stroke of the whole wheel changes between -150mm-175mm, and the camber angle of the wheel changes between ; The variation of the initial value of in relative static balance is, which meets the design requirements of variation . The initial value of kingpin caster angle is 3.0°. It can be seen from the figure that the runout stroke of the whole wheel changes between - 125mm-150mm, and the kingpin caster angle changes between which shows a monotonic change curve; The variation of the initial value of in relative static balance is, which meets the design requirements of variation.

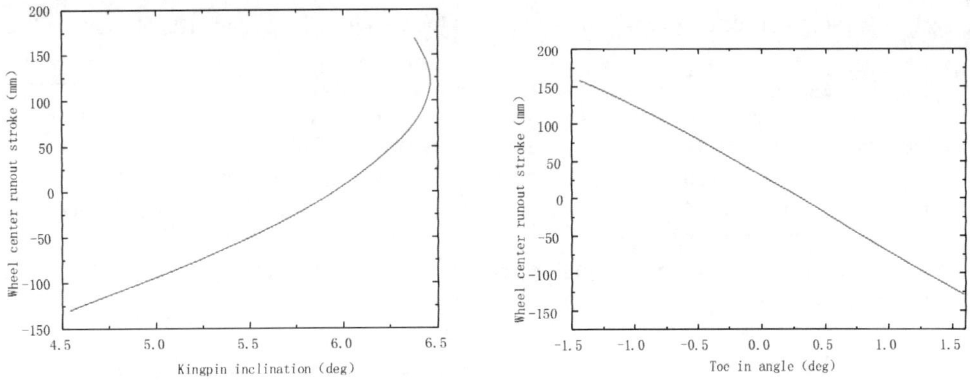

Fig. 6 variation curve of kingpin inclination and toe in with wheel jump displacement

When the established suspension virtual prototype model is in static equilibrium, the initial value of kingpin inclination is 5.9°, and the variation relationship of kingpin inclination with wheel center runout is shown in Figure 6. It can be seen from the figure that the runout stroke of the whole wheel changes between - 125mm-175mm, and the inclination angle of the wheel kingpin changes between 4.53°~6.47°, which is in the shape of left opening parabola; The variation of the initial value of 5.9° in relative static balance is -1.37°~1.57°, which meets the design requirements of variation -1.5~2°. The initial value of front wheel toe in is 12mm. It can be seen from the figure that the runout stroke of the whole wheel changes between - 125mm-160mm, and the front wheel toe in changes between -

Journal of Physics: Conference Series **2365** (2022) 012007 doi:10.1088/1742-6596/2365/1/012007

62.2mm-68.8mm, with a monotonic change curve; The variation of 12mm initial value in relative static balance is -74.23mm-56.81mm.

5.2 Influence analysis of rear suspension wheel alignment parameters

When the established suspension virtual prototype model is in the static equilibrium state, the camber angle maintains the initial value, which is 1.0°, and the variation relationship of the camber angle value with the runout stroke of the wheel center is shown in Figure 7. It can be seen from the figure that the runout stroke of the whole wheel changes between - 125mm and 175mm, and the camber angle changes between, which is in the shape of left opening parabola; The variation of initial value in relative static balance is, which meets the design requirements of variation. The initial value of rear wheel toe in is 6mm. It can be seen from the figure that the runout stroke of the whole wheel changes between - 125mm-160mm, and the front wheel toe in changes between - 1.9mm-11.6mm, with a monotonic change curve; The variation of initial value of 6mm in relative static balance is -4.1mm-5.6mm.

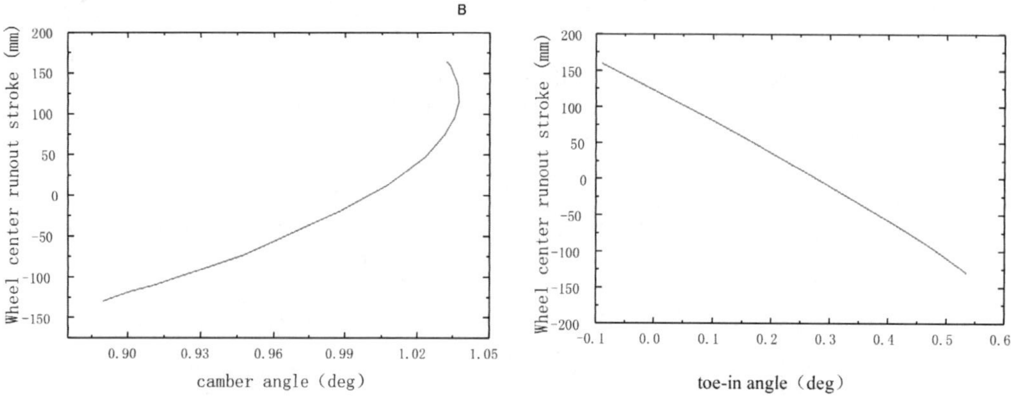

Fig. 7 variation curve of rear wheel camber and toe in with wheel runout displacement

5.3 Analysis on the influence of front and rear suspension axle track

When the established suspension virtual prototype model is in the static equilibrium state, the initial value of the front wheel track is 2550.46mm, and the variation relationship of the front wheel track value with the runout stroke of the wheel center is shown in Figure 8. It can be seen from the figure that the runout stroke of the whole wheel changes between - 125mm-175mm, and the track width changes between 2537.30mm~2557.29mm, which is in the shape of left opening parabola; The variation of the initial value of 2550.46mm in relative static balance is -13.16mm-6.83mm. In the design process, the variation range of the relative initial value of the wheel track is required to be within ±4mm, so it has exceeded the reasonable requirements.

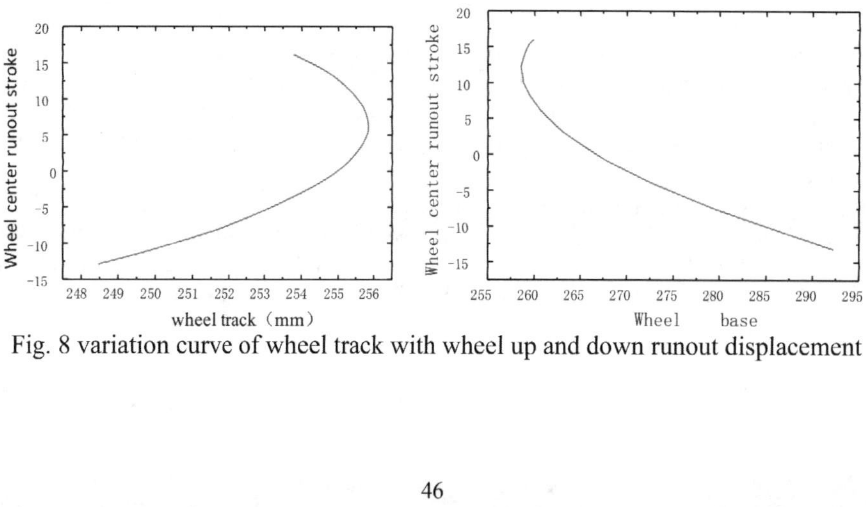

Fig. 8 variation curve of wheel track with wheel up and down runout displacement

 IOP Publishing
2365 (2022) 012007 doi:10.1088/1742-6596/2365/1/012007

The initial value of the rear suspension axle track is 2578.35mm. It can be seen from the figure that the runout stroke of the whole wheel varies from - 130mm to 160mm, and the rear suspension axle track varies from 2577.67mm to 2580.74mm, with a monotonic change curve; During relative static balance, the variation of the initial value of 2578.35mm wheelbase is -0.68mm-2.39mm. In the design process, the variation range of the relative initial value of wheelbase is required to be included, so it meets the reasonable requirements.

5.4 Influence analysis of suspension stiffness

5.4.1 Influence analysis of front suspension stiffness

Influence analysis of front suspension stiffness when the established suspension virtual prototype model is in static equilibrium, the initial value of left suspension stiffness is 118N/mm and the initial value of right suspension stiffness is 110N/mm. The variation relationship between left and right suspension stiffness with wheel center runout is shown in Figure 9. It can be seen from the figure that the runout stroke of the whole wheel varies from - 130mm to 160mm, the stiffness value of the left suspension varies from 105N/mm to 150N/mm, and the stiffness value of the right suspension varies from 96N/mm to 142N/mm, with a monotonic change curve; When the stiffness value of the left suspension is relatively static balance, the change of the initial value of the suspension stiffness is 118N / mm, which is 13N/mm-32n/mm; When the stiffness value of the right suspension is relatively static balanced, the variation of the initial value of the suspension stiffness of 110N/mm is 14N/mm-32N/mm.

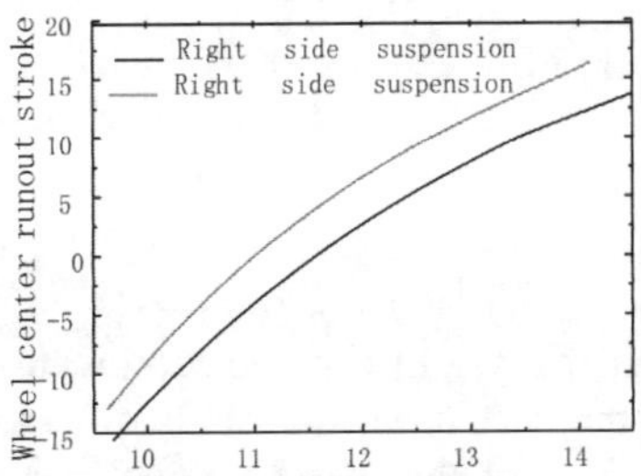

Fig. 9 variation curve of suspension stiffness with wheel up and down runout displacement

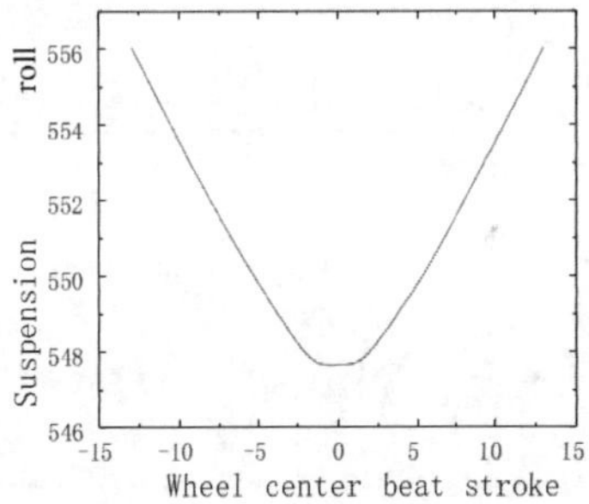

Fig. 10 variation curve of suspension roll angle stiffness

The roll angle stiffness of McPherson independent suspension can be calculated by the calculation method given in ADAMS / car software. Take the upper and lower connection points of the suspension elastic elements as the displacement reference points, calculate the relative displacement between the upper and lower connection points during the suspension movement, calculate the spring deformation and the elastic value of the deformation based on the relative displacement, and finally convert it into the elastic recovery torque T transmitted by the suspension system to the vehicle body, and calculate the corresponding vehicle body angle Φ_r, The calculation method of suspension roll angle stiffness is shown in formula (8):

$$K_{\Phi r} = dT / d\Phi_r \quad (8)$$

Draw the calculated suspension roll stiffness value into curves as shown in Figure 9 and figure 10.

Compared with FIG.9 and FIG.10, it can be seen that in the initial position, the left wheel is located at the point of - 130mm in the coordinate system and the right wheel is located at the point of 130mm. At this time, the suspension roll stiffness reaches the peak, which is about 5560N·m/° With the up and down movement of the wheels on both sides (the left wheel moves up and the right wheel moves down), the suspension roll stiffness begins to show a downward trend, When the left and right wheels are moved to 0mm, the roll stiffness of the suspension also reaches the minimum value, which is about

5470N·m/°; As the wheels on both sides continue to move, the suspension roll stiffness value begins to rise again until the left wheel moves to the point of 130mm and the right wheel moves to the point of - 130mm, and the suspension roll stiffness value returns to the peak value of 5560N·m/° again.

5.4.2 Influence analysis of rear suspension stiffness

When the established suspension virtual prototype model is in the static equilibrium state, the initial value of the left suspension stiffness of the rear axle is 65N/mm, and the initial value of the right suspension stiffness is 68N /mm. The relationship between the left and right suspension stiffness and the runout stroke of the wheel center is shown in Figure 12. It can be seen from the figure that the runout stroke of the whole wheel varies from - 130mm to 160mm, the stiffness value of the left suspension varies from 25N/mm to 234N /mm, and the stiffness value of the right suspension varies from 28N/mm to 243N/mm, with a monotonic change curve; When the stiffness value of the left suspension is relatively static balanced, the variation of the initial value of 65N/mm suspension stiffness is -40n/mm -169n/mm; When the relative stiffness of the right suspension is -175n/ 175mm.

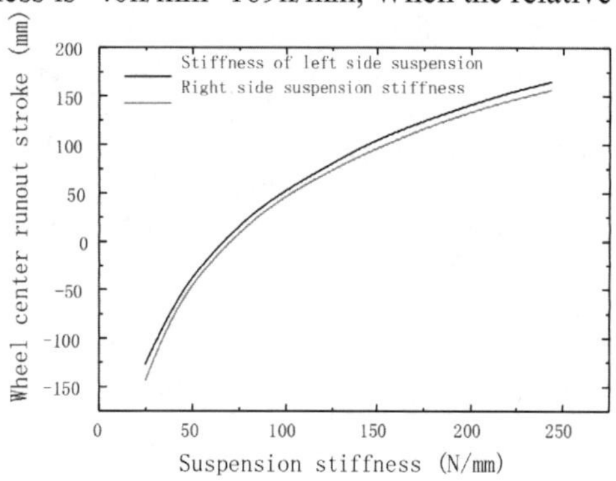

Fig. 11 Curve of suspension stiffness with wheel jump

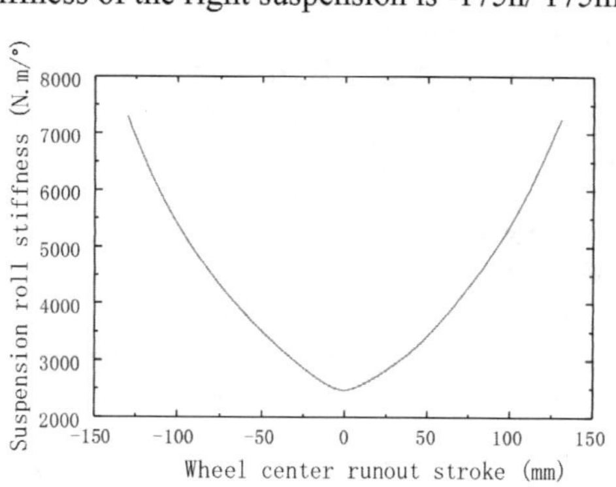

Fig. 12 Change curve of suspension roll Angle stiffness

In the initial position, the left wheel is located at the point of - 130mm in the coordinate system and the right wheel is located at the point of 130mm. At this time, the suspension roll stiffness reaches the peak, about 7250N·m/° With the up and down movement of the wheels on both sides (the left wheel moves up and the right wheel moves down), the suspension roll stiffness begins to show a downward trend. When the left and right wheels move to 0mm, The roll stiffness of the suspension also reaches the minimum value, about 2440N·m/°; As the wheels on both sides continue to move, the suspension roll stiffness value begins to rise again until the left wheel moves to the point of 130mm and the right wheel moves to the point of - 130mm, and the suspension roll stiffness value returns to the peak value of 7250N·m/° again.

6. Conclusion

This chapter mainly carries out kinematic simulation on the established vehicle suspension model in ADAMS simulation software, mainly analyzes the changes of wheel alignment parameters such as camber angle, front toe in, kingpin inclination angle and kingpin caster angle with the runout stroke of the wheel center, and then analyzes the motion characteristics of the suspension. The changes of wheel alignment parameters and track width of single trailing arm independent suspension with wheel jump meet the corresponding design indexes. The lower limit of the variation range of wheel kingpin inclination of Mai independent suspension exceeds the design indexes, and its track width value also exceeds the range specified in the design indexes when the wheel jump range is ± 40mm.

Acknowledgments

Young and middle-aged scientific research backbone training project OF Nantong Institute of Technology (ZQNGG407)

Reference

[1] Zhang L J.(2016)Modeling and Semi-active Control Research of the Vehicle Fitted with Hydraulically Interconnected Suspension System[D]. Hunan University. Changsha 11-20
[2] Gong H Q.(2015)Research of Light Bus Ride Comfort Based on Rigid-flex Coupled Virtual Prototype[D].Wuhan University of Science and Technology. Wuhan 26-35
[3] Sun Z M.(2014)Kinematics Simulation and Optimization Design of a Vehicle Suspension System Based on ADAMS[D]. Changchun University of Technology. Chang chun 46-50
[4] Peng P P.(2015)Studies of 8×8 Multi-axle Vehicle's Handling and Stability.Beijing Institute of Technology[D]. Beijing 44-56
[5] Feng Y J.(2014)Fatigue life analysis of automobile wishbone based on rigid-flexible coupling simulation [D]. Jilin University.Chang chun 40-47
[6] Yong X U.(2013)The optimal design of the steering mechanism and the torsion bar for the front double. wishbone suspension[D].HuNan University. Changsha 43-54

A positioning correction method for turbine blade film cooling holes

Pengfei Li[1], Jianmei Guo[2], Yanhua Cai[2], Dongbo Wu[1*], Jiawei Liang[1], Hui Wang[1], and Xiao Han[2]

[1]Department of Mechanical Engineering, Tsinghua University, Beijing, China

[2]Beijing Institute of Electro-machining Co.,Ltd., Beijing Key Laboratory of Electro Discharge Machining Technology, Beijing, China

[*]Corresponding author's e-mail: wdb4451253@163.com

Abstract. It is significant to improve the machining accuracy of film cooling holes on aero-engine turbine blades. This paper presents a method to correct the position of film cooling holes on turbine blades. In this method, the pre-positioning algorithm and the precise positioning algorithm are proposed. The pre-positioning algorithm introduces the blade structure information. In the precise positioning algorithm, 3D coordinate transformation, ∠B and ∠C of 5-axis machine tool updating, iteration points creating based on the normal vector are designed. The operation of the gas film cooling hole correcting by the method is given. Experiment shows that, after 7 iterations, the error is reduced to 0.0344mm, achieving high-precision positioning of turbine blades. Furthermore, the coordinates are corrected to realize the high-precision machining of the film cooling holes, which can improve the cooling performance of turbine blades.

1. Introduction

Turbine is one of the most important components of aero-engine, which bears great thermal and mechanical loads, and its working environment is particularly bad. In the process of aero-engine operation, turbine blades are subjected to high temperature and high-pressure gas for a long time. Reducing the temperature of turbine blades and improving the heat dissipation efficiency of turbine blades are key to improving the performance of aero-engines [1].

Film cooling technology is one of the most important cooling technologies for aero-engine turbine blades, which is mainly realized through film cooling holes on turbine blades. The higher the machining accuracy of film cooling holes in turbine blades, the better the cooling effect will be. Therefore, it is of great significance to improve the machining accuracy of film cooling holes on turbine blades for improving the performance and service life of turbine blades of aeroengines [2].

Turbine blades are usually made by one-time investment casting. Due to the influence of thermal stress and casting precision limitation in the casting process, there will be a certain profile error between the actual surface structure and the theoretical model of the turbine blade. Therefore, it is necessary to locate the actual position of the turbine blade during the processing of the film cooling hole [3].

The traditional positioning of turbine blades is mainly to manually position the position and angle of the film cooling hole, which is low accurate and reduces the production efficiency of turbine blades.

The existing film cooling hole positioning method using CMM to measure the turbine blade, then manually uses the data to position on the computer, and then inputs the positioning results to the CNC machine tool. Although this method has high positioning accuracy, the operation process is complicated and cannot be completed automatically [4]. The existing methods are limited for accurate positioning of turbine blades, which is difficult to achieve high precision machining of film cooling holes.

In the study of the positioning correction method for turbine blade film cooling holes, the definitions are shown in Table 1.

Table 1. Definitions.

Name	Symbol	Form	Coordinate system	Description
Theoretical point	O	(X,Y,Z)	Blade	Points on the theoretical model of the blade
Calculated point	S	(X,Y,Z)	Machine tool	Points output after iteration of the method
Converted point	S'	(X,Y,Z,B,C)	Machine tool	Points get from convert S to the five-axis machine tool coordinate system
Machine tool point	Q	(X,Y,Z,B,C)	Machine tool	Actual blade points get from the machine tool probe to measure
Measured point	P	(X,Y,Z)	Machine tool	Points get from convert Q to the blade coordinate system

2. Pre-positioning algorithm

In the design of the pre-positioning algorithm for turbine blades, the initial positioning of turbine blades is proposed by introducing the structural information of turbine blades, which will improve the positioning efficiency of the algorithm.

In the CAD model of turbine blade, six non-coplanar points are selected. In order to improve the positioning, the following conditions should be met when six points are selected: a) The normal vectors of points 1, 2 and 3 are roughly along the y axis; b) The normal vectors of points 4 and 5 are roughly along the x axis; c) The normal vector of points 6 is roughly along the z axis. The Pre-positioning algorithm is shown as follows.

1. Determine ***xOz*** plane by conditions as follows.
   a) Convert the points *1*, *2* and *3* by formula (1).

$$\begin{cases} y_1' = y_1 - y_0 \\ y_2' = y_2 - y_0 \\ y_3' = y_3 - y_0 \end{cases} \tag{1}$$

   b) Points *1'*, *2'* and *3'* are on the plane.
2. Determine ***yOz*** plane by conditions as follows.
   a) Convert the points *4* and *5* by formula (2).

$$\begin{cases} x_4' = x_4 - x_0 \\ x_5' = x_5 - x_0 \end{cases} \tag{2}$$

   b) Points *4'* and *5'* are on the plane.
   c) The ***yOz*** plane is perpendicular to the ***xOz*** plane.
3. Determine ***xOy*** plane by conditions as follows.
   a) Convert the point *6* by formula (3).

$$z_6' = z_6 - z_0 \tag{3}$$

   b) Point *6'* is on the plane.
   c) The ***xOy*** plane is perpendicular to the ***xOz*** plane and the ***yOz*** plane.
4. Determine *x* axis, *y* axis and *z* axis by conditions as follows.
   a) The intersecting line of the ***xOy*** and ***xOz*** planes is the *x* axis.
   b) The intersecting line of the ***xOy*** and ***yOz*** planes is the *y* axis.
   c) The intersecting line of the ***yOz*** and ***xOz*** planes is the *z* axis.
5. Determine the origin *O'*.
6. Determine the rotation matrix ***R*** and translation matrix ***t*** as follows.
   a) Generate a sphere with the origin *O'* as center and radius *r*=1.
   b) The intersections of *x*, *y* and *z* axis and sphere are corresponding to (1,0,0), (0,1,0) and (0,0,1) respectively, where $x, y, z > 0$.

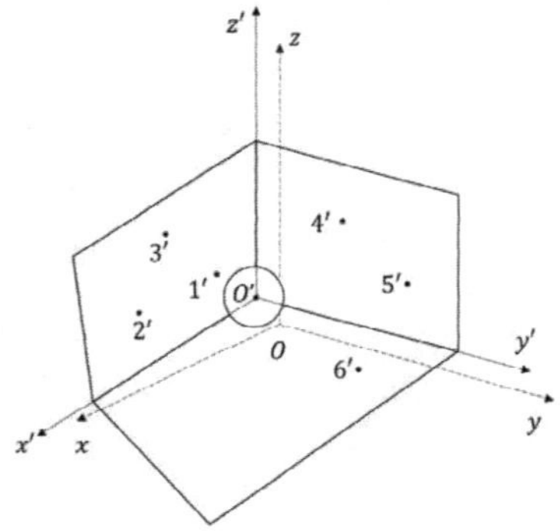

Figure 1. Pre-positioning algorithm.

3. Positioning algorithm

3.1. 3D coordinate transform

In the design of precise positioning algorithm for turbine blades, the algorithm is applied to calculate the measurement results iteratively for many times, so that the positioning error of turbine blades can meet the accuracy requirements.

d_{max}, the maximum distance between calculated point *S* and measured point *P*, is adopted as $error$ to measure the positioning effect of the algorithm. The smaller the value of d_{max} is, the better the positioning effect of the algorithm is. Actually, the positioning error is less than d_{max}.

Error equalling to zero means the calculated point coincides with the measured point completely, at which the turbine blade achieves absolutely accurate positioning and the positioning error is 0. However, since the actual shape of turbine blades cannot be the same as the theoretical shape, the value of $error$ in actual measurement cannot be 0.

For the positioning of turbine blades, the following formula (4) can be obtained,

$$\begin{gathered} \boldsymbol{R}A + \boldsymbol{t} = B \\ error = \sum_{i=1}^{N} \left\| \boldsymbol{R}A^i + \boldsymbol{t} - B^i \right\|^2 \end{gathered} \tag{4}$$

where, *A* refers to the theoretical point *O*, *B* refers to the measured point *P*.

The 3D coordinate transformation is shown as follows.

1. Calculate the centroids of two 3D point sets.

$$\text{centroid}_A = \frac{1}{N}\sum_{i=1}^{N} A^i$$
$$\text{centroid}_B = \frac{1}{N}\sum_{i=1}^{N} B^i \tag{5}$$

2. Solve the rotation matrix $\boldsymbol{R}$ by SVD [5].

$$\boldsymbol{H} = (A - \text{centroid}_A)(B - \text{centroid}_B)^T$$
$$[\boldsymbol{U},\boldsymbol{S},\boldsymbol{V}] = SVD(\boldsymbol{H}) \tag{6}$$
$$\boldsymbol{R} = \boldsymbol{V}\boldsymbol{U}^T$$

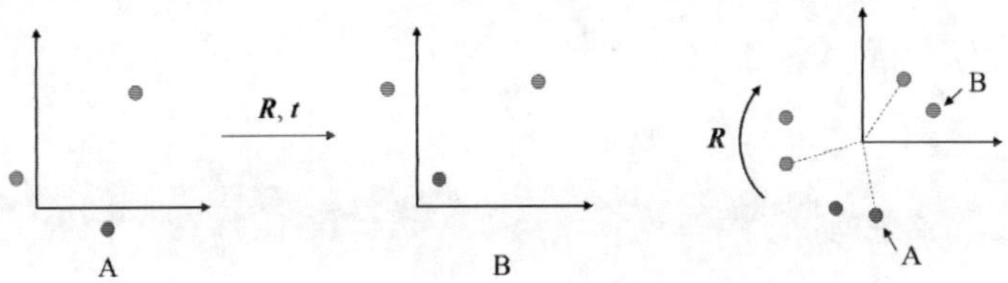

Figure 2. Solve the rotation matrix $\boldsymbol{R}$.

3. Solve the translation matrix $\boldsymbol{t}$.

$$\boldsymbol{R}A + \boldsymbol{t} = B$$
$$\boldsymbol{R}\cdot\text{centroid}_A + \boldsymbol{t} = \text{centroid}_B \tag{7}$$
$$\boldsymbol{t} = \text{centroid}_B - \boldsymbol{R}\cdot\text{centroid}_A$$

3.2 Update ∠B and ∠C of 5-axis machine tool

In the practical operation, in order to improve the positioning efficiency, the five-axis machine tool can be designed to always feed along the normal vector direction of the blade.

Design idea of the method is as follows: a) According to the solved R matrix, the theoretical normal vector N is updated. Rotate the current normal vector to the vertical direction of the machine tool to obtain the new $\angle B$ and $\angle C$ and the new X, Y, Z coordinates; b) Ensure that the feed direction of the machine tool spindle is always along the normal vector direction of the blade.

Update algorithm is shown as follows.

1. Obtain the new normal vector direction $\boldsymbol{N'} = \boldsymbol{R}\cdot N$.
2. The rotation mode of the machine tool spindle is shown in Figure 3, the calculation expressions of $\angle B$ and $\angle C$ are shown as formula (8).

$$\angle C = \arcsin\left(\frac{|y|}{\sqrt{x^2+y^2}}\right)$$
$$\angle B = \arccos\left(\frac{|z|}{\sqrt{x^2+y^2+z^2}}\right) \tag{8}$$

3. Modify ∠B and ∠C, the modify expression is as formula (9) and formula (10).

$$\angle C = \begin{cases} C, x \geq 0, y \leq 0 \\ 180^\circ - C, x < 0, y \leq 0 \\ 180^\circ + C, x \leq 0, y > 0 \\ 360^\circ - C, x > 0, y > 0 \end{cases} \tag{9}$$

$$\angle B = \begin{cases} -B, z \geq 0 \\ -\left(180^\circ - B\right), z < 0 \end{cases} \tag{10}$$

4. Convert the calculated point S to obtain the converted point S' with the new $\angle B$ and $\angle C$.

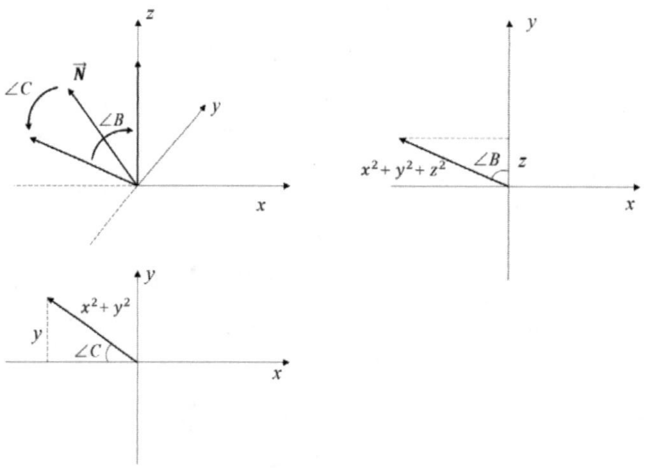

Figure 3. Rotation mode of the machine tool spindle and calculation principle of $\angle B$ and $\angle C$.

3.3. *Create iteration points based on the normal vector*

In order to further improve the positioning efficiency, it can be created more iteration points based on the normal vector of turbine blades.

Create solution is shown as follows.

1. Set distance l;
2. Solve $\boldsymbol{R}$ and $\boldsymbol{T}$ by measured point P and theoretical point O.
3. Obtain the new normal vector direction $\boldsymbol{N'} = \boldsymbol{R} \cdot N$.
4. Create new 6 points by $l \cdot N' + P$ to constitute 12 points with P.
5. Create new 6 points by $l \cdot O + P$, to constitute 12 points with O.
6. Solve the new $\boldsymbol{R}$ and $\boldsymbol{T}$ by two groups of 12 points.
7. Repeat 4, 5 and 6 to create new 6 points by the new $\boldsymbol{R}$ and $\boldsymbol{T}$.
8. Stop until $\boldsymbol{R}$ and $\boldsymbol{T}$ no longer change to obtain the final $\boldsymbol{R}$ and $\boldsymbol{T}$.

4. Method Application

4.1. *Correction design*

The process of correcting the coordinates of the turbine blade film cooling holes is as follows:

1. Calculate the theoretical normal vector of gas film holes.
2. Calculate X, Y, Z coordinates of gas film holes when $\angle B$ and $\angle C$ are set to zero.
3. Correct the X, Y, Z coordinates of the gas film hole by the $\boldsymbol{R}$ and $\boldsymbol{t}$ obtained by the method.
4. Calculate the $\angle B$ and $\angle C$ when the normal vector is rotated to the vertical direction.

5. Convert the X, Y, Z coordinates of gas film holes to the X, Y, Z, B, C, in which the $\angle B$ and $\angle C$ values are obtained in step 4.
6. Finally, the modified gas film hole coordinates is obtained.

4.2. System design

According to above positioning algorithms and correction design, the system is shown as figure 4.

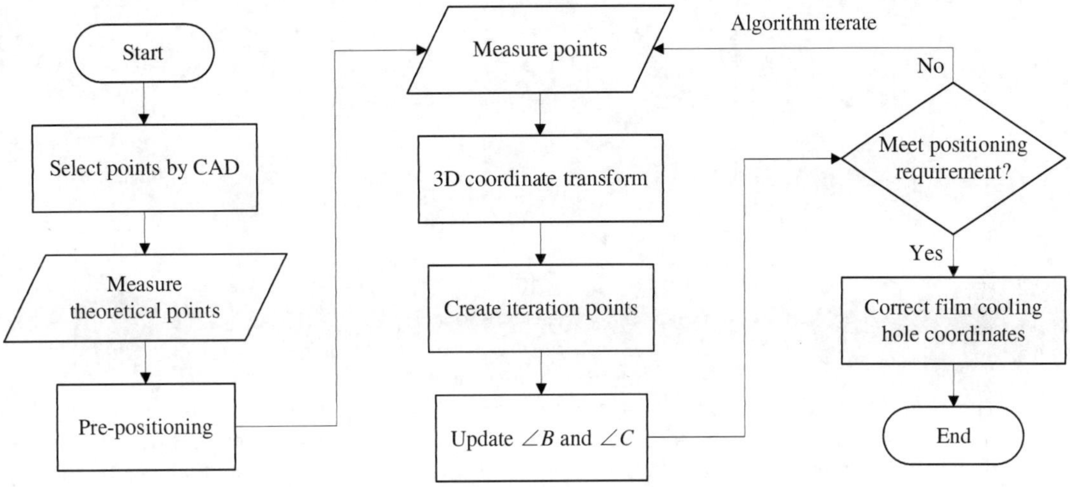

Figure 4. System flowchart.

5. Experiment

The five-axis machine tool is used to conduct a measurement and positioning experiment on a turbine blade. The positioning accuracy of machine tool is 4 μm and the repeat positioning accuracy is 3 μm. The probe is Pioneer TP300.

Firstly, six points are selected on the CAD model, and then the blade is fixed on the machine tool workbench by the fixture. Then, the measurement and positioning experiment is carried out. Measuring points is shown as figure 5.

The final average error and maximum error data are shown as figure 6.

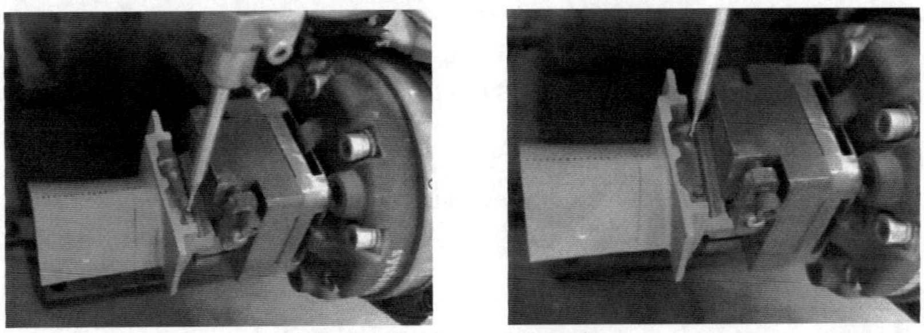

Figure 5. Measuring points.

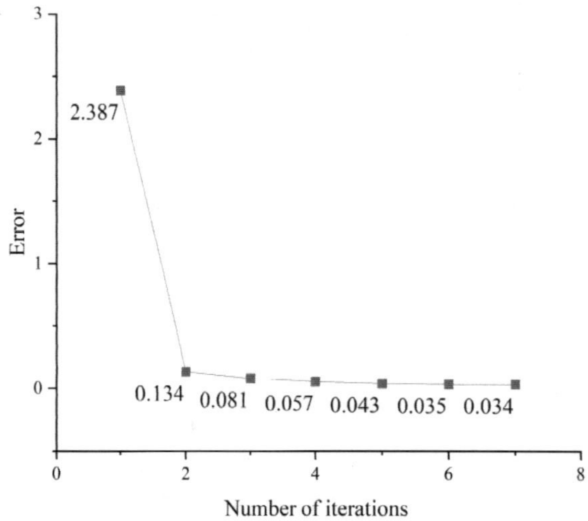

Figure 6. Experimental data.

The result show that, after the first pre-positioning algorithm processing, the error significantly decreased, and then after several iterations, the error tended to be stable. Since the actual shape of the turbine blade cannot be exactly the same as the theoretical shape, the optimal positioning of the turbine blade is achieved at this time.

6. Conclusions

The positioning correction method proposed in this paper has been verified by actual machine tool and turbine blade experiments. The conclusions can be summarized as follows:

1) Experimental result show that the error converges to 0.0344mm after 7 iterations by the positioning algorithm, which meets the technical requirements.

2) The algorithm can achieve high-precision positioning of turbine blades through rapid iterative detection, and obtain the transformation matrix to establish the coordinate system of the actual blades.

3) According to the measured results of the algorithm, the coordinates of the gas film cooling holes are corrected by the method. Therefore, the method realize the high-precision processing of the gas film cooling holes, which can improve the cooling performance of turbine blades.

References

[1] Wei Z, Zhang S, Jafari S, Nikolaidis T. Gas turbine aero-engines real time on-board modelling: A review, research challenges, and exploring the future. Progress in Aerospace Sciences. 2020 Feb 1;121:100693.

[2] Zhang J, Zhang S, Chunhua WA, Xiaoming TA. Recent advances in film cooling enhancement: a review. Chinese Journal of Aeronautics. 2020 Apr 1;33(4):1119-36.

[3] Huang PH, Shih LK, Lin HM, Chu CI, Chou CS. Novel approach to investment casting of heat-resistant steel turbine blades for aircraft engines. The International Journal of Advanced Manufacturing Technology. 2019 Oct;104(5):2911-23.

[4] Wang H, Zhang K, Wu D, Yu T, Yu J, Liao Y. Analysis and optimization of the machining fixture system stiffness for near-net-shaped aero-engine blade. The International Journal of Advanced Manufacturing Technology. 2021 Apr;113(11):3509-23.

[5] Sorkine-Hornung O, Rabinovich M. Least-squares rigid motion using svd. Computing. 2017 Jan 16;1(1):1-5.

Research on Transient Torsional Vibration Test of Harmonic Reducer

Tianhao Xu[1], Jing Tan[2], Dishan Huang*[3] , Tong Tong[4] and Jingjun Gu[5]

[1,2,3]School of Mechatronic Engineering and Automation, Shanghai University, Shanghai, P.R. China

[4,5]Nantong Zhenkang Machinery Co. Ltd,Jiangsu,China

*hdishan@shu.edu.cn

Abstract. The torsional vibration of harmonic reducer is related to the positioning accuracy and Trajectory Accuracy of industrial robots. Based on this, this paper studies and experiments the impact response of harmonic reducer. By establishing the torsional vibration model with periodic variable stiffness of harmonic reducer, the unit impulse response and characteristics are analyzed; The time-varying torsional stiffness of the harmonic reducer is verified by setting up an experimental platform for the harmonic reducer, and the transient torsional vibration characteristics of the harmonic reducer are obtained The research will provide basic data for the optimal application of harmonic reducer.

1. Introduction

The torsional characteristic of the harmonic reducer is a very important transmission characteristic, which directly affects the positioning accuracy and trajectory accuracy of the industrial robot.

Xu LiZhong[1] and others studied the forced influence of harmonic drive system on excitation, and found that the torsional vibration response frequency depends on the excitation frequency and the parameter frequency. Huang Dishan[2] et al. studied a method for predicting the free and forced responses of parametric systems. Based on the concept of frequency cracking, a special triangular series approximation response form is proposed, and the parametric vibration equation is transformed into an infinite number of homogeneous linear equations by using the harmonic balance method. Harmonic coefficients, which give an analytical representation of the parametric vibrational transient response.

Gu Jingjun[3,4] and others studied the periodic time-varying phenomenon of stiffness caused by the manufacturing error of the harmonic reducer through the analysis of the transmission mechanism, and established the parametric vibration model of the harmonic drive system. The parametric vibration response of the system is simulated, and the bispectral characteristics and combined harmonics of the steady-state response of the parametric system are studied theoretically and experimentally.

In this paper, the parametric vibration will be simulated and analyzed, the torsional vibration characteristics of the harmonic reducer will be studied and analyzed by the impact experiment, and the transient vibration law will be explored.

2. Periodically time-varying torsional stiffness

Factors such as design, manufacturing, and assembly can cause fluctuations in torsional stiffness. Torsional stiffness due to machining errors Periodicity as an example: Suppose an ideal flex wheel meshes with the teeth of a rigid wheel with accumulated errors in the circumferential pitch. When the rigid wheel is fixed, as the wave generator rotates, the flex wheel rotates and slips. In the rigid wheel in Figure 1, it is assumed that the left tooth pitch is sparse, while the right tooth pitch is dense. At the position of AA' point, the meshing of the flexible pulley and the rigid pulley is close to normal; when the flexible pulley rotates with the wave generator to the BB' point, due to the difference in the tooth pitch between the flexible pulley and the rigid pulley, as shown in Figure 2, the rigid pulley and the rigid pulley The two ends of the long shaft of the flexible wheel interfere, and additional elastic deformation occurs.

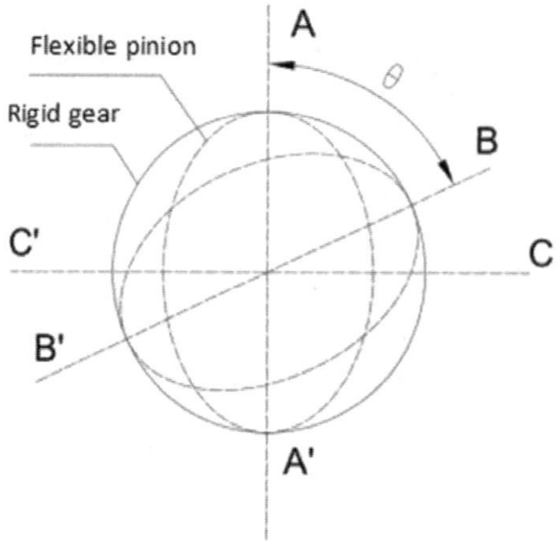

Figure 1. Schematic diagram of meshing position of rigid wheel and flexible wheel

When the cam is rotated by the θ angle, the elastic deformation of the meshing teeth increases, and additional torque is required compared to the original meshing in the AA'position.Therefore, as the meshing point goes from AA' to BB', the interference elastic deformation of meshing between the teeth increases, and the meshing stiffness increases, resulting in an increase in torsional stiffness. When the meshing point reaches the CC'point, the meshing stiffness reaches its maximum. When the flex wheel continues to rotate with the cam, the meshing stiffness gradually decreases, and the tooth meshing stiffness returns to the original state at the point A'A. When the cam makes one revolution, the tooth mesh stiffness, i.e. the torsional stiffness, will fluctuate twice. Based on this, shock vibration experiments will be carried out on the vibration model of the harmonic reducer under static and dynamic conditions, respectively.

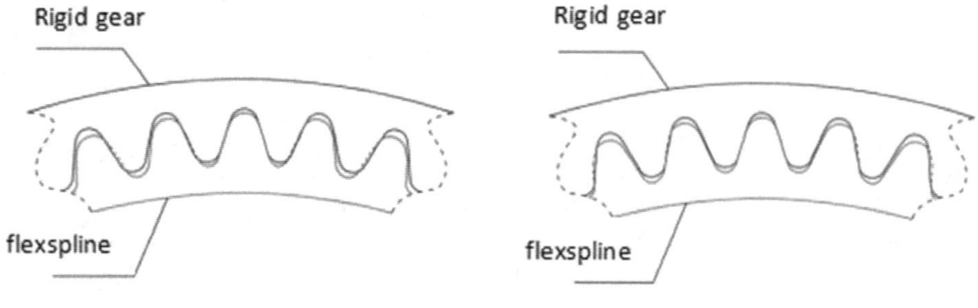

(a) Point B engagement status (b) Point B' engagement status

Figure 2. BB' point interference meshing state

3. Simulation and experiment

3.1 Theoretical Simulation

Based on the time-varying characteristics of torsional stiffness, the torsional stiffness of the harmonic reducer is set as:

$$K = K_o(1 + \beta\cos\omega_o t) \qquad (1)$$

where K_o is the average torsional stiffness, β is the torsional stiffness fluctuation index, and ω_o is the torsional stiffness variation frequency.

The harmonic reducer is driven by a servo motor, and the output is usually loaded with a moment of inertia. Therefore, the normalized parameter torsional vibration equation of the harmonic reducer is established as:

$$\frac{d^2\theta}{dt^2} + 2\zeta\omega_o\frac{d\theta}{dt} + \omega_n^2(1 + \beta\cos\omega_o t)\theta = 0 \qquad (2)$$

Among them, $\theta(t)$ is the rotational angular displacement, ω_o is the input cam rotational frequency , ζ is the damping ratio, β is the torsional stiffness modulation index, ω_n is the system natural frequency, and $\omega_o$is the stiffness fluctuation frequency. Put the parameters ω_n=25 , ω_o=5, ζ=0.025, β=0.3 into the formula 2, given the initial conditions $x|_{t=0}$=0 and $\frac{dx}{dt}|_{t=0}$=1 and use the Ronge-Kutta algorithm to simulate, the obtained spectrum is shown in Figure 3, It can be seen that the first sudden peak is the oscillation frequency of the system, and the quadrilateral frequency is distributed around the first sudden peak.

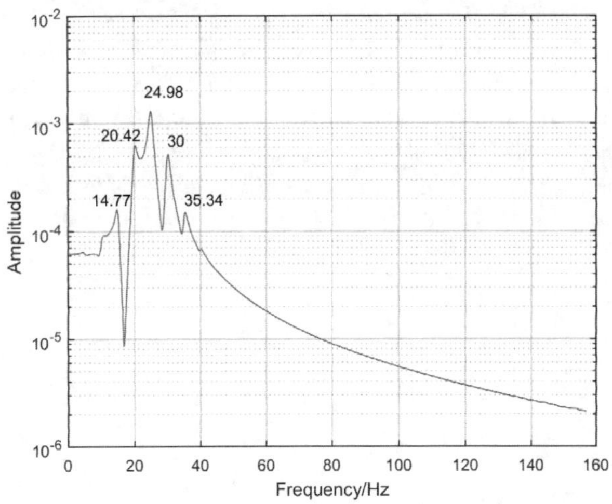

Figure 3. Spectrum diagram of shock vibration response per unit of parameter system

3.2. Impact test

A torsional vibration testing device was established. As shown in figure 4, it includes harmonic reducer, support frame, motor, load, etc.

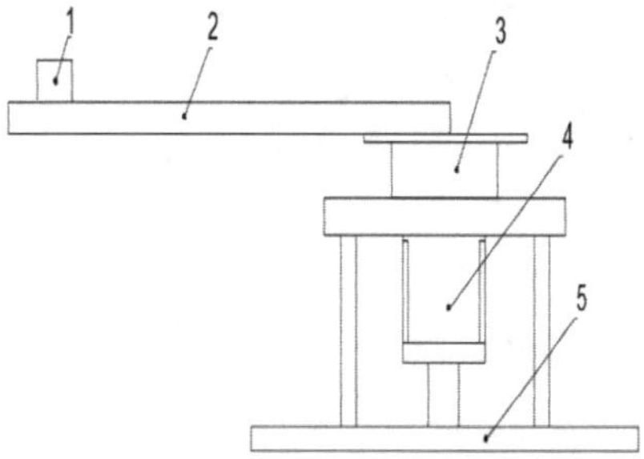

(a) Schematic diagram of the harmonic reducer test device

(b) The picture of theharmonic reducer test device

Figure 4. Torsion experiment device of harmonic reducer: 1—load block, 2—load arm, 3—harmonic reducer, 4—motor, 5—support frame

Vibration detection is performed using a wireless sensor located 70cm away from the rotating shaft. When the end of the load wall is impacted, the sensor collects the torsional vibration signal of the system.

3.2.1 Static impact test of reducer

When the system is static, the stiffness is constant, then the dynamic normalization equation

$$\ddot{\theta}(\mathrm{t}) + 2\zeta\omega_{\mathrm{n}}\dot{\theta}(\mathrm{t}) + \omega_n^2\theta(t) = 0 \quad (3)$$

on the static system , the end of the load arm is impacted once by a rubber hammer, the vibration signal is collected once by the collector, and the collected signal is processed and analyzed.

Perform self-power analysis on the obtained data. Figure 5 below is the self-power spectrum obtained at the 0°position . The first inrush peak 25.31Hz in the figure is the oscillation frequency of the harmonic reducer under static conditions.

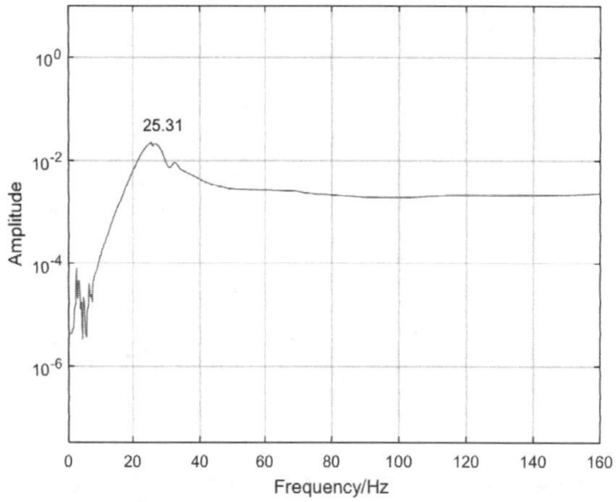

Figure 5. The response spectrum of the harmonic reducer under the shock in the static state

The same experiment was performed for the other seven positions, changing the impact position. Table 1 shows the oscillating frequency measured at each position in eight circumferential positions. The oscillating frequency of the system changes with the change of the angular position, that is, the torsional stiffness of the harmonic reducer fluctuates about two times for each cycle of fluctuation.

Table 1. Oscillation frequencies at eight positions under static shock

angle	Numerical value
0°	25.31 Hz

45°	23.44 Hz
90°	20.94 Hz
135°	21.88 Hz
180°	24.38 Hz
225°	20.63 Hz
270°	21.56 Hz
360°	21.88 Hz

3.2.2 Impact test when the reducer rotates

The shock response experiment was carried out while the reducer was rotating, and the shock was performed on the designated position of the load arm, and the shock position was the same as the static shock position. The obtained data is also analyzed by self-power spectrum. Figure 6(a) is the experimental diagram obtained at the 0° position at the driving speed of 300 rpm, where the highest peak at 25Hz is no longer, and there are peaks on both sides, and the peak frequency Symmetrical with respect to the frequency at 25Hz, the experimental results are similar to the simulation results.

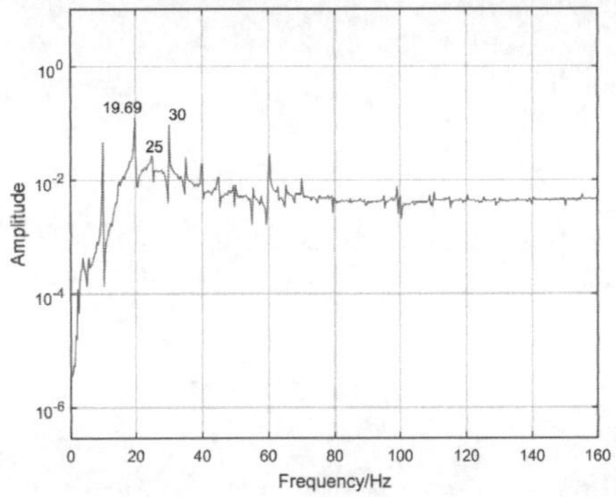

(a) Self power spectrum at 0°position at 300 rpm

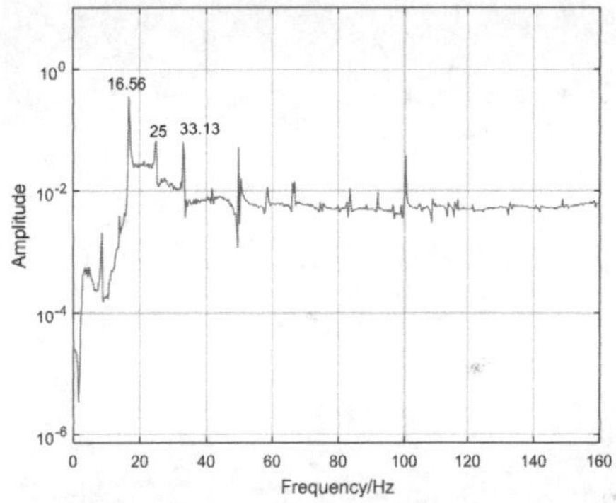

(b) Self power spectrum at 0°position at 500 rpm

Figure 6. Self-power spectrum obtained by dynamic shock at different rotational speeds

The driving speed of the motor was changed, and the speed was set at 500 rpm, and the same experiment was carried out. When the rotational speed increases, the spacing between the symmetrical peaks increases. Comparing the difference between the symmetrical peak frequencies at each position, it can be inferred that when the rotational speed increases, the side frequency spacing Δf will increase from 10.31 to 16.57.

4. Conclusions

Impact experiment of the harmonic reducer, some laws of transient torsional vibration are obtained. The specific conclusions are as follows:

(1)harmonic reducer varies with position, and its stiffness fluctuates about twice for every cycle.

(2)Through the impact test under harmonic rotation, it is concluded that the torsional vibration response spectrum is composed of the oscillation frequency and the combination of the oscillation frequency and the stiffness fluctuation frequency. The side frequency is symmetrical with respect to the oscillation frequency, and as the rotational speed increases, the side frequency spacing also increases.

References

[1] Xu L, Zhu C, Qin L 2007 Parametric Vibration for Electromechanical Integrated Electrostatic Harmonic Drive [J]. *Mechatronics*, **17 (1)** :31-34.

[2] Huang DS, Zhang YY, Hexi Shao 2017 Free response approach in a parametric system [J]. *Mechanical Systems and Signal Processing*, **91** : 313-325.
[3] Gu JJ, Tong T, Huang DS, Li M 2021 Study on torsional vibration of a harmonic driver based on time-varying stiffness caused by manufacturing error [J]. *Journal of Vibroengineering*, Vol. 23, Issue 3, p 619-631.
[4] Gu JJ 2021 Study on torsional stiffness and dynamics of double rigid wheel harmonic reducer [D]. (Shanghai: Shanghai University).

Research on free vibration characteristics of axially moving functionally graded plates under the action of temperature field

YongHeng Mao, YouQi Tang, JunHao Wu

Shanghai Institute of Technology, College of Mechanical Engineering, Shanghai 201418, China

myh8035@163.com

Abstract: As a new type of material, functionally graded materials are widely used in industry, aerospace and other fields due to their special physical properties. Among them, there are also many in the thermal environment, and the existence of the thermal environment will also have a great impact on the stability of the moving object. Therefore, it is of great significance to study the axial motion functionally graded materials in the temperature field. In this paper, taking the axially moving functionally graded plate as the research object, the dynamic control equation of the axially moving functionally graded plate in the thermal environment is established, and the simplified linear dimensionless control equation and boundary conditions of the axially moving functionally graded , using the multi-scale method to find the governing equation, and using the complex mode method to solve the equation to obtain the natural frequency of the system. Furthermore, the effects of axial velocity, functional gradient index, viscoelastic coefficient and thermal parameters on natural frequency are analyzed by Matlab.

1. Introduction

Functionally graded plates are widely used in daily life and industrial fields, such as the casings of moving rockets in aerospace, the casings of highly moving satellites in space, and so on, which can be simplified as axial motion plate models. The stability of the axial moving plate and the stationary plate are different, and the axial speed will bring complex lateral vibration problems to the moving plate. Therefore, many scholars have not stopped their research on the sports board. Gupta and Talha[1] gave an extensive review of the structural responses of functionally graded materials and structures, focusing on the structural characteristics of FGM plates/shells under thermomechanical loading under various boundary and environmental conditions, giving an overview Different fabrication procedures and future research directions required to properly implement these materials in design and analysis. Tahir [2] presented the wave propagation analysis of porous functionally graded (FG) sandwich panels in a hot and humid environment. The effects of moisture and temperature on wave propagation in porous FG sandwich panels and their effect on material expansion were investigated. Dastjerdi[3] performed a physically consistent analysis of the nanostructures, interpreting thermal and temperature effects as a 3D problem, and after calculating the deflection relationship, conducted a systematic study of the bending response of nanoporous FGMs in a hot and humid environment. Mao and Zhang [4] analyzed the dynamic response of in-plane motion multilayer functionally graded piezoelectric plates. Tang [5] studied the dynamic properties of in-plane moving viscoelastic plates in a fluid environment. Hao[6] used the exact solution method to study the free vibration of Mindlin annular plate under arbitrary boundary conditions.

2. Mathematical model of axial motion functionally gradient plate

In this chapter, we will study the free vibration of axially moving functionally graded plates under the action of a temperature field. Firstly, the dimensionless governing equations and boundary conditions of the simplified axial motion functionally graded plate in the temperature field are given.

As shown in Figure 1, it is assumed that the functional ladder is composed of metal and ceramics, which is a thin plate model that moves in the axial direction. The length and width of the functionally graded plate are a and b, respectively, and, $a>b$, the thickness of the plate is h, and the functionally graded plate moves with a speed of Γ(t) in the in-plane x-direction. The coordinate circle is assumed to be in a corner of the midplane of the plate. For the functionally graded material plate, it is assumed that its lower side is composed of ceramic material, the upper side is composed of metal material, and the middle is composed of a mixture of the two materials. Because the Poisson's ratio of the ceramic material and the Poisson's ratio of the metal material are similar, let their Poisson's ratio be the same μ.

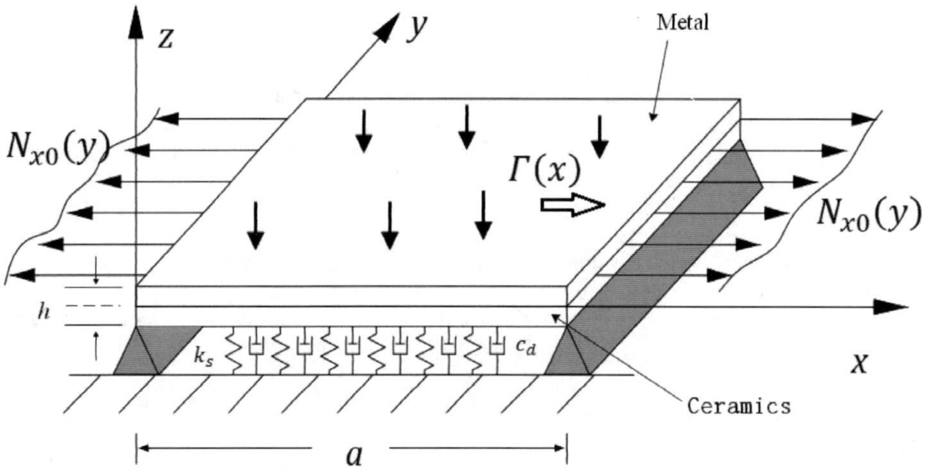

Figure 1. Physical model of axial motion functional gradient plate

The governing equation in dimensionless form is

$$
\begin{aligned}
& w_{,tt} + 2\gamma_0 w_{,xt} + \left(\gamma_0^{\,2} - 1\right) w_{,xx} + K_{\mathrm{f}}\left(w_{,xxxx} + 2\xi^2 w_{,xxyy} + \xi^4 w_{,yyyy}\right) + k_{\mathrm{s}} w \\
& + q\left(w_{,xx} + \xi^2 w_{,yy}\right) + \varepsilon c_{\mathrm{d}}\left(w_{,t} + \gamma_0 w_{,x}\right) = -\varepsilon \sin[\omega t](2w_{,xt} + c_{\mathrm{d}} w_{,x} + 2\gamma_0 w_{,xx})
\end{aligned}
\tag{1}
$$

In the above formula, γ_0 is the average axial velocity of the functional gradient plate; γ_1 is the amplitude of the in-plane axial velocity pulsation; ω is the frequency of the in-plane axial velocity pulsation. Since the simply supported boundary is often used as the boundary condition in engineering applications, it can be regarded as a four-sided simply supported plate, and the boundary condition is in the form of

$$
w\Big|_{x=0}^{x=1} = 0, \quad w_{,xx}\Big|_{x=0}^{x=1} = 0; w\Big|_{y=0}^{y=1} = 0, \quad w_{,yy}\Big|_{y=0}^{y=1} = 0 \tag{2}
$$

3. Complex mode analysis

The first-order approximate solutions of equations (1) and (2) can be written as

$$w(x,y,t;\varepsilon)=w_0(x,y,T_0,T_1)+\varepsilon w_1(x,y,T_0,T_1)+O(\varepsilon^2) \tag{3}$$

where T0=t is the fast time scale due to some natural frequencies of the translational plate; T1=εt is the slow time scale with slow amplitude and phase changes caused by small perturbations in viscoelasticity and velocity.

Substitute equation (3) into (1) and (2), separate the ε^0 order quantity, and get

$$\begin{aligned}&\varepsilon^0: w_{0,T_0T_0}+2\gamma_0 w_{0,xT_0}+\left(\gamma_0^2-1\right)w_{0,xx}+K_{\mathrm{f}}\left(w_{0,xxxx}+2\xi^2 w_{0,xxyy}+\xi^4 w_{0,yyyy}\right)\\&+k_{\mathrm{s}}w_0+q\left(w_{0,xx}+\xi^2 w_{0,yy}\right)+c_{\mathrm{d}}\left(w_{0,T_0}+\gamma_0 w_{0,x}\right)=0\end{aligned} \tag{4}$$

$$\varepsilon^0:\ w_0\big|_{x=0}^{x=1}=0,\quad w_{0,xx}\big|_{x=0}^{x=1}=0;\quad w_0\big|_{y=0}^{y=1}=0,\quad w_{0,yy}\big|_{y=0}^{y=1}=0 \tag{5}$$

The solution of the above equation (4) can be expressed as

$$w_0(x,y,T_0,T_1)=\sum_{m=1}^{\infty}\sum_{n=1}^{\infty}A_{mn}(T_1)\psi_{mn}(x,y)e^{i\omega_{mn}T_0}+cc \tag{6}$$

Among them, A_{mn} is an undetermined complex function; ψ_{mn} is the *mn*-th order modal function of the linearly derived system, and ω_{mn} is the *mn*-th order natural frequency of the linearly derived system. cc represents the complex conjugate of the preceding terms on the right-hand side of equation (6).

ψ_{mn} in equation (6) is

$$\psi_{mn}(x,y)=C_{1n}\left(e^{i\beta_{1n}x}+C_{2n}e^{i\beta_{2n}x}+C_{3n}e^{i\beta_{3n}x}+C_{4n}e^{i\beta_{4n}x}\right)\sin(m\pi y) \tag{7}$$

$\beta_{jn}(j=1,2,3,4)$ is the four characteristic roots of equation (4). Substitute equations (6) and (7) into equation (4), and get

$$\beta_{jn}^4+f_1\beta_{jn}^2+f_2\beta_{jn}+f_3=0\,(n=1,2,...) \tag{8}$$

In

$$\begin{aligned}&f_1=\frac{-\gamma_0^2+2m^2\pi^2K_f\xi^2+1-q}{K_f},\ f_2=\frac{(-2\omega_{mn}+ic_d)\gamma_0}{K_f},\\&f_3=\frac{m^4\pi^4K_f\xi^4+ic_d\omega_{mn}+k_s-m^2\pi^2\xi^2q-\omega_{mn}^2}{K_f}\end{aligned} \tag{9}$$

The solution of equation (9) can be expressed as

$$\begin{aligned}&\beta_{1n}=g_1-\frac{1}{2}\sqrt{g_2-g_3},\ \beta_{2n}=g_1+\frac{1}{2}\sqrt{g_2-g_3},\\&\beta_{3n}=-g_1-\frac{1}{2}\sqrt{g_2-g_3},\ \beta_{4n}=-g_1+\frac{1}{2}\sqrt{g_2-g_3}.\end{aligned} \tag{10}$$

In

$$g_1 = \frac{1}{2}\sqrt{-\frac{2f_1}{3} + \frac{\sqrt[3]{2}g_4}{3\sqrt[3]{g_5 + \sqrt{g_5^2 - 4\sqrt[3]{g_4}}}} + \frac{\sqrt[3]{g_5 + \sqrt{g_5^2 - 4\sqrt[3]{g_4}}}}{3\sqrt[3]{2}}}, \tag{11}$$

$$g_2 = -2f_1 - 4g_1^2, g_3 = \frac{f_2}{g_1}, g_4 = f_1^2 + 12f_3, g_5 = 2f_1^3 + 27f_2^2 - 72f_1f_3.$$

For the boundary condition of simply supported four sides of axial motion, the frequency equation is

$$\begin{aligned}&\left[e^{i(\beta_{1n}+\beta_{2n})} + e^{i(\beta_{3n}+\beta_{4n})}\right]\left(\beta_{1n}^2 - \beta_{2n}^2\right)\left(\beta_{3n}^2 - \beta_{4n}^2\right) + \left[e^{i(\beta_{1n}+\beta_{3n})} + e^{i(\beta_{2n}+\beta_{4n})}\right]\left(\beta_{3n}^2\right.\\ &\left.-\beta_{1n}^2\right)\left(\beta_{2n}^2 - \beta_{4n}^2\right) + \left[e^{i(\beta_{2n}+\beta_{3n})} + e^{i(\beta_{1n}+\beta_{4n})}\right]\left(\beta_{2n}^2 - \beta_{3n}^2\right)\left(\beta_{1n}^2 - \beta_{4n}^2\right) = 0\end{aligned} \tag{12}$$

4. Numerical analysis

In this section, the single variable method will be used, and other parameters remain unchanged, the direct multi-scale method will be used to analyze the natural frequency of the functional gradient plate, the actual value will be brought in, and the graph will be plotted using Matlab. The effects of axial velocity, viscoelastic coefficient, temperature field and gradient function index on the natural frequency are considered respectively. Since the parameter is dimensionless, the coordinate unit is 1.

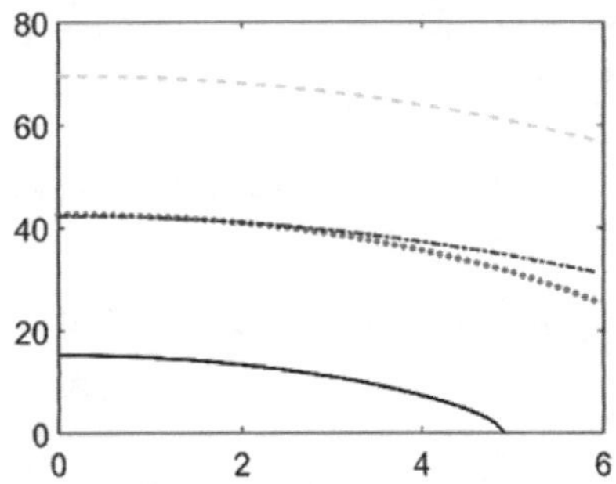

Figure 2. The relationship between the first fourth-order frequency and the in-plane translational velocity of functionally graded plates

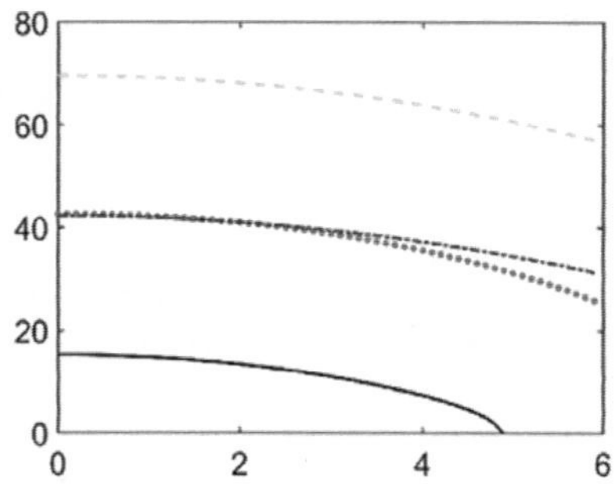

Figure 3. Variation of the first-order natural frequency of functionally graded plates with viscoelasticity

Figure 2 shows the relationship between the natural frequency and the average velocity of the first four-order functionally graded plate in the thermal field. The parameters are set as T_0=20K, T=25K, n=1, μ=0.1, ξ=1, k_s=0.5, c=2, a=1m, h=0.01m and N_{x1}=27000N. The first, second, third, and fourth order natural frequencies ω_{11}, ω_{12}, ω_{21}, and ω_{22} are represented by solid lines, dashed-dotted lines, dotted lines, and dashed lines, respectively. As can be seen from the figure, the natural frequency decreases as the average in-plane velocity increases. Internal work resonance may occur when the in-plane average velocity is equal to a special value. As can be seen from the figure, when γ=1.8, the second-order natural frequency is equal to the third-order natural frequency, that is, $\omega_{12} \approx \omega_{21}$, and a 1:1 internal resonance occurs at this time.

Figure 3 shows the change of the first-order natural frequency and the change of the average speed under the change of the damping stage. The parameters are set as T_0=20K, T=25K, N_{x1}=27000N, a=1m, h=0.01m, n=1, μ=0.1, ξ=1, ks=0.5 and γ=1. The solid line, the dotted line, and the dashed line represent the change of the first-order natural frequency with the translational velocity under the damping c=0, c=6 and c=12, respectively. It can be seen that with the increase of damping, the first-order natural frequency of the functionally graded plate gradually decreases.

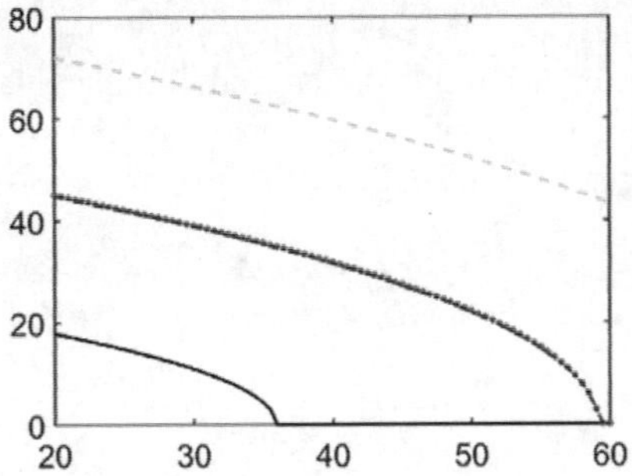

Figure 4. The relationship between the first four natural frequencies of functionally graded plates and temperature

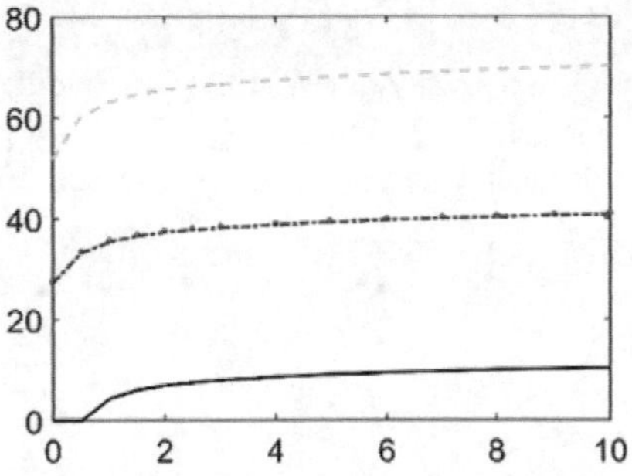

Figure 5. The relationship between the natural frequency of the functionally graded plate and the functional gradient index

Figure 4 shows the relationship between natural frequency and temperature. The parameters are set as T_0=20K, n=1, μ=0.1, ξ=1, k_s=0.5, c=2 and γ=1. The first, second, third, and fourth order natural frequencies ω11, ω12, ω21, and ω22 are represented by solid lines, dashed-dotted lines, dotted lines, and dashed lines, respectively. It can be seen from the figure that the natural frequency of the functionally graded plate decreases as the temperature T increases. And, when T=36K, the first-order frequency decreases to zero, and instability occurs.

Figure 5 shows the relationship between the first-order natural frequency and the gradient index of the functionally graded plate. The parameters are set as μ=0.1, ξ=1, k_s=0.5, c=2 and γ=1. The first, second, third, and fourth order natural frequencies ω_{11}, ω_{12}, ω_{21}, and ω_{22} are represented by solid lines, dashed-dotted lines, dotted lines, and dashed lines, respectively. It can be seen from the figure that when the functional gradient index varies between 0 and 4, the fourth-order natural frequency has an obvious increasing trend, and then with the increase of the index, the natural frequency increases slowly.

5. Conclusion

In this paper, the free vibration of axially moving functionally graded plates in temperature field is studied. First, the partial differential equation of lateral vibration of the axially moving functionally graded plate in the temperature field, and the boundary conditions of four sides simply supported are derived. The partial differential equations and boundary conditions were linearized, and the natural frequencies of the axially moving functionally graded plates were analyzed by the direct multiscale method. And some actual values are given, and the single variable method is used to analyze the effects of viscoelastic coefficient, functional gradient index, axial velocity and temperature field on the natural frequency under the condition of controlling other parameters unchanged.

References

[1] Gupta A , Talha M . Recent development in modeling and analysis of functionally graded materials and structures[J]. Progress in Aerospace Sciences, 2015, 79:1-14.

[2] Tahir S I, Chikh A, Tounsi A, et al. Wave propagation analysis of a ceramic-metal functionally graded sandwich plate with different porosity distributions in a hygro-thermal environment[J]. Composite Structures, 2021, 269: 114030.

[3] Dastjerdi S, Malikan M, Dimitri R, et al. Nonlocal elasticity analysis of moderately thick porous functionally graded plates in a hygro-thermal environment[J]. Composite Structures, 2021, 255: 112925.

[4] Mao J J, Zhang W. Buckling and post-buckling analyses of functionally graded graphene reinforced piezoelectric plate subjected to electric potential and axial forces[J]. Composite Structures, 2019, 216: 392-405.

[5] Tang Y Q, Yan C H, Jiang S Y. Dynamic Behaviors of an Axially Accelerating Viscoelastic Plate Immersed in Fluid[J]. Available at SSRN 4022888.

[6] Hao Q, Chen Z, Zhai W. Free Transverse Vibration of Mindlin Annular and Circular Plate with General Boundary Conditions[C]//International Conference on Computational & Experimental Engineering and Sciences. Springer, Cham, 2019: 1291-1302.

Analysis of Dynamic Characteristics of Hydraulic Driven System for Weight Bearing Lower Limb Exoskeleton

QiYing Su, ZhongCai Pei and ZhiYong Tang

School of Automation Science and Electrical Engineering, Beihang University, Beijing, China

suqiying@buaa.edu.cn

Abstract. A new kind of weight-bearing lower limb exoskeleton driven by hydraulic cylinder is designed, and its structural characteristics and driving principle are analysed. The mathematical model and control block diagram of the valve controlled unsymmetrical cylinder system for the controlled rotary joint are established. The effects of structural parameters such as servo valve bandwidth, load mass and hydraulic cylinder size on the dynamic characteristics of the system are studied, and the force tracking curve optimized by structural parameters is given to prove the correctness of the theoretical analysis of the system, which provides theoretical support for the design of the weight-bearing lower limb exoskeleton.

1. Introduction

Wearable exoskeleton is a typical parallel man-machine integrated mechanical device [1], which uses biological concepts for reference to integrate machine and human. It is used to improve human physical ability, carry heavy objects, reduce the damage to human body caused by load and make the human body has large bearing capacity and good rigidity of the machine without losing its own mobility.

The active drive [1] is usually selected to assist the flexion and extension freedom of the lower limb exoskeleton knee joint and hip joint, which is driven by the actuator. This is because the daily activities such as walking and standing are completed on the sagittal plane, which is consistent with the bionic biology. However, the degrees of freedom of the ankle are mostly driven passively [2], which are realized by energy storage elements such as springs.

At present, the exoskeleton active drive joint has many kinds of modes, such as hydraulic drive, motor drive, pneumatic drive, etc. Compared with other drive modes, hydraulic drive has the advantages of high-power density, high reliability, stable operation, easy overload protection and step less speed regulation [3]. For the weight-bearing exoskeleton that needs large joint driving torque and power, the hydraulic drive makes the structure simpler, the volume smaller and the weight lighter. Under the same load, the hydraulic drive mode shows good power assisted performance [4-8].

To ensure the comfort, safety and flexibility of the exoskeleton, it is particularly important to improve the control accuracy, response speed and anti-disturbance ability of the hydraulic control system. In this paper, a hydraulic assisted lower limb exoskeleton is designed, the mathematical model of the hydraulic valve-controlled cylinder system is established. The control performances and influencing factors are theoretically analysed and simulated to find the best control strategy and parameter matching relationship. The research results provide a theoretical basis for the design of high-performance assisted walking exoskeleton.

2. Composition and working principle of hydraulic exoskeleton

The designed three-dimensional model of the assisted lower limb exoskeleton is shown in figure 1. The lower limb exoskeleton adopts the anthropomorphic structure. According to the actual parameters of the human body, such as the length of thigh and shank, waist circumference and joint degrees of freedom, the overall design is carried out on the basis of analysing the joint structure of the lower limb and the main limb size parameters of the human body, so as to avoid motion interference and ensure the freedom and safety of the operator's activities to the greatest extent.

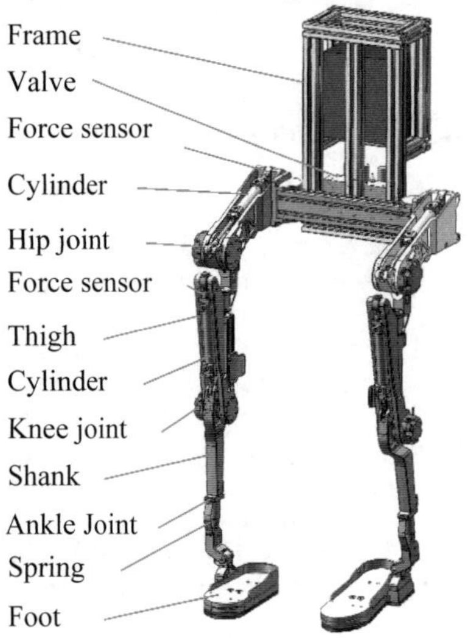

Figure 1. Three-dimensional model

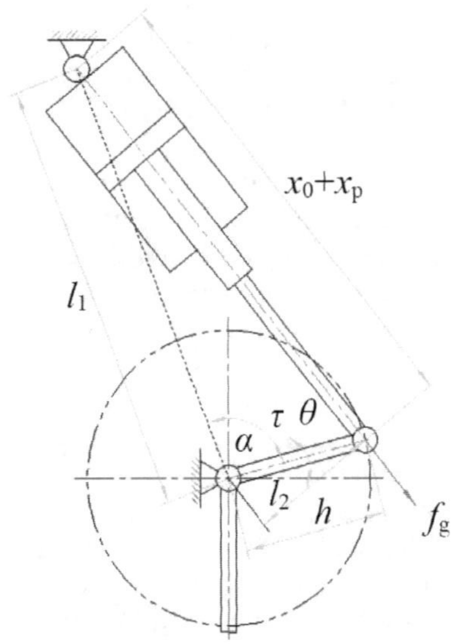

Figure 2. Schematic diagram of joint driven

The lower limb exoskeleton mechanical structure mainly includes seven parts: waist, hip, thigh, knee, shank, ankle and foot. The degree of freedom of the exoskeleton is preserved as follows: the hip joint retains 2 degrees of freedom, namely, adduction/abduction and flexion/extension; the knee joint retains 1 degree of freedom, namely, flexion/extension; the ankle joint retains 2 degrees of freedom, which are flexion/extension and adduction/abduction. These five degrees of freedom ensure that the lower limb exoskeleton robot can help the wearer with normal operation. Considering the human-machine integrated walking needs of the human and the exoskeleton, the hip joint retains the rotation and flexion/extension movement. The knee joint is the hinge of lower limb movement, and also the joint that bears the largest weight and supporting torque during weight-bearing support so that only one degree of freedom of flexion/extension is reserved and the minor rotation movement is excluded. The ankle joint receives the maximum instantaneous moment when the human body is walking, but the stress time is short and the rotation angle is small so that the elastic metal sheet is added to the ankle joint to provide moment support when the ankle joint has small rotation, and play a damping and buffering role to reduce the stress on the human ankle joint. Length adjustment mechanisms are designed at the thighs and legs of the exoskeleton to meet the needs of wearers with different height and body types.

The driving of exoskeleton hip joint and knee joint in the sagittal plane is completed by hydraulic cylinder. Figure 2 shows the working principle of a joint driving. One end of the hydraulic cylinder is fixed by a hinge shaft and can rotate in the sagittal plane. The piston rod is connected to the connecting rod through the hinge shaft, the output force f_g and displacement x_p generate the joint torque τ and angle θ through the force arm h to drive the joint to rotate.

3. Model of hydraulic control system

The hydraulic control system of exoskeleton is mainly composed of hydraulic cylinder, electro-hydraulic servo valve, servo amplifier and force sensor. The principle of the hydraulic valve-cylinder system is shown in figure 3. The system uses four side slide valves to control the single rod double acting hydraulic cylinder, which belongs to the force control system with asymmetric cylinder controlled by symmetric valve [9]. Set the valve spool displacement $x_v > 0$ and hydraulic cylinder displacement $x_p > 0$ as positive direction when moving to the right.

Defining load pressure as $p_L = p_1 - p_2$, load flow as $q_L = (q_1 + q_2)/2$, and taking the ratio of the two chamber areas of the hydraulic cylinder as $\eta = A_2 / A_1$, after deriving, the following dynamic equations can be obtained.

Linearized flow equation of servo valve port:

$$q_L = K_q x_v - K_c p_L \quad (1)$$

Flow continuity equation of hydraulic cylinder:

$$q_L = A_m \dot{x}_p + C_t p_L + V_c \dot{p}_L / 4\beta_e \quad (2)$$

Force balance equation of hydraulic cylinder:

$$f_g = A_c p_L = M_t \ddot{x}_p + B_p \dot{x}_p + K_p x_p \quad (3)$$

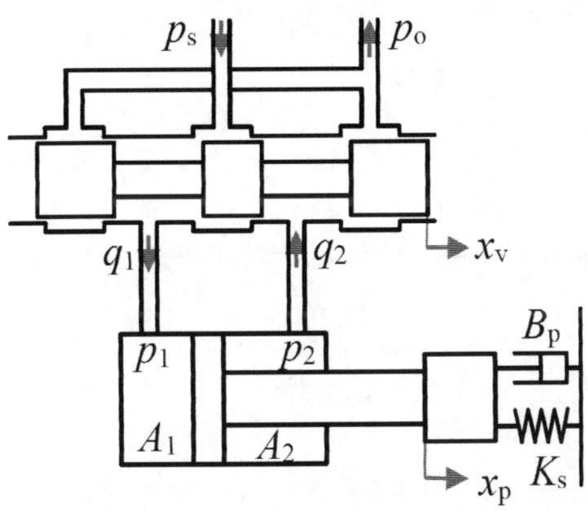

Figure 3. Hydraulic valve control cylinder system

Where q_1, q_2 ——flow of servo valve port, m^3/s

p_1, p_2 ——pressure of servo valve port, N/m^2

K_q ——flow gain of servo valve, m^2/s

K_c ——flow pressure coefficient of servo valve, $m^5/N \cdot s$

A_c ——equivalent action area of cylinder, m^2

A_m ——average area of two chambers of hydraulic cylinder, m^2

V_c ——equivalent volume of cylinder, m^3

β_e ——effective bulk modulus of elasticity, N/m^2

C_t ——leakage coefficient of cylinder, $m^5/N \cdot s$

f_g ——output force, N

M_t ——mass, Kg

B_p ——viscous damping coefficient, $N \cdot m/s$

K_s ——load spring stiffness, N/m

Combined with equations (1), (2) and (3), a simplified block diagram of the hydraulic control system can be obtained as shown in figure 4.

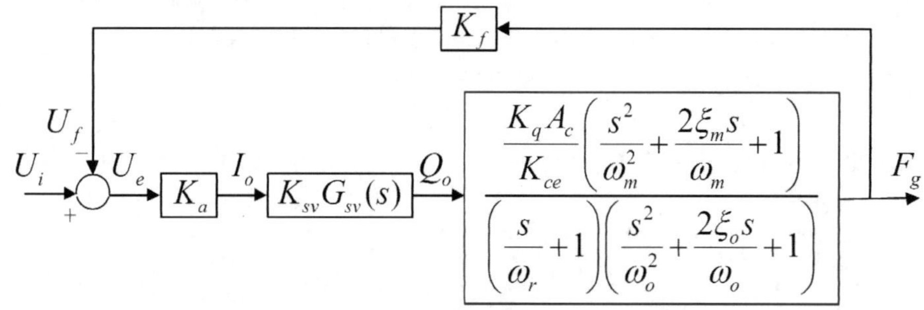

Figure 4. Block diagram of hydraulic control system

where U_i, U_f ——input and feedback voltage, V

K_a ——servo amplifier gain, A/V

K_f ——force sensor coefficient, A/m

K_q ——flow gain of servo valve, m^2/s

$G_{sv}(s)$ ——transfer function of Servo valve

ω_m ——load natural frequency, rad/s

ξ_m ——load damping ratio

ω_r ——frequency of Hydraulic and load coupling in series

ω_o ——frequency of Hydraulic and load coupling in parallel

ξ_o ——hydraulic damping ratio

4. Analysis of dynamic characteristics and influencing factors

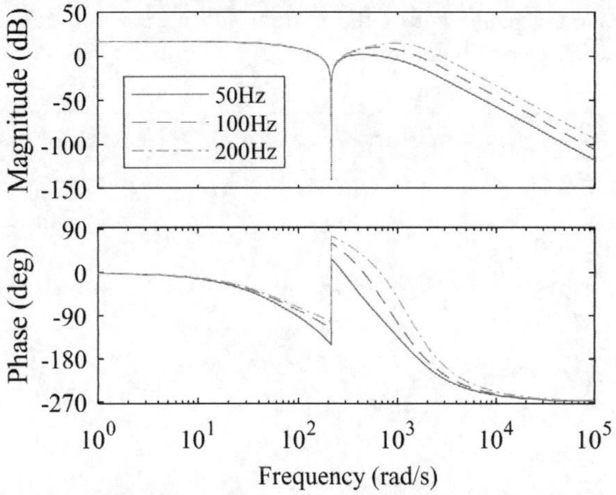

Figure 5. Open loop Bode diagram of different servo valve bandwidth

In order to study the dynamic characteristics of the hydraulic control system, taking the hydraulic control system of knee joint as an example, the system parameters are substituted into the block diagram in figure 4. The effects of parameters such as electro-hydraulic servo valve, load mass and hydraulic cylinder action area on the system control performance are analysed, so as to lay a foundation for obtaining the best control performance.

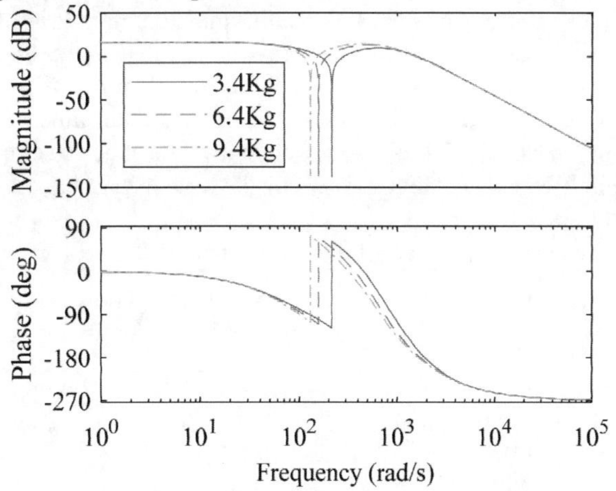

Figure 6. Open loop Bode diagram of different load masses

Servo valves with 50Hz, 100Hz and 200Hz bandwidth are used respectively. The open-loop Bode diagram of the system is shown in figure 5. It can be seen from the figure that the servo valve with a bandwidth above 50Hz has only an impact on the high-frequency band of the system, but has little impact on the amplitude and phase of the low-frequency band. Therefore, if the servo valve with 100Hz bandwidth is adopted, the influence of its dynamic characteristics can be ignored.

The load mass is taken as 9kg, 6kg and 3kg respectively. The open-loop Bode diagram of the system is shown in figure 6. It can be seen from the figure that the bandwidth of the system will decrease with the increase of load mass. Therefore, on the basis of ensuring the strength of exoskeleton, weight reduction of exoskeleton legs is not only the requirement of structural design, but also the requirement of control performance.

The hydraulic cylinder diameter/rod diameter are taken as 20/10mm, 30/15mm and 40/20mm respectively. The open-loop Bode diagram of the system is shown in figure 7. It can be seen from the figure that the change of hydraulic cylinder diameter/rod diameter has a great impact on the control performance of the system, especially in the low frequency band, the increase of cylinder diameter /rod diameter will increase the phase lag of the system and reduce the tracking accuracy. Therefore, the hydraulic cylinder diameter/rod diameter should be reduced as much as possible under the allowable conditions of structural design.

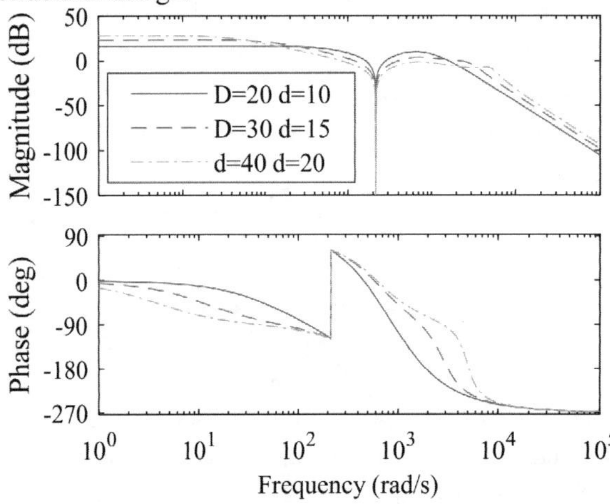

Figure 7. Open loop Bode diagram of different cylinder diameter

From the above analyses, it can be seen that the system is a type 0 system with a second-order unstable term. If only the dynamic characteristics of the servo valve are ignored, it is a third-order system. The natural frequency of mechanical load ω_m divides the open-loop frequency characteristics into two areas. The system is not allowed to work beyond ω_m that is the upper limit of the working frequency. As the frequency of exoskeleton walking is far less than ω_m, there will be no out of range operation. The dynamic performance of the system is mainly determined by the parallel coupling frequency of hydraulic and mechanical. In the low frequency band, the structural parameters have little influence on the amplitude frequency characteristics of the system, but have a greater influence on the phase frequency characteristics.

Adding PID controller to the system and taking the control coefficient $K_p = 3, K_I = 0.01$, $K_D = 2$, the force tracking curves at 2Hz and 10Hz are shown in figure 8 and figure 9 under the condition of cylinder size $D = 20$ $d = 10$, servo valve $\omega_{sv} = 628\text{rad/s}$ and load quality $M_t = 3.4\text{K}_g$.

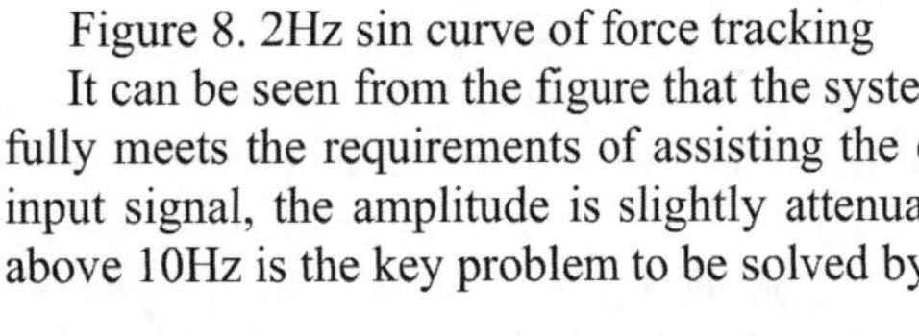

Figure 8. 2Hz sin curve of force tracking

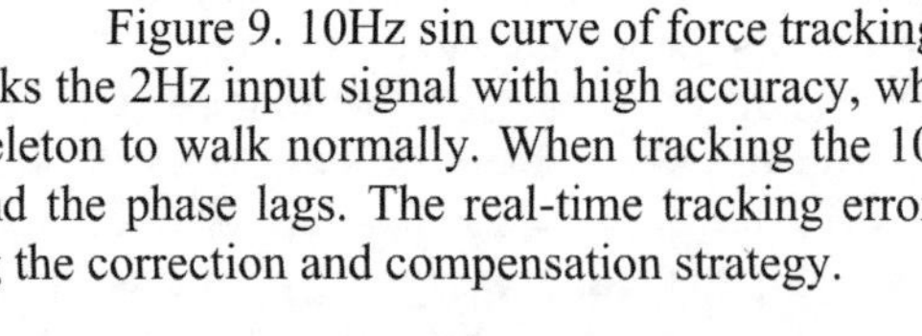

Figure 9. 10Hz sin curve of force tracking

It can be seen from the figure that the system tracks the 2Hz input signal with high accuracy, which fully meets the requirements of assisting the exoskeleton to walk normally. When tracking the 10Hz input signal, the amplitude is slightly attenuated and the phase lags. The real-time tracking error in above 10Hz is the key problem to be solved by using the correction and compensation strategy.

5. Conclusions

It is very important to optimize the structural parameters of the hydraulic driven lower limb exoskeleton. High frequency response servo valve can be simplified as proportional link. Reducing the load quality is of great significance to lightweight and control performance. Under the condition of ensuring the structural strength of the system, the use of smaller hydraulic cylinders is the key to improve the dynamic characteristics in low-frequency band. In the high frequency band, the force tracking accuracy also needs to be realized by optimizing the design of the compensation controller.

Acknowledgments
This work was supported in part by the project funded by National Nature Science Foundation of China 51075017.

References

[1] Anam K, Al-Jumaily A A 2012 Active Exoskeleton Control Systems: State of the Art [J]. Procedia Engineering, 41:988-994
[2] Walsh C J, Endo K and Herr H A 2007 Quasi-passive Leg Exoskeleton for Load-carrying Augmentation [J]. International Journal of Humanoid Robotics, 4(3):487-506.
[3] Wang C X 2011 Hydraulic control system [M]. Mechanical industry Press, Beijing.
[4] Zoss A B 2006 Kazerooni H and Chu A Biomechanical Design of the Berkeley Lower Extremity Exoskeleton (BLEEX) [J]. IEEE/ASME Trans. Mechatronics, 11(2):128-138
[5] Ghan J, Steger R and Kazerooni H 2006 Control and System Identification for the Berkeley Lower Extremity Exoskeleton (BLEEX) [J]. Advanced Robotics, 20(9): 989-1014
[6] Racine J L 2003 Control of a Lower Extremity Exoskeleton for Human Performance Augmentation [D]. Berkeley, CA: U.C.Berkeley.
[7] Steger J R 2006 A Design and Control Methodology for Human Exoskeletons [D]. Berkeley, CA: U.C.Berkeley.
[8] Amundson K, Justin R and Nathan H 2006 Development of hybrid hydraulic-electric power units for field and service robots [J]. Advanced Robotics, 20(9): 1015-1034
[9] Liu C N 1989 Optimization design theory of hydraulic servo system [M]. Metallurgical Industry Press, Beijing.

Journal of Physics: Conference Series **2365** (2022) 012012 doi:10.1088/1742-6596/2365/1/012012

Analysis and Verification of Phase Current Detection Method of Permanent Magnet Synchronous Motor Based on Three Sampling Resistors Configuration

Ziwei Zhu[1], Jiayong Cao[1], Qianli Ma[1], Fazhe Li[1]

[1]School of Mechanical Engineering, Shanghai Institute of Technology, 201418, China

E-mail: caojy@sit.edu.cn

Abstract. In the permanent magnet synchronous motor (PMSM) control system, the phase current sampling unit is an important part of the whole motor control system. Traditionally, the hall-effect components are used to detecting the phase currents of the PMSM. However, the size of the hardware with Hall-effect components are often large. In this paper, a scheme of detecting the phase currents of PMSM by using three sampling resistors on the main circuit was discussed. In order to overcome the phenomenon that phase currents flow discontinuously across the three sampling resistors, a method for reconstructing the phase currents was proposed. Both simulation results and experimental results show that the phase currents can be reconstructed successfully and the high sampling accuracy can be obtained.

1. Introduction

With the rapid development of power electronics technology and computer technology in recent years, permanent magnet synchronous motors (PMSM) are characterized by high reliability, low energy consumption, high efficiency, and high-power density. PMSMs are widely used in aerospace, industrial servo system, CNC machine tools and other high-tech field. In the PMSM driven system, the closed-loop control of the entire currents can be realized by feeding back the collected phase current to the current PI regulators for improving the control accuracy of the entire control system. Therefore, collecting the information of motor phase currents is an important problem in the whole control system. The traditional voltage source inverter controller uses a current sensor on the AC side of the inverter circuit to detect the motor phase current. Generally, the current sensor adopts Hall sensor to sample the motor phase current. When using the Hall sensor to detect the current, two current sensors [1-2] can be used to detect the DC bus current or a single current sensor can be used [3-5] Sampling the motor phase current. However, using these methods, the acquisition and detection must be isolated, so that the sampling current without interference can be better obtained, which will also increase the volume and cost of the entire control system. For occasions with high volume requirements and more space constraints, the implementation of this method is not appropriate, this drawback is highlighted in today's increasingly demanding industrial integration requirements. Therefore, resistors detection is an important development trend of precision driving in the future, which suitable for production cost-sensitive industries. The literature adopts the detection method of a single resistor [6-8] to detect the DC side bus current of the inverter circuit and solves the problem of current reconstruction through the method of voltage vector phase shifting.

In this paper, three sampling resistors are used to detect the motor current. In the scheme of using three-resistor sampling, the main problem is that none of the sampling resistor terminal voltages can fully reflect the full waveform of the phase current in a PWM cycle. Therefore, how to reasonably set the sampling instant of the digital controller and reconstruct the three-phase circuit of the motor through a reasonable algorithm has become a key issue, this paper mainly studies the above problems.

2. Principle of current detection method

2.1. Mathematical model in synchronous coordinate system

The mathematical model of a surface-mounted PMSM in the d-q frame can be expressed as follows[9].

$$\frac{d}{dt}i_d = \frac{1}{L_d}u_d - \frac{R}{L_d}i_d + \frac{L_q}{L_d}p\omega_m i_q \tag{1}$$

$$\frac{d}{dt}i_q = \frac{1}{L_q}u_q - \frac{R}{L_q}i_q + \frac{L_d}{L_q}p\omega_m i_d \tag{2}$$

$$T_e = 1.5p\lambda i_q \tag{3}$$

Equations (1) and (2) are known as voltage equations. Equation (3) is named as equation of electromagnetic torque equation. The quantities in (1)-(3) refer to Table 1.

Table 1. Quantities in the PMSM model

Variables	Description	Variables	Description
L_q, L_d	q-axis and d-anis inductances	R	Resistance of the stator windings
i_q, i_d	q-axis and d-axis currents	u_q, u_d	q-axis and d-axis voltages
ω_m	Angular velocity of the rotor	λ	Amplitude of the flux induced by the permanent magnets of the rotor in the stator phases
p	Number of pole pairs	T_e	Electromagnetic torque

2.2. Main circuit of PMSM inverter and PWM strategy

A three-phase bridge inverter shown in Figure 1 was adopted in this paper, every bridge contains two IGBT devices which are named higher IGBT and lower IGBT. The middle points of three phase bridge are connected to the U, V and W ports of the motor respectively. There are two pulse width modulation (PWM) strategies named sinusoid PWM and space-vector PWM (SVPWM). We used SVPWM strategy for driving the IGBTs considering that the SVPWM has the advantages of lower switch device loss and less harmonic components.

In SVPWM method, the inverter divides the voltage vector into 6 non-zero vectors $V_1, V_2, V_3, V_4, V_5, V_6$ and 2 zero vectors V_0, V_7, and each output voltage vector is composed of these 8 voltage vectors. Therefore, the reference voltage vector is composed of two effect voltage vectors and two zero vectors in each PWM cycle.

2.3. Principles of Current detection and reconstruction

Three sampling resistors $R_i (i = 1,2,3)$ which connect the emission ports of the lower IGBTs and the electrical ground are configured. The basic principle of current reconstruction is when all the three-phase lower tubes are turned on, removing the phase with the shortest conduction time, only the currents of the other two phases are collected, and then the currents of the other phase is obtained according to Kirchhoff's law, during a control cycle. Thereby the current reconstruction is realized.

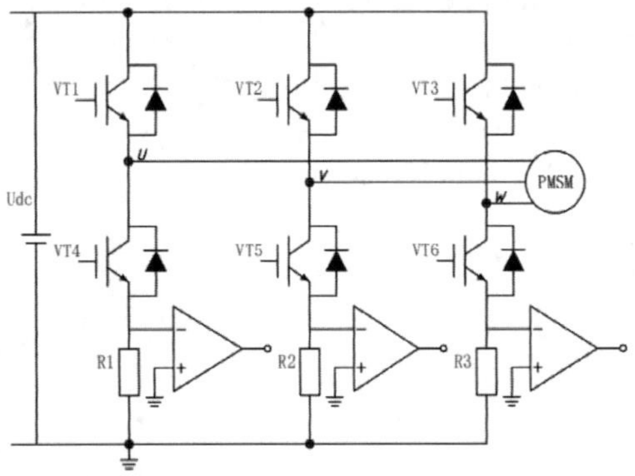

Figure 1. Three-resistor sampling circuit diagram

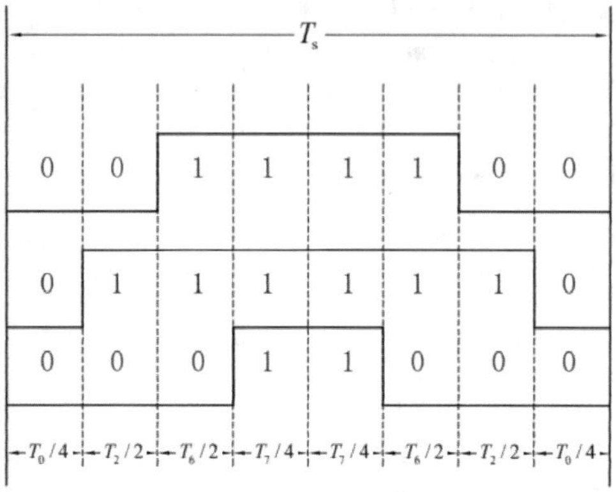

Figure 2. Three-phase bridge arm conduction diagram

The phase current reconstruction method adopted in this paper is to complete the current sampling during the period of zero voltage vector action. In order to facilitate analysis and research, the direction of the current flowing into the motor is defined as the positive direction, and the direction of the current flowing out of the motor is negative direction.

Taking the reference voltage vector in sector II as an example, the sequence of the voltage vector is $U_0 \rightarrow U_2 \rightarrow U_4 \rightarrow U_7 \rightarrow U_7 \rightarrow U_4 \rightarrow U_2 \rightarrow U_0$, as shown in the Figure 2, the upper switch tube conduction time of phase B is longest, so in order to improve the accuracy of the sampling currents, only the currents of A-phase and C-phase are collected during the PWM cycle.

At the sampling moment, due to the effect of inductance, the currents' direction of motor will not change abruptly. The current is freewheeled by the freewheeling diode in anti-parallel with the lower arm of phase B. Then the direction of i_b and i_c are positive, the direction of i_a is negative. By analogy, the corresponding relationship of motor phase current in other sectors can be obtained, as shown in Table 2.

Table 2. Current acquisition diagram

Sector where the reference voltage is located	I	II	III	IV	V	VI
The current collected by the sampling resistor	i_b、i_c	i_a、i_c	i_a、i_c	i_a、i_b	i_a、i_b	i_b、i_c

3. Simulation of the current detection method

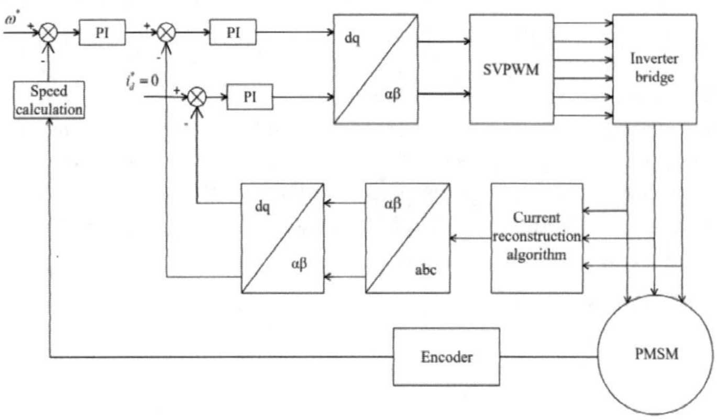

Figure 3. System control block diagram

In order to verify the feasibility and accuracy of this method, using the MATLAB/Simulink platform, a PMSM based on line current reconstruction is built as a closed-loop control variable for simulation. The system control block diagram is shown in Figure 3. The inverter switching frequency is 10KHz, the DC busbar voltage is 311V.

Main motor parameters are as follows, the number of pole pairs $P_n = 4$, stator resistance $R = 2.785\Omega$, stator d, q axis inductance $L_d = L_q = 0.0085H$, permanent magnet flux linkage $L_d = L_q = 0.0085H$, moment of inertia $J = 0.001kg \cdot m^2$.

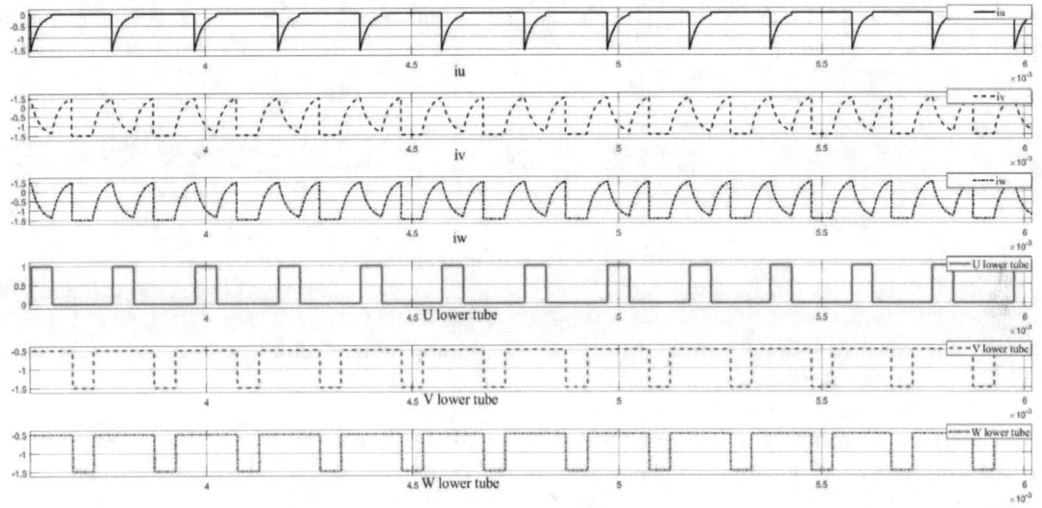

Figure 4. Three-phase current waveform and three-phase upper arm conduction time waveform

The three-phase current waveform and the three-phase upper tube conduction time waveform obtained by simulation is shown in Figure 4.

4. Experiment verification

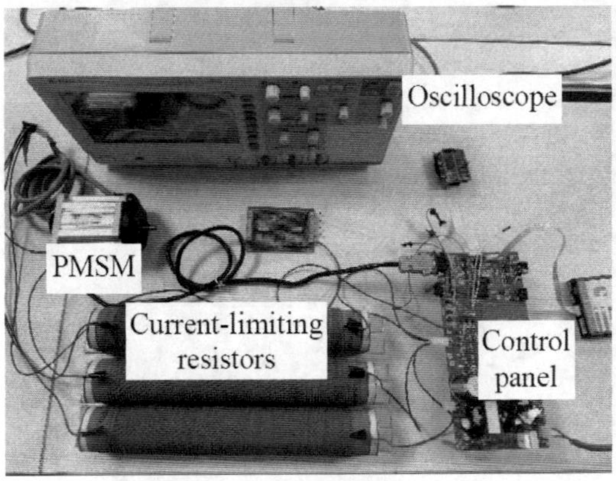

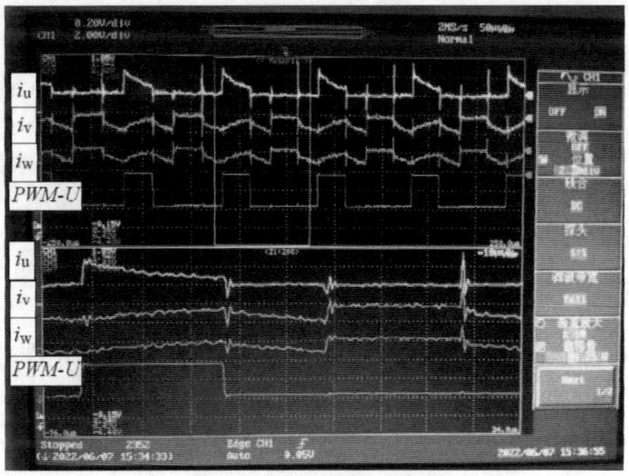

Figure 5. Test bench Figure 6. Experimental results

In order to prove the reliability of the proposed method, a three-phase voltage inverter test bench as shown in Figure 5 was built in this study as the driver for the PMSM, in order to prevent the current from being too large during debugging, the motor is externally connected with the current limiting resistors.

$$m = index(\min(i_1, i_2, i_3)) \tag{4}$$

$$i'_k = \begin{cases} i_k & k \neq m \\ \sum_{j=1, j\neq m}^{3} i_j & k = m \end{cases} \tag{5}$$

The entire sampling flow chart in the software program is shown in the Figure 7. When the microcontroller has an underflow flag, the microcontroller starts to sample the motor current through the ADC. The minimum value of the current collected in the three-phase circuit is discarded, and only the two-phase current is retained. Then another phase current is obtained by calculation as shown in Equation (4) and Equation (5).

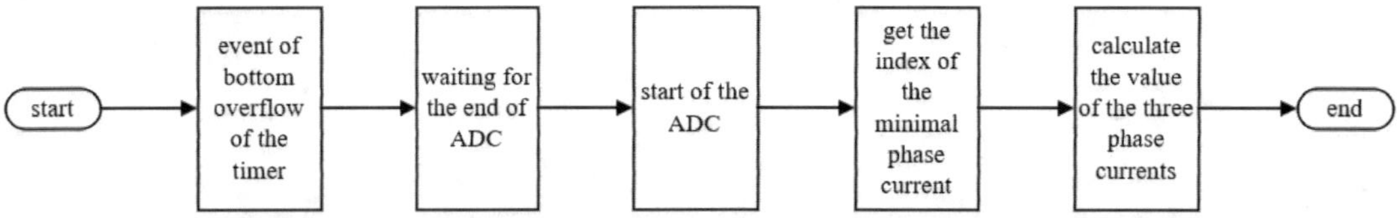

Figure 7. Current sampling process

The experimental results are shown in Figure 6. Comparing the experimental results with the simulation results, it is found that the detected current waveform is basically consistent with the simulation waveform. The experimental results also prove the correctness and reliability of the proposed method.

Figure 8 shows the current waveform collected by the ADC of STM32F407, which is directly obtained. There are many harmonic components in the current waveform, so a digital filter is used to filter the current wave. Figure 9 gives the waveform after filtering.

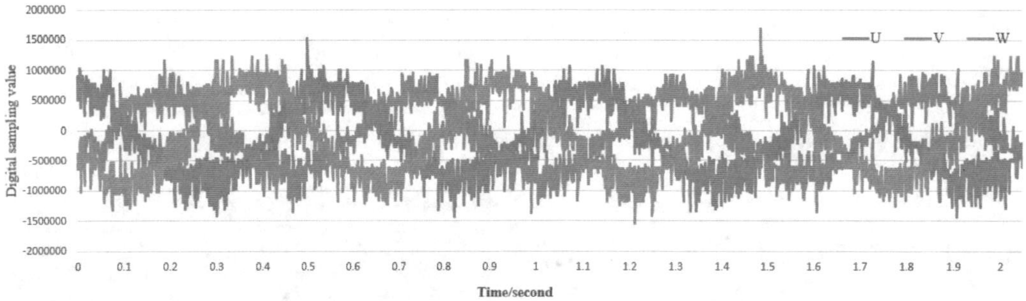

Figure 8. Current obtained from ADC of single chip microcomputer

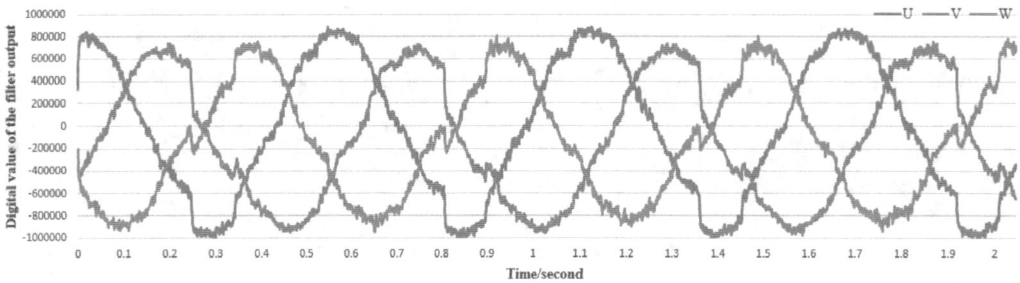

Figure 9. Waveform of current collected by ADC of single chip microcomputer after passing through filter

5. Conclusions

This paper analyzes the method based on phase current sampling and reconstruction, the essence of this method is to use the conduction of lower switch tubes of the inverter to collect the two-phase current of the motor through three sampling resistors, then calculate the current of another phase. The three-resistors sampling method can effectively achieve the purpose of current detection, realize the high-precision control of the motor, and greatly reduce the size and production costs of the motor driver.

References

[1] WX Huang, YW Hu. 2007 J. Cage Induction Generator Used in 270V DC Power System Generator System Based on Space Vector Modulation Strategy. *Transactions of China Electrotechnical Society*. **22** (in Chinese).

[2] SZ Jiang, XH Li. 2007 J. Current Sensor in Motor Driver. *Electric Machines & Control Application*. **34**:29-32(in Chinese).

[3] YK Gu, FL Ni, et al. 2009 J. Novel method for phase current reconstruction using a single DC-Link current sensor. *Electric Machines and Control*. **13** 811-816(in Chinese).

[4] BX Chu, YW Hu, et al. 2010 J. Phase Current Sampling Reconstruction for Inverter. *Transactions Of China Electrotechnical Society*. **25** 111-117(in Chinese).

[5] JQ Xu, JJ Li. 2018 J. Synchronous Rectification and Braking Control for Brushless DC Motor With Single Current Sensor. *Journal of Beijing University of Technology*. 44 848-854(in Chinese).

[6] XB Cheng, WX Huang, et al. 2010 J. Single Resistor Current Sensing and Reconstruction Method for Phase Currents in Inverters. *Electric Drive*. **40** 3-6(in Chinese).

[7] YG Zhang, S Qin and P Liu. 2021 J. A motor phase current reconstruction technology based on multiple adjustment strategies. *Power System Protection and Control*. **49** 63-72(in Chinese).

[8] KY Huang, RZ Wu, et al. 2018 J. Phase Current Reconstruction Strategy for PMSM Using One-shunt Current Sampling. *Proceedings of the CSU-EPSA*. **30** 114-120(in Chinese).

[9] TL Guo, ZX Sun, et al. 2019 J. A Simple Current-Constrained Controller for Permanent-Magnet Synchronous Motor. *IEEE Transactions on Industrial Informatics*.**15** 1486-1495.

[10] Candelo-Zuluaga C, Riba JR and Garcia A. 2021 J. PMSM Parameter Estimation for Sensorless FOC Based on Differential Power Factor. *IEEE Transactions on Instrumentation and Measurement*. **70**.

Thermal Analysis and Static Lubrication Performance Study of Liquid Metal Journal Bearing

Haifeng Wang, Yazhen Wang*, Zhaoxiang Xu

School of Mechatronic Engineering and Automation, Shanghai University, Shanghai, P.R. China.

*meyzwang@shu.edu.cn

Abstract. The temperature distribution of liquid metal journal bearing system in X-ray tube of medical CT machine was simulated by using simulation software, and the working temperature of the herringbone groove liquid metal journal bearing was initially determined. On this basis, the lubricating properties of the liquid metal lubricant were analyzed. The lubrication model of herringbone groove liquid metal journal bearing with Galinstan liquid metal as lubricant was established, and the influence of herringbone groove structure parameters on the static lubrication performance of liquid metal journal bearings was analyzed. The results show that the Galinstan liquid metal has good high temperature stability and can act as a reliable lubricant in the high temperature working environment of the medical CT machine X-ray tube.

1. Introduction

The rotating anode X-ray tube is the core component of X-ray Computed Tomography Equipment (CT), which consists of three major parts: the rotating anode, the cathode that emits the electron beam, and the vacuum glass cover. Among them, the rotating anode also includes the target plate and bearing system and other components.

When the CT machine is working, a large amount of heat is generated when the electron beam strikes the target plate, and the target plate material will melt partially if it is on a fixed anode target, so it is necessary to rotate the anode to make the heat spread to a ring area and dissipate it through the cooling tube inside the shaft. In conventional X-ray tubes, deep groove ball bearings are usually used to support the rotating body, which has the disadvantages of limited life, unstable current supply, weak heat transfer, high vibration and noise, and low load capacity[1]. In order to improve the working performance of CT machines, a new type of bearing capable of stable current supply, large thermal capacity as well as load-bearing capacity and stable working performance is needed to replace the traditional deep groove ball bearing. Muijderman E [2] used herringbone groove bearings to replace the ball bearings in the diagnostic X-ray tube and found that not only the bearing load carrying capacity was improved, but also the problems of limited life of the ball bearings, difficulties in anode cooling, and bearing noise were solved. Gerkema J [3] conducted an experimental study on smaller size helix groove journal bearings using liquid gallium-based alloys as lubricant and found that gallium-based alloys can be used as a reliable lubricant for helix groove bearings under non-oxidizing conditions. Prokhorenko V Y[4] found that gallium-based alloys have high electrical conductivity, thermal conductivity and high temperature stability. Kezik V Y [5] used gallium as a lubricant and investigated the lubricating properties of gallium in the temperature range of 30-300°C using different methods and found that gallium-based liquid metal is mainly mixed laminar and turbulent lubrication in high temperature environment. Bo G[6] found that gallium-based liquid metals not only have good thermal conductivity, electrical conductivity, low viscosity, nontoxicity, biocompatibility and low melting point, but also have good flexibility, ductility and recombination ability at room temperature. Xu M[7] established a lubrication performance calculation model of herringbone groove bearing for the low viscosity of liquid metal gallium, and analyzed the bearing lubrication film pressure distribution, load capacity magnitude, temperature rise and stiffness under different eccentricity. However, the above literature does not consider the effect of the high temperature operating environment in the X-ray tube of CT machine on the lubrication performance of liquid metal lubricants, which further affects the static performance of herringbone groove liquid metal journal bearings.

In this paper, Galinstan is used as the lubricant of herringbone groove journal bearing, and the structure of CT machine X-ray tube with herringbone groove liquid metal journal bearing is shown in Figure 1. Firstly, the temperature distribution of liquid metal journal bearing system in X-ray tube of medical CT machine was simulated, and the working temperature of the bearing was initially determined. The physical properties of the liquid metal Galinstan under high temperature environment were analyzed. Secondly, the lubrication model of herringbone groove liquid metal journal bearing with Galinstan liquid metal as lubricant was established, and the influence of herringbone groove structure parameters on the static lubrication performance of liquid metal journal bearings was analyzed.

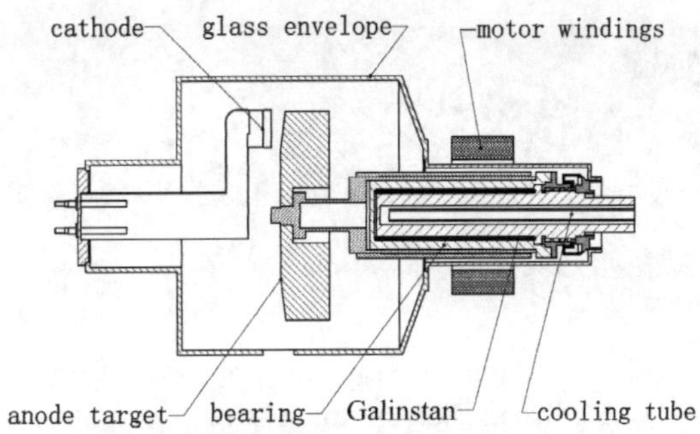

Figure 1. CT machine x-ray tube structure

2. Thermal simulation of X-ray tube bearing system

The structure of X-Ray ray tube with liquid metal journal bearing is simplified and modeled by SolidWorks modeling software, and the sketch of the model is shown in Figure 2 below.

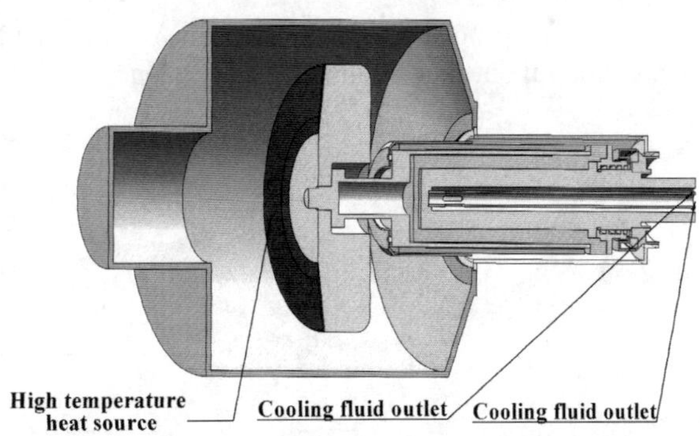

Figure 2. X-ray tube 3D model

The boundary conditions are the combination of finite element analysis and actual working conditions. The boundary conditions of the X-ray tube of the CT machine studied in this paper are based on actual measurements, and the basic parameters are as follows: coolant inlet temperature: 50°C; inlet flow rate: 1.7m/s. The high temperature surface of the target plate is 1100°C.

The overall temperature distribution of the rotating anode target plate and the temperature distribution of the shaft diameter and bearing bush are shown in Figure 3-4.

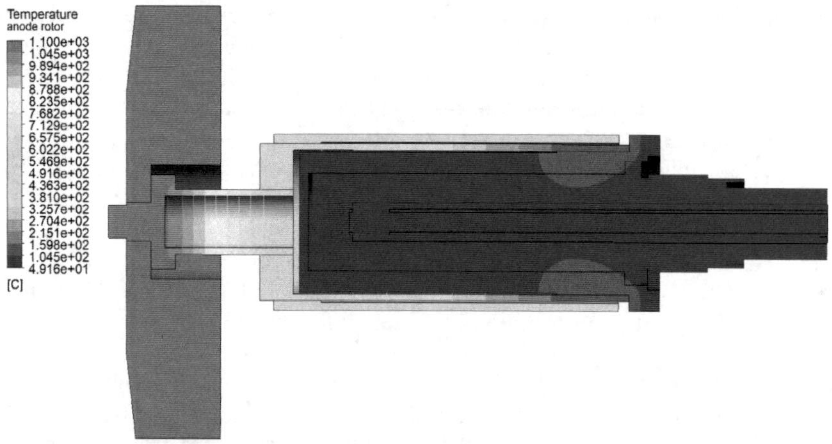

Figure 3. Rotating anode temperature distribution

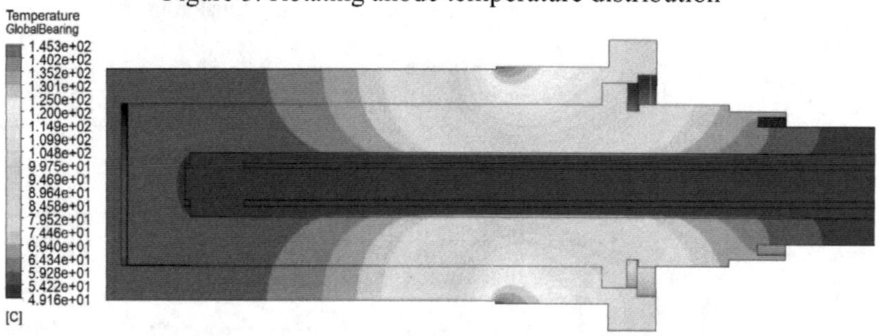

Figure 4. Journal bearing temperature distribution

From the above figure, it can be seen that the maximum temperature of the lubrication interface between the shaft diameter and the bearing sleeve can reach 130°C. At this temperature, the viscosity of liquid metal Galinstan is about 0.0016 Pa·s.

3. Mathematical model and calculation method

3.1. The establishment of the mathematical model

The basic structure of the herringbone groove liquid metal journal bearing and the schematic diagram of the expansion along the circumferential Φ are shown in Figure 5 below.

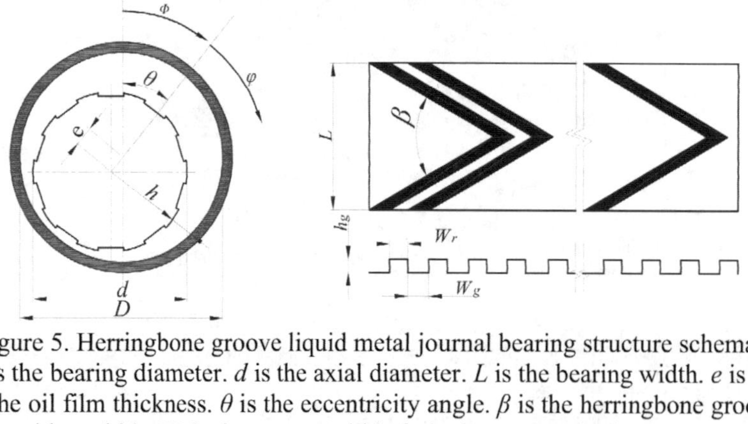

Figure 5. Herringbone groove liquid metal journal bearing structure schematic

Where, D is the bearing diameter. d is the axial diameter. L is the bearing width. e is the eccentricity distance. h is the oil film thickness. θ is the eccentricity angle. β is the herringbone groove helix angle. W_r is the convex ridge width. W_g is the groove width. h_g is the groove depth.

The expression of Reynolds equation for herringbone groove liquid metal journal bearing considering turbulent flow in right-angle coordinate system is shown in Eq. 1

$$\frac{\partial}{\partial x}\left(\frac{h^3}{k_x\mu}\frac{\partial p}{x}\right)+\frac{\partial}{\partial y}\left(\frac{h^3}{k_y\mu}\frac{\partial p}{y}\right)=\frac{U}{2}\frac{\partial h}{\partial x} \tag{1}$$

Where: U is the bearing linear velocity. h is the lubricant film thickness. μ is the lubricant dynamic viscosity. p is the lubricant film pressure. x is the circumferential coordinates($x=\Phi r$). y is the axial coordinates. k_x, k_y is the turbulence coefficient.

Dimensionless Eq. 1, and the dimensionless Reynolds equation is shown in Eq. 2.

$$\frac{\partial}{\partial \varphi}\left(K_x H^3\frac{\partial P}{\partial \varphi}\right)+\left(\frac{D}{L}\right)^2\frac{\partial}{\partial \varphi}\left(K_y H^3\frac{\partial P}{\partial \lambda}\right)=3\frac{\partial H}{\partial \lambda} \tag{2}$$

Where: $\varphi=\Phi-\theta$, $\lambda=y/(L/2)$ (L is the bearing width). $H=h/s$(s *is* bearing radius clearance). $P=(c^2/2U\mu r)p$. $K_x=12/k_x$, $K_y=12/k_y$.

Due to the special structure of herringbone groove liquid metal journal bearing, there are a large number of boundary conditions to be handled when calculating in right angle coordinates, and the calculation efficiency and accuracy will be affected. In order to improve the calculation efficiency and accuracy, the calculation will be carried out in oblique coordinate system. The schematic diagram of the oblique coordinate transformation is shown in Figure 6 below.

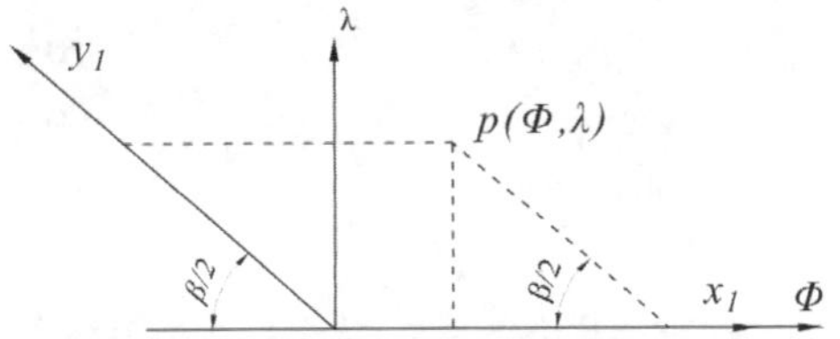

Figure 6. Coordinate transformation diagram

The coordinate transformation equation can be obtained from Figure 5, as shown in Eq. 3.

$$\begin{cases}\Phi=x_1-y_1\cos(\beta/2)\\ \lambda=y_1\sin(\beta/2)\end{cases}\Leftrightarrow\begin{cases}x_1=\Phi+\lambda/(\tan(\beta/2))\\ y_1=\lambda/\sin(\beta/2)\end{cases} \tag{3}$$

3.2. Numerical calculation method

In this paper, the finite difference method is used to solve the lubricating film pressure distribution, and we divide the lubricating film into finite meshes along the helix angle of the herringbone groove in the circumferential and axial directions. Since the thickness of the lubricating film at the junction of the groove and the convex ridge is discontinuous, the difference expression of Reynolds equation is established in oblique coordinates for the continuous lubricating film on the groove and the convex ridge and the discontinuous lubricating film thickness respectively, and the expression is shown in the following Eq. 4.

$$\begin{cases}A_{i,j}P_{i+1,j}+B_{i,j}P_{i-1,j}+C_{i,j}P_{i,j+1}+C_{i,j}P_{i,j-1}-F_{i,j}=E_{i,j}P_{i,j}\\ A'_{i,j}P_{i+1,j}+B'_{i,j}P_{i-1,j}+C'_{i,j}P_{i,j+1}+C'_{i,j}P_{i,j-1}-F'_{i,j}=E'_{i,j}P_{i,j}\end{cases} \tag{4}$$

Where: P is the dimensionless pressure of lubricating film. A to F are the difference coefficients related to lubricating film thickness and other structures at lubricating film continuity. A' to F' are the difference coefficients related to lubricating film thickness and other structures at lubricating film discontinuity.

The expression for calculating the dimensionless bearing capacity is shown in Eq. 5.

$$\begin{cases} \overline{F}_x = -\int_0^{2\pi}(\int_{-1}^{1} Pd\lambda)\sin\Phi d\Phi \\ \overline{F}_y = -\int_0^{2\pi}(\int_{-1}^{1} Pd\lambda)\cos\Phi d\Phi \end{cases} \tag{5}$$

The total dimensionless bearing capacity of liquid metal journal bearings is shown in Eq. 6.

$$\overline{F} = \sqrt{\overline{F}_x + \overline{F}_y} \tag{6}$$

4. Influence of the structural parameters on lubrication film pressure and bearing capacity

4.1. Initial structure parameters

The initial structural parameters of the herringbone groove liquid metal journal bearing and the physical parameters of the lubricant are shown in Table 1 below.

Table 1. Initial parameters table

Parameters	Value
Bearing diameter/(mm)	30
Width to diameter ratio	1
Radius clearance/(mm)	0.02
Groove depth/(mm)	0.02
Number of grooves	10
Helix angle	60
Eccentricity	0.5
Declination angle	30
Viscosity/(Pa·s)	0.0016
Density/(kg/m3)	6440
Rotational speed /(r/min)	11000

4.2. Influence of eccentricity

From Fig. 7.a and 7.b, we can find that the pressure at the groove is larger than that at the convex ridge, which is caused by the wedge effect at the groove. The distribution of the lubricating film pressure at the top of the herringbone groove in the circumferential direction is shown in Fig. 7.c. It can be seen that the variation of the pressure at the top of the herringbone groove increases gradually with the increase of the eccentricity. In addition, with the increase of eccentricity, the wedge effect of bearing lubrication gap is more significant, and the lubricating film pressure and load-bearing capacity also gradually increase.

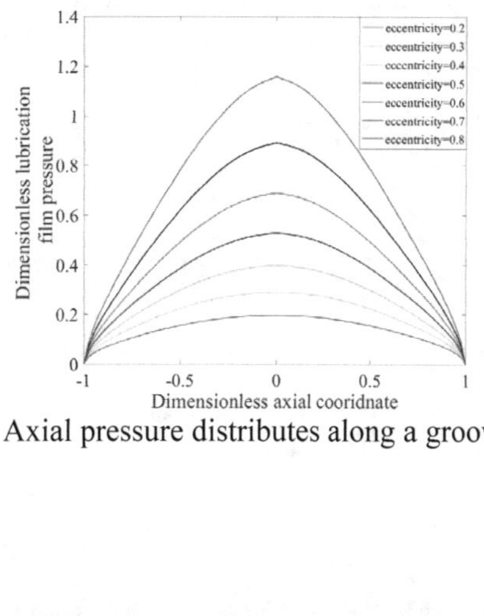

(a) Axial pressure distributes along a groove

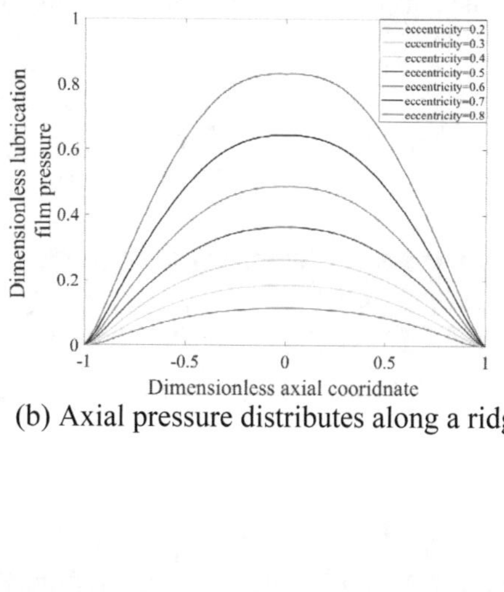

(b) Axial pressure distributes along a ridge

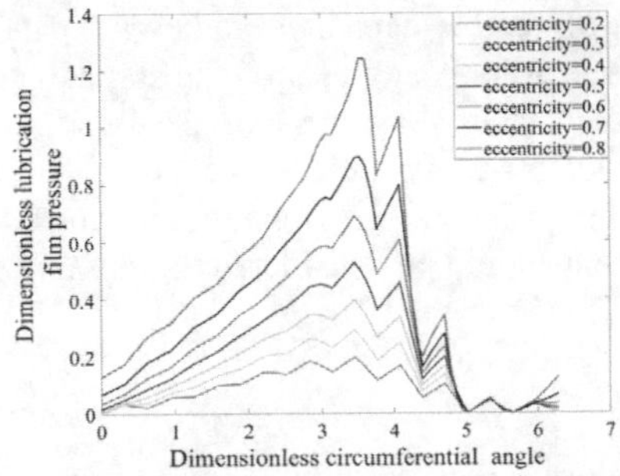

(c) Herringbone groove top circumferential dimensionless pressure distributes

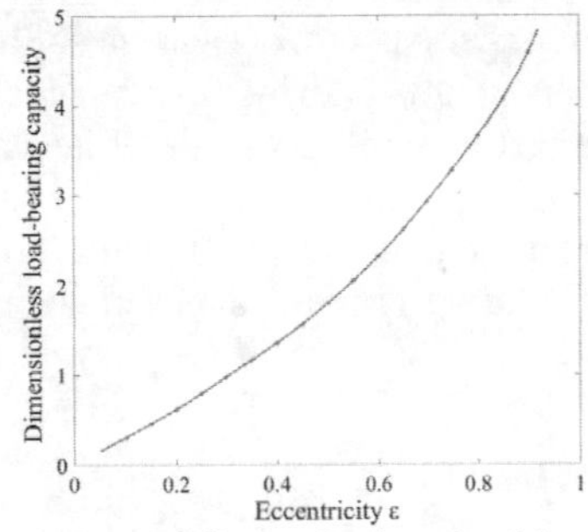

(d) Variation of dimensionless load-bearing capacity with eccentricity

Figure 7. Variation of lubricating film pressure and bearing capacity with eccentricity

4.3. Influence of width to diameter ratio

From Figure 8, we can see that with the increase of the width to diameter ratio, the pressure at the top of the herringbone groove along the circumferential direction also increases, and the bearing capacity gradually increases. There are two main reasons for this: first, with the increase of the width-to-diameter ratio, the herringbone groove becomes longer and the pumping effect is more obvious; second, with the increase of the width-to-diameter ratio, the lubricating film bearing area increases.

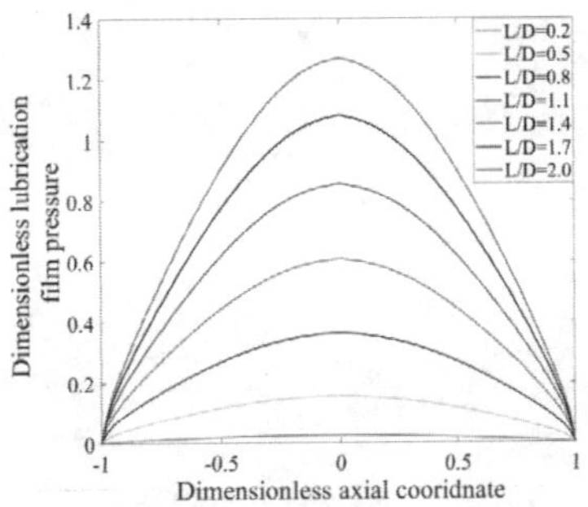

(a) Axial pressure distributes along a groove

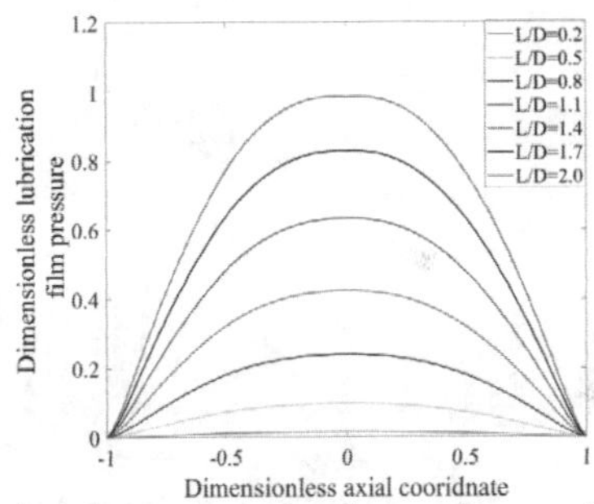

(b) Axial pressure distributes along a ridge

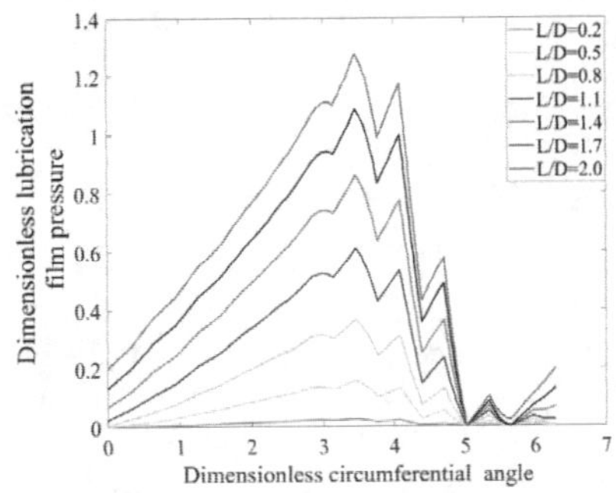

(c) Herringbone groove top circumferential dimensionless pressure distributes

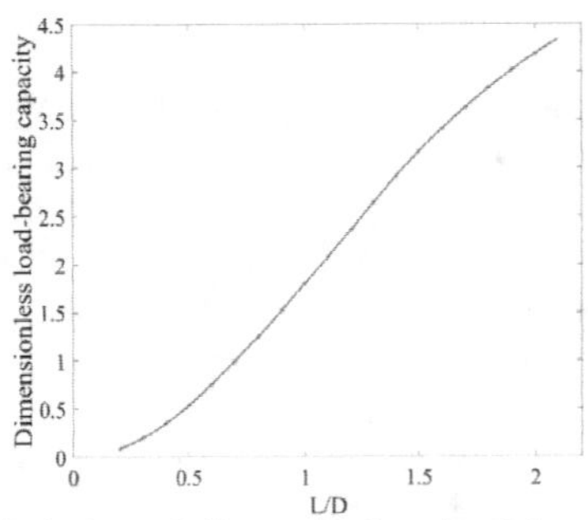

(d) Variation of dimensionless load-bearing capacity with width to diameter ratio

Figure 8. Variation of lubricating film pressure and bearing capacity with width to diameter ratio

4.4. Influence of helix angle of herringbone groove

From Figure 9, we can see that with the increase of herringbone groove helix angle, the pressure at the top of the herringbone groove along the circumferential direction first increases, when reaching a certain angle, the pressure at the top of the herringbone groove with the increase of herringbone groove helix angle no longer has obvious changes. From the viewpoint of bearing capacity, the bearing capacity of liquid metal journal bearing firstly increases with the increase of helix angle, and then starts to decrease with the increase of helix angle of herringbone groove after reaching a certain angle. In addition, the helix angle of the herringbone groove has a certain influence on the friction resistance coefficient and stiffness of the liquid metal journal bearing. Therefore, when designing herringbone bearings, bearing capacity, bearing stiffness and frictional resistance coefficient should be considered at the same time.

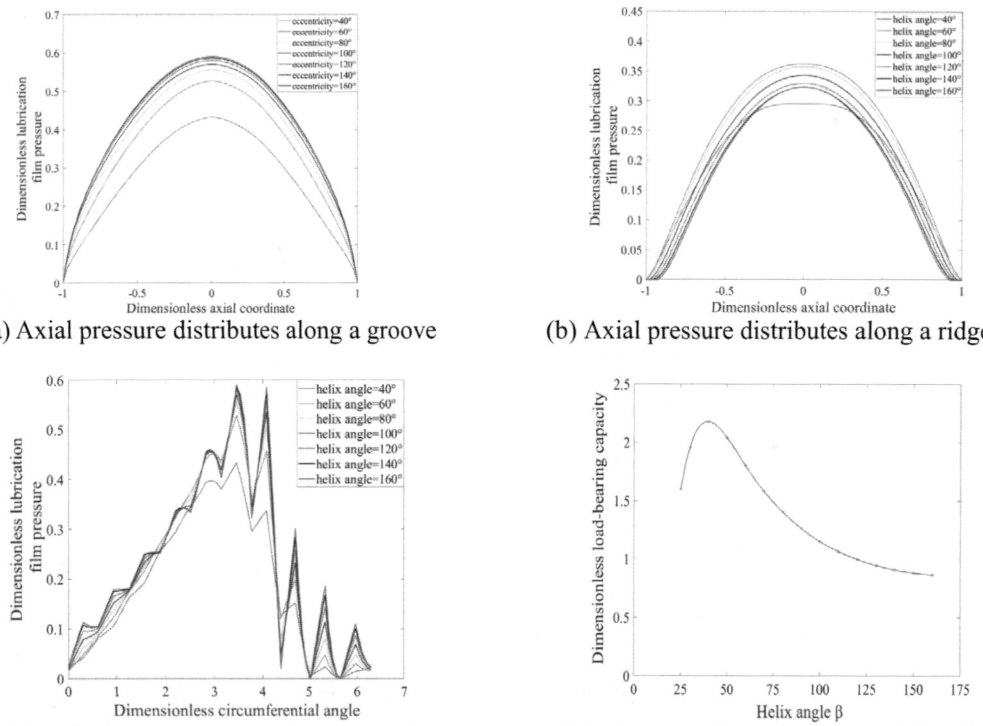

(a) Axial pressure distributes along a groove (b) Axial pressure distributes along a ridge

(c) Herringbone groove top circumferential dimensionless pressure distributes (d) Variation of dimensionless load-bearing capacity with helix angle

Figure 9. Variation of lubricating film pressure and bearing capacity with helix angle

4.5. Influence of herringbone groove depth

In Figure 10, we can see that with the increase of groove depth, the pressure at when it acts as lubricant, the groove the top of the herringbone groove along the circumferential direction gradually decreases, and the bearing capacity also decreases. However, because the viscosity of liquid metal is small depth is too shallow will adversely affect the formation of lubricating film. However, due to the small viscosity of liquid metal, the groove depth is too shallow will have a negative impact on the formation of lubricating film. Therefore, the groove depth is an important factor in the design of the bearing, and needs to be selected after considering various factors to ensure the stability of the bearing and sufficient load carrying capacity.

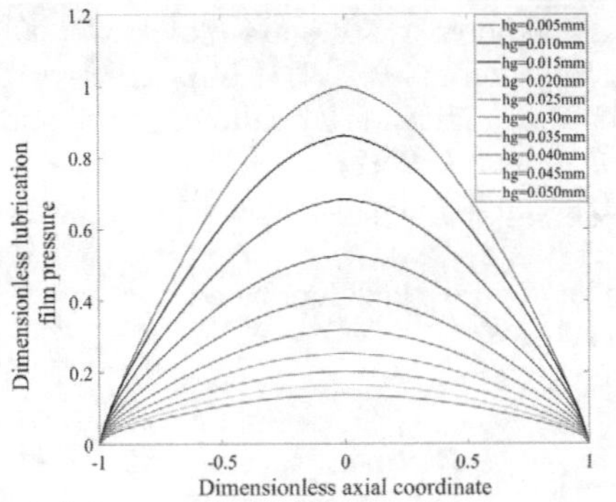

(a) Axial pressure distributes along a groove

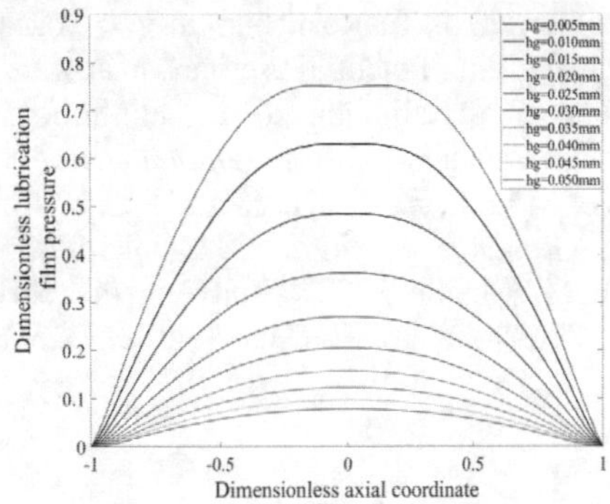

(b) Axial pressure distributes along a ridge

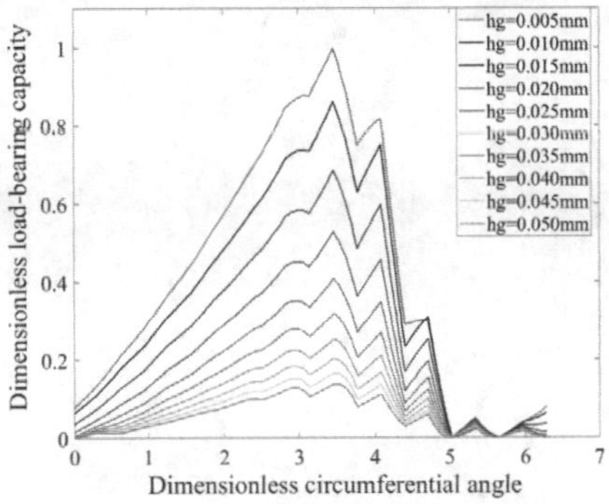

(c) Herringbone groove top circumferential dimensionless pressure distributes

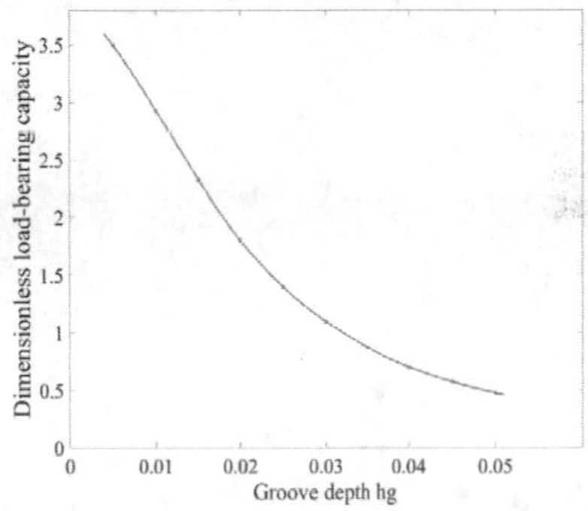

(d) Variation of dimensionless load-bearing capacity with groove depth

Figure 10. Variation of lubricating film pressure and bearing capacity with groove depth

5. Conclusion

In this paper, the influence of the basic structural parameters of herringbone groove liquid metal journal bearing on the bearing capacity is studied, and the main conclusions obtained are: the pressure of lubricating film and the bearing capacity increase with the increase of eccentric distance, but the eccentric distance is too large will lead to the uneven distribution of axial lubricating film pressure, resulting in the decrease of rotor stability. The lubricating film pressure and bearing capacity increase with the increase of bearing width to diameter ratio. The lubricating film pressure and bearing capacity increases with the increase of herringbone groove helix angle and then decreases, there is a helix angle that makes the best bearing capacity. The lubricating film pressure and bearing capacity decreases with the increase of herringbone groove depth, but because the viscosity of liquid metal lubricant is small, the groove depth is too small will affect the distribution of lubricant on the whole lubrication interface, so we must choose the herringbone groove depth very carefully.

References

[1] Hattori H, Fukushima H, Yoshii Y, Nakamuta H, Iwase M, and Kitade K 2009 Proposal of a high rigidity and high speed rotating mechanism us-ing a new concept hydrodynamic bearing in X-Ray tube for high speed computed tomography [J]. *Journal of Advanced Mechanical Design, Systems, and Manufacturing*. **3(1)**: 105-114.

[2] Muijderman E, Roelandse C, Vetter A, Schreiber P 1989 A diagnostic X-ray tube with spiral-groove bearings [J]. *Philips Tech Rev*. **44(11/12)**: 357-363.

[3] Gerkema J 1985 Gallium-based liquid-metal full-film lubricated journal bearings [J]. *ASLE transactions*. **28(1)**: 47-53.

[4] Prokhorenko V Y, Roshchupkin V V, Pokrasin M A, Prokhorenko S V, and Kotov V V 2000 Liquid gallium: Potential uses as a heat-transfer agent [J]. *High Temperature*. **38(6)** : 954-968.
[5] Kezik V Y, Kalinichenko A, Kalinitchenko V 2003 The application of gallium as a liquid metal lubricant [J]. *International Journal of Materials Research*. **94(2)**: 81-90.
[6] Bo G, Ren L, Xu X, Du Y, Dou S 2018 Recent progress on liquid metals and their applications [J]. *Advances in Physics: X*. **3(1)**: 1446359.
[7] Xu M, Chen W 2021 Study on the performance of liquid metal lubricated V-groove bearing considering turbulence [J]. *Applied Sciences*. **11(3)**: 940.

Design of actuator of marine garbage cleaning vessel based on ocean wave driving

JiangMeng Zou[a], Xi Zhu[a], Zhi Han[a], Chun Li*[a]

[a]School of Industrial Automation, Beijing Institute of Technology Zhuhai, Zhuhai, 519088, China

*Corresponding author: 351005011@qq.com

ABSTRACT: There are usually two ways to salvage and clean the garbage floating in the ocean, one is relying on manual and the other one is dependent on garbage salvage vessel driven by fuel oil. It is known that human collection is time-consuming and laborious, and machinery driven by fuel produces pollutant and pollutes the ocean environment in return. Taking these issues into consideration, this study designed a marine garbage cleaning vessel based on ocean wave driving which could gather and collect the floating garbage on the sea. The structure of the vessel actuator mainly constructed of a buoy-ratchet mechanism, the gear set-differential mechanism, the incomplete teeth-gathering mechanism, and a conveyor belt for transporting garbage. Through innovative structural design, this device could drive the whole transmission system by wave energy, which not only improves the work efficiency, increases reliability of structures but also reduces the consumption of manpower and material resources. Thus, the mechanical vessel is in line with sustainable development strategy, making full use of natural resources to clean ocean waste, and has a very high practical application value.

1. Introduction

With the development of the marine industry and the increase of various maritime activities, more and more garbage are floating on the sea [1]. In order to protect the Marine environment, the problem of floating garbage cleaning has become an important issue in the field of international Marine environmental protection. The existing Marine garbage collection mainly relies on manual regular salvage of Marine garbage, which will spend a lot of time, but the effect is not good, and it is also dangerous to the salvage personnel [2].

Many institutions at home and abroad have studied the surface garbage cleaning ship [3-5]. Australian scientists Pete Ceglinski and Andrew Turton invented Seabin [6]. It is a garbage can powered by water pump source to collects the surrounding garbage. Compared with a lot of manual fishing, it saves a lot of human resources, but it needs to rely on a lot of electricity to drive, and the effective range is small and low efficiency.

In addition, a sea surface garbage cleaning robot designed and manufactured by Ran Marin in the Netherlands [7]. Using the automatic control system, although the overall efficiency is improved, but the price is too expensive, the high cost makes the product not easy to promote. Therefore, this paper designs a cheap, green device, only rely on the waves to complete the sea garbage gathering, transfer and collection device.

2. Wave-drive principle

2.1 Characteristics and driving principle of sea waves

In a set of waves, the highest point is called the peak, the lowest point is called the trough, the difference between the peak and the trough is called the wave height, Wave length as shown in Figure 1, the distance between the two adjacent peaks is called the wavelength, and Wave height as shown in Figure 1.Wave energy exists in the form of the kinetic energy and potential energy contained in the waves on the ocean surface. The water quality point moves around the center of its trajectory, with the kinetic energy [8]. Through the China Ocean Forecast website, this paper selected the Zhuhai Kou Nei Bay for research, in the 72-hour forecast, the wave height is 1.5-1.9m. The buoy size design of this design will be calculated based on this data.

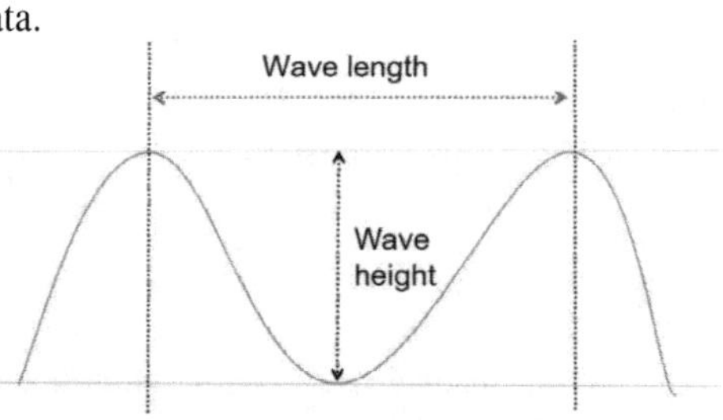

Figure 1 Schematic Sea wave diagram

Using the characteristics of the periodic fluctuations of the waves, the continuous rotation of the driving shaft is designed with the continuous fluctuation conversion cost, so as to convert the position energy of the waves into the rotating mechanical energy of the device. When the wave comes, the buoy front is the first to contact the peak of the wave. When the wave comes, the front end of the buoy first contacts the crest of the wave. At this time, driven by the crest, the buoy rotates $\alpha°$ counterclockwise around the O point, driving the principal axis to rotate counterclockwise, as shown in FIG. 2. When the wave continues to move and exceeds the front end of the buoy, the buoy rotates $\beta°$ clockwise around point O under the action of gravity, and the principal axis does not move at this time, as shown in FIG. 3. Therefore, in the course of a wave, the float of this design rotates $(\alpha+\beta)$ ° counterclockwise, therefore, driving the axis of the buoy to rotate $(\alpha+\beta)°$.

Through design calculation, multiple floats on one spindle, and multiple floats drive the driving shaft continuously counterclockwise. Therefore, the continuous counterclockwise rotation of the driving shaft can be realized under the continuous push of the sea waves, and the rotation serves as the driving power source of the whole device.

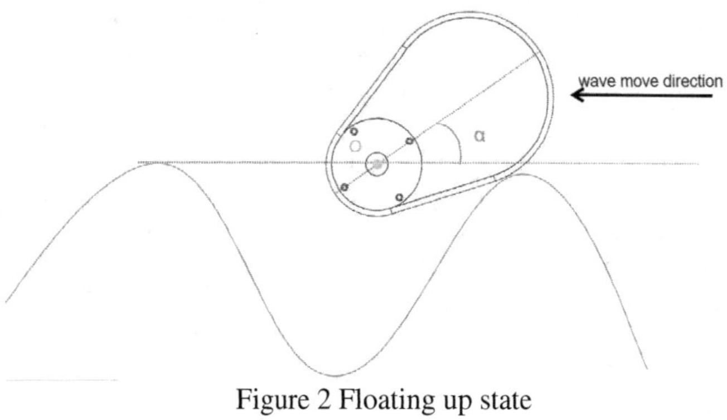

Figure 2 Floating up state

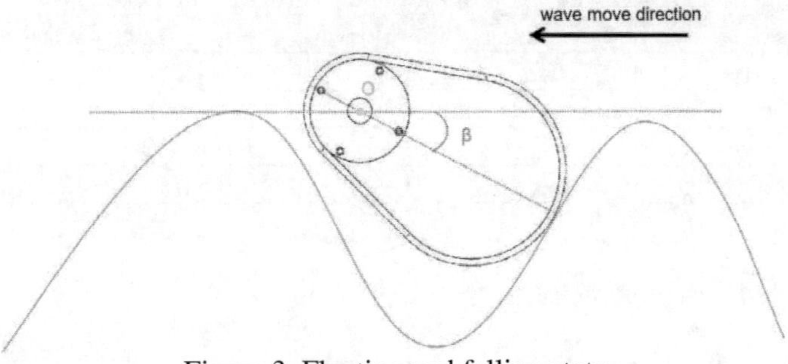

Figure 3. Floating and falling state

α —— angle of buoy lift (°);
β —— angle of buoy fall (°).

2.2 Characteristics of the sea surface garbage

Marine garbage has a wide range of pollution sources, many kinds of toxic and harmful substances, a large range of diffusion, far-reaching harm, complex control and difficult to control. Eighty percent of Marine waste comes from land and 20 percent from maritime activities. For example, abandoned fishery production facilities damaged by typhoons, Marine floating waste brought by river basins, household waste in coastal areas, coastal sea aquaculture and ship dumping waste, and waste generated from coastal industrial enterprises are all the main sources of Marine floating waste. Statistics show that in the sea surface garbage, plastic class accounted for 31%, foam class accounted for 46%, wood products accounted for 16%. The waste is soft and light, which can leave the air inside, with a certain buoyancy, discarded on the sea will change the position with the waves, and finally accumulate on the shore, as shown in Table 1. At the same time, the Technical Regulations for Marine Waste Monitoring and Evaluation (Trial) [9] is issued by for garbage classification, as shown in the Table 2, R is the maximum size of the floating garbage cutting center. In the national offshore floating garbage, medium and small pieces of floating garbage accounted for 98% of the overall waste, the overall size design of this design organization will be calculated based on these statistics [10].

Table 1. The common type of ocean garbage

Garbage source	Types of garbage
Human coastal activity	Plastic bottles, plastic bags, toys, plastic tableware, food packaging containers, clothes, shoes, metal drink cans and related materials, paper waste, glass bottles, etc.
Shipping/fishing activities	Buoys, beacons, fishing gear, etc.
Medical/Hygiene Supplies	Plastic syringes, diapers, cotton swabs, sanitary napkins, toothpicks
Smoking supplies	Lighters, cigarettes, cigarette butts, filters
Other waste	Home appliances and electronic products, batteries processed wood, rubber tires, metal cans, barrels, gas cvlinders (>4L, glass bulbs, lampshades and tubes, glass and ceramic building materials, cloth carpets and cloth funiture, etc.

Table 2. The size of garbage floating in the ocean

Classification	Maximum size ofthe centroid ofan object to be cut(R/cm)
small pieces of garbage	R<2.5
medium rubbish	2.5<R≤10
chunks of garbage	10<R≤100
extra large piece of garbage	R>100

3. Design scheme and design principle

3.1 Overall scheme design

In order to realize the purpose of using wave can collect garbage, first design the wave energy into mechanical energy mechanism, namely buoy-ratchet mechanism, because the wave fluctuations bring direct rotation is too slow, it is difficult to meet the demand of drive, so designed the gear set-differential mechanism, garbage collection is designed incomplete teeth-gathering mechanism floating garbage together, finally through a garbage conveyor together, the top view as shown in figure 4. The waves drive the buoy up and down. Using the ratchet characteristics to convert the upward swing of the drift to a rotation in one direction of the driving shaft, Via two cardan joint the transmission into the growth system, with a transmission ratio of 1 / 9. Then, through the differential, the speed of both sides is synthesized into one direction, drive a drive shaft to transfer to the incomplete gear shaft. The removable toothed slats is driven by incomplete gears [11]. Then drive the connected gathering baffle to achieve reciprocating push and pull movement, to achieve the purpose of gathering garbage, last, a conveyor belt drive shaft is then driven by the gear set-differential mechanism, use a special conveyor belt to hook up and transport the garbage to the garbage basket. According to the data collected in this paper, the actuator is mainly aimed at cutting medium and small pieces of garbage with the maximum size of the centroid of the cutting object less than 10cm. Therefore, the maximum size width of the design mechanism is 1.5m and 3.5m long.

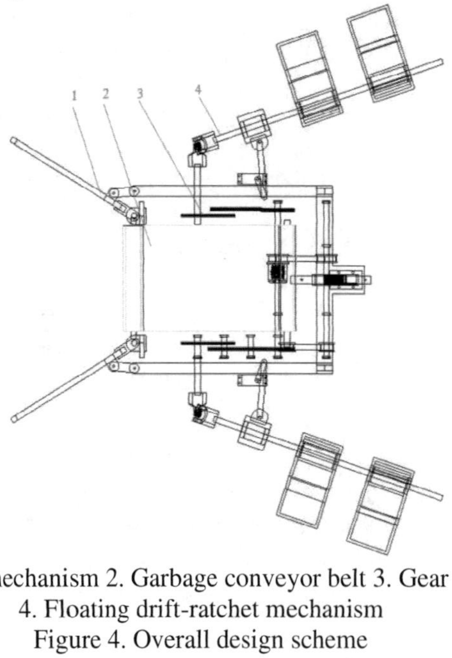

1. Incomplete teeth-gathering mechanism 2. Garbage conveyor belt 3. Gear set-differential mechanism 4. Floating drift-ratchet mechanism

Figure 4. Overall design scheme

3.2 Buoy-ratchet mechanism

According to the principle of wave drive, to realize the drive, it is necessary to drive the driving wheel to rotate when the buoy is lifted, and the driving shaft does not rotate when the buoy is falling, so that the buoy-ratchet mechanism is designed, as shown in Figure 5. The exterior consists of a shell 7 fixed by rivets and ratch, and the interior is composed of the spine claw 3 cooperating with the spring 5 and is connected to the active wheel 4 by the pin 2. When the waves through, shell 7 and ratch go up, claw 3 meshes with ratch, make the inside and outside two parts into a whole, thus completes the buoy up drive the driving shaft 4 rotation, when the shell 7 and ratch fall, spine claw 3 pressure, spring 5 contraction, external cannot transfer force to internal, the buoy and falling are realized, and the driving shaft 4 does not rotate[11].

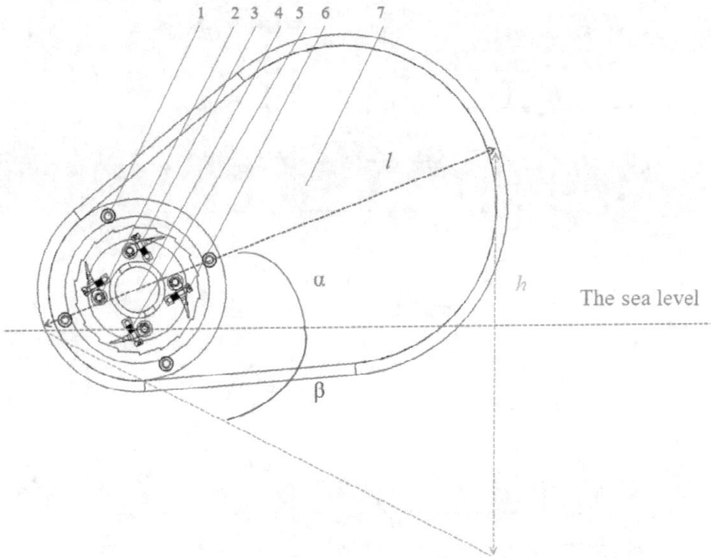

1. Rivet 2. pin 3. spine claw 4. active shaft 5. spring 6. bolt 7. Shell

Figure 5 Mooring buoy structure

Based on the survey statistics, wave height is assumed 1.5-1.9m, and taking the average value assumed 1.7m to calculate the dimensions of the mooring buoy. Ideally, for a drift elevation angle of 60°, the drift length out of the formula (1) is about 1.96m, and the actual drift length is designed as 2m to facilitate processing.

$$\alpha + \beta = 60° \tag{1}$$

and

$$\begin{aligned} l &= h \div \sin 60° \\ &= 1.7 \div \sin 60° = 1.96m \end{aligned} \tag{2}$$

3.3 Gear sets-differential mechanism

In order to achieve the purpose of increasing the speed, First, the speed direction is turned to the lateral direction through the speed steering device composed of two universal joints, as shown in Figure 6. Set up the large gear$Z_1 = Z_{2'} = 60$, smaller gear$Z_2 = Z_3 = 20$, $Z_{3'} = 50$, $Z_5 = 50$, other gear parameters are shown in Table 3. The principle of using the big gear to drive the pinion can increase the speed, increase the speed twice, since the waves on both sides do not necessarily arrive at the same time, the transmission shaft does not necessarily rotate at the same time, so use four bevel gears and a planet carrier to form a speed composite system, combine the different rotations of the two axes into the rotation of one axis. Finally, the gear ratio could be calculated as:

$$i_{13} = \frac{z_2}{z_1} \times \frac{z_3}{z_2'} = \frac{1}{9} \tag{3}$$

By medium wheel Z_4 and $Z_{4'}$ is then transmitted to the gear 5 to drive the conveyor belt drive shaft, so that it drives the conveyor belt to rotate.

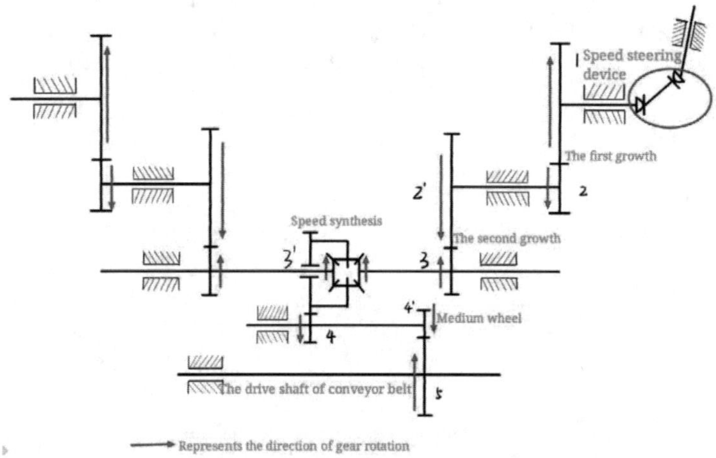

Figure 6. Overall structure of the gear group

Table 3. Gear parameters

Gear Name	Gear Modulus	Tooth Numbers
1	3.5	60
2	3.5	20
2'	3	60
3	3	20
3'	2.5	50
4	2.5	20
4'	2.5	20
5	2.5	50

3.4 Incomplete teeth-gathering mechanism

The gathering mechanism is composed of incomplete gear and linkage mechanism. First, the tooth frame 3 is fixed together with the horizontal push rod 6 and the long bar 8, the fixing seat 10 relates to the hull, and the horizontal push bar is fixed together with the slider, which is directly coordinated with the guide rail 2 to make the movement silkier. The speed synthesis device 7 transmits the power transmitted by the float to the output shaft 4 by the synchronous pulley 5, driving it to rotate counterclockwise. Then, the incomplete teeth 1 cooperate with the tooth frame 3 to drive the horizontal push rod 6 for the forward and backward reciprocating movement[13], so that the short rod 11 can drive the baffle 12 for the opening and closing movement to gather the garbage[14].In order to reduce the resistance when the baffle 12 is opened and closed, a 10mm*10mm hole is designed on the baffle, as shown in Figure7, to reduce the resistance and ensure that too small garbage is not missed. Through reasonable mechanical structure design, the structure of the marine garbage collection vessel is shown in the Figure 8.

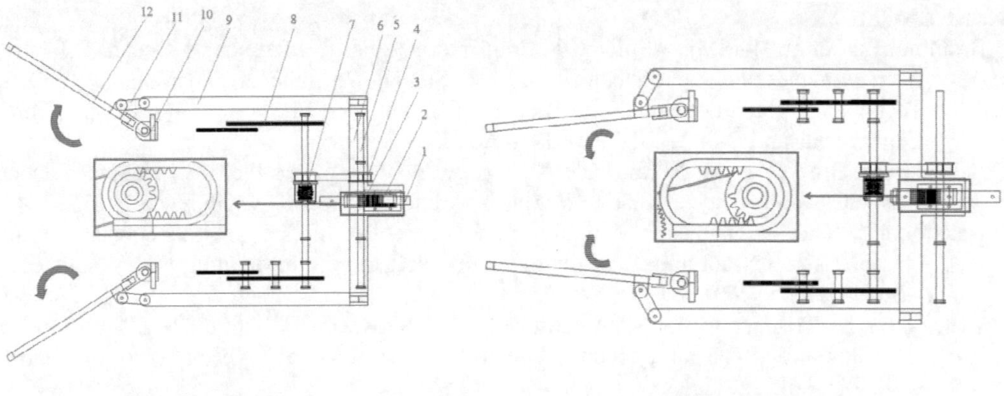

(a) Open state (b) Close state

1. Incomplete teeth, 2. guide rail, 3. tooth frame, 4. output shaft, 5. synchronous pulley 6. Horizontal push rod
7. Speed synthesis device, 8. Long rod 9. gear set, 10. fixed seat, 11. connecting rod, 12. baffle

Figure 7 Diagram of gathering mechanical structure

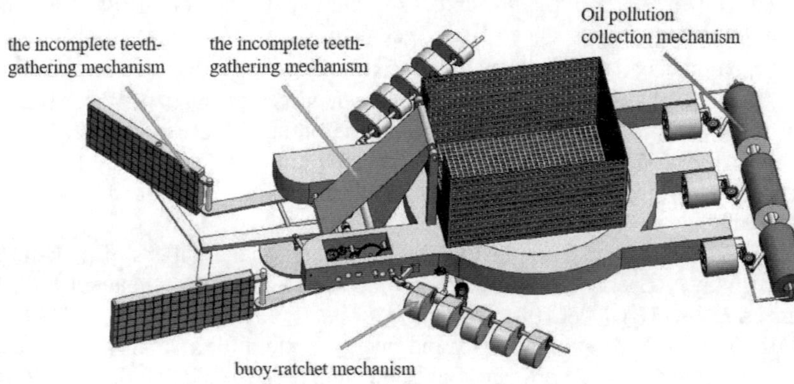

Figure 8 Overall structure diagram

4. Conclusion

The innovations of this study are summarized and concluded as follows:

① In view of the shortage of the existing garbage collection devices, this paper proposes that the green environmental protection sea surface waste cleaning device that uses waves for power can better complete the garbage collection work under the premise of energy saving and green environmental protection.

② The actuator uses a pure mechanical structure, by using wave energy to drive the whole structure, to realize the function of collecting offshore garbage, clean and environmental protection, and can also replace artificial, has a very high application value.

③ The float-ratchet mechanism converts wave fluctuations into rotation and is then supplied to the entire actuator.

④ The gear set-differential mechanism increases the efficiency of the wave drive, achieved growth rate, so that the wave energy can be effectively used to drive.

⑤ The incomplete tooth-gathering mechanism is also driven by wave energy to bring floating garbage together for easy collection.

REFERENCES

[1] Brambini, R. , et al. "[ASME ASME 2017 36th International Conference on Ocean, Offshore and Arctic Engineering - Trondheim, Norway (Sunday 25 June 2017)] Volume 1: Offshore Technology - Hydrodynamics and Capture Efficiency of Plastic Cleanup Booms: Part I — Experiments and Dyn." (2017):V001T01A052.

[2] Sainte-Rose, Bruno , et al. "Multi-Scale Numerical Analysis of the Field Efficiency of an Ocean Plastic Cleanup Array." Offshore Marine & Arctic Engineering Conference 2016.

[3] Nakamura, S., and K. Fukuzaki. "Method of removing, collecting, and treating substances floating on water surface and apparatus for removing, collecting, and treating the substances." US, US6190570 B1. 2001.

[4] Walczyk, S. L. "Trash collection skimmer boat." US, US7045058 B2. 2006.

[5] Hines, D. "Skimmer, barge and methods for recovering and transferring heavy oil or bitumen." US, US8273250 B2. 2013.

[6] MOLE, BETH, and AUTHOR. "Sea trash defies ocean boundaries. " Science News 186.7(2014):13-13.

[7] A display of a sea surface cleaning robot, Robot Technology and Application, 2016(05):8.

[8] Zhao, W., Y. Liu, and W. Wang. "Current Situation and Development of Marine Renewable Energy Power Generation." Smart Grid (2015).

[9] Department of Ecological and Environmental Protection, State Oceanic Administration. Notice of the Department of Ecological and Environmental Protection of the State Oceanic Administration on the Issuance of the Technical Regulations for Marine Waste Monitoring and Evaluation (For Trial Implementation) [EB / OL]. （2015-12-18）.

[10] Zhang, C., S. Luo, and L. Li. "Current Situation and Countermeasures of Water Body Pollution in Rural Areas of China. " Meteorological and Environmental research 6.1(2015):6.

[11] Su, C., et al. "An Autonomous Ship for Cleaning the Garbage Floating on a Lake." 2009 Second International Conference on Intelligent Computation Technology and Automation (ICICTA 2009) 2009.

[12] Wang, G., C. Zheng, and K. Chen. "Modeling and Numerical Analysis of a Novel Anti-reverse Ratchet Wave Energy Collector." Zhong Guo Jixie Rongcheng/China Mechanical Engineering 28.11(2017):1319-1325.

[13] Zhang W, Jiang D, Li P. "Design of Power and Energy System for a Small Garbage Cleaning Ship. " IOP Conference Series: Earth and Environmental Science, 2020, 603(1):012046 (7pp).

[14] Li Zehan, Yang Jie. "A multifunctional and sustainable water area waste cleaning device. " Shandong Industrial Technology, 2021(05):123-127.

Study on Calibration of Medical Analgesia Pump

Huijuan Wang, Mengdie Xiang, Chuandong Xie, Yang Xu and Han Luo

Center for Medical Metrology, Chongqing Academy of Metrology and Quality Inspection, No. 1, Yangliu North Road, Yubei District, Chongqing, China 401123;

email: 540063612@qq.com

Abstract. The medical analgesia pump is a liquid infusion device that can maintain a stable concentration of drugs in the blood and achieve better analgesic treatment with fewer drugs. It generally allows the patient to self-press to add an additional infusion based on the ongoing infusion, known as patient-controlled analgesia (PCA). It makes the treatment more individualized and consistent with the large variation in pain sensation. It is very important to the analgesic effect and medical safety whether the infusion of medical analgesia pump is accurate and the flow rate is stable. However, the current standard testing methods of the pharmaceutical industry are only applicable to the laboratory testing of the medical analgesic pump in the factory inspection, and not applicable to the evaluation of the accuracy of the medical analgesia pump used in medical institutions. In this paper, an infusion pump quality analyzer was used to calibrate the constant speed and the additional PCA infusion of medical analgesia pump, which ensured the accuracy and reliability of analgesic drug infusion.

1. Introduction

Patients after surgery often have pains that affect their quality of life and even endanger normal signs. In this situation, taking analgesic drugs or postoperative local anesthesia are generally used to relieve the pain. In recent years, postoperative local anesthesia is being used more and more frequently because of its various advantages. Postoperative local anesthesia is the use of the analgesic pump, through continuous micro infusion of anesthetic to achieve the function of local anesthesia.

The flow of anesthetic infusion is very low and less than 10 ml/h or even below 2 ml/h. The time of anesthetic infusion is very long and lasts for even more than 24 hours. The requirement for the precision of the anesthetic infusion is higher than normal infusion. Therefore, gravity infusion is not applicable, instead, special medical analgesia pumps are used to ensure a continuous micro infusion of anesthetic into the body. Whether the infusion of medical analgesia pump is accurate and the flow rate is stable are particularly important to the analgesic effect and medical safety, which is a small but important item in the precision medical.

2. Principle and management of medical analgesia pump

2.1 Principle of medical analgesia pump

At present, the commonly used medical analgesia pump includes the electronic analgesia pump which can be used only once or repeatedly and the mechanical analgesia pump. According to the actual needs of medical institutions, only calibration for reusable electronic analgesia pumps is carried out. Electronic medical analgesia pumps generally have the functions of constant speed infusion and patient-controlled analgesia（PCA) injection, which can set a constant flow rate to infuse analgesic drugs. When patients feel severe pain, they can increase a certain amount of analgesic drugs by pressing the PCA button. In order to ensure the safety of analgesia, PCA flow is usually fixed, and it is necessary to press again after a certain time interval to for PCA to take effect.

The electronic medical analgesia pump is generally composed of electronic controller, peristaltic pump, liquid storage box (bag), PCA touch device and other components. The Schematic diagram is shown in Figure 1 below:

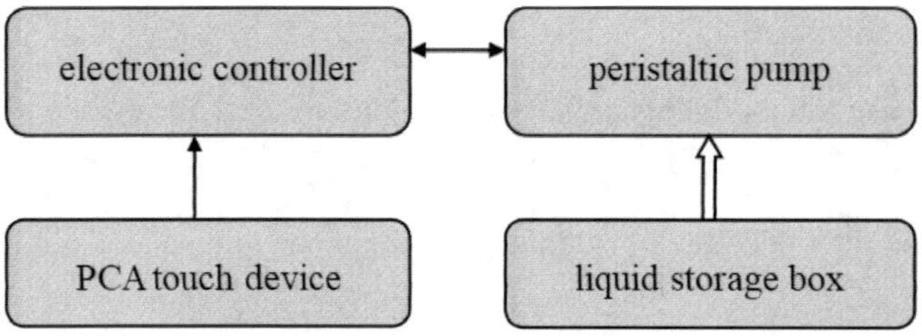

Figure 1. The schematic diagram of the electronic medical analgesia pump

2.2 Current measurement and calibration status of medical analgesia pump

At present, there is no national standard or calibration specification for medical analgesia pumps, but there is an applicable pharmaceutical industry standard YY/T1469-2016 Portable Electric Infusion Pump. In this standard, there are clear provisions for the accuracy of continuous infusion mode and PCA infusion mode of portable electric infusion pump. However, the main standard equipment for its detection is the electronic balance. The detection method is to calculate the error by converting the measured quality of output liquid into the flow rate. Because the density of liquid is affected by temperature and pressure, and the electric balance is not suitable for moving, the measurement needs to be carried out in a constant temperature laboratory, which is only applicable to the factory inspection of the medical analgesia pump, and not applicable to the measurement calibration of the medical analgesia pump already used in clinical wards by the equipment management department of medical institutions or the metrology department. It is necessary to calibrate the accuracy of the medical analgesia pump that has been used for a period of time.

This paper intends to study a more practical and convenient calibration method for medical analgesia pumps by using conventional equipment existing in the metrology department: IDA4 Plus infusion equipment analyzer, so as to provide technical support for precision medicine.

3. Measurement and calibration method of medical analgesia pump

- Calibration standard equipment: infusion equipment analyzer, model: IDA4 Plus; electronic stopwatch, minimum division value 0.1s;
- Accessories: manufacturer's original disposable liquid storage box and pipeline;
- Medium: sterilized water for injection.

3.1 Calibration Preparations

Ensure that the environment meets the requirements of the instrument: the temperature is 15-30 ° C, the humidity is ≤80%, and no mechanical vibration or electromagnetic interference affecting the test. Install the drainage pipe and waste liquid collection container for the infusion equipment analyzer in advance, put 30ml sterilized water for injection into the original disposable drug storage bag of the manufacturer, connect the liquid storage box with the analgesic pump, and connect the infusion tube to the test port of the infusion equipment analyzer.

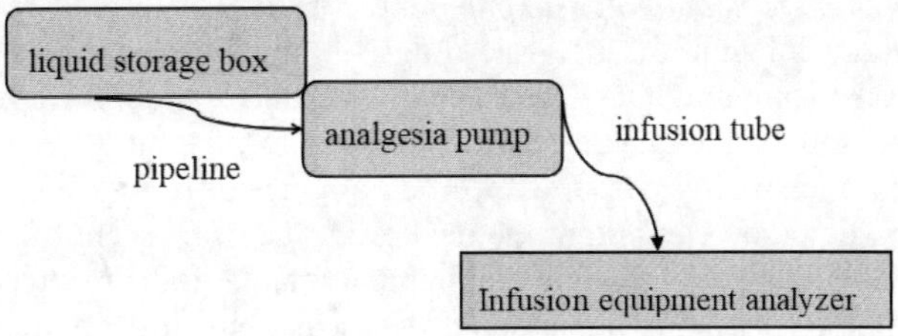

Figure 2. Connection diagram of electronic medical analgesia pump velocity calibration

3.2 Calibration of the constant flow rate of medical analgesia pump

Since the principle of the medical analgesia pump is similar to that of the medical infusion pump, we calibrated the constant flow rate of the medical analgesia pump by referring to JJF1259-2018 Calibration Specification for Medical Injection Pump and Infusion Pump.

3.2.1 Selection of calibration points

JJF1259-2018 Calibration Specification for Medical Injection Pump and Infusion Pump requires to determine the calibration points according to user requirements, and it is recommended to select one point within the range of low, medium and high speed as the calibration point. The flow rate of the analgesic pump is lower than that of the ordinary injection pump and infusion pump, so 2ml/h, 5ml/h and 10ml/h are selected as calibration points in the common range of analgesic drug use. Take three measurements at each calibration point.

3.2.2 Calibration method of flow rate

After the equipment is connected according to Figure 2 and the preparation is completed, the flow rate of the analgesia pump is adjusted to the corresponding speed, and then the pipeline in rapid swith and the liquid storage box are filled with liquid for 30 seconds. Then the analgesia pump starts infusion, and press the "Start" button of the infusion equipment analyzer to observe the displayed value of flow rate. After the infusion was performed for 20 minutes, read the flow rate displayed on the infusion equipment analyzer after the flow rate stabilizes. The interval of each reading is 5 minutes, and each calibration point is read 3 times.

The relative error and repeatability were calculated according to formulas 1 and 2.

The relative error is calculated by the formula

$$\delta = \frac{Q-\overline{Q}}{\overline{Q}} \times 100\% \tag{1}$$

In the formula:

δ -- The error of the flow rate relative to the indicated value of the calibrated instrument;

Q -- The flow set value of the calibrated instrument, ml/h;

$\overline{Q}$ -- Arithmetic average value of three flow rate measurements on the infusion equipment analyzer, ml/h

The repeatability calculation formula is

$$b = \frac{R}{1.69\overline{Q}} \times 100\% \tag{2}$$

In the formula:

b -- The repeatability of the flow indication value of the calibrated instrument;

R -- Range of three flow rate measurements of the infusion equipment analyzer, ml/h

3.3 PCA flow calibration of medical analgesia pump

In addition to the constant infusion of analgesic drugs, the medical analgesia pump can also let the patient infuse the prescribed amount of analgesic drugs according to their own pain level. For the safety of the patient, the PCA flow of the analgesic pump is fixed. Every time the patient presses the PCA button, the analgesic pump can only output a specified amount of drugs. Only after a certain locking time interval, the analgesic pump can output the next drug. It can be seen that whether the amount of medicine output after each pressing meets the requirements stated in the manual is related to the safety of analgesia, and it is one of the main indicators of the measurement calibration of medical analgesia pump.

It is stipulated in YY/T1469-2016 Portable Electric Infusion Pump to continuously measure the infusion quantity of 25 PCA and weigh the total amount of liquid output. However, for the safety of infusion, the current medical analgesia pumps used by medical institutions set the PCA locking time interval to prevent patients from being infused too many analgesics. Generally, the minimum setting value is 15 mins, so the throughput of 25 PCA needs 6.25 hours, which is not conducive to the on-site detection of the medical analgesia pump. Considering the actual needs of on-site detection, since the flow measurement accuracy of the IDA4plus infusion equipment quality analyzer used in the measurement calibration can reach 1%, we determined the throughput of PCA to be 5, and the measurement only takes about 1 hour.

According to the requirements of YY/T1469-2016 Portable Electric Infusion Pump, the relative error of PCA flow error shall be measured when PCA single injection amount is the minimum and PCA injection speed is the maximum.

Connect medical analgesia pump and infusion equipment analyzer according to figure 2. The constant flow rate of the medical analgesia pump was adjusted to the lowest (usually not 0), the PCA flow rate was adjusted to the minimum and the injection speed was adjusted to the highest. At first, the infusion was performed in rapid swith for 30 seconds to fill the pipeline and the liquid storage box with liquid, and then the analgesia pump started the infusion. After the infusion was performed for 20 minutes, the "Start" button of the infusion equipment analyzer was pressed to record the flow, and then the PCA button was pressed once to record the start time. Then press the PCA key every 15 minutes. After the fifth press and PCA input is completed, press the "End" key of the infusion equipment analyzer and record the end time to calculate the total infusion time. Read the flow measurement value displayed on the infusion equipment analyzer, measure it 3 times in the same way, and calculate its average value.

PCA flow relative error is:

$$\sigma = \frac{N-\overline{N}}{\overline{N}} \times 100\% \qquad (3)$$

In the formula:

σ -- the relative error of PCA flow of the calibrated instrument;

N -- the theoretical flow of the calibrated instrument (constant flow rate × total duration of infusion + single PCA flow rate of the calibrated instrument ×5), ml;

$\overline{N}$ -- Arithmetic average value of flow measurement value of infusion equipment analyzer, ml.

4. CALIBRATION EXAMPLES

The calibration test of measurement parameters of 15 medical analgesia pumps of a certain brand was carried out with the above measurement calibration method. According to the brand manual, the minimum constant infusion flow rate of the 15 analgesic pumps was 0.01ml/h, the minimum PCA flow rate was 1ml/ time, and the maximum PCA input flow velocity was 10ml/h.

The measurement results obtained after detection are shown in Figure 3-Figure 9

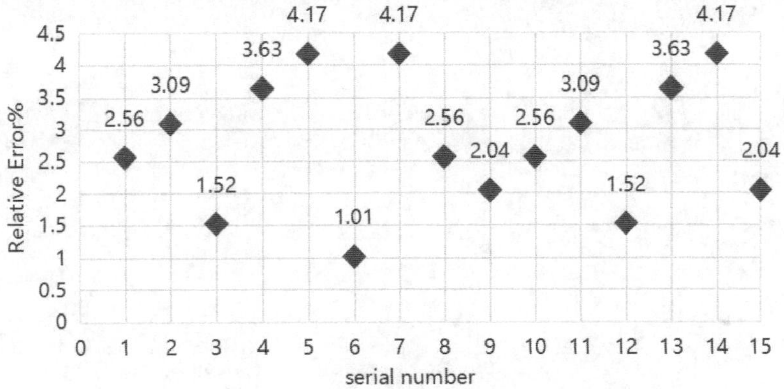

Figure 3. Relative error of flow velocity set at 2mL /h

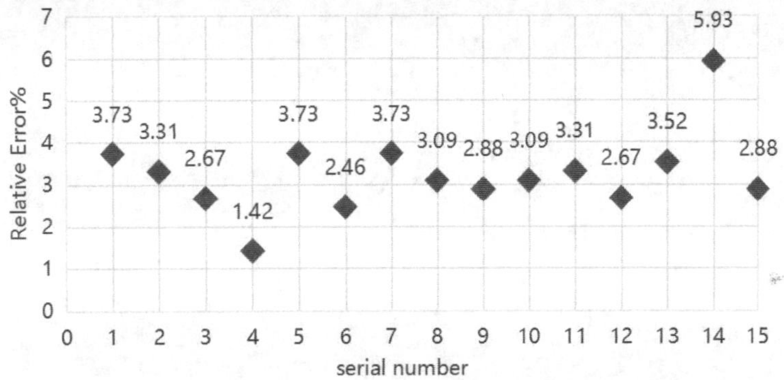

Figure 4. Relative error of flow velocity set at 5mL /h

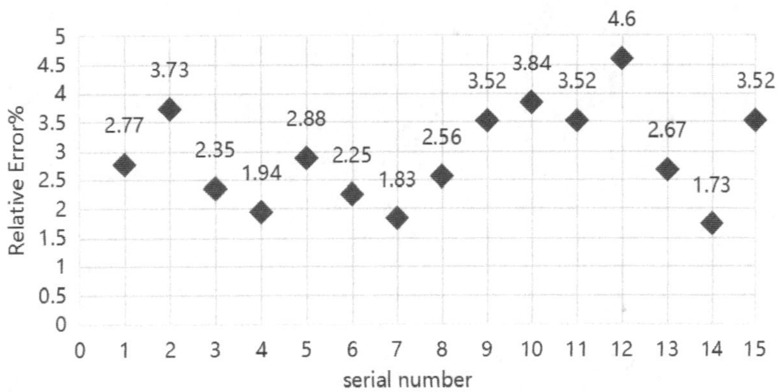

Figure 5. Relative error of flow velocity set at 10mL/h

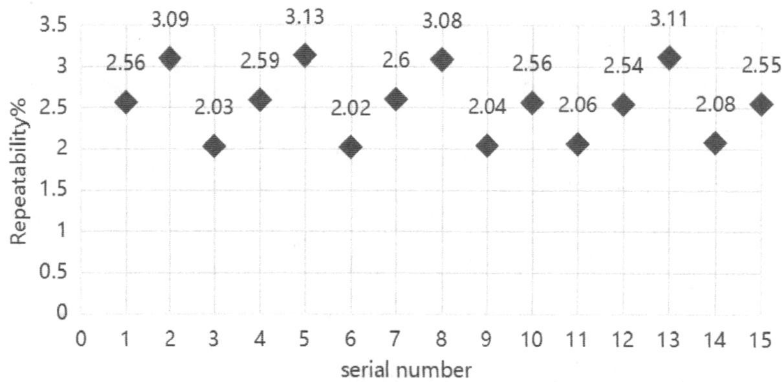

Figure 6. Flow velocity repeatability at a flow velocity of 2ml/h

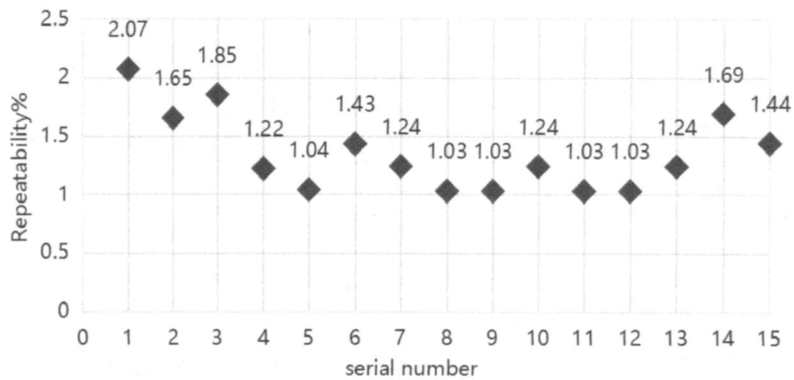

Figure 7. Flow velocity repeatability at a flow velocity of 5ml/h

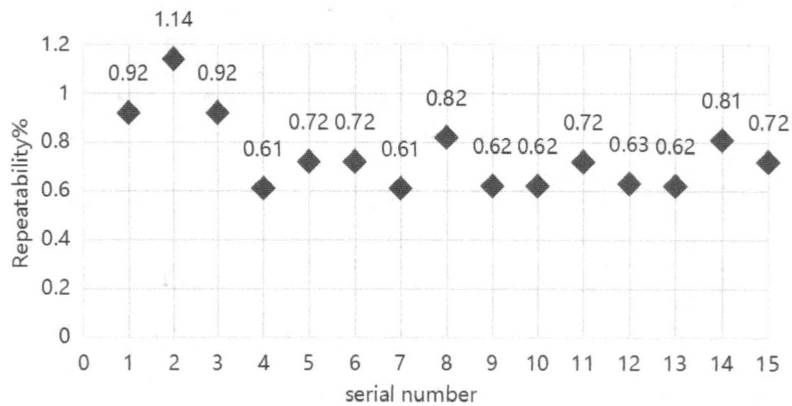

Figure 8. Flow velocity repeatability at a flow velocity of 10ml/h

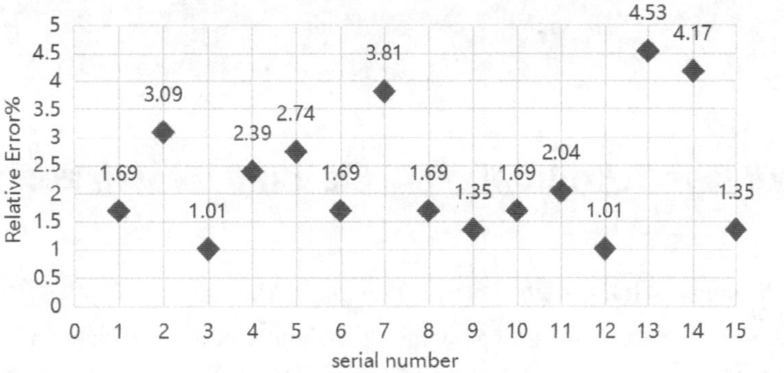

Figure 9. PCA flow relative error

According to the above calibration test, it can be found that:

(1) The relative velocity errors of this measurement calibration are all within the range specified in JJF1259-2018 Calibration Specification for Medical Injection Pump and Infusion Pump and the pharmaceutical industry standard YY/T1469-2016 Portable Electric Infusion Pump.

(2) The flow velocity repeatability of this measurement calibration is within the range specified in JJF1259-2018 Calibration Specification for Medical Injection Pump and Infusion Pump.

(3) The PCA flow of measurement calibration is less than 5%, meeting the accuracy requirements stated in YY/T1469-2016 Portable Electric Infusion Pump and the manual of medical analgesia pump. The PCA flow measured in this method is the sum of the PCA flow and the constant velocity accumulative flow. However, since the constant velocity has been adjusted to the minimum, the constant velocity flow of 1h is several orders of magnitude smaller than the PCA flow of 5 times, so the error caused by the constant velocity flow error can be ignored.

5. Conclusions

In this paper, the IDA4plus infusion equipment analyzer was used to measure and calibrate the flow rate and PCA flow of the medical analgesia pump, which solved the problem of accuracy control of the medical analgesia pump used by medical institutions. The IDA4plus infusion equipment quality analyzer used in this method directly measures liquid volume rather than quality and is not affected by ambient temperature and air pressure. The measurement department has high accuracy and traceability, and has been used for infusion pump and injection pump measurement calibration for years. Its reliability and convenience have been verified. This method can guarantee the accurate and safe use of analgesic drugs and provide technical support for precision medicine.

References

[1] JJF1259-2018 Calibration Specification for Medical Injection Pump and Infusion Pump [S].
[2] YY/T1469-2016 Portable Electric Infusion Pump [S].
[3] Xu Shuxing, Liu Hui 2022 Calibration method and Uncertainty evaluation of electronic injection pump automatic dosage . China Metrology, 4:95-96.
[4] Di Wei, Yan Zhengguang , Jiang Xuandong 2021 .Quality Control Data Analysis of Medical Infusion Pump . China Medical Equipment, 18:175-178.
[5] Yao Junming 2020 Study on Quality Control and Testing of Infusion Pump .China Strategic Emerging Industries (4):163-165.
[6] Shang Hongtao, Xu Tao, Tang Hui 2013 Discussion on Detection Technology of Infusion Pump and Injection Pump .Chinese medical equipment,10:42-44.
[7] Lin Jinnan, Jin Wei, Yang Jun. 2010 Analysis and Discussion on quality test results of in-service micro-injection pump.Annual Conference of Chinese Medical Association Medical Engineering Branch and Chinese Clinical Medical Engineering and Medical Informatization Conference,Qingdao.165-170.

Design and Application of New On-board Hydraulic Jacks

Hongyan Guan, Xizhe Du and Wenqiang Yang

School of Mechanical Engineering, Lanzhou Jiaotong University, Lanzhou, Gansu, China 730070

Corresponding author: 940982655@qq. com

Abstract. Vehicle hydraulic jacks have been widely used in vehicle repair and vehicle lifting support, etc. Aiming at the problems of single control, low position accuracy, and poor safety of existing vehicle-mounted hydraulic jacks, a new on-board vehicle hydraulic jack was proposed and designed. The handle and anti-reversal mechanism are designed to achieve automatic control of the hand, and the displacement sensor is added to realize the feedback control system that monitors the jack elevation in real time, thereby improving the accuracy and safety of position control; then using SOLODWORKS—the three-dimensional software to establish the overall model of the hydraulic jack and perform related performance checks; Finally, the hydraulic system of the hydraulic jack is simulated to show the working characteristics of the equipment based on the AMEsim simulation software.

1. Introduction

The vehicle jack is an important maintenance auxiliary tool and it is necessary for daily driving, and it is widely used in daily driving. Commonly used portable jacks are divided into mechanical jacks and hydraulic jacks the two serious of products. Of which, hydraulic jacks based on Pascal's Law and it passes through hydraulic transmission, with the help of the oil tubing. Flexible arrangement of transmission mechanism with the help of tubing connection, it has the advantages of light weight, compact structure and small volume. Vehicle hydraulic jacks are widely used in vehicle repair and vehicle lifting support, etc. And hydraulic jacks are used in vehicle tire replacement and vehicle chassis. It has the overall light weight of the device, compact structure,small size and other advantages. And the various types of sensors can be use in hydraulic jacks to form a closed-loop system to achieve precise and reliable control,that is the hydraulic control system possesses automation characteristics[1]. The hydraulic jacks includes into electric jacks and manual jacks. And the manual hydraulic jack is more laborious to use, which_it is very inconvenient for people with low strength to use; And the electric hydraulic jack is more dependent on the power supply, in the case of insufficient power supply or abnormal motor can not be used. In view of these two problems, a hydraulic jack was designed that it combines manual operation with automatic operation,and a labor-saving structure is adopted in the design of the manual device , which satisfies the user with small force, and the two operating modes can be freely selected when using, so as to make up for their respective deficiencies.

1. 1 The design of the hydraulic system

As shown in Figure 1 that the model of hydraulic jacks was built in Solid Works ,which can visually and integrally show the whole construction of this equipment . As shown in Figure 2,the diameter is rotated from large to small when the motor or handle drives the eccentric wheel 4, and the reset spring forces the piston rod of the piston cylinder 5 to extend, and the oil in the fuel tank 1 is sucked into the rodless cavity of the piston cylinder through the oil filter 2、the check valve 3. And when the eccentric wheel 4 of its diameter is rotated from small to large, the piston rod will be pushed back to the rightmost position of the hydraulic cylinder, and the hydraulic oil will be pressured to the system. Meanwhile the piston cylinder 9 will act. Then the plunger extends to lift the weight. When the limit height is reached, the signal sent by the displacement sensor makes the electromagnetic relief valve 7 electromagnet 1YA get power, and the electromagnetic relief valve will opened, the oil is returned to the tank through the electromagnetic relief valve, which keeps the jacking height constant, so that to avoid the explosion of the cylinder accident; When the weight falls back, the electromagnet 2YA of the two-position two-way electromagnetic directional valve 11 gets electricity, and the upper position works ,then the oil flows out from the plunger cylinder 9, and it directly returns to the oil tank through the two-position two-way electromagnetic directional valve 11.

Then a displacement sensor was added this design in order to make the use of the hydraulic jack more reliable, and the displacement of the plunger can be monitored real-timely. When the plunger of the piston cylinder 9 reaches the working height, the displacement sensor will monitor the physical signal into electrical signals, using the control system to make the electromagnetic relief valve 7 electromagnet 1YA electricity,which will realize pump unloading ; If due to the leakage problem, resulting in the plunger bars 9 slide, the electromagnetic relief valve 7 electromagnet 1YA loss of power, prevent oil back to the tank, continue to supply oil to the system, so that the plunger rises to the target position again .

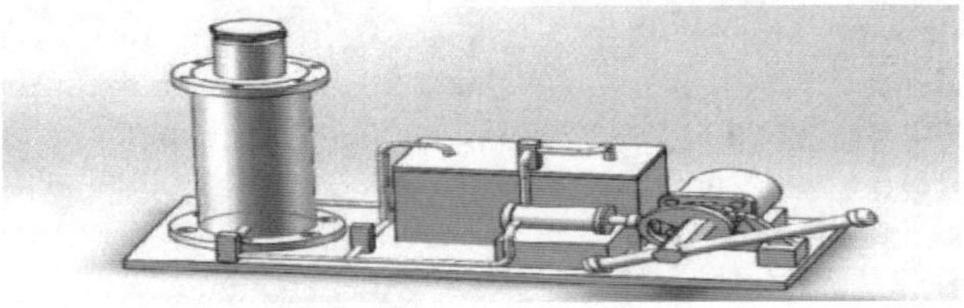

Figure 1 the SolidWorks model of the hydraulic jack

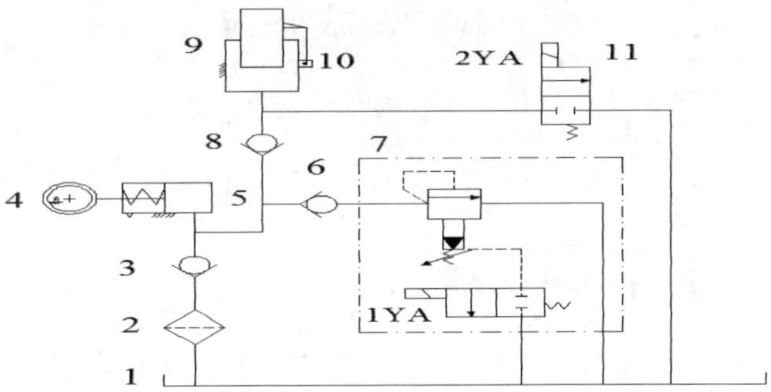

Figure 2 the principle of the hydraulic system

1-Fuel tank; 2-Oil filter; 3,6,8-Check valve; 4-eccentric; 5-Piston cylinders; 7-Electromagnetic relief valve; 9-Plunger bars; 10-Displacement sensor; 11-Two-position, two-way solenoid reversing valve;

Due to the general family car does not exceed 2. 5t, so the system load is F=25000N, according to the literature [2], the maximum working pressure of the system is P=16Mpa, the ascending velocity of the weight is 0. 01m/s.

The plunger diameter of the plunger cylinder is calculated according to the system load size and the selected system pressure, which is obtained by formula $\frac{\pi}{4}D^2p = F$.

$$D = 1.13 \times 10^{-3}\sqrt{\frac{F}{P}} = 44.7mm \tag{1}$$

The plunger diameter is 50mm. According to the actual needs , which selects the piston cylinder stroke S=150mm. According to the speed of the plunger movement, the flow rate is

$$q = v \times \frac{\pi D^2}{4} \approx 2 \times 10^{-5} m^3/s \tag{2}$$

So the time for the plunger to rise to the highest point is obtained

$$t = \frac{V}{q_z} = \frac{As}{q_z} = 14.8s \tag{3}$$

thereinto, v is the volume, m^3; A for the area of action of the plunger, m^2.

The mechanical energy is converted into liquid pressure energy. According to the mechanical design manual [2], the maximum working pressure of the piston cylinder can be determined P=20Mpa .The calculated piston cylinder flow rate is $2.6 \times 10^{-5} m^3$. According to the literature [2] Table 21-1-3, the piston cylinder bore diameter is 32mm, and according to the relationship between the working pressure of the hydraulic cylinder and the piston rod diameter . the piston rod diameter can be obtained [3]is $22.4mm$. Thus,the diameter of the piston rod is rounded to 22mm. According to the mechanical design manual [2], the thrust of the piston cylinder is $16085N$

1.2. Design of mechanical structures

The manual oil supply action is intermittent, in order to ensure the safety of the use of the equipment that in the handle rotation intermittent time it will not be reversed, thereby it injures the operator. Therefore, the power transmission shaft added a ratchet structure to prevent the handle and eccentric wheel reversal. It can also stabilize the system pressure to a certain extent .

According to the overall size of the equipment, the eccentric wheel diameter is $50\,mm$ and it sets $10\,mm$ to the thickness, and the hole processed at the distance from the center of the circle $10\,mm$, and the regular hexagon of the outer circle $\phi = 18mm$, and the countersink thickness is $5\,mm$ that its use for manual rotary slots. A blank sector is left in the eccentric wheel to reduce the weight of the equipment, and the motor is connected to the eccentric wheel using the ordinary flat key[4]. Due to the large thrust of the piston rod,and in order to ensure the normal operation of eccentric circle deformation, the static stress analysis of the eccentric circle is performed by using the simulation module in SolidWorks software. The static stress analysis of an eccentric circle is based on the finite element method [5][6]. The results are shown in Figure 3. The maximum stress of the eccentric circle is 3. 442eMpa , which is less than the yield stress of 2. 206e^{2}Mpa . Finally the strength of eccentric circle can meet normal operation.

This design leaves a hole in the tail of the rotating rod, so that the handle can slide freely as shown in figure 4, to maximize the extension of the moment arm and reduce the difficulty of operation. Considering the size of the piston cylinder, the length of the handle is 800MM when the driving force only needs 200N. With the increase of driving force, the handle length can be reduced. This design solves the disadvantage of higher driving force of conventional manual jacks, which can be operated by women as long as the handle length is reasonably designed.

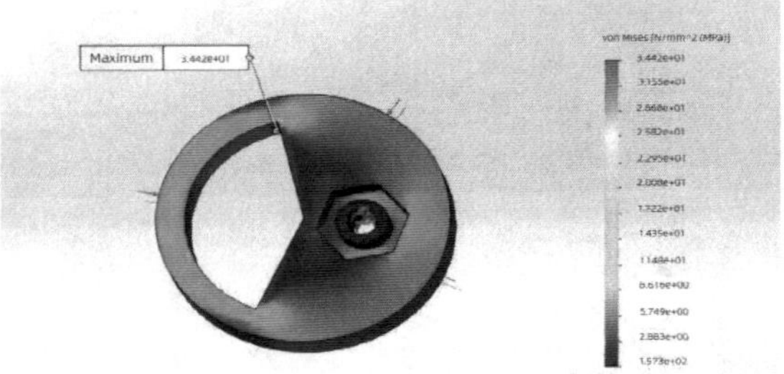

Figure 3 Eccentric wheel stress analysis result

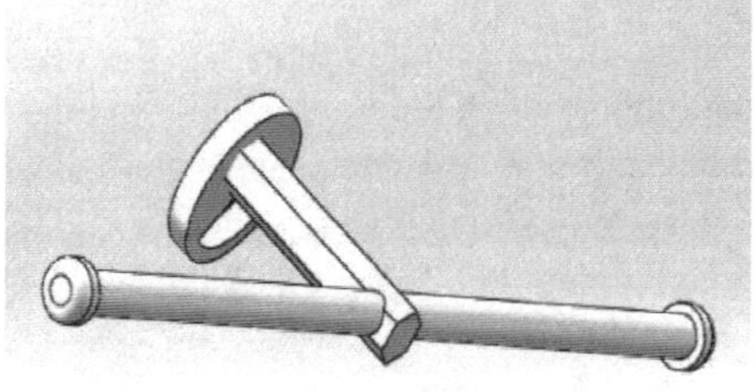

Figure 4 Ratchet model

1.3. Auxiliary equipment
The displacement sensor is installed on the piston cylinder of the jacking device of the hydraulic jack, and its role is to realize real-time monitoring of the position of the plunger rod to ensure the safety and stability of the use of the equipment. When the equipment completes the ascent task, the displacement sensor monitors the plunger extension displacement of the plunger cylinder , and it feeds the physical information into an electrical signal. When the height of the jack reaches the set value, the electromagnet of the control solenoid relief valve is charged, and the hydraulic pump unloads, so that the plunger of the plunger cylinder no longer rises. Considering the leakage of the hydraulic system, the motor does not stop at this time,and the motor will continue to drive the hydraulic pump to complete the oil supply - unloading - oil supply cycle, that is,to achieve a dynamic balance process.

2. AMEsim simulation experiments

2.1.AMEsim Model Establishment
AMEism software is used to build the model of the hand-operated hydraulic jack, as shown in Figure5. The whole system includes the piston cylinder, plunger cylinder, external signal input, electromagnetic relief valve[8], two-position two-way valve, displacement sensor and other major components .

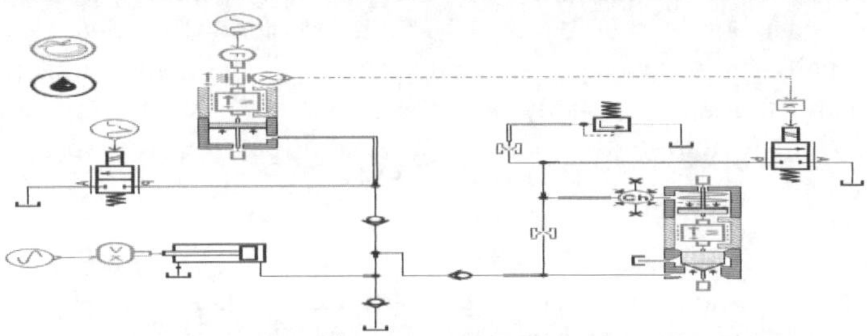

Figure 5 The AMEsim model of the hydraulic jack

2.2. Simulation parameter settings
After the hydraulic jack model is established by AMEsim, the parameters of each component are set, the piston diameter is 32mm;the piston rod diameter is 22mm ;the piston cylinder stroke is 35mm;the plunger diameter is 50mm;the plunger cylinder stroke is 150mm; the external load is 25000N;the pressure on the system is 16Mpa. In the case of the hydraulic jack which bears the maximum external load, it is set according to the calculated value, and it can be seen from equation (3) that it takes 14. 8s to lift the heavy object to the highest place, and the simulation time is set here to 15. 5s .

2.3. Simulation result analysis
When the hydraulic jack is working, the eccentric wheel is driven by the motor to rotate, and the piston rod makes a regular telescopic movement, so it similarly generates a cosine signal. The input signal of the piston cylinder is defined as a cosine signal with a frequency of 1Hz and an amplitude of 0. 017[9], as shown in Figure 6. After entering the cosine signal, the piston cylinder makes a straight-line reciprocating motion with a period of 1s, with an amplitude of 0. 035m and the system operates normally.

The displacement of the lifting device is stepped, as shown in Figure 7, the main reason is that the piston cylinder movement process is divided into oil absorption and oil pressing process. The horizontal section of the curve indicates that it is in the oil absorption process, and the rising section of the curve indicates that it is in the oil pressing process. When the system runs to the 15s, the weight reaches the top, and the average speed of the weight rise is 0. 01m/s, which is consistent with the calculation working state. The pressure of the plunger cylinder inlet is maintained at 12. 7MPa from the beginning of the oil inlet until the heavy object which is lifted to the highest place. the pressure rises to 13. 1MPa after 15s and it maintained until the end of the simulation, at this time the curve indicates that after the heavy object reaches the highest point, and the check valve holds the pressure for the system to prevent the weight from falling back, and due to the extremely small compressibility of the oil, the inlet pressure of the plunger cylinder is slightly increased, but the pressure is not greater than the maximum working pressure of the plunger cylinder. At this time, the hydraulic system of the equipment is operating normally.

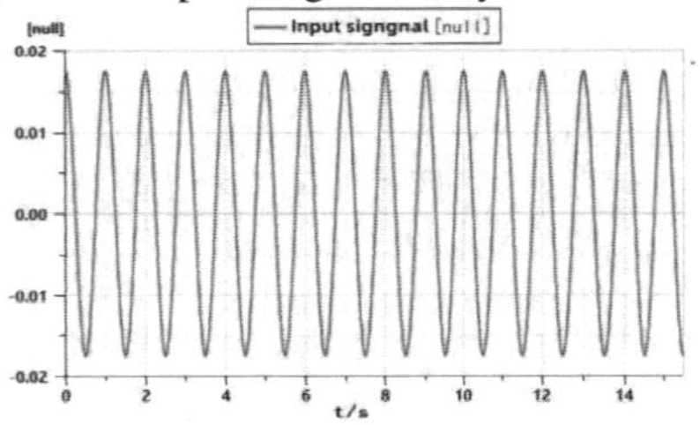

Figure 6 Eccentric wheel stress analysis

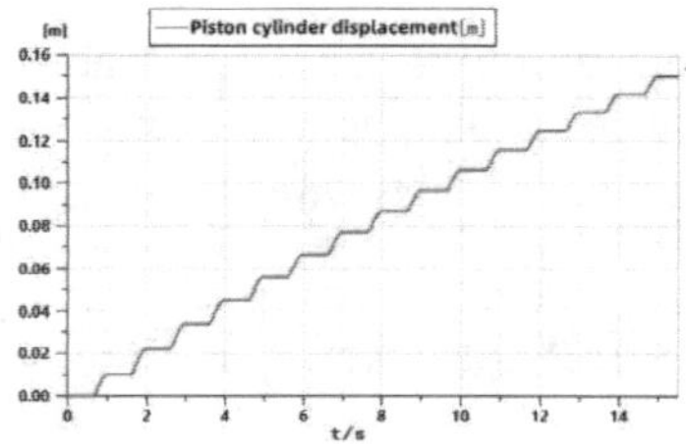

Figure7 Plunger bar displacement

3. Conclusion

In this paper, taking the jack as the research object, a self-contained hydraulic jack is designed, which solves the following problems: the problem of slow drop caused by leakage of the jacking mechanism, the problem of reversal of the eccentric wheel, and the problem of excessive driving force during manual control. Finally, by setting up a simulation model of the control system on AMESim, the entire hydraulic control system runs stably, and the simulation results are consistent with the theoretical calculation, which proves that the equipment can be used normally.

Acknowledgments

The project is funded by the Science and Technology Department of Gansu Province.The name of the project is"Design and Development of Intelligent Multifunctional Dust Suppression Vehicle Fog Cannon Structure" . And the project number is 21JR7RA302 .

References

[1] ZHONG J. (2019)Talk about Hydraulic Transmission Control System and Application Research in Mechanical Design and Manufacturing [J]. Information Systems Engineering：(09)89 (in Chinese).

[2] CHENG D X . (2008) Mechanical Design Handbook (5th Edition)[M]. Beijing:Chemical Industry Press (in Chinese).

[3] GUO L, GONG X. (2015)Hydraulic System Design[M]. Beijing:Chemical Industry Press(in Chinese).

[4] GUO J C, WU W, ZHENG J, et al. (2020) Calculation and Analysis of Static Characteristics of a Jack Assembly[J]. Mechanical Research & Application, 33(02): 38-39+43(in Chinese).

[5] CHEN Z X, CHEN L S , LIU J Y. (2019) Finite Element Analysis and Design of Cam-rotor Vane Motor Stator[J]. Machine Tool & Hydraulics. 47(21):130-134(in Chinese).

[6] Ding J S, Yang L. (2014) Measurement and Analysis of Some MaximumIsometric Muscle Strengths[J]. Industrial Health and Occupational Diseases, (03)142-145(in Chinese).

[7] ZHAO Y B, ZHANG L, WU Y, et al. (2014) Modeling and Simulation of Electromagnetic Relief Valve Based on AMESim[J]. Coal Mine Machinery , 35(07): 45-47(in Chinese).

[8] LIN Q. (2016) AMEsim Software in the Application of Jacks, Hydraulic Circuit Dynamic Performance Simulation Analys[J]. Equipment Manufacturing Technology, (01): 155-157(in Chinese).

Experimental study on the ozonation of transformer oil in wastewater enhanced by hydrocyclone and cavitation

Zihan Qu[1], Xuan Cai[1], Ying Zhang[1], Chi Zhang[1], Jing Wang[2], Danping Yu[2]

[1] State Grid Hubei Electric Power Research Institute, Wuhan, Hubei, P.R.China, 430071

[2] PowerChina Equipment Research Institute, Shanghai, 200233, China

quawater@163.com

Abstract: The treatment of transformer oil wastewater in the accident oil pond with ozone has no secondary pollution to the environment. The cavitation technique was used to enhance the oxygenation process and hydrocyclone separator to do the pretreatment. After transformer oil sewage enters the hydrocyclone, the separation efficiency reached up to 60%, which reduces the processing load of subsequent cavitation device. Through 10min cycle treatment of ozone in a cavitation device, the oil concentration of the imported wastewater was reduced from 50mg/L to below 5 mg/L, which has reached the Class-I discharge standard of industrial water. The optimal operation parameters of cavitation were studied by adjusting the inlet pipe gap and gas-liquid ratio of cavitation.

1. Introduction

Transformer oil leakage in accident oil pool, as one of the main water environment pollution risks in Power plants[1], will bring great environmental responsibility and dispute risk. Existing accident oil pools lacks consideration of environmental risk prevention and accident handling capacity.

Transformer oil, containing mostly mineral insulating oil, is the main pollutant in the substation accident oil pool. It is a distillate product of petroleum, whose main components are alkanes and naphthenic saturated hydrocarbons. Physical separation, biodegradation, biofilm filtration reaction and chemical oxidation or a combination of the above methods[2-4] are generally used for aromatic unsaturated hydrocarbons and other compounds. The reason why the ozone oxidation method contributes to the removal of organic bond fracture is the oxidation effect of O_3, O_2 and free radical particles (OH·), which are extremely high energy and unstable. Oxidizing substance can react with the organic matter quickly, destroy its chemical bonds and eventually generate CO_2 and H_2O. Advanced oxidation process, such as cavitation enhanced ozone-generated hydroxyl radical has been widely used in wastewater treatment[5-6], such as cabin ballast water, surface water and industrial pesticide wastewater[7-9]. Cavitation enhanced ozone has a better effect on the treatment of pollutants in water than the single process or the combined process of photocatalytic oxidation[10]. It is the best synergistic effect of the two phases in current water treatment and has certain advantages in energy consumption, treatment efficiency and process cost[11]. The throat part and air inlet of cavitation are the structures that have an important influence on cavitation effect, and the inlet pressure is also an important factor affecting cavitation[12-13].

According to the characteristics of transformer oil, the principle of cavitation enhanced ozone process and the principle of cavitation device, relevant experiments are designed to make the wastewater containing transformer oil reach the discharge standard. In this work, it is mainly aimed to treat the sewage containing transformer oil in the substation accident oil pool using combined advanced process of vortex cavitation, ozone and hydrocyclone. The discharging wastewater will reached Class-I discharge standard of industrial water (below 5 mg/L).

2. Experimental

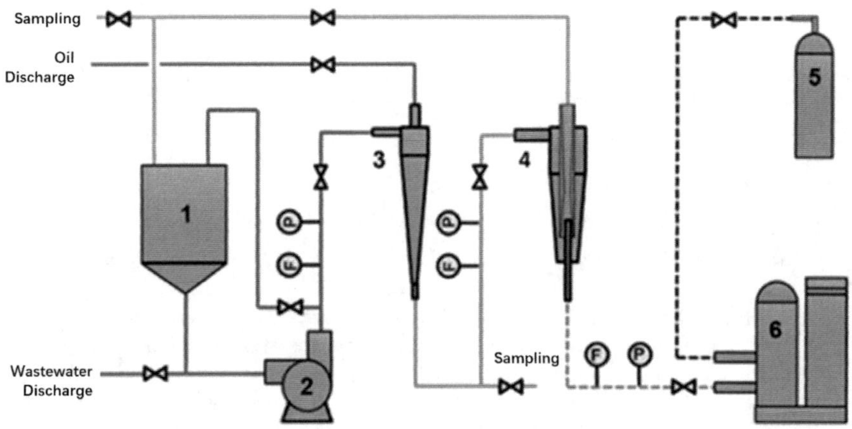

1. material tank; 2. pump; 3. hydrocyclone; 4. Cavitation pipe; 5. Oxygen container; 6. Ozone generator

Figure 1. Flow chart of experimental System

The hydrocyclone-cavitation-ozone oil removal experimental system was established, and the flow chart of the process is shown in Figure 1. The effect of reaction time on degradation rate and outlet concentration of transformer oil was studied.

The hydrocyclones and hydrocavators used in this study are independently developed, and their basic process parameters are shown in table 1.

Table 1. Main parameters of hydrocyclone and cavitation pipe

	Design pressure / MPa	Operating pressure / MPa	Temperature / ℃	Pressure loss / MPa
Hydrocyclone	1.0	0.5	10～40	0.1～0.2
Cavitation pipe	1.0	0.5	10～40	0.2～0.3

Transformer oil and tap water were used in the experiment to configure oily sewage at a concentration of 50mg/L. Transformer oil is colorless or light yellow liquid, not miscible with water, can be miscible in ether, chloroform, glycerin and other organic solvents. Its main physical properties are as follows:

Table 2. Main physical properties of 25# transformer oil

Kinematic viscosity /mm^2s^{-1}	density /kgm^{-3}	flammability	Flash point
22.6 (40℃)	0.89(20℃)	combustible	135°

3. Results and discussion

3.1. Analysis of hydrocyclone pretreatment

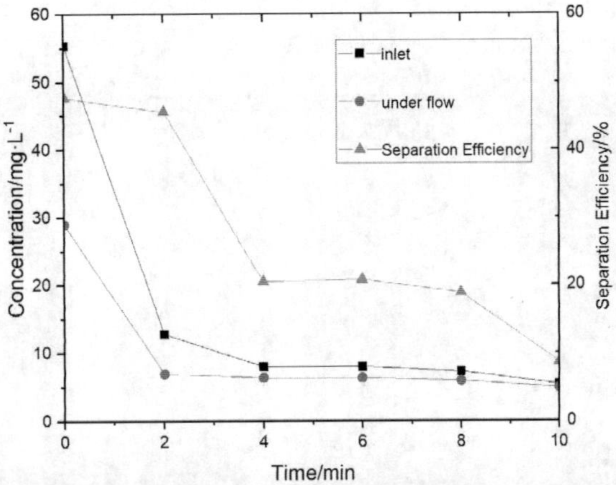

Figure 2. Oil removal efficiency curve of hydrocyclone

Hydrocyclone is an important pretreatment and separation equipment, which occupies a small area, easy to maintain with large processing capacity and low cost. At a circulating flow rate of $1m^3/h$, the cyclone pressure loss is about 1-2bar and the split flow ratio is about 10%. After a 10min cycle, when the inlet concentration gradually decreases, the inlet and outlet concentration and separation efficiency of the cyclone are calculated. The results show that the separation efficiency decreases with the decrease of inlet concentration. This is because at low concentration, transformer oil mainly exists in the form of micro-emulsified oil droplets and dissolved oil below 5μm. The influence of micro-characteristics such as Brownian motion is more significant than that of macro-force (centrifugal force), and the centrifugal migration effect is not obvious. When the inlet concentration is about 50mg/L, the separation efficiency can reach more than 60, which reduces the inlet wastewater concentration of subsequent cavators. 10% of the pretreatment overflow drainage can be discharged directly or after concentration as hazardous waste, reducing the discharge of hazardous waste in the entire accident oil pool.

3.2. Cavitation performance analysis

Cavitation number is a dimensionless parameter defined as the ratio of the pressure drop of saturated vapor pressure at the throat of the cavitation device and the corresponding temperature of the liquid medium here to the kinetic energy head at the throat. Cavitation number is defined by the following formula:

$$C_V = \frac{P_\infty - P_V}{\frac{1}{2}\rho\mu_\infty^2} \tag{1}$$

For throttling cavitation elements, cavitation effect can be generated when the cavitation number is below 1, stable cavitation can be generated when the cavitation number is about 0.5, and the space-time effect is better when the cavitation number is about 0.29. However, the cavitation number can only guide the performance design of the cavitation device macroscopically, because the cavitation process is affected by multiple factors such as model structure, surface condition and material properties. Generally, specific structural parameters need to be optimized through experiments.

Pressure difference is the main driving force of ozone inhalation. The mixture of gas-liquid two phase process is affected by adjusting the annulus between the intake pipe and the swirl shell. In the case of no ozone inhalation, the annulus is adjusted. As shown in figure 3, the annulus between the intake pipe and the swirl shell is above 3mm to form a stable bubble column, and about 2mm bubbles are evenly dispersed and fill the whole window space. The air bubble about 1mm is a violently swinging air column, indicating when the ring gap is about 2mm, the gas-liquid contact effect is better. After ozone inhalation at the flow rate of about 0.8 m^3/h, gas liquid two phase mixture is relatively uniform, the bubble-liquid contact is the most effective.

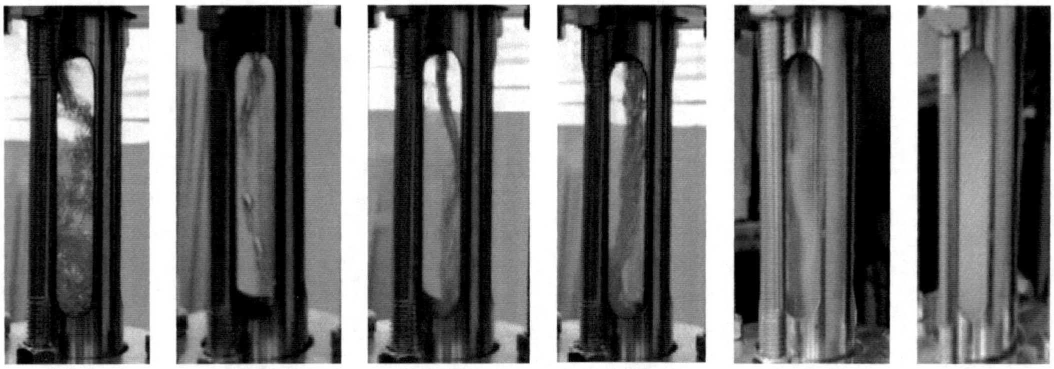

dr=5mm	dr=2mm	dr=3mm	dr=2mm	dr=2mm	dr=2mm
Q_0=0 m3	Q_0=0 m3	Q_0=0 m3	Q1=0.2m^3	Q2=0.5m^3	Q3=0.8m^3

Figure 3. Bubble column with ozone intake

3.3. Oil removal efficiency using ozone and cavitation

Under the condition of good mixing, the circulating flow rate of the system is 1m^3/h, the sewage treatment flow rate of the cavitation is about 0.8-0.9m^3/h, and the pressure loss is about 2-3bar. After 10min of circulating treatment, the overall oil content is successfully reduced to the level I industrial water discharge standard (5mg/L). The treatment effect is especially obvious in the first half (4 minutes ago), and the data in the second half fluctuate greatly and the degradation rate is low. It can be seen that the transformer oil removal effect of cavitaion-ozone oil removal is closely related to the inlet concentration. When the inlet concentration is about 50mg/L, the oil removal rate of higher oil concentration is also higher. When the concentration is below 10mg/L, the oil removal efficiency of cavitation-ozone decreases sharply. Therefore, in the case of low concentration, ozone consumption of oil removal by ozone is huge, and other auxiliary oil removal technologies should be explored to enhance the treatment effect of low concentration oily wastewater.

3.4. Oil-removal efficiency curve of air/oxygen source ozone generato

As is shown in figure 5, the results of experiments using oxygen source ozone generator show that oil concentration reaches 7 mg/L after 30 minutes. Due to the partial pressure of oxygen in the air is only 20%, the ozone concentration is low, after 30 minutes, only 1 mg/L oil content. On the one hand, lower oxygen partial pressure lead to export the ozone concentration is low, On the other hand, some oil slicks entrap in the circulation process also worsen the reaction effect.

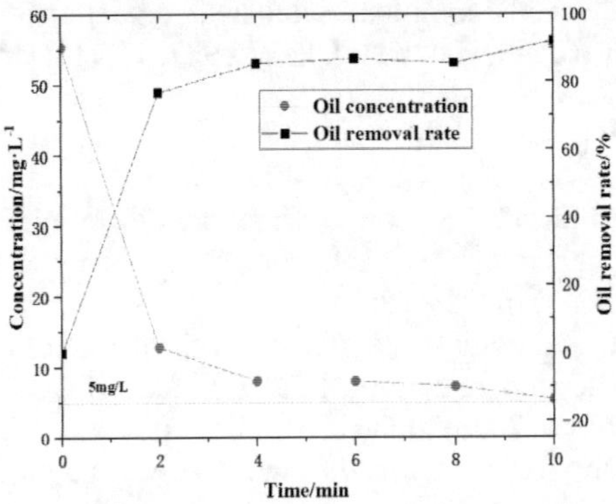

Figure 4. Ozone oil removal efficiency curve using cavitaion-ozone process

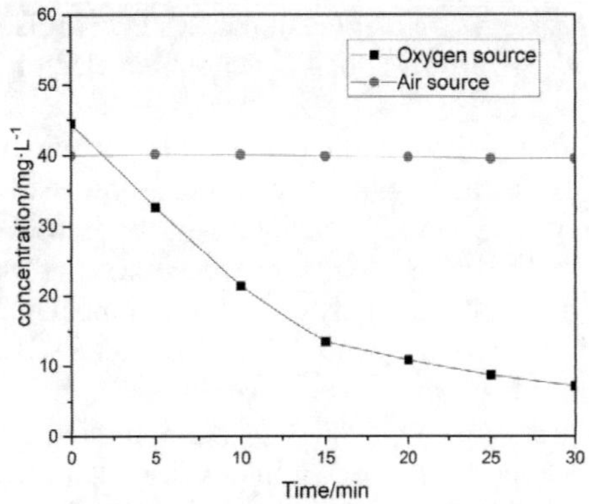

Figure 5. Oil-removal efficiency curve of air/oxygen source ozone generator

4. Conclusion

(1) The hydrocyclone-cavitation-ozone experimental system for the treatment of sewage containing transformer oil was established, whose treatment capacity was up to $2m^3/h$.

(2) After transformer oily sewage entered the cyclone separator, the treatment effect of the underflow outlet (oily sewage entering the cavitation device) changed with time (i.e., the oil concentration at the inlet). The results show that when the concentration was relatively higher, the separation efficiency can reach up to 60%, which reduced the inlet sewage concentration of subsequent cavitation device.

(3) The optimal operation parameters of cavitation were studied by adjusting the inlet pipe clearance and gas-liquid ratio of cavitation. The optimized inlet pipe clearance was about 2mm and the optimized inlet gas was about $0.8m^3$.

(4) The effect of air source and oxygen source ozone generator on oil removal performance was studied. The oil removal effect of air source ozone was significantly lower. After 30 minutes of circulation, only 1mg/L oil content was decreased.

(5) The influence of reaction time on the degradation rate and outlet concentration of transformer oil was studied. When the inlet concentration is about 50mg/L, the oil removal rate of higher oil concentration is also higher, and when the concentration is below 10mg/L, the oil removal efficiency of cavity-ozone decreases sharply. Through the cycle treatment of 10min, the advanced treatment of imported oily wastewater with a concentration of 50mg/L has been realized, which has reached the Class-I discharge standard of industrial water.

Acknowledgement

This work was sponsored by Science and Technology Project of State Grid Corporation of China: Study on key technologies of water environment risk assessment, prevention and control in substation/converter Station (No. 5500-202022425A-0-0-00).

References

[1] Larin, B.M., Bushuev, E.N., Larin, A.B., Karpychev, E.A., Zhadan, A.V. Improvement of water treatment at thermal power plants. Thermal Engineering, 2015, 62(4): p.286-292.

[2] Alexander J.C, Ramirezcortina C.R. A comparative study: degradation of 2, 5-dichlorophenol in wastewater and distilled water by ozone and ozone-UV. Ozone Science & Engineering. 2015, 38(3): p.181-193.

[3] Wang H, Yuan S, Zhan J, Wang Y, Yu G, Deng S, Hang J, Wang B. Mechanisms of enhanced total organic carbon elimination from oxalic acid solutions by electro-peroxone process. Water Res. 2015, 80: p.20-29.

[4] Moiseeva N.I., Gekhman A.E., Moiseev I.I. Metal complex catalyzed oxidations with hydroperoxides: Inner-sphere electron transfer. J. Mol. Catal. A: Chem. 1997, 117(1-3): p.39-55.
[5] Gogate, P.R. and A.B. Pandit, A review and assessment of hydrodynamic cavitation as a technology for the future. Ultrason Sonochem, 2005. 12(1-2): p. 21-7.
[6] Černigoj, U., U.L. Štangar, and P. Trebše, Degradation of neonicotinoid insecticides by different advanced oxidation processes and studying the effect of ozone on TiO2 photocatalysis. Applied Catalysis B: Environmental, 2007. 75(3-4): p. 229-238.
[7] Badve, M.P., M.N. Bhagat, and A.B. Pandit, Microbial disinfection of seawater using hydrodynamic cavitation. Separation and Purification Technology, 2015. 151: p. 31-38.
[8] Barik, A.J. and P.R. Gogate, Degradation of 4-chloro 2-aminophenol using a novel combined process based on hydrodynamic cavitation, UV photolysis and ozone. Ultrasonics Sonochemistry, 2016. 30: p. 70-78.
[9] Raut-Jadhav, S., et al., Treatment of the pesticide industry effluent using hydrodynamic cavitation and its combination with process intensifying additives (H 2 O 2 and ozone). Chemical Engineering Journal, 2016. 295: p. 326-335.
[10] Li, W.X., et al., Eutrophic water purification efficiency using a combination of hydrodynamic cavitation and ozonation on a pilot scale. Environ Sci Pollut Res Int, 2015. 22(8): p. 6298-307.
[11] Raut-Jadhav, S., et al., Effect of process intensifying parameters on the hydrodynamic cavitation based degradation of commercial pesticide (methomyl) in the aqueous solution. Ultrason Sonochem, 2016. 28: p. 283-293.
[12] Wang, B., H. Su, and B. Zhang, Hydrodynamic cavitation as a promising route for wastewater treatment – A review. Chemical Engineering Journal, 2021. 412: p. 128685.
[13] Pawar, S.K., et al., Sonochemical effect induced by hydrodynamic cavitation: Comparison of venturi/orifice flow geometries. AIChE Journal, 2017. 63(10): p. 4705-4716.

A New Derailment Detection Method of Rail Vehicles

Qiang Tao[1] , Xiaojie Sun[1,*] , Yi Liu[2] , Chenxi Dai[1] , Yulun Zhao[1]

1. School of Railway transportation, Shanghai Institute of Technology, Shanghai 201418, China
2. China Railway Wuhan Bureau Group Co.,Ltd, Research Institute of Science and Technology, Wuhan, 430070, China

*Corresponding author: sxjlm2003@163.com

Abstract. Aiming at the problem of train derailment detection, researchers at domestic and abroad mainly focus on the derailment coefficient and wheel load reduction rate, but the derailment coefficient and wheel load reduction rate can't accurately reflect the status of rail vehicle derailment in time. In this paper, a new derailment detection method of rail vehicles is designed based on the lateral acceleration and roll angle as the key characteristic physical quantities of wheelset derailment. Firstly, the virtual simulation method is used to grasp the characteristics of the key physical quantities. Then, the appropriate scheme of derailment detection system is proposed to measure the key characteristic physical quantities and judge whether the rail vehicle is derailed. Finally, the simulation data is used to prove the effectiveness of the detection method, which can lay the foundation for the practical engineering application.

1. Introduction

Rail vehicles in China have developed rapidly, among which derailment has always been the hot issue. Over the past 100 years, scholars have carried out basic research on train derailment mechanism and evaluation index. French scholar Nadal[1] proposed the calculation method of the critical derailment coefficient Q / P according to the relationship between the normal force and tangential friction force at the wheel-rail contact point and the lateral force and vertical force of the wheel when the climbing side wheel is suspended, which is used as the criterion of wheel derailment ;British scholar Gilchrist and Brickle[2] studied the quasi-static derailment of a single wheel under steady rolling conditions, and analyzed the effects of attack angle, longitudinal and rotational creep, contact elliptical spot shape and wheel load on derailment. Due to the insufficient Nadal criterion, the wheel load reduction rate is proposed to evaluate the safety of derailment of one side wheel due to severe load reduction when the lateral force of the wheelset is 0 or close to 0.

At present, countries around the world have adopted different derailment evaluation criteria.The basic indicators for judging rail vehicle derailment in China and abroad are derailment coefficient and wheel load reduction rate. The safety of the rail vehicle is evaluated by measuring wheel-rail force and calculating derailment safety index. However, subsequent studies found that the Nadal criterion was quite conservative for small attack angle and large longitudinal creep, which was confirmed by Japanese researchers in laboratory and full-scale tests[3-4]. Chen[5] pointed out by analyzing the mechanism of rail vehicle derailment, can't simply according to the wheel load reduction rate to detect whether derailment. Many variables of derailment safety index can't be directly measured, such as the lateral force and wheel load acting on the flange. In addition, due to the high cost of instrumented wheelset and the difficulty in maintenance, there are many limitations, which are not suitable for the on-line derailment detection of daily operating rail vehicle[6-7]. Therefore, the selection of appropriate physical quantities for rail vehicle monitoring and derailment risk assessment has become a hot issue in the field of derailment research.

There are many derailment detection methods and application cases at home and abroad. The typical monitoring physical quantities include wheel-rail relative displacement, wheel lift and axle box vibration. The test methods, sensor selection and installation requirements of these derailment characteristic physical quantities are different. How to select simple and easy-to-measure rail vehicle derailment characteristic physical quantities, and then design a new derailment detection method for derailment detection of rail vehicles, which will be of great significance to improve the safety of rail vehicle operation and reduce the huge losses caused by derailment.

2. Derailment detection scheme

2.1. Selection of characteristic quantities

The selection of derailment detection physical quantity is various, which can be divided into wheel-rail relative displacement and vibration acceleration.

Based on the derailment detection scheme of wheel-rail relative displacement, Zhai[8] proposed the derailment judgment principle of wheel lifting displacement; Tang[9] installed an electromagnetic displacement sensor to detect the lateral and vertical relative displacement of the wheel-track. When the relative displacement reaches the limit value, the early warning and emergency braking system will be started.

Based on the derailment detection scheme of vibration acceleration, Zhao[10] developed the derailing monitoring system of the running gear, which is based on the composite sensor installed in the axle box to detect the periodic impact of the wheel rolling on the sleeper or fastener once the rail vehiclehad derailed.Yuefu and Wang[11] installed acceleration sensor on the bogie. He compares the collected acceleration data with the pre-established acceleration statistical feature database data to determine whether the derailment is based on the waveform characteristics of vibration acceleration. Based on the vibration acceleration, the derailment is judged. When the periodic vibration of the rolling sleeper or fastener is used, the vehicle has derailment to alarm. When the measured acceleration is compared with the statistical database data, it is difficult to establish the database.

Combined with the above data, on the basis of previous studies, the author comprehensively considers the vibration acceleration and wheel-rail relative displacement of two derailment characteristic quantities, and uses the derailment simulation test of rail vehicles to analyze the derailment conditions of straight and curved lines. Combined with the difficulty of test, the lateral acceleration and roll angle of wheelset are finally selected as the derailment characteristic quantities.

2.2. Derailment detection method

2.2.1. Principle

An acceleration sensor and a roll angle sensor are installed near the wheelset to sense the lateral acceleration and roll angle signals. With the help of the data acquisition module, the data are collected. Then determine whether the rail vehicle is in the straight line or curve section by positioning device. In the derailment detection module, the lateral acceleration and roll angle data are compared with the safety critical value, and then the data are transmitted to the host computer to analyzeand diagnose whether the rail vehicle has derailment risk. When it exceeds the safety critical threshold, the alarm device will be triggered and the train will be emergency braked.

2.2.2. Overall flow chart

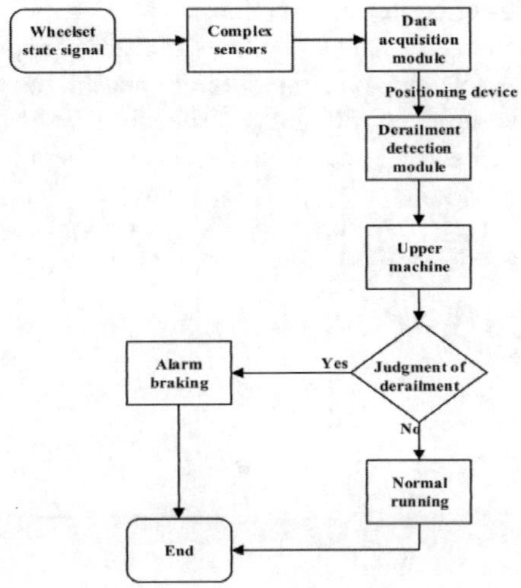

Fig.1 Overall flow chart

The overall flow chart of derailment detection method is shown in Fig.1. The state signal of the wheel is sensed by the composite sensor, and the data are collected andconverted by the data acquisition module. Whether the train is in the straight line or curve section is judged by the positioning device; the characteristic quantities were analysed by derailment detection device,and the operation of the rail vehicle is diagnosed by the host computer. Finally, the results are fed back by the execution device.

2.2.3. The design of detection system

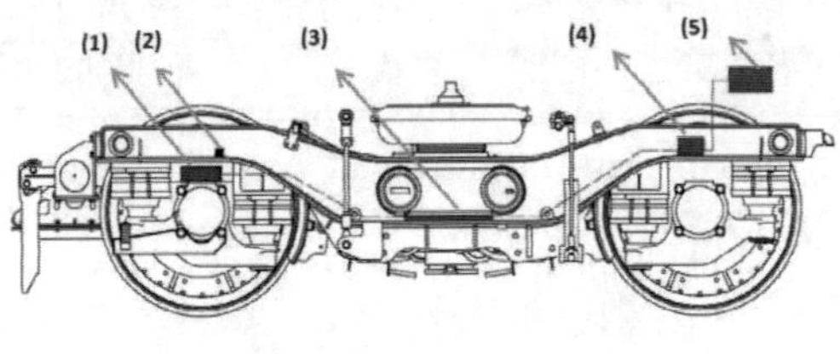

(1): Complex sensors
(2): Positioning device
(3): Data acquisition module
(4): Derailment detection module
(5): Host computer

Fig.2 Derailment detection system

The designed bogie derailment detection system is shown in Fig.2. It includes composite sensor, positioning device, data acquisition module, derailment detection module and host computer. The acceleration and angle composite sensors are installed at the axle box of the wheelset, and the data acquisition module is used to monitor the lateral acceleration and roll angle of the wheelset at the same time. Then, the location of the current train is determined based on the positioning device. Finally, the derailment detection module compares the lateral acceleration and roll angle data with the derailment safety threshold set in the upper computer system to diagnose whether the rail vehicle has derailment risk. When the lateral acceleration and roll angle exceed the safety threshold, the alarm device will be triggered, the derailment state indicator light will become red, and the rail vehicle will be emergency braked.

3. Obtain the physical quantities of contrast detection

Using the above detection method, the physical quantities of the detection characteristics are obtained by simulation. Taking the rail vehicle as an example, the Simpack software is used to build the rail vehicle model. The lateral acceleration and roll angle of the wheelset of the rail vehicle under straight and curve conditions are simulated and compared under normal operation conditions and derailment conditions, as shown in Fig.3 below :

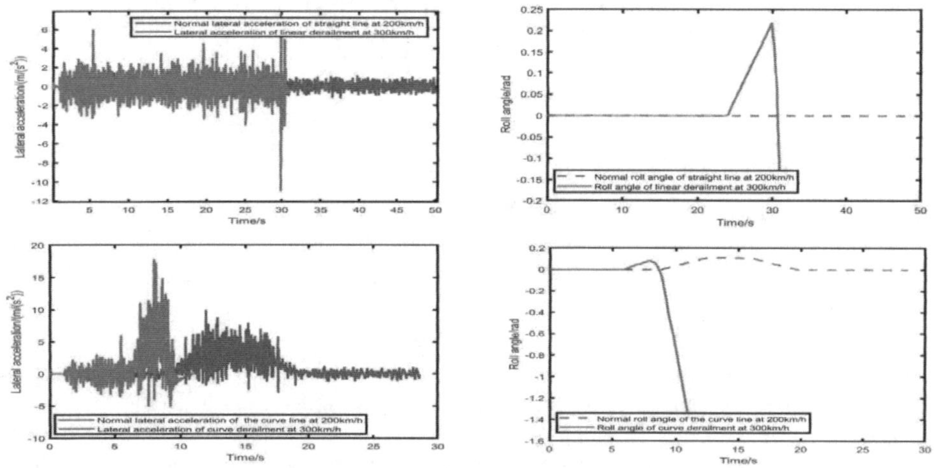

Fig.3 Comparison of lateral acceleration and roll angle curves

Fig.3 shows that when the rail vehicle is running in a straight line, with the increase of speed, the lateral acceleration at the wheelset will suddenly increase when derailment occurs, and the inclination angle will also change at the same time. When the rail vehicle runs normally on the curve, it first enters the moderate curve, then enters the circle curve, and finally the moderate curve. At this time, the lateral acceleration and roll angle change with the rail vehicle entering the curve of different sections; If derailment occurs, as shown in the above figure, in the moderate curve, so the values of lateral acceleration and roll angle will change suddenly at the moderate curve. The effectiveness of the detection method is proved by comparing the physical quantities obtained by simulation, which can effectively detect the change of the characteristic physical quantities when the rail vehicle derailment occurs.

4. Conclusion

By using a new derailment detection method of rail vehicles designed in this paper, the comparative characteristic physical quantities are obtained through simulation. It is found that when the rail vehicle is in the normal state of straight track, the lateral acceleration and roll angle at the wheelset change little. When derailment occurs, it will change significantly. When the rail vehicle is in the normal state of curve line, the values of lateral acceleration and roll angle at the wheelset will change with the change of curve superelevation compared with that in the linear state. When derailment occurs, it will change suddenly at the moment of derailment. The above conclusions prove the effectiveness of the above detection methods.

Due to the superelevation influence of the rail vehicle running in the curve section, there are some deviations in the lateral acceleration and roll angle with straight one. The future work can combine the rail vehicle derailment detection method and derailment early warning, which lays a deeper foundation for the practical significance of the project.

References

[1] Nadal M J 1896 Theorie de la stabilite des locomotives Part II: Mouvement de lacet. Annales des Mines p 232
[2] Gilchrist A O and Brickle B V 1976 A re-examination of the proneness to derailment of a railway wheel-set ARCHIVE Journal of Mechanical Engineering Science 1959-1982 (vols 1-23) pp 131-141
[3] Arai S and Yokose K 1975 Simulation of Lateral Motion of 2-Axle Railway Vehicle in Running Vehicle System Dynamics pp 165-169
[4] Yokose K 1966 A theory of the derailment of wheelset Quarterly Report of RTRI pp 30-34
[5] Chen G X and Luo Y 1990 The preliminary analysis of wheel derailment judgment principle Railway vehicles p 13-18
[6] Wei L et al 2014 Indirect method for wheel-rail force measurement and derailment evaluation Vehicle System Dynamics pp 1622-1641
[7] Wei Let al 2015 Derailment safety evaluation for railway vehicles based on wheelset modle Journal of the China Railway Society pp 25-31
[8] Zhai W M and Chen G 2001 Method and criteria for evaluation of wheel derailment based on wheel vertical rise Journal of China Railway Science pp 17-26
[9] Tang Y K 2012 Development of derailment detection system for no-load freight train Industrial control computer pp 52-53
[10] Zhao Z P 2020 Application of Running Part and Derailment Detection System for Urban Railway Vehicles Shanghai Energy Conservation pp 1082-1085
[11] Yuefu C Q and Wang G M 2014 Research on the method of monitoring vehicle fault by bogie vibration acceleration Foreign railway vehicles pp 36-41

Journal of Physics: Conference Series **2365** (2022) 012019 doi:10.1088/1742-6596/2365/1/012019

An anti-collision system for tower cranes considering lifting heavy objects

Dongfeng He[1] , Junping He[1] , Zhibin Li[2], Bo Huang[2] and Hongliang Cheng[2*]

[1]Gansu First Construction Group Co., Ltd. Gan su, China

[2]School of electronics and control engineering, Chang'an University, Xi'an, China

*hlcheng@chd.edu.cn

Abstract. Most of the traditional tower crane anti-collision solutions only consider the ideal working environment of the tower crane without lifting heavy objects or fail to consider the collision under the condition of lifting heavy objects, which has certain limitations in practical engineering applications. This paper proposes an anti-collision system for tower cranes that considers lifting heavy objects. The system introduces the concept of the dangerous area of cylindrical heavy objects, and explains in detail the anti-collision principle of tower cranes when lifting heavy objects. At the same time, the specific scheme of hardware controller and software design of the system is introduced. The system makes the tower crane anti-collision system closer to the real working situation, and is a supplement to the research and application field of tower crane anti-collision.

1. Introduction

Tower crane is a kind of construction machinery equipment commonly used in the construction field. The tower crane is generally composed of four major mechanisms: hoisting, luffing, slewing and operation. The fixed tower crane has no operating mechanism.

With the development of urbanization, the demand for tower cranes is also gradually increasing. On larger construction sites, the number of tower cranes has increased significantly, which puts forward higher requirements for safe operation between tower cranes [1]. In response to this situation, a large number of scholars have proposed different anti-collision schemes for tower crane groups. Most of these schemes only consider the ideal working environment of the tower crane without a load or fail to consider the collision when the load is lifted. Based on this, this paper proposes an anti-collision scheme for tower cranes considering the lifting of heavy objects.

This paper mainly studies three aspects of tower crane anti-collision controller scheme, including tower group anti-collision algorithm, software design of tower crane anti-collision system and draws corresponding conclusions.

2. Hardware design of anti-collision system for tower crane

The hardware design of the tower crane anti-collision system is shown in figure 1. The MCU in the system is used for calculation and control; the power module is used to supply power to the system and the sensor group; the boost module can increase the voltage from 5V to 15V to supply power to the sensor group; the sound and light alarm module provides sound and light alarm information; zigbee wireless communication module is used for information exchange between sensors, actuators and controllers between tower cranes [2]; the LoRa wireless communication module transmits the current tower crane information back to the ground control center in real time for relevant personnel on the ground to view and monitor; the display module provide a visual tower crane operation monitoring platform for tower crane drivers. The sensor group consists of weight sensor, height sensor, amplitude sensor and angle sensor. The weight sensor measures the weight of the heavy object; the height sensor and the luffing sensor realize the measurement of the height of the tower crane and the amplitude of the trolley; the angle sensor is used to measure the rotation angle of the slewing mechanism of the tower crane. The actuator includes a starting circuit, a hoisting circuit, a slewing circuit, and a luffing circuit. These circuits work in coordination to ensure the safe operation of the tower crane.

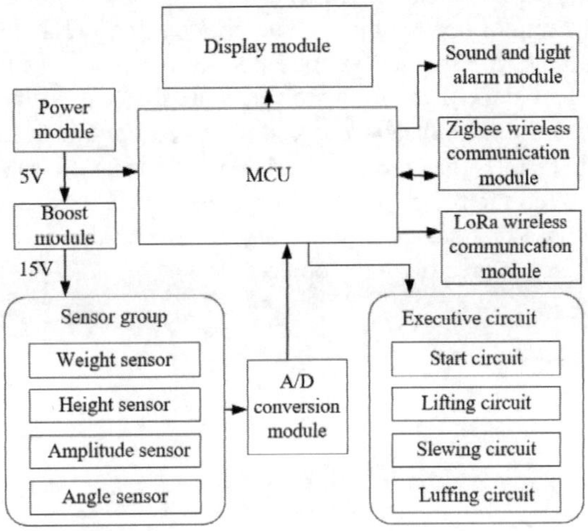

Figure 1. Hardware design of anti-collision system for tower crane.

3. Research on anti-collision algorithm of tower group

3.1. Establishment of mathematical models of tower groups and buildings

Within the construction site, the true north direction is the positive direction of the x-axis, and the counterclockwise direction perpendicular to it is the positive direction of the y-axis, and a plane rectangular coordinate system as shown in figure 2 is established. The coordinate system should ensure that each tower crane and its rotation range as well as the buildings on the construction site are included in the first quadrant of the coordinate system. It is stipulated that when the boom of the tower crane points to the positive direction of the x-axis, the rotation angle is 0°, and the counterclockwise direction is the direction of increasing the rotation angle, and the maximum is 360°. When the boom rotates more than 360°, the calculation of the rotation angle starts from 0° again. As shown in figure 2, the center coordinates of the tower crane $TC0$ can be obtained by measurement, expressed as (x_0, y_0), the coordinates of the farthest end of the tower crane boom are (x_0', y_0'), and the coordinates of the luffing trolley is (x_{c0}, y_{c0}), and the other tower crane coordinates are expressed in a similar way. The shadows of A1 and A2 in the figure are the building models.

After the coordinate system is established, the relevant coordinates are represented according to the existing conditions. Take the tower crane $TC0$ as an example, the boom length is represented by R_{l0}, and the trolley width is represented by R_{c0}. The coordinates of the end point of the tower crane $TC0$ boom and the coordinates of the luffing trolley are calculated as follows:

$$x_0' = x_0 + R_{l0} \cos\theta \quad (1)$$

$$y_0' = y_0 + R_{l0} \sin\theta \quad (2)$$

$$x_{c0} = x_0 + R_{c0} \cos\theta \quad (3)$$

$$y_{c0} = y_0 + R_{c0} \sin\theta \quad (4)$$

3.2. Heavy hazardous area model

The operation process of tower crane will be affected by many factors, including the tower crane vibration caused by Inertia of heavy objects, the lateral and vertical vibration of the tower crane boom caused by the movement of the tower crane luffing trolley, and the inertia of the hoisted weight itself. These factors may cause the heavy objects lifted by the tower crane to rotate [3-6]. figure 3 shows the model diagram during the lifting of heavy objects. The oblique dotted line is the wire rope for mooring the heavy objects, and the thick line segment represented by AB is a mass line segment formed by the two end points of the heavy objects lifted by the tower crane that are farthest apart. This line segment is used to approximately represent the heavy objects, $A'B'$ is a certain position during the rotation of the heavy object. The shaded part is the circular plane formed during the rotation of the weight, and H_0 is the distance from the hook to the rotation plane of the weight. It is stipulated that the heavy objects in the figure and the wire ropes that hoist the heavy objects are rotated to form a cylinder as shown in the figure, which is the dangerous area of heavy objects.

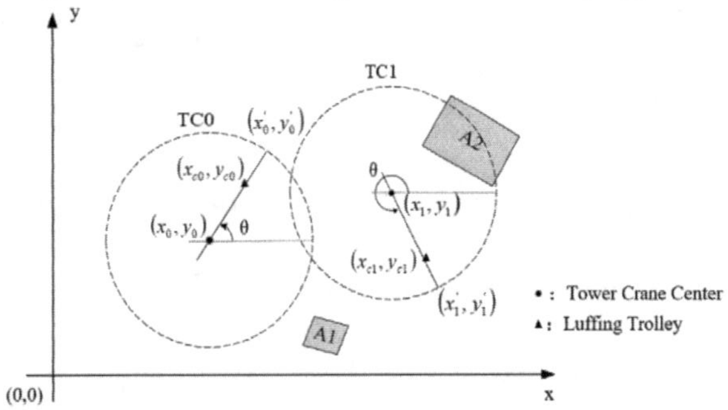

Figure 2. Tower crane plane coordinate system.

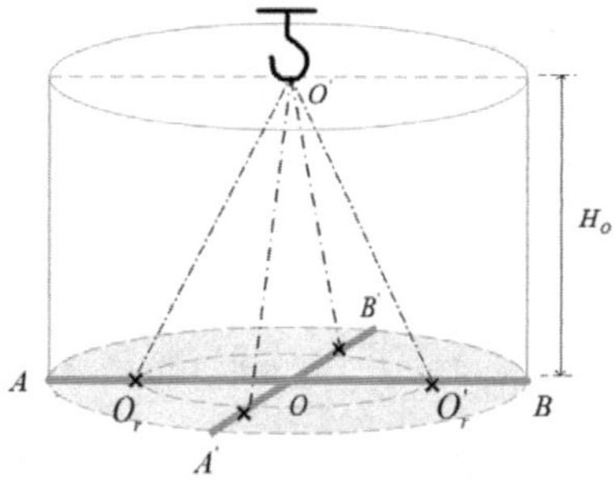

Figure 3. Cylindrical heavy object hazardous area model.

3.3. Research on anti-collision algorithm for tower group

Anti-collision between tower cranes:

Figure 4 and figure 5 are schematic diagrams of lifting heavy objects between tower cranes of equal height and tower cranes of unequal height, respectively. Among them, H_{r1} and H_{r2} are the lengths of the slings, H_{01} and H_{02} are the heights of the cylindrical danger zone of heavy objects.

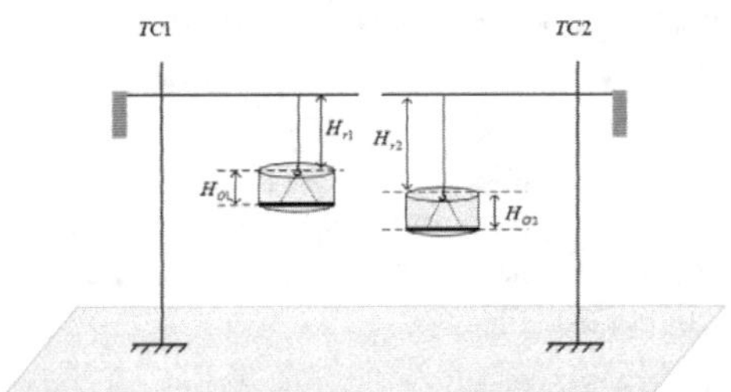

Figure 4. Equal height tower crane lifting load model.

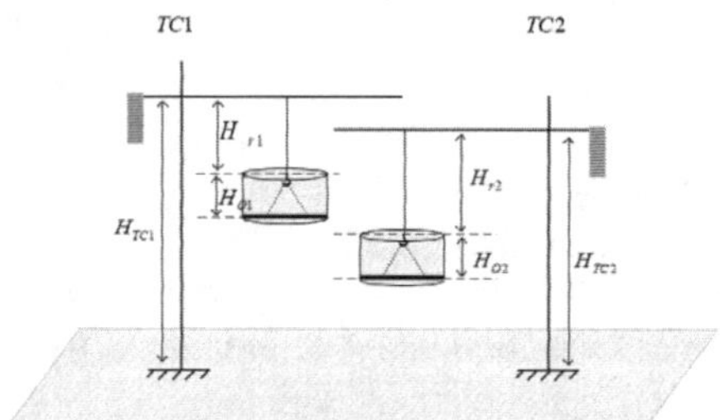

Figure 5. Unequal height tower crane lifting load model.

The conditions for non-collision between tower cranes of equal height are as follows:

First of all, it should be satisfied that the end point of the boom of the tower crane will not collide with the boom of another tower crane, that is:

$$\left[(x_2^{'}-x_1^{'})^2+(y_2^{'}-y_1^{'})^2\right]^{1/2}+\left[(x_2^{'}-x_1)^2+(y_2^{'}-y_1)^2\right]^{1/2}>d_1 \tag{5}$$

$$\left[(x_1^{'}-x_2^{'})^2+(y_1^{'}-y_2^{'})^2\right]^{1/2}+\left[(x_1^{'}-x_2)^2+(y_1^{'}-y_2)^2\right]^{1/2}>d_2 \tag{6}$$

In the formula, d_1 and d_2 are the lengths of the booms of the $TC1$ tower crane and the $TC2$ tower crane, respectively.

Secondly, when at least one tower crane is hoisting heavy objects, the dangerous area of cylindrical heavy objects of the tower crane that is hoisting heavy objects should be prevented from colliding with the suspending rope of another tower crane that is not hoisting heavy objects during the operation; or it should prevent from colliding with the suspending rope of another tower crane that is hoisting the weight and at the same time it should prevent from entering the cylindrical danger area formed by this tower crane's heavy load.

The conditions for non-collision between tower cranes of different heights are as follows:

When the high-level tower crane satisfies $H_{r1}+H_{O1}<H_{TC1}-H_{TC2}$, there is no risk of collision between the two tower cranes. If at least one of the two tower cranes lifts load, the boom of the low tower crane should avoid collision with the hoisting rope of the high tower crane that is hoisting and the boom of the low tower crane should not enter the cylindrical danger zone of the heavy load of the high tower crane; the cylindrical dangerous area of the heavy object lifted by the high-level tower crane should prevent collision with the sling rope of the low-level tower crane that is not hoisting or the cylindrical dangerous area of the heavy object lifted by the high-level tower crane should prevent collision with the sling rope of the low-level tower crane that is hoisting and at the same time should not enter the cylindrical danger zone formed by this low-level tower crane.

Anti-collision between tower crane and building:

Figure 6 is a projection of the positional relationship between the tower crane and the building. Figure 7 is a model diagram when the tower crane height is greater than the building height.

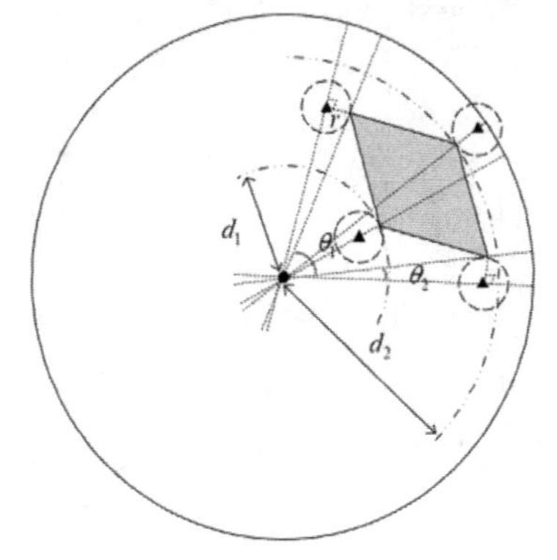

Figure 6. Projection of collision relationship between tower crane and building.

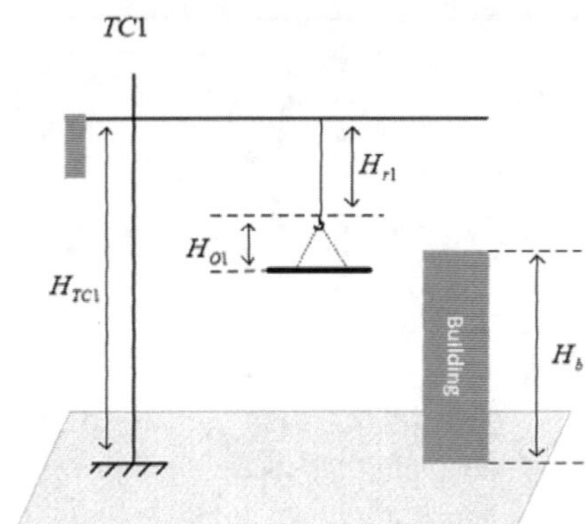

Figure 7. The tower crane height is greater than the building height model diagram.

When the height of the tower crane is less than or equal to the height of the building and the tower crane is not lifting heavy objects, if the tower crane does not enter the area with an angle of θ_1, the tower crane will not collide with the building; when the tower crane is lifting heavy objects, if the tower crane does not enter the area with an angle of $\theta_1+2\theta_2$ area, the tower crane will not collide with the building.

If H_{TC1}-(H_{r1}+H_{O1})> H_b, there is no collision risk between the tower crane and the building; when the tower crane is not loaded with heavy objects, if H_{TC1}-H_{r1}> H_b, there is no collision risk between the tower crane and the building, if H_{TC1}-H_{r1}< H_b, then the precondition of non-collision between the tower crane and the building is:

$$\left[(x_1 - x_{c1})^2 + (y_1 - y_{c1})^2\right]^{1/2} < d_1 \tag{7}$$

$$\text{or} \left[(x_1 - x_{c1})^2 + (y_1 - y_{c1})^2\right]^{1/2} > d_2 \tag{8}$$

When the tower crane is loaded with heavy objects, if $H_{TC1} - (H_{r1} + H_{O1}) < H_b$, when the tower crane does not enter the area with an angle of θ_1+2θ_2, there is no risk of collision between the tower crane and the building. When the tower crane enters the area with an angle of θ_1+2θ_2 area, then the precondition of non-collision between the tower crane and the building is:

$$\left[(x_1 - x_{c1})^2 + (y_1 - y_{c1})^2\right]^{1/2} + r < d_1 \tag{9}$$

$$\text{or} \left[(x_1 - x_{c1})^2 + (y_1 - y_{c1})^2\right]^{1/2} - r > d_2 \tag{10}$$

where r is the radius of the bottom surface of the cylindrical hazardous area of the heavy object.

4. Software design of anti-collision system for tower crane

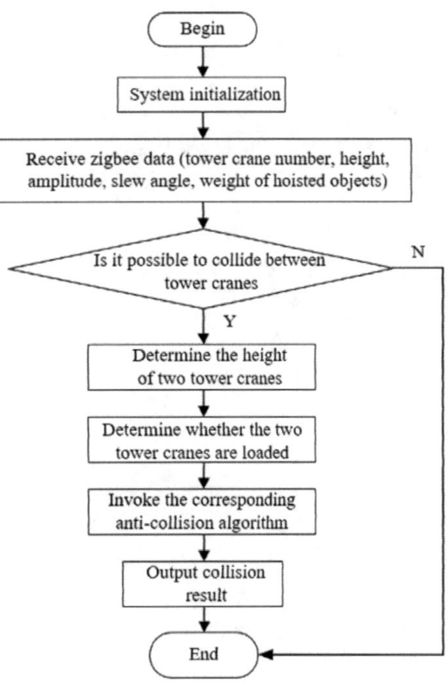

Figure 8. Flow chart of anti-collision procedure between tower cranes.

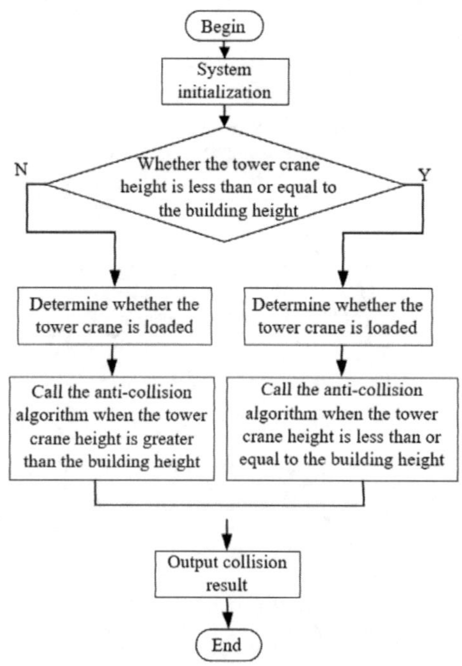

Figure 9. Anti-collision procedure flow between tower crane and building.

Figure 8 shows the program flow chart of anti-collision between tower cranes. The system compares and analyzes the tower crane parameters transmitted by zigbee and its own tower crane parameters to determine whether there is a collision risk between the tower cranes. If there is a risk of collision, the corresponding anti-collision algorithm is invoked according to the height of the two tower cranes and whether they are lifting heavy objects to achieve anti-collision between tower cranes.

Figure 9 shows the flow chart of the collision avoidance procedure between the tower crane and the building. After initialization, the system compares the height of the tower crane and the building, and invokes the corresponding anti-collision algorithm to achieve anti-collision between the tower crane and the building.

5. Conclusions

Most of the current research on anti-collision systems of tower cranes only considers the ideal working environment of the tower crane without lifting heavy objects or fails to consider the collision under the condition of lifting heavy objects. Aiming at this situation, this paper proposes an anti-collision system for tower cranes considering lifting heavy objects. The system considers the working situation of the tower crane under the condition of lifting heavy objects, and establishes a corresponding model, which makes the anti-collision system of the tower crane closer to the real working situation. It supplements the research and application of tower crane anti-collision system, and has certain practical significance in engineering application.

References

[1] Khodabandelu Ali, JeeWoong Park, and Cristian Arteaga 2020 Crane operation planning in overlapping areas through dynamic supply selection. *Automation in Construction*. 117: 103253.

[2] Koumboulis Fotis N, Kouvakas Nikolaos D, Giannaris George L, et al 2016 Independent motion control of a tower crane through wireless sensor and actuator networks. *Isa Transactions*, 60, 312-320.

[3] Jean-Pierre Sleiman, Emile Zankoul, Hiam Khoury, et al 2016 Sensor-based planning tool for tower crane anti-collision monitoring on construction sites. *Construction Research Congress*. 2624-2632.

[4] Lianghai Jin, Han Liu, Xiazhong Zheng, et al 2020 Exploring the impact of wind loads on tower crane operation. *Mathematical Problems in Engineering*. 2020.

[5] Runhui Feng, En Zhang, and Mingxiao Dong 2021 Jib vibration and payload swing of tower cranes in the case of trolley motion. *Arabian Journal for Science and Engineering*. 46(12), 12179-91.

[6] Mingxiao Dong, Jiyong Wang, Bo Pang, et al 2021 Maximum Oscillation Amplitude of Cargo Caused by Inertia Forces for Tower Cranes. *International Conference on Education, Management, Computer and Society*. Atlantis Press, 1410-13.

Scenario modeling and operation path planning of autonomous driving tower cranes

Xuan DU[1,2,a], Xian-zhe YU[1,2,b], Xing-yu YUN[1,2,c]

[1]Hubei Key Laboratory of Hydroelectric Machinery Design & Maintenance, Yichang 443000, China;

[2]College of Mechanical & Power Engineering,China Three Gorges University, Yichang 443000, China;

[a]xdu@ctgu.edu.cn; [b]956449611@qq.com; [c]1370780559@qq.com

Abstract. Combined with the operation specification of building tower cranes, the three-dimensional path planning technology of tower cranes and the optimization of path points are studied. The improved RRT-Connect algorithm is proposed. By introducing heuristic information in the path search, the path search efficiency and quality of the algorithm are improved. By eliminating invalid points in the path, the control burden of tower crane is reduced, and the optimized lifting point movement path is transformed into tower crane joint movement, and finally the calculation result is uploaded to PLC through LAN. The running trajectory of the tower crane lifting point is visually displayed in the scene model, and the validity and safety of the path are verified. The real tower crane is controlled. The operation is completed and the automatic driving of the tower crane is realized..

1. Introduction

Tower crane, also known as tower crane or tower crane, consists of tower body, boom, lifting trolley, sling and hook. Tower cranes are used for the vertical and horizontal transportation of materials and the installation of building components in the construction of houses, so they play an important role in the construction of medium and high-rise buildings. Affected by factors such as labor efficiency and complex site environment, the operating efficiency and reliability of tower cranes need to be improved urgently. With the development of robot technology, through the research of tower crane automatic control technology, sensing and detection technology, scene modeling and motion planning technology, the realization of remote control and automatic driving of tower crane can effectively reduce the labor intensity of tower crane operators, improve work efficiency and ensure the safety of tower crane operation.

In the industrial application of automatic assembly and welding production lines, the operation control of the manipulator has many similarities with the construction tower crane. The current mainstream intelligent control method for robotic arms is to use 3D lidar or camera to extract the geometric information of the surrounding scene and the position information of the target point, then plan a motion trajectory of the end of the robotic arm in the scene model, and finally calculate the joints of the robotic arm. The motion trajectory is driven by the motor to work, so as to realize the intelligent control of the manipulator within a certain range [5-10]. Compared with the robotic arm, the tower crane has a large working space and a complex environment. At the same time, it is necessary to strictly abide by the operating specifications to ensure safety. This paper studies the key technologies such as the operation scene modeling in the unmanned tower crane, the operation path planning and data transmission of the tower crane in the complex scene, and lays the foundation for the realization of the unmanned tower crane.

2. Construction of tower crane operation scene model

The operating scene of the tower crane is large and complex. In order to ensure that the created scene model meets the motion planning requirements of the tower crane, this paper chooses to use the 3D lidar scanning modeling method to obtain the 3D model of the environment where the tower crane is located. In order to prevent the lack of information on the 3D point cloud map due to obstruction by obstacles, install the 3D lidar in the middle of the tower crane boom, with the laser emission system of the lidar facing down, and control the tower crane boom to rotate once to obtain the points of the tower crane operation scene. Cloud model. Since the operation of the tower crane will cause the lidar to vibrate during the scanning process, there will be more noise in the point cloud obtained by scanning, and because the laser emission frequency of the lidar is high, the rotation speed of the tower crane boom is slow, and the scanning obtained The point cloud contains a large number of duplicate point clouds. In order to obtain a point cloud model with moderate density and less noise, Gaussian filtering is used to remove the noise in the point cloud, and the influence of the noise in the point cloud on path planning is eliminated, then the voxel filter is used to reduce the density of the point cloud, and the processing The resulting 3D point cloud model is converted into an octree map, which reduces the amount of environmental information and improves the stability of the simulation environment while ensuring that key point cloud data will not be lost.

Figure 1. Tower crane operation scenario model

3. Tower crane lifting point motion path planning

The RRT-Connect algorithm is an improved algorithm of the RRT (Rapidly Expanding Random Tree) algorithm. It not only has the good adaptability of the RRT algorithm to 3D complex scenes, but also has faster computing speed and higher efficiency. The difference between the RRT-Connect algorithm and the RRT algorithm is that the RRT-Connect algorithm introduces a double tree expansion link. The RRT-Connect algorithm is a two-way incremental construction method. It generates a new state point at the start and end point of the path. If the point is in a collision-free position, find the node closest to the node in the search tree as the reference node. Starting from the reference node, it extends towards the random node with a certain step size, and the position of the end point of the extension line is added to the search tree as a valid node. The growth process of this search tree continues until the distance between the target node and the search tree is within a certain range and terminates. The search algorithm then searches the search tree for a shortest path connecting the starting point to the ending point. The advantage of the RRT-Connect algorithm is that two trees are sampled at the same time, and two sampling points are obtained for each sampling, which speeds up the connection between the starting point and the end point. However, when the RRT-Connect algorithm is applied to the path planning of the lifting point of the building tower crane, there is a key problem: the algorithm converges slowly in a large three-dimensional space. Therefore, the path planning algorithm of the tower crane is improved.

In order to ensure that the final motion path of the tower crane hoisting point conforms to the operating specifications, the motion path of the tower crane hoisting point is divided into three sections: 1) The lifting path of the hoisting point: the tower crane hoisting point is raised to a safe height. 2) The joint motion path of the hook, the hoisting trolley and the boom: take the current position of the lifting point as the starting point and the predetermined position above the target point as the ending point, and use the algorithm to plan a path. 3) The descending path of the lifting point: the tower crane is dropped to the target point. Splicing the three paths together is the motion path of the tower crane hoisting point.

The RRT-Connect algorithm uses random expansion to generate nodes. When the scene model is too large, the random nodes generated by the algorithm have a high probability of heading to the invalid area, which affects the efficiency of path search. In order to solve the problem of low efficiency of the algorithm in large scenes, this paper introduces the heuristic information of the direction from the current node to the end point in the improved algorithm. In path planning that introduces heuristic information, the direction of generating path points is no longer completely random, but toward the end point based on a certain probability, which improves the efficiency of the algorithm while ensuring the global search capability. The algorithm steps are as follows:

1) The starting point and ending point of the input are used to generate the root node of the algorithm.

2) Generate a heuristic information q_{bias} or direction information q_{rand} for a random direction. where the probability of generating heuristic information is p. Assuming that the random value in [0, 1] is rand, the determination of the heuristic information q_{bias} and the random direction information q_{rand} is shown in formula (1):

$$\begin{cases} q_{\mathrm{bias}} = \frac{S-S_g}{\|S-S_g\|} & \mathrm{rand} < p \\ q_{\mathrm{rand}} = \frac{S-S_{\mathrm{rand}}}{\|S-S_{\mathrm{rand}}\|} & \mathrm{rand} > p \end{cases} \tag{1}$$

In the formula: S is the coordinate of the current point; S_g is the coordinate of the target point; S_{rand} is the randomly generated coordinate.

3) Generate a new node S_{new} from the existing node S along the direction of the information in 2), where the step size of generating S_{new} is ρ.

4) Obtain the nearest node S_{nearst} of the new node S_{new}, connect the two points to generate a hanging point motion path, and then Perform collision detection on this path. If the detection is passed, the expanded new node S_{new} is valid, otherwise step 2) is re-executed.

5) Repeat steps 2) to 4) for each end node between the two trees until the two trees are connected together.

The probability p of generating heuristic information in the algorithm has a greater impact on the generation of the path. Too large p value will make the randomness of the algorithm too low and reduce the global search ability of the algorithm; too small p value will cause the algorithm to generate The path is easy to go to the invalid area, which reduces the efficiency and quality of the algorithm search. In the tower crane operation scenario, the influence of the algorithm with different p values on the path planning results is shown in Table (1):

Table 1. Influence of P value on path search

Influence factor	Different value					
P	0.0	0.10	0.20	0.30	0.4	0.5
Average search time	9.12	6.17	5.98	5.80	5.85	5.99
Average number of search nodes	652.1	527.6	499.6	464.9	482.7	496.7
Number of iterations	1321.3	914.2	561.6	510.7	550.3	591.3

4. Optimization and transmission of tower crane joint running path data

The tower crane uses PLC to control the inverter and the motor to complete the operation. Too many waypoints during the operation of the tower crane will increase the workload of the motor, so it is necessary to reduce the invalid waypoints in the path; in order to realize the automatic driving of the tower crane, it is necessary to convert the motion of the tower crane lifting point into different joints of the tower crane and upload it to the PLC, and finally control the motor work according to the joint operation. Specific steps are as follows:

1) Read the motion path of the tower crane lifting point;

2) Determine whether the slope of the two line segments AB and BC connected to the three adjacent path points A, B, and C on the path is approximately 180°. If so, it is considered that there is an invalid path point B, and it is eliminated.

3) Calculate the motion path of the tower crane joint, and the calculation method is shown in formula (2);

$$\begin{cases} \theta_i = arctan\,(y/x) \\ d_i = \sqrt[2]{x^2 + y^2} \\ z_i = z - h_0 \end{cases} \tag{2}$$

In the formula (x, y, z) are the three-dimensional coordinates of the hoisting point of the tower crane in the operation scene of the tower crane; θ_i is the rotation angle of the tower crane boom, d_i is the running distance of the hoisting trolley relative to the tower body; z_i is the distance of the hook Lifting height; h_0 is the vertical distance between the hoisting trolley and the ground in the tower crane operating scene model.

4) Connect the PLC and the host to the same local area network, and write the motion of the tower crane joint calculated in 2) into the address specified by the PLC;

5) The frequency converter reads the movement of the joints in the PLC, controls the motor work, and realizes the unmanned driving of the tower crane.

5. Analysis of experimental results

Based on the operating scene model of the tower crane, use ROS to simulate the path planning algorithm in the 3D scene, set the step size of the algorithm to generate new nodes ρ=1, the safe distances D_{safe}=3 and R_{safe}=3, and the probability p of generating heuristic information =0.3. The statistical data of the RRT-Connect algorithm before and after the improvement after multiple tests are shown in Table 2: the average search time is reduced by 13.5%, the average number of search nodes is reduced by 12.9%, and the average number of nodes for generating paths is reduced by 9.2%. This shows that the improved RRT-Connect algorithm has higher computational efficiency and better path search results in large scenes, and maintains a safe distance from obstacles, which is in line with the operating specifications of tower cranes.

Table 2. Comparison of path search efficiency before and after improvement

	Average time	**Average number of search nodes**	**Path average number of nodes**
RRT-Connect	6.12s	681.6	421.3
Improved RRT-Connect	5.29s	593.3	382.7

In order to realize the unmanned driving of the tower crane in the real scene, the three-dimensional lidar is used to build a model of the environment where the tower crane is located, and the improved RRT-Connect algorithm is used to quickly plan a path that conforms to the operating specifications of the tower crane. Lift the hook first, then transport the target from the ground on one side of the building to the top on the other side of the building, and finally lower the hook. The movement path of the hanging point is shown as the red solid line in Figure 3:

Figure 2. Motion path diagram of tower crane lifting point

After the tower crane runs safely in the simulation environment, the motion data of each joint of the tower crane is extracted. The joint motion data is replaced by the motion of the tower crane joints, and the sensor data is transmitted to the tower crane's PLC control system, so as to realize the unmanned driving of the tower crane.

6. Conclusion

In order to realize unmanned tower crane, the working path planning of tower crane and the data transmission technology between PLC and host are studied. First, the improved RRT-Connect algorithm is used to realize the motion path planning of the tower crane lifting point. The improved RRT-Connect algorithm not only improves the efficiency of path search, but also generates a path that conforms to the operating specifications of the tower crane, which improves the efficiency and safety of the automatic operation of the tower crane. Finally, after verifying the feasibility of the path in the simulation environment, the path points are extracted and saved in turn, and the invalid path points are eliminated and sent to the specified address of the PLC. At the test site, the tower crane control system reads the waypoints in the PLC. Complete the automatic control of the tower crane and realize the unmanned driving of the tower crane.

Reference

[1] Wangz. (2021) Obstacle Detection and SLAM Path planning based on Lidar [D]. Beijing Jiaotong University,2021.(03) p79

[2] Qinpl,Zhangcw,Zhoulb,Wangjl. (2022) Research on 3D target detection of unmanned trackless rubber wheel vehicle in coal mine [J]. Industry and Mine Automation,2022,48(02):35-41.

[3] Baosj,Zhangmy,Gesr,Liuq,Yuanxm,Senwm,Yany,Zaol. (2022) Underground driverless path planning of trackless rubber tyred vehicle based on improved A* and artificial potential field algorithm [J/OL]. Journal of China Coal Society:1-14[2022-03-21].

[4] Guoyc,Liuxx,Zhangwg,Yany. (2020) 3D Path Planning Method for UAV Based on Improved Artificial Potential Field [J]. Journal of Northwestern Polytechnical University,2020,38(05):977-986.

[5] Xiel,Lius. (2018) Dynamic obstacle-avoiding motion planning for manipulator based on improved artificial potential filed [J]. Control Theory & Applications,2018,35(09):1239-1249.

[6] Zhanjw,Huangyq. (2021) Research on dynamic obstacle avoidance of robot by combing safety A* algorithm and dynamic window approach [J/OL]. Computer Engineering:1-11[2021-11-14]

[7] Zengdh. (2021) Collision-free Path Planning for Overhead Crane System Using Modified Ant Colony Algorithm[C]// 2021 9th International Electrical Engineering Congress (iEECON). 2021. Pullman Pattaya Hotel G, Pattaya, Thailand. p 1108-1115

[8] Liyz,Wangst,Jianglq,Mengj,Xieyl. (2021) Motion Planning of Mobile Manipulators Based on RRT with Sparse Nodes [J]. China Mechanical Engineering,2021,32(12):1462-1470.
[9] Zhangyl,Lipf,Wangxh,Wangwj. (2021) Motion planning method of manipulator based on improved rrt combined with B-spline [J/OL]. Computer Integrated Manufacturing Systems:1-17[2021-10-29].
[10] Qureshi A H , Ayaz Y . (2015) Intelligent bidirectional rapidly-exploring random trees for optimal motion planning in complex cluttered environments[J]. Robotics & Autonomous Systems, 2015, 68:1-11.
[11] Puxp,Zouw,Wangp,Zhouh. (2017) Smooth Path Planning Based on Non-uniformly Modeling and Cubic Bezier Curves [J]. Acta Automatica Sinica,2017,43(05):710-724.
[12] Zhangmh,Zhangyf, Ouyanghm, et al. (2020) Modeling and Adaptive Control for Tower Crane Systems with Varying Cable Lengths[M]. 2020.

Path Planning Method of Rolling Brush of Car Washing Machine Based on Fractional Differentiation

[1st]Hao Yan[1,a], [2nd] Yafeng Qiu[1,b*], [3rd] Xiaoming Hou[1,c], [4th] Zhao Jin[1,d]

[1]School of Mechanical Engineering, Nanjing University of Science and Technology, Nanjing 210094, China

[a]yh_mail@njust.edu.cn; [b*]njlghcn@sina.com; [c]1320279372@qq.com; [d]1229101532@qq.com

Abstract. The roller brush cleaning path of the gantry reciprocating automatic car washing machine is the key to determining the quality of the car wash. For the issue that the traditional method cannot accurately and reliably calculate the roller path, a rolling brush path planning scheme combined with fractional differential algorithm is proposed. By analysing the working form of the car washing machine and the appearance characteristics of the car, a multi-view car detection system is designed, and the car images are collected by using multiple cameras. The possibility of the path of each roller brush of the car washing machine is extracted from the contour information, and the contour curve is converted into the roller brush path in the form of coordinate points through software. Finally, the actual size of the car to be washed is calculated by the multi-view camera, and the average error between the calculation result and the actual result is 2.63%, which meets the operating conditions of the gantry reciprocating automatic car washing machine.

1. Introduction

With the rapid development of today's society and the continuous improvement of the national economic level, people gradually pay attention to the cleaning and maintenance of vehicles. In order to improve the efficiency of car washing and reduce costs, unattended automatic car washing machines have begun to develop instead of traditional manual car washing. The roller brush of an automatic car washer is used as a scrubbing actuator, and its motion trajectory affects the operation efficiency and cleaning quality of the car washer. It is of great significance to study how to formulate its motion route efficiently and accurately. The traditional automatic car washing machine rolling brush motion control method at home and abroad adopts current detection technology. The control method based on current detection is to detect the load current of the rolling brush rotary motor. The load current corresponds to the contact distance between the rolling brush and the car body surface, and monitors the current. The size of the roller brush changes the direction of movement [1]. This method controls the trajectory by detecting the current, and has a slow response to current mutation and poor anti-interference ability. In recent years, control methods based on ultrasonic technology have also been gradually used in automatic car washers. Chu ZB et al. [2] proposed a positioning control strategy for the two-link manipulator arm of a car washing machine, whose motion trajectory is controlled by the distance fed back by the ultrasonic sensor. This method is superior to the route controlled by current detection, but it cannot accurately control the body shape in complex situations, especially when detecting an irregular plane or a plane with a certain inclination angle on the body, it is easy to see that the rolling brush of the car washer cannot be cleaned or is not cleaned enough.

At present, the problems existing in the trajectory planning of the rolling brush of the automatic car washing machine can be converted into the problems of car body positioning and contour extraction. Zhu M et al. [3] proposed a car positioning algorithm based on orthogonal dual cameras, and obtained the coordinates of the feature points on both sides. Based on the Fast-RCNN method, Zhang Q et al. [4] realized the accurate measurement of the vehicle distance in the range of 5m through the principle of binocular ranging. The car washing detection method based on image processing can significantly increase the cleaning accuracy of the car washing machine, has strong adaptability to complex situations, and has good versatility for different cars. This paper proposes a solution that can quickly and accurately extract the outer contour of the vehicle. Multiple sets of cameras are used to extract the contour of the vehicle, calculate the position of the vehicle being washed, and convert the contour into a set of rolling brush motion trajectory points.

2. Extraction scheme of vehicle contour position information

This scheme is developed based on the gantry reciprocating automatic car washer, and the information extraction scheme is designed according to the shape of the car and the cleaning methods of each roller brush. As shown in Figure 1, the gantry reciprocating automatic car washing machine washing system includes 1-vertical brush, 2-horizontal brush, 3-skirt brush, 4-wheel brush and 5-gantry frame. Each set of roller brushes cleans different surfaces of the car. The middle, upper side and front and rear of the car are cleaned, the horizontal brush is used to clean the top surface of the car (including front and rear glass, hood, trunk lid, etc.), the skirt brush is used to clean the lower side of the car, and the wheel brush is used to clean the car hub. 4 sets of roller brushes work together to complete the cleaning of the car. The vehicle shape information that needs to be extracted can be divided into: the outer contour of the car side (vertical brush and skirt brush), the outer contour of the car top surface (horizontal brush), the position of the wheel hub (wheel brush), the upper side of the car Slope (the vertical brush inclination, the slope of the upper side of different types of cars is different based on the characteristics of the car).

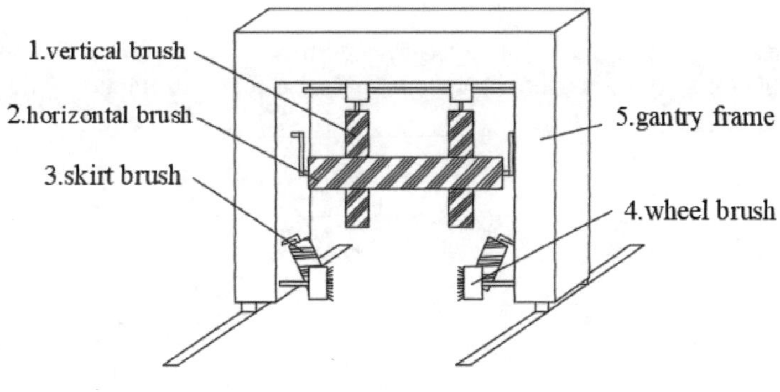

Figure 1. Gantry reciprocating automatic car washing machine

According to the shape of the car and each brush cleaning method, the visual inspection system as shown in Figure 2 is established. The system includes three sets of cameras, light sources, etc., the car is washed and parked in the to-be-washed area of the car washer, and the side camera 3 is the main camera, which captures the side of the car. The image extracts its top surface contour (horizontal brush cleaning path) and calculates the position of the hub (wheel brush cleaning position), the top camera 1 collects the top image of the car to extract the side contour (vertical brush cleaning path), and the camera 2 collects the front image of the car Extract the slope information of the upper side of the car (the vertical brush inclination).

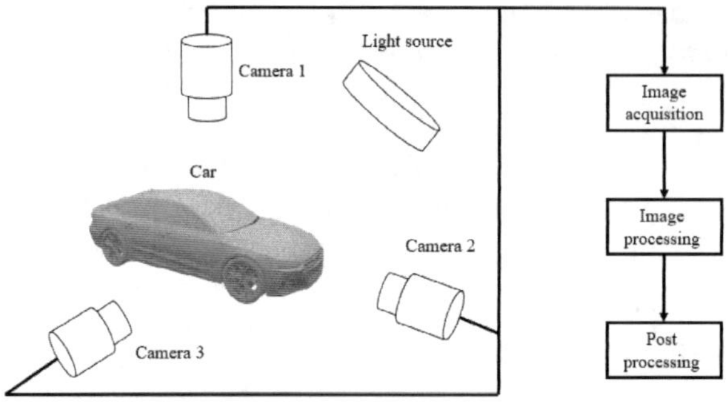

Figure 2. Schematic of vision inspection System

3. Detection algorithm

Since the roller brush moves along the outer contour of the car during cleaning, the key to the algorithm is the complete extraction of the outer contour of the vehicle. Most of the traditional contour edge extraction algorithms use differential operators to perform convolution operations on the image. Common operators include Prewitt algorithm, Sobel operator, Laplacian operator, Canny operator, etc. [5]. Compared with other operators, the Canny operator is an extension based on the common method, so the detection effect of the edge is better, but the traditional Canny algorithm uses Gaussian filtering, which will blur the edge of the image and lose some information. The gradient calculation method is relatively simple, and it is difficult to retain the real edge, and the double threshold needs to be adjusted manually, which is not suitable for automatic car washing machine detection systems. For these problems, this paper designs a method to detect the edge of the car contour in the automatic car washing machine.

3.1. bilateral filter

Different from Gaussian filtering, bilateral filtering adds the influence factor of grayscale difference between pixels to the model, which can better preserve image edge information. The bilateral filtering model is as follows

$$Gray(i,j)=\frac{\sum_{(k,l)\in\Omega_{r,i,j}} f(k,l)\omega_r(i,j,k,l)\omega_s(i,j,k,l)}{\sum_{(k,l)\in\Omega_{i,j}} \omega_r(i,j,k,l)\omega_s(i,j,k,l)} \tag{1}$$

$$\omega_s(i,j,k,l)=\mathrm{e}^{-\frac{(i-k)^2+(j-l)^2}{2\sigma_s^2}} \tag{2}$$

$$\omega_r(i,j,k,l)=\mathrm{e}^{-\frac{\|f(i,j)-f(k,l)\|^2}{2\sigma_r^2}} \tag{3}$$

Where: $f(i,j)$ is the pixel value of the original image at (i,j), $f(k,l)$ is the pixel value of the filter window center (i,j), $Gray(i,j)$ is the pixel value at (i,j) after image processing, ω_s and ω_r are the spatial domain weight coefficient and value domain weight coefficient respectively, σ_s and σ_r are the standard deviation of spatial domain and value domain respectively. σ_s determines the distribution of pixels in the spatial domain. The larger its value, the better the noise reduction effect, but the worse the image sharpness. In order to ensure the noise reduction effect and image sharpness at the same time, σ_s is taken as

$$\sigma_s=(r-C)/2 \tag{4}$$

Where: C is a constant and r is the radius of the filter template. Generally, The relationship between range standard deviation σ_r and noise standard deviation σ_n is linear, $\sigma_r = 1.95\sigma_n$ [6]

$$\sigma_n = 6\sum_{imagel} |I(x,y) * N| \sqrt{\pi/2} / (W/3)(H/3) \tag{5}$$

Where: W, H is the size of the image, N is the template of discrete transformation.

3.2. Edge detail enhancement

In order to make the vehicle contour and edge details more prominent and clear in the image, the image is enhanced with details, and the multi-scale method is used to enhance the vehicle contour edge information. The multi-scale image detail enhancement model [8] is as follows:

$$B_1 = G_1 * I, B_2 = G_2 * I, B_3 = G_3 * I \tag{6}$$

Where: I is the filtered image; B_i is the image after Gaussian filtering at different scales; G_i represents kernel functions of different scales, and the standard deviation of the three kernel functions is taken as 1.0, 2.0 and 4.0.

The pixel points of I and B_i are subtracted one by one to obtain the edge details of different scales of the image, and the enhanced image is formed by adding them to the original image according to different weight distribution.

$$D_1 = I - B_1, D_2 = B_1 - B_2, D_3 = B_2 - B_3 \tag{7}$$

$$I^* = (1 - k_1 \times \mathrm{sgn}(D_1)) \times D_1 + k_2 \times D_2 + k_3 \times D_3 + I \tag{8}$$

$$\mathrm{sgn}(Gray(x,y)) = \begin{cases} 1, Gray(x,y) > 0; \\ -1, Gray(x,y) < 0; \\ 0, Gray(x,y) = 0; \end{cases} \tag{9}$$

Where: I^* is the enhanced image, D_i is the edge details of different scales, k_i is the weight, taking 0.5, 0.5, 0.25 respectively. In addition, in order to prevent the image gray supersaturation, special processing is added to the coefficient of D_1.

3.3. Gradient amplitude calculation

The improved Sobel operator is used to calculate the gradient. The improved Sobel operator refers to the idea of the literature [9]. It is a method that implicitly combines the fractional differential operator and the Sobel operator. When fractional differentiation processes an image, it can enhance and improve the medium and high frequency areas, and retain the low frequency area, while the edges and noises in the image belong to the high frequency area, details and textures belong to the intermediate frequency area, and the low frequency area is generally the smooth part [10]. The algorithm in this paper is to extract the contour of the car. The fractional differential operator can improve the details of the contour and suppress the influence of noise, so this method is selected to achieve.

Common fractional differential definitions include Riemann-Liouville (R-L) definition, Caputo definition and Grünwald-Letnikov (G-L) definition, G-L definition is more suitable for digital image processing than the other two definitions, G-L definition [11, 12] as follows:

$$^{GL}_{a}D^{v}_{x} f(x) = \lim_{h\to 0} h^{-v} \sum_{n=0}^{\frac{x-a}{h}} \frac{\Gamma(n-v)}{\Gamma(n+1)\Gamma(-v)} f(x-nh) \tag{10}$$

Where: v represents the differential order of the function, a represents the lower boundary; x represents the upper boundary, h is the sampling interval, $\Gamma(\bullet)$ is the Gamma function, if $t \in [a,x]$ is in

$f(t)$, and the sampling interval is set to $h=1$,and then $n=x-a$ is obtained, the differential expression of the v-order differential of $f(t)$ can be obtained:

$$\frac{d^v f(t)}{dt^v} \approx f(t)+(-v)f(t-1)+\frac{(-v)(-v+1)}{2}f(t-2)+\ldots+\frac{\Gamma(-v+1)}{n!\Gamma(-v+n+1)}f(t-n) \tag{11}$$

The model of the traditional Sobel operator is as follows:

$$\text{Sobel}(x,y)=\left[\frac{\partial f(x,y)^2}{\partial x}+\frac{\partial f(x,y)^2}{\partial y}\right]^{\frac{1}{2}} \tag{12}$$

$$\frac{\partial f(x,y)}{\partial x}=\left|f(x+1,y-1)+2f(x+1,y)+f(x+1,y+1)\right|-\left|f(x-1,y-1)+2f(x-1,y)+f(x-1,y+1)\right| \tag{13}$$

$$\frac{\partial f(x,y)}{\partial y}=\left|f(x-1,y-1)+2f(x,y-1)+f(x+1,y-1)\right|-\left|f(x-1,y+1)+2f(x,y+1)+f(x+1,y+1)\right| \tag{14}$$

Where: $f(a,b)$ represents the Gray-scale value of point (a,b) of the image

According to equations 9 and 10, the calculation formula of the gradient magnitude of the Sobel operator in the direction and in the direction of the fusion G-L fractional derivative is as follows:

$$D^v\left[\text{Sobel}(x,y)\right]=\frac{\partial^v f(x,y)}{\partial x^v}+\frac{\partial^v f(x,y)}{\partial y^v} \tag{15}$$

$$\begin{aligned}\frac{\partial^v f(x,y)}{\partial x^v}=&\left[\frac{\partial f(x,y)^2}{\partial x}+\frac{\partial f(x,y)^2}{\partial y}\right]^{\frac{1}{2}}+(-v)\left[\frac{\partial f(x-1,y)^2}{\partial x}+\frac{\partial f(x-1,y)^2}{\partial y}\right]^{\frac{1}{2}}\\&+\frac{(-v)(-v+1)}{2}\left[\frac{\partial f(x-2,y)^2}{\partial x}+\frac{\partial f(x-2,y)^2}{\partial y}\right]^{\frac{1}{2}}+\ldots\end{aligned} \tag{16}$$

$$\begin{aligned}\frac{\partial^v f(x,y)}{\partial y^v}=&\left[\frac{\partial f(x,y)^2}{\partial x}+\frac{\partial f(x,y)^2}{\partial y}\right]^{\frac{1}{2}}+(-v)\left[\frac{\partial f(x,y-1)^2}{\partial x}+\frac{\partial f(x,y-1)^2}{\partial y}\right]^{\frac{1}{2}}\\&+\frac{(-v)(-v+1)}{2}\left[\frac{\partial f(x,y-2)^2}{\partial x}+\frac{\partial f(x,y-2)^2}{\partial y}\right]^{\frac{1}{2}}+\ldots\end{aligned} \tag{17}$$

Take the first three terms of equations 16 and 17 and write them as g_x and g_y, the gradient magnitude $M(x,y)$ and gradient direction $\theta(x,y)$ at the pixel (x,y) in the image can be obtained as follows

$$M(x,y)=\sqrt{g_x^{\,2}+g_y^{\,2}} \tag{18}$$

$$\theta(x,y)=\arctan(g_x/g_y) \tag{19}$$

3.4. post-processing

(1) After the gradient processing, since the collected car image is more complex and does not require much information in the image, the threshold segmentation image is used to fill the gradient processed image. Although the threshold segmentation cannot accurately identify the edge, it is not suitable for the car and the background. The segmentation effect is good. The traditional threshold selection methods based on histogram area division include the maximum entropy method, the minimum cross entropy method, the OTSU method and the minimum error method. This paper adopts the OTSU algorithm, which is fast and simple in operation.

(2) After the image is filled, many connected holes and isolated small connected domains will be formed. The seed filling method is used to fill the holes, and the unneeded small connected domains are removed by judging the scale of the connected domain.

(3) In order to make the vehicle contour smoother, the mathematical morphology processing method is adopted. First, the operator structure is established, and the operator performs the first closing operation and then the opening operation on the connected domain of the vehicle contour.

As a result, the algorithm flow of extracting the contour of the car is shown in the figure 3.

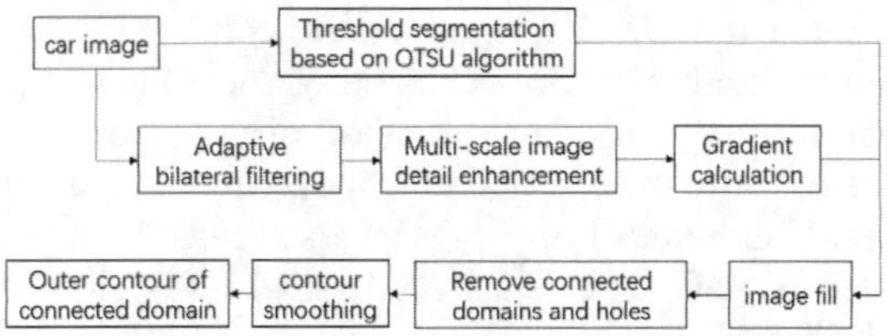

Figure 3. Flow diagram of algorithm

4. Experimental results and analysis

In this paper, the HBV-1780-2 camera is used, and the image resolution is 1280×720. The camera calibration adopts the Zhang Dingyou calibration method. In order to facilitate the establishment of the experimental environment, a 1:32 car model is used to conduct simulation experiments to evaluate the algorithm effect. For the side image of the car shown in Figure 4(a), the image is grayed out to reduce the amount of calculation. The effect after processing is shown in Figure 4(b), adaptive bilateral filtering, multi-scale detail enhancement effect. As shown in Figure 4(c), the gradient calculation of the fusion fractional differential is then performed on the image, as shown in Figure 4(d), and the OTSU threshold segmentation image is filled in the gradient map, as shown in Figure 4(e), The result after removing unwanted connected domains and voids and smoothing the contour is shown in Figure 4(f).

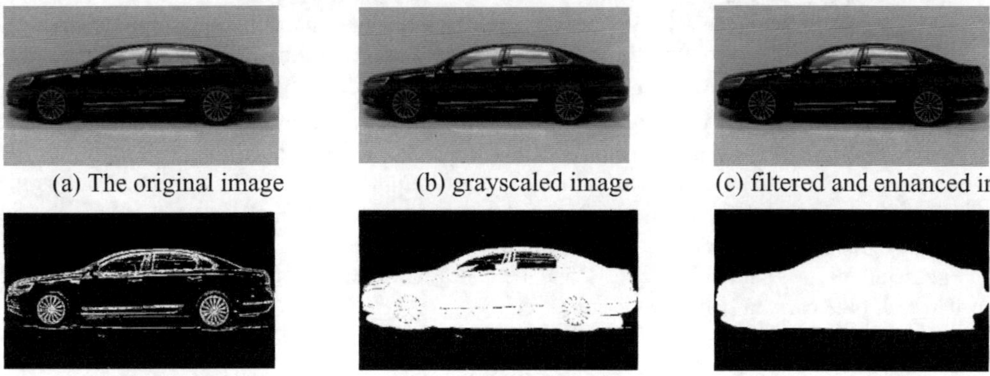

(a) The original image (b) grayscaled image (c) filtered and enhanced image

(d) gradient image (e) filled image (f) Smooth rendering

Figure 4. Results of Side image processing

The contour of the side of the car is obtained as shown in Figure 5 (a). The outer contour of the top surface, that is, the horizontal brush cleaning track, can be obtained by dividing the contour. The segmentation point is the farthest before and after the contour, as shown in Figure 7. In addition, the gradient image is transformed by Hoff circle, which can detect the radius and centre position of the circle in the two-dimensional image space. Because the size of the car tire is 60cm-80cm, and the connecting line between the front and rear axle centres of the car is horizontal, after screening the detected circle, the hub position can be found. The outer contour of the top surface and the hub position are added to the original image, and the effect is shown in Figure 5 (b).

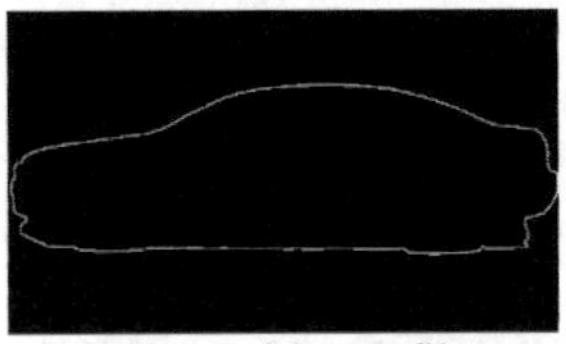

(a) car side profile

(b) The outer contour and hub are superimposed to the original image

Figure 5. Effect of the contour on the original image

The original image and processing results of the images collected by the top camera are shown in Figures 6(a) and 6(b). The extracted outer contour is the vertical brush cleaning trajectory of the car washer. As shown in c) and 6(d), the calculation method of the slope of the upper side of the outer contour of the front of the car is as follows. Search downward from the extreme value of the top surface of the contour. When the rear-vision mirror of the vehicle is searched, the gradient of the contour curve will change drastically, and stop at this time. Search and calculate the slope of the search path to get the slope of the upper side of the car.

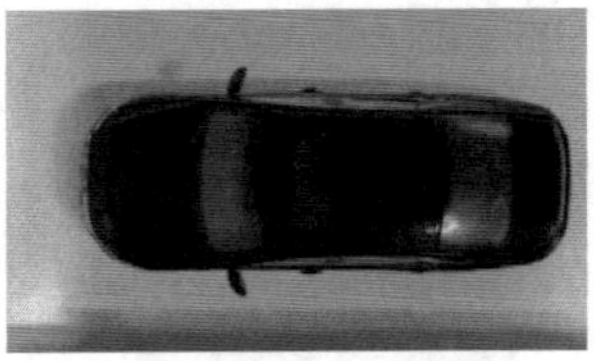

(a) car top image

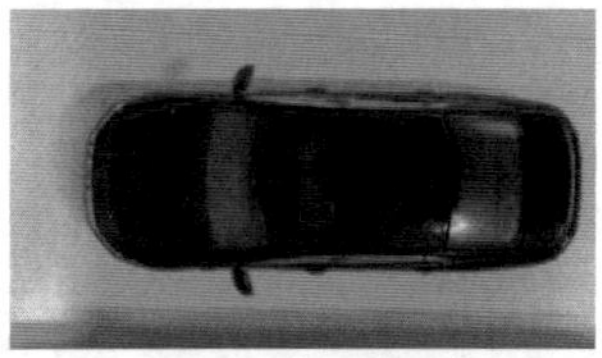

(b) The outer contour is superimposed to the original image

(c) car front image

(d) The outer contour is superimposed to the original image

Figure 6. Results of top and front image processing

After the algorithm is processed, the outer contour information of the car is obtained, but the contour information at this time is only a description of the boundary of the car without specific position information. It is necessary to perform contour tracking calculation on the existing information. Position information and coordinate it, and use software for processing. Because the amount of contour coordinate data is too large, the coordinates are sampled at equal intervals, and the sampling interval is 50. The processing result is shown in Figure 7, and the coordinates after sampling can be seen. The outer contour features of the car can still be well preserved.

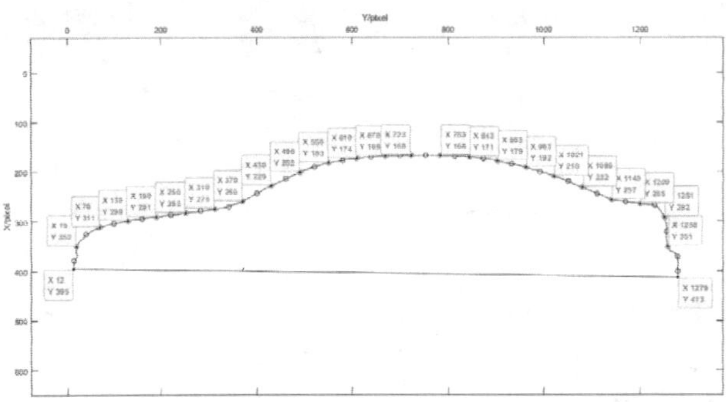

Figure 7. Results of side track sampling processing

Finally, by transferring the contour coordinates of each surface of the car from the camera coordinate system to the world coordinate system, the distance between the car and the car washer and the actual size of the car can be obtained. This system uses multiple cameras, and the relationship between the cameras is known, and the car can be directly calculated. Relative to the positional relationship of the camera, the side camera is the main camera, and the pixel coordinates of the contour in the top camera and the side camera are recorded. The distance between the top camera and the side camera is known, and the distance between the side camera and the car is obtained by coordinate conversion. The car size can be obtained by calculating the result. The output of the result is the vehicle length L, the vehicle width W, the vehicle height H and the wheel hub size R. As shown in Table 1, it can be seen that the error of the vehicle size calculated by this scheme is 2%-4%. It is verified that the algorithm can obtain accurate vehicle contour data and meet the operating requirements of the gantry reciprocating automatic car washer.

Table 1. Results of car size calculation

vehicle data	*Calculated data/mm*	Measured data/mm	*Error rate*
Vehicle length L	150.96	154.15	2.06%
Vehicle width W	54.87	57.38	4.37%
Vehicle height H	44.92	45.79	1.99%
Wheel size R	15.50	15.18	2.11%

The above experiments prove the practicability of the algorithm in the built measurement system, which can ideally extract the outer contours of the car.

5. Conclusion

In this paper, a new planning scheme is proposed to solve the problem of lack of reliability and accuracy of the traditional cleaning path planning method of the roller brush of the car washing machine. A multi-view detection system was established in the car washing environment, and images were collected from all sides of the car. A method of adaptively detecting the edge of the car contour was designed, and the Sobel operator implicitly combined with fractional differential was proposed to extract the image gradient. Combined with operations such as adaptive bilateral filtering and edge detail enhancement, the contour edges of each side of the car are obtained. Finally, the coordinate points of the car contour are obtained through coordinate conversion, and 4 kinds of data of the vehicle shape are calculated. It is verified that the scheme can be applied to gantry reciprocating automatic car washing machine and has a good application prospect.

Reference

[1] Xu R, Ge YP.(2012) Design of Roll-over Automatic Car Wash Profiling System. Machine Building & Automation.35-38.

[2] CHU ZB, WANG H, HU YX.(2017) Positioning control strategy and realization of two-link manipulator for washing machine. Manufacturing Automation.142-145.

[3] ZHU M, WU ZW.(2021) Research on positioning algorithm of car wash based on orthogonal dual camera. Manufacturing Automation.147-150.

[4] Zhang Q, Hu GD, Li YS, et al.(2018) Binocular vision vehicle detection method based on improved Fast-RCNN. Journal of Applied Optics.832-838.

[5] HE Q, LIU BY.(2021)Review of Infrared Image Edge Detection Algorithms. Infrared Technology. 199-207.

[6] BIRADAR N, DEWAL M L, ROHIT M, et al. (2016)Echocardiographic image denoising using extreme total variation bilateral filter. Optik - International Journal for Light and Electron Optics, 127(1).

[7] LI DJ, GUO S, YANG L. (2021) Yarn defect detection based on improved image threshold segmentation algorithm. Journal of Textile Research. 82-88.

[8] QIN YY, CUI W, LI Q, et al. (2019)Traffic Sign Image Enhancement in Low Light Environment. Procedia Computer Science, 154(C).
[9] BALOCHIAN S, BALOOCHIAN H. (2022)Edge detection on noisy images using Prewitt operator and fractional order differentiation. Multimedia Tools and Applications, 81(7): 9759-70.
[10] JIANG ZH, WU QQ, GUI WH. et al. (2017)Adaptive Detection of Blast Furnace Surface Contour with Fractional Multi-directional Differential Operator. Acta Automatica Sinica. 2115-2126.
[11] ZHANG XF, HE H. (2021) Multi-focus Image Fusion with Multi-scale Based on Fractional Order Differentiation. Journal of Northeastern University(Natural Science). 1071-1078.
[12] WHARMBY A W. (2016)The application of the fractional calculus model for dispersion and absorption in dielectrics II. Infrared waves. International Journal of Engineering Science, 104: 62-74.

Design and Verification of Simulation Data Generator for Rail Transit Switch Machine

Shiyan WEI[1], Jinbai ZOU[1], Quan SHA[1], Guoyi JI[1]

[1]Shanghai Institute of Technology, China

Corresponding author: 1437192148@qq.com

Abstract. Because the fault data of rail transit switch machine are difficult to obtain and the site fault is difficult to reproduce, it is difficult to diagnose or predict the switch machine. In this paper, the power fault data of S700K switch machine is divided into creeping fault and mutation fault, and a simulation data generator for generating massive fault data is developed. The generators involve the synthesis of minority over-sampling techniques and generative confrontation networks. Finally, the long-term memory neural network is used to predict the generated gradual fault data to verify the authenticity and reliability of the simulation data generator. The experimental results show that the generated data can predict the future power trend of the switch machine, which proves the authenticity and feasibility of the simulation data generator.

1. Introduction

The switch machine is the core component of the rail transit switch device. It is necessary to monitor and diagnose the fault of the switch machine in order to improve the safety of railway transportation. However, the fault frequency of the switch machine in rail transit is low, which leads to the scarcity of fault data and is unable to diagnose or predict the fault of the switch machine. The development of simulation data generator can effectively solve the above problems. The simulation data generator can simulate some characteristics of the fault representation of the switch machine, and provide massive fault data for various fault diagnosis or prediction technologies. Hou used the normal probability data of the switch machine to predict the degradation performance index of the slowly changing fault, and conducted fault location and identification for the power data after the fault[1], so as to achieve the health management of the switch machine and provide theoretical support for the maintenance and repair work of the subway operation and maintenance site. In this paper, the power curve of the switch machine is analyzed, and three data generation intelligent algorithms in the simulation data generator are introduced. Finally, the intelligent model of fault prediction is used to verify the multi-features of the generated data, which proves the authenticity and feasibility of the simulation data generator.

2. S700K switch machine fault data classification

The S700K switch machine is widely used as the first generation of speed-raising turnouts. Whether it is the action principle or the control circuit design, the S700K switch machine is leading. In the current research on the switch machine, the power in the switch machine conversion process is regarded as an important data that can reflect the real-time working condition of the switch[2] . The following is the fault classification of the switch machine.

2.1. Creeping fault

The definition of gradual fault is that fatigue fault occurs in the mechanical equipment related to the switch machine over time. In order to prevent the switch machine fault in advance, this kind of fault can be predicted. The causes of this kind of failure are as follows: sanding of slide bed, oil shortage of slide bed, loose of fixed parts of slide bed. These three reasons will increase the friction resistance between the slide bed and the track, and will eventually affect the normal switch.

When the S700K switch machine works normally, the power response is that the time required for a switch machine to work normally is 7s, which is divided into three stages: start-up, conversion and locking. Figure 1 is the normal power time curve of the switch machine. It can be seen from figure 1 that the power suddenly rises at the beginning, and then the power will suddenly drop to about 0.5 kW and remain near this value, and then the small step phenomenon occurs. Finally, the representation circuit is automatically cut off, and the power is reduced to 0. Figure 2 is the power data curve of gradual failure. It can be seen from figure 2 that the power is normal in the start-up phase of the previous seconds, but the power will suddenly increase at a certain moment in the conversion phase, and then it will remain stable at a higher power and cannot enter the locking phase until the phase breaker starts the protection device to cut off the circuit, and the power will suddenly drop to 0.

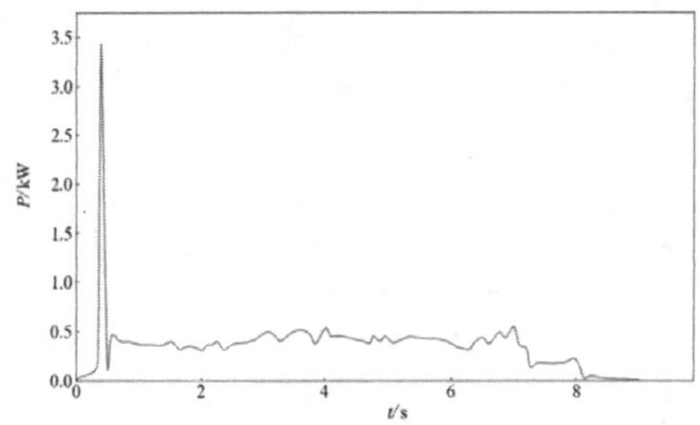

Figure 1. Normal w3orking power curve of switch machine

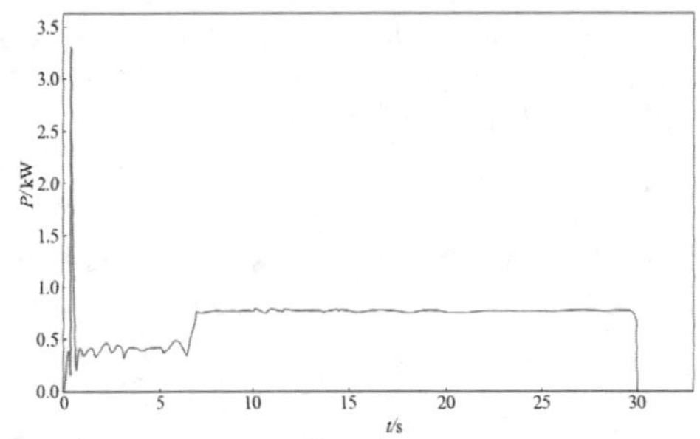

Figure 2. Power fault power curve of switch machine

2.2. Mutation fault

Mutant faults can't be predicted because of their suddenness, which can only be analyzed afterwards, such as fault diagnosis or classification. The reasons for this kind of failure are as follows: S700K switch machine detection rod is not normal into the gap of the representation rod; card gap; work slide plate damage; diode breakdown causes short circuit.

3. Simulation data generation method

In order to obtain a large number of these types of fault data, this paper develops a simulation data generator. The main principle is to use synthetic minority oversampling technique (SMOTE) or generative adversarial networks (GAN) to simulate and generate a large number of fault data based on a small amount of fault data and normal data.

3.1. SMOTE algorithm
SMOTE is an oversampling method designed by Chawla et al. in 2002. This technique is used to connect a few fault samples and their ' k ' nearest line segments to artificially generate a few samples. Depending on the oversampling rate set, the adjacent elements in the 'k' nearest neighbor are randomly selected. One of the shortcomings of SMOTE algorithm is the overgeneralization of a few class spaces without considering most classes, which may increase the overlap between classes. So, this paper uses one of the improved SMOTE algorithm, Borderline-smote. Firstly, the minority samples of the boundary, namely the normal data samples, are identified, and then the synthetic samples are generated by using the SMOTE algorithm to oversample the minority classes. This technique is used to connect line segments of a few fault samples and all training samples to generate a few samples artificially. The design of Borderline-smote strengthens the sampling of a few examples of boundary[3].

3.2. GAN
GAN has also been widely used in dealing with data imbalance[4]. Unlike smote algorithm, GAN is a method based on neural network. GAN was proposed by lan J. Goodfellow et al. in 2014, similar to the generation model of confrontation and game between two neural networks[5]. As shown in figure 3, the noise Z outputs G (z) through the generator G, and the discriminator D judges the real data P_{data} and the simulated sample G (z) of the switch machine. If the data is real, the output is close to 0, otherwise close to 1. When the discriminator cannot judge the authenticity of the input data, the output probability of each judgment of the discriminator is 0.5, which reaches the global optimal solution.

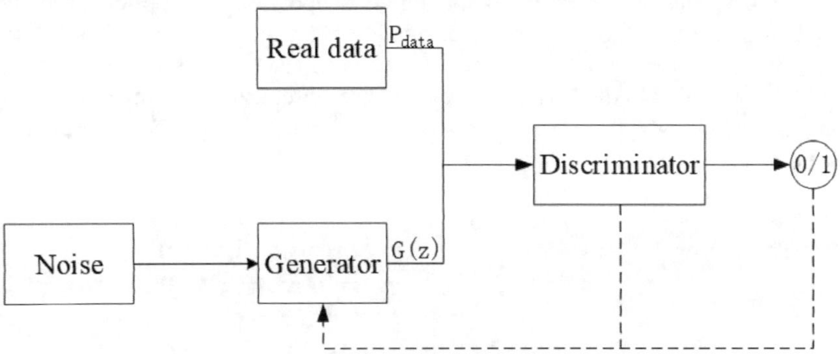

Figure 3. Principle block diagram of GAN algorithm

4. Power prediction of switch machine based on LSTM
In the second chapter, the method of S700K switch simulator data generator is described. In this chapter, a generation method is selected to design the switch simulator data generator, and the generated fault power data are used to train the long-term and short-term memory neural network prediction model. The prediction effect can verify the authenticity of the generated data, so as to prove the feasibility of the simulator data generator.

4.1. Analog data generator based on Borderline-Smote
In this experiment, the power data of gradual fault and normal working described in the first chapter are selected. The original data samples are three groups of gradual fault power data and three groups of normal working power data. The simulation data generator method selects the Borderline-Smote method in the second chapter. The original data sample generates 203 sets of power data through the simulation data generator. Figure 4 shows the generated 203 sets of data. These data include normal power, fault power and intermediate state data from normal to gradual faults.

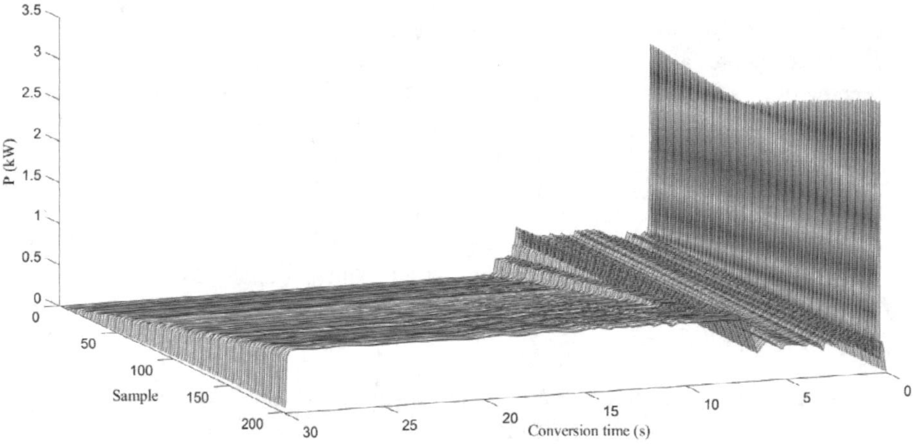

Figure 4. 203 groups of generated data

4.2. Gradient fault power prediction based on LSTM
Long short term memory (LSTM) is an improved form of convolutional neural network model[6]. Long-term dependent information can be learned, and the power time data of switch machine have the characteristics of time series, which provides theoretical support for the application of LSTM in the prediction of fault time series of switch machine. As shown in figure 5, the horizontal line above the figure is the key to the LSTM model running information interaction, a bit like a conveyor belt that runs through the entire model. In the LSTM model, there are three gate structures Ω, namely forgetting gate, selection gate and output gate. Their functions are to discard unimportant information, determine updated information and output final information.

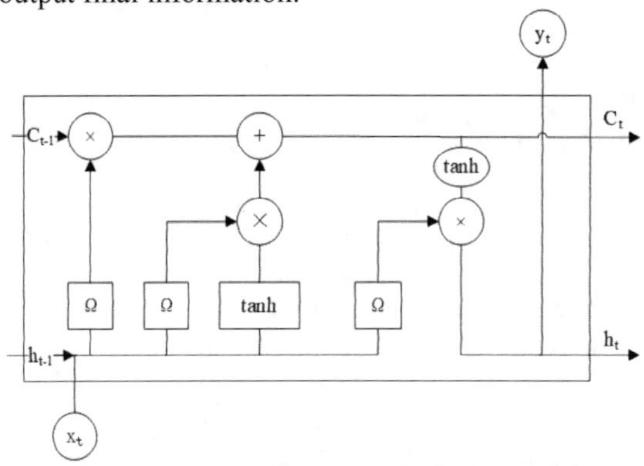

Figure 5. Frame diagram of LSTM model

In order to verify the authenticity and feasibility of the simulated data generator designed in this paper, experiments are carried out on the gradual fault of the switch machine. The 203 groups of generated data in Section 3.1 are used as the input of LSTM, and each group of data is used as the input of a moment. Divide 203 sets of data into 160 training sets and 43 test sets. The main steps of the experiment are to train the training set, predict the latter value by the specified step length, and then test the test set. Since the sampling time interval of the microcomputer monitoring system of the switch machine is 40millisecond, a group of 30 s fault power data has 750 data points. The step size of this experiment is set as 25, the value of the 26th step size is predicted, and 500 rounds of training are conducted.

4.3. Results and analysis

In this experiment, LSTM is used to predict the time series of the power data of the gradual fault of the switch machine. As shown in figure 6. It is the training result of the power data of the gradual fault of the switch machine. The dashed line represents the input of the LSTM model, and the solid line is the output. It can be seen from the figure that the training effect of the model is roughly consistent with the original data. Figure 7 is the test result of gradual fault power data, the virtual line is the power data of the test set, and the real line is the power data of the predicted output. It can be proved from these two figures that the power data generated by the data generator can well train the LSTM model, and can predict the approximate change trend of power. It also proves the availability of generated data, indicating the authenticity and feasibility of the simulated data generator designed in this paper.

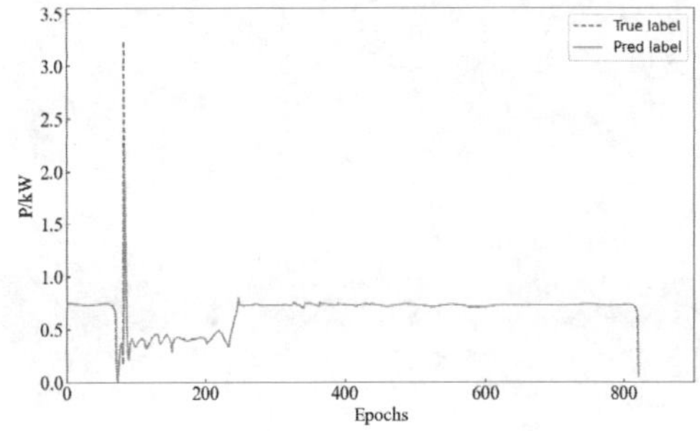

Figure 6. Training results of gradual fault power data

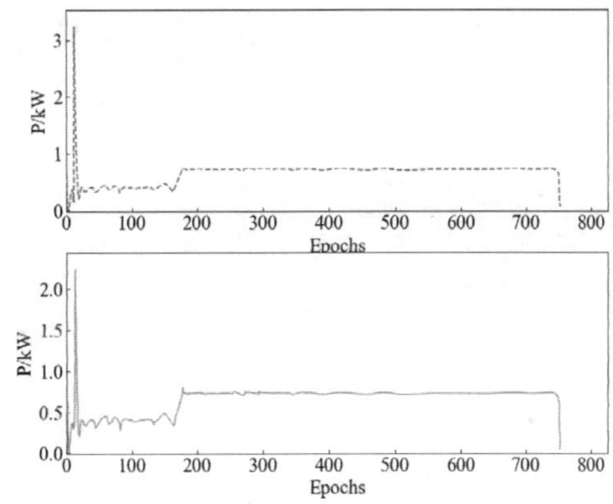

Figure 7. Test results of gradual fault power data

5. Conclusions

In order to solve the problem of scarce fault data of rail transit equipment, this paper first proposes and designs a simulation data generator. Taking rail transit switch machine as the object, this paper expounds the intelligent algorithms involved in the generator, and selects Borderline-smote algorithm to generate gradual fault power data, generating 203 sets of data. The LSTM prediction model is used to train and test 203 groups of generated data. Experiments show that the generated data can well train the prediction model, and the predicted results are also consistent with the actual data, which proves the authenticity and feasibility of the generated data.

References

[1] Hou Dashan. Points Fault Prognosis and diagnosis based on neural network[D]. Beijing: Beijing Jiaotong University. 2019: 28−30.
[2] Zhang Yao. Research on Fault Prediction and Fault Diagnosis of Railway Turnout[D]. Beijing:

Beijing Jiaotong University. 2021.
[3] Shi Hongbo, Chen Yuwen, Chen Xin. Summary of research on SMOTE oversampling and its improved al-gorithms[J]. CAAI transactions on intelligent systems, 2019, 14(6): 1073−1083.
[4] Shi, J., Ding, Y. and Lv, Z., 2021. An Intermittent Fault Data Generation Method Based on LSTM and GAN. In: 2021 Global Reliability and Prognostics and Health Management. Nanjing. 1-4.
[5] Huo, Y., Shang, Y., Xu, B., Li, Y. and Y. Yang, 2021. A Fault Data Generation Algorithm Based on GAN and Policy Gradient Mechanism. In: 2021 IEEE International Symposium on Broadband Multimedia Systems and Broadcasting. Chengdu. 1-6.
[6] Lu, H. and Yang, F., 2018. A Network Traffic Prediction Model Based on Wavelet Transformation and LSTM Network. In: 2018 IEEE 9th International Conference on Software Engineering and Service Science. Beijing. 1-4.

Development of an Adaptive Wall-pressing Mechanism for an In-pipe Cleaning Robot

[1st] **Zeyu Yang**[1,a], [2st] **Yuanxun Fan**[1,*], [3st] **Yu Yan**[1,b]

[1]School of Mechanical Engineering, Nanjing University of Science and Technology, Nanjing, Jiangsu Province, China

[a]yangzeyu@njust.edu.cn; *fanyx@njust.edu.cn; [b]noel.yan@outlook.com

Abstract. This paper proposes a novel adaptive wall-pressing mechanism for an in-pipe cleaning robot (ICR) which combines two type of self-adaptive mechanisms. Parametric modelling and optimization for the mechanism is proceeded in order to improve the carrying capacity and adaptability of the robot in the water supply pipeline. Force in the compression spring due to the self-weight of ICR is reduced by 31% after optimization. A virtual simulation platform is established to analyse the stability and obstacle-surmounting capacity of ICR under different pre-loading forces.

1. Introduction

Water supply network is an important part of water supply system, it needs to meet the requirements of users for quantity, quality and pressure of the water. The underground water supply network is significant and complicated, its safety is the top priority of public health in the 21st century [1]. At present, the aging water supply pipelines in many cities are seriously corroded, which leads to the deterioration of drinking water quality, the reduction of pipeline water delivery capacity and the increase of water supply energy consumption. Finding a method to prevent and control the pollution of water quality in the water supply network, ensure the water pressure of water supply, reduce the energy consumption of water supply, and improve the safety of water supply has become an important subject of urban water supply.

Up to now, plenty of in-pipe robots have been presented. Ho Moon Kim et al. design an in-pipe robot which can realize the differential-speed function [2]. Young-Sik Kwon et al. develop a two-moduled inspection robot that moves in horizontal and vertical pipelines [3] and a small-sized in-pipe robot with diameter of 40mm to 70mm [4]. Zhao W et al. design a large inspection robot for long-distance pipeline [5]. Zhang Y et al. develop an inspection robot for gas pipe that an actively adapt the pipe-diameter and automatically adjust tractive force [6]. Park J et al. design a tracked robot that can estimate the relative attitude between the robot's main body and the pipe using the angular sensors attached to a pantograph mechanism [7]. Lee D et al. develop an inspection robot with novel mechanism that can travel through various pipe configurations including vertical, elbow, and branch pipes [8]. Most of them are used for maintenance and inspection, there is almost no in-pipe robot designed for cleaning water supply pipelines. The ICR introduced in this paper is a robot equipped with high-pressure water jet cleaning system to clean the water supply pipeline, it shall be applied to the pipeline whose diameter is range from 250mm to 300 mm, and figure 1 is the cleaning system we proposed. Since ICR needs to work in long-distance, wet pipelines and maintain a constant moving speed, choosing motor as driving force will face problems such as poor power supply, insufficient driving force when slipping and waterproof measures. Take the problems above into consideration, ICR is powered by winch with the wirerope. The water jet used for cleaning is provided by a high-pressure pump, and the real-time image taken during the cleaning process will be transmitted to the image processing system on the ground.

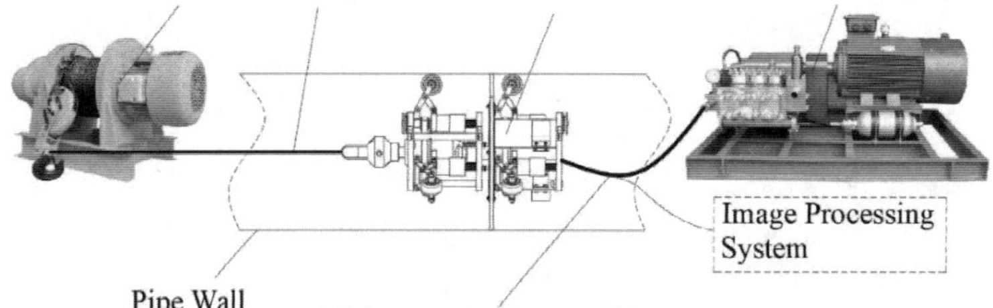

Figure 1. The cleaning system of ICR

2. Design of the adaptive wall-pressing mechanism

2.1. General structure of ICR

Now, the mechanisms of in-pipe robots developed by researchers can be classified into eight types [9], the ICR introduced in this paper is a combination of wheeled and wall-pressing type mechanisms, and its general structure is shown in figure 2. The ICR consists of a cleaning module and a detection module, the cleaning module is equipped with a rotary cleaning nozzle for cleaning the water supply pipeline. The detection module is equipped with three cameras in the circumference of ICR in order to capture real-time images inside the pipeline and transmit them to the image processing system outside the pipeline via data transmission line. There are elastic board between the two modules to block the spray generated by the cleaning nozzle as much as possible for clearer images. The ICR has a radially symmetric structure, each module is surrounded by 3 supporting frames at 120 degree apart circumferentially.

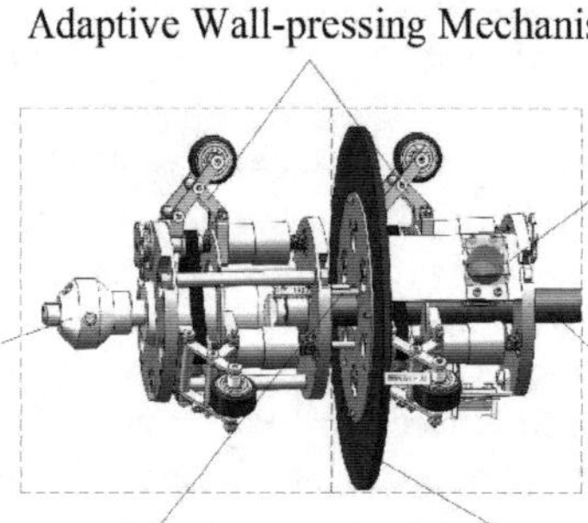

Figure 2. General structure of ICR

2.2. Principle of the adaptive wall-pressing mechanism

The ICR is required to work in pipelines with diameters range from 250mm to 300mm. At the same time, during the ICR's travel along the pipe, it often faces the sudden changes of pipeline's diameter due to corrosion, rust and breakage. In order to ensure the adaptability in different pipe sizes and make sure the ICR does not suffer from rigid impact during the changes of pipeline's diameters, it requires the ICR to have a suitable wall-pressing mechanism.

The adaptive wall-pressing mechanism developed by researchers can be divided into two types: synchronous adjusting mechanism and independent adjusting mechanism. The synchronous adjusting mechanism means that all the supporting frames of the in-pipe robot are adjusted by a single adaptive mechanism, and the radial dimensions of all supporting frames will be changed at the same time. This type of adjusting mechanism can ensure that robot's centre align with pipeline's centre when the pipeline's diameters remain unchanged or changes synchronously, but the adaptability is poor when the diameter change partly.On the contrary, the independent adjusting mechanism is able to adjust every supporting frame independently. However, due to the self-weight of the robot, the radial dimension of the adaptive wall-pressing mechanism at the bottom of the robot is different from that of the top, which results in the misalignment or even a large deviation between the robot's centre and pipeline's centre. This can lead to the poor cleaning effect or even the crush of the robot. To avoid these consequences, giving the system an initial pre-loading force is necessary.

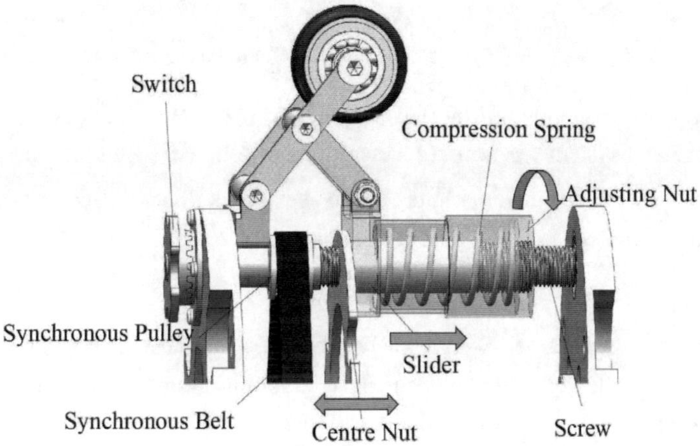

Figure 3. Structure of the adaptive wall-pressing mechanism

Considering the problems of the two adjusting mechanisms above, a new adaptive wall-pressing mechanism is proposed in this paper to allow ICR to work properly. As shown in figure 3, the adaptive wall-pressing mechanism is composed of two parts: a rigid adjusting mechanism and a flexible adaptation mechanism.

The rigid adjusting mechanism consists of centre nut, screw and synchronous belt drive system. Before ICR enters the pipe, adjust the switch located at the front of ICR depending on the diameter and the scaling situation of the pipe. The synchronous belt drive system drives three screws synchronously, and the central nut will slide back and forth to switch among different size of pipelines, the switch will be locked up with robot's frame afterwards. The ICR will be set a suitable initial position by the rigid adjusting mechanism.

The flexible adaptation mechanism consists of slider, compression spring and adjusting nut. When facing the changes of pipeline's diameter during the process, the slider will compress the spring to offset the impact. The spring is also designed to neutralize the self-weight of ICR in different postures, the pre-loading force in the spring can be adjusted by rotating the adjusting nut in order to meet the demand. The compression spring is surrounded by slider and adjusting nut to prevent rusting.

Compared to the traditional structure of wall-pressing mechanism, the new mechanism will ensure the alignment between ICR and pipelines while maintaining better adaptability.

2.3. Optimization of the adaptive wall-pressing mechanism

To maximize the carrying capacity of ICR, reduce the pre-loading force it needs to neutralize the self-weight, it is necessary to optimize the dimensions of components in the mechanism. The simplified free-body diagrams of ICR in pipelines with diameter of 300mm and 250mm is shown in figure 4.

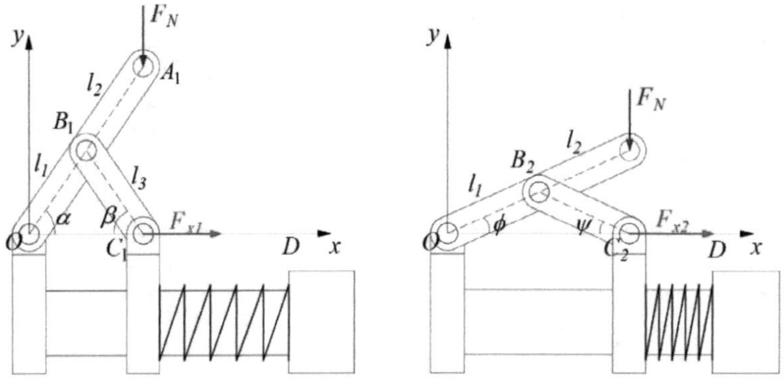

(a) (b)

Figure 4. Free-body diagram of adaptive wall-pressing mechanism:
(a)-300mm pipeline; (b)-250mm pipeline

The two situation shown above are the limit positions of ICR in pipelines of different diameters Establish the coordinate system at point O, the force F_x on the spring is determined mainly by $l_1, l_2, l_3, \alpha, \beta, \phi$ and ψ, therefore the variable is chosen as $x = \{l_1, l_2, l_3, \alpha, \beta, \phi, \psi\}$.

The objective function can be expressed as follows:

$$f(x) = \min F_x \tag{1}$$

F_N is the normal force of the self-weight of ICR on the adaptive wall-pressing mechanism, when the ICR is in position (a), the vertical displacement y_A at point A_1 and the horizontal displacement x_c at point C_1 can be determined by the following equation:

$$\begin{bmatrix} y_A \\ x_C \end{bmatrix} = \begin{bmatrix} l_1 + l_2 & 0 & 0 \\ 0 & l_1 & l_3 \end{bmatrix} \begin{bmatrix} \sin\alpha \\ \cos\alpha \\ \cos\beta \end{bmatrix} \tag{2}$$

Differentiating equation (2), the instantaneous velocity matrices at points A_1 and C_1 with α and β as variables are obtained as follows:

$$\begin{bmatrix} \delta_A \\ \delta_C \end{bmatrix} = \begin{bmatrix} l_1 + l_2 & 0 & 0 \\ 0 & l_1 & l_2 \end{bmatrix} \begin{bmatrix} \cos\alpha \\ -\sin\alpha \\ -\sin\beta \end{bmatrix} \tag{3}$$

According to the principle of virtual work, ignoring friction, the sum of work done by all external forces is zero, i.e.:

$$[F_N \quad F_{x1}] \begin{bmatrix} \delta_A \\ \delta_C \end{bmatrix} = 0 \tag{4}$$

Where F_{x1} is the force on spring when the diameter of pipeline is 300mm, solving equation (4):

$$F_{x1} = \frac{F_N (l_1 + l_2)}{l_1 \cdot \tan\alpha + l_3 \cdot \dfrac{\sin\beta}{\cos\alpha}} \tag{5}$$

According to the geometric conditions:

$$l_1 \cdot \sin\alpha = l_3 \cdot \sin\beta \tag{6}$$

Substituting equation (6) into equation (5), the force F_{x1} can be obtained as follows:

$$F_{x1} = \frac{F_N (l_1 + l_2)}{2l_1 \cdot \tan\alpha} \tag{7}$$

Similarly, F_{x2} can be obtained as follows:

$$F_{x2} = \frac{F_N (l_1 + l_2)}{2l_1 \cdot \tan\phi} \tag{8}$$

According to the dimensional relationships of the structure of ICR, the constraints can be obtained as follows:

$$\begin{cases} (l_1 \cdot \cos\alpha + l_3 \cdot \cos\beta) \geq 32 \\ 115 - (l_1 \cdot \cos\phi + l_3 \cdot \cos\psi) \geq 60 \\ 49 = (l_1 + l_2)\sin\alpha \\ 49\sin\phi = 24\sin\alpha \end{cases} \tag{9}$$

Since that $F_{x2} > F_{x1}$, it is more efficient to choose F_{x2} as the optimization objective function. The initial point of optimization is the original designing dimensions of the components in the mechanism, i.e.: $x = \{35, 35, 35, 44°, 44°, 20°, 20°\}$ Substituting the constraints into the objective function, the relationship between the value of the function and the number of iterations is shown in figure 5. According to the result, force on the spring due to self-weight of ICR reduces by about 31%.

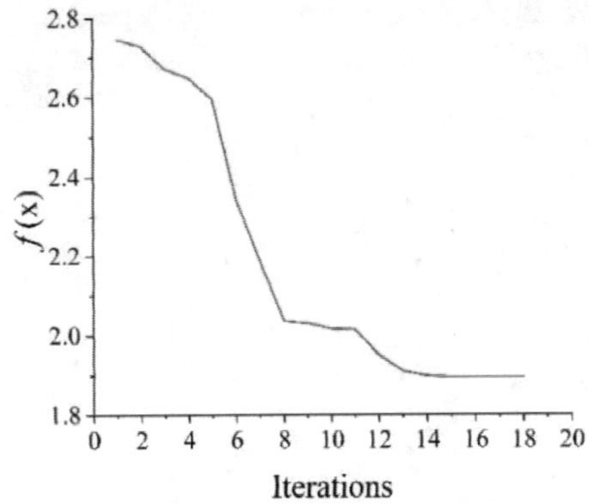

Figure 5. Result of optimization

3. Simulation Analysis

Some simulations have been conducted in order to testify the alignment between ICR and pipelines while traveling and obstacle-surmounting, the virtual simulation platform includes a simplified ICR model, a 300mm diameter pipe with obstacles, and a simulated winch.

As shown in figure 6, the pipe is fixed to the ground while there are three different obstacles inside the pipe, which are obstacle for the lower supporting frames only (obstacle 1), obstacle for all supporting frames (obstacle 2), and obstacle for the upper supporting frames only (obstacle 3), respectively. The height of the obstacle are all 10mm, these obstacles are used to simulate different situations that ICR may encounter in pipelines. A translational joint was set between the wirerope and the ground, while a constant velocity *v* was applied along the *x*-direction to simulate the traction process of ICR by the winch.

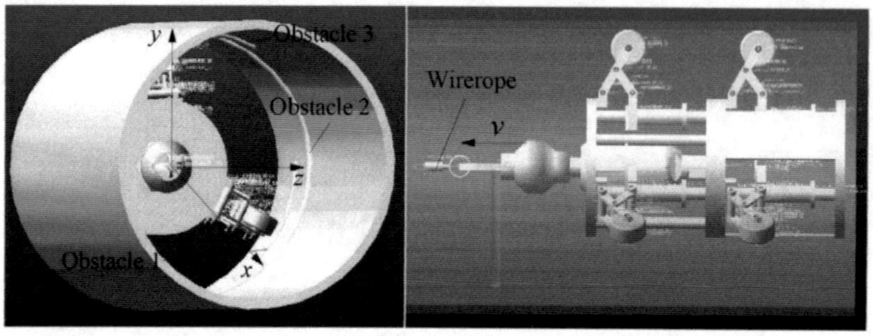

Figure 6. The virtual simulation platform of ICR

In order to analyse the stability of ICR while surmounting obstacles under different pre-loading forces, the pre-loading forces in the spring are set to 0N, 30N, 60N and 120N respectively, where the force in spring caused by the self-weight of ICR is about 60N. The displacement of the centre of the nozzle in y-direction and z-direction is shown in the figure 7.

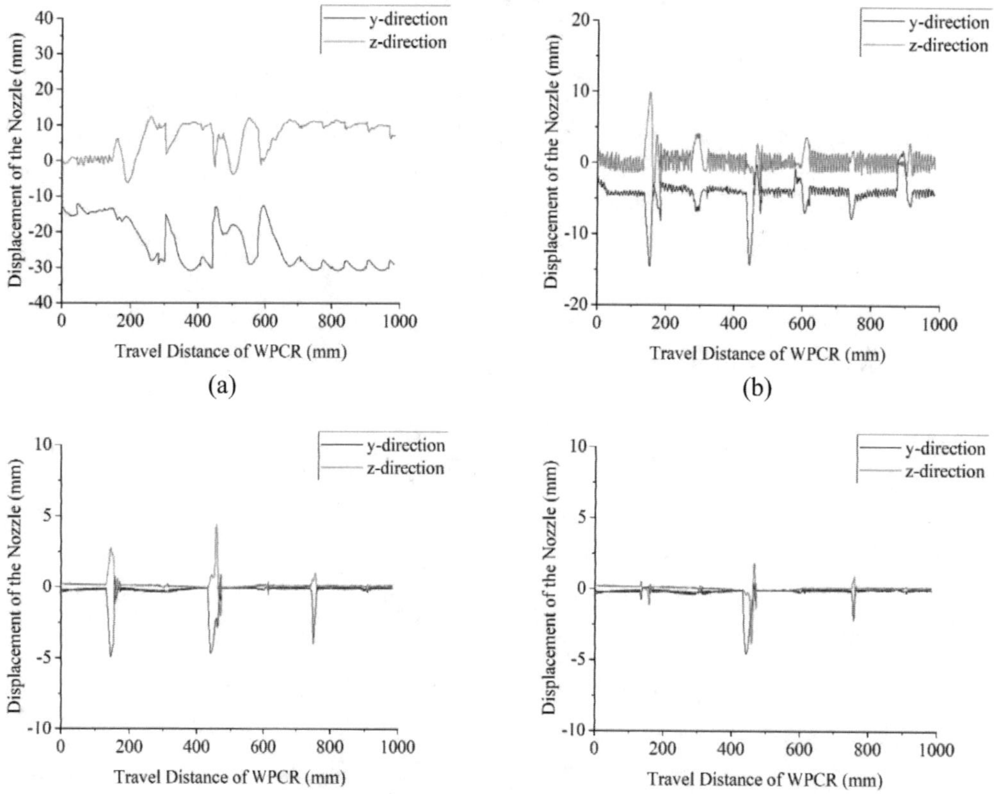

Figure 7. Displacement of the nozzle with pre-loading force of:
(a)-0N; (b)-30N; (c)-60N; (d)-120N.

As shown in figure 7, the ICR has a poor alignment with the pipeline due to its self-weight without pre-loading force, and there is large fluctuation while traveling. After increasing the pre-loading force to 30N, the fluctuation in ICR is significantly less than the situation above. However, the nozzle's centre is still below the pipe's centre in y-direction because that the pre-loading force fails to neutralize the self-weight of ICR, and the vibration in z-direction is obvious at the same time. When the pre-loading force just neutralizes the ICR's self-weight, the ICR has a good alignment with the pipeline while traveling without obstacles, the fluctuation of the nozzle is also significantly smaller than the previous two situations when surmounting obstacles. Continuing to increase the pre-loading force, it can be seen that the fluctuation is decreased further when crossing obstacle 1 and obstacle 3, the displacement is within 2mm. The displacement in y-direction when crossing obstacle 2 is still about 5 mm, this is because the spring on the lower side is more easily compressed by ICR's self-weight than the spring on the upper side. The results shown in the simulation indicates that increasing pre-loading force is an effective measure to increase the stability of ICR.

4. Conclusions

This paper proposes a new adaptive wall-pressing mechanism for an in-pipe cleaning robot with high-pressure water jet devices. By optimizing the adaptive wall-pressing mechanism, the force in compression spring caused by its self-weight reduces by about 31%, which enables ICR to have better carrying capacity. The stability of ICR is verified through simulation experiments, and it is proven that the fluctuation caused by obstacle-surmounting can be decreased by adding more pre-loading force to the springs. The results will provide more theoretical basis for the development of the physical prototype in the future.

References

[1] Edwards, M. (2004) Controlling corrosion in drinking water distribution systems: a grand challenge for the 21st century. *Water science and technology*, **49**: 1-8.

[2] Kim H M, Suh J S, Yun S C, et al., 2013. An In-pipe robot with multi-axial differential gear mechanism. In: *IEEE/RSJ International Conference on Intelligent Robots & Systems*. Tokyo. 252-257.

[3] Kwon Y S, Lim H, Jung E J, et al., 2008 Design and motion planning of a two-moduled indoor pipeline inspection robot. In: *IEEE International Conference on Robotics and Automation.* Pasadena. 3998-4004.

[4] Kwon Y S, Yi B J., 2010 Development of a pipeline inspection robot system with diameter of 40mm to 70mm (Tbot-40). In: *2010 IEEE International Conference on Mechatronics and Automation.* Xian. 258-263.

[5] Zhao W, Zhang L, Kim J. (2020) Design and Analysis of Independently Adjustable Large In-Pipe Robot for Long-Distance Pipeline. *Applied Sciences*, **10**:3637.

[6] Zhang Y, Yan G. (2007) In-pipe inspection robot with active pipe-diameter adaptability and automatic tractive force adjusting. *Mechanism & Machine Theory*, **42**:1618-1631.

[7] Park J, Hyun D, Cho W H, et al. (2010) Normal-Force Control for an In-Pipe Robot According to the Inclination of Pipelines *IEEE Transactions on Industrial Electronics*, **58**: 5304-5310.

[8] Lee D, Park J, Hyun D, Yook G, Yang H S. (2012) Novel mechanisms and simple locomotion strategies for an in-pipe robot that can inspect various pipe types. *Mechanism and Machine Theory*, **56**: 52-68.

[9] Kakogawa A, S Ma. Mobility of an in-pipe robot with screw drive mechanism inside curved pipes. In: *IEEE International Conference on Robotics & Biomimetics*. Tianjin. 1530-1535.

Numerical simulation analysis of the flow field in the bore of a two-stage light gas gun

Jinlin Dong[1,a] and Bing Cao[1,b]

[1]School of Mechanical Engineering, Nanjing University of Science & Technology, Nanjing 210094 China

[a]E-mail: dongjinlin231@163.com

Corresponding author: [b]E-mail:caobing202815@163.com

Abstract. In order to study the effect of different launch tube diameters of gas-driven two-stage light gas guns on the flow field during the launch process, a model of the piston movement in the pump tube to the projectile launch was established using Fluent. Using the dynamic mesh technology and 6DOF model, a numerical simulation of the projectile launch process of the two-stage light gas gun was carried out, and the effects of different launch tubes on the projectile velocity and the flow field in the bore were obtained. The results show that when the mass difference between the piston and the projectile is large, appropriately increasing the diameter of the launch tube is conducive to the flow of gas in the high-pressure cone and increases the projectile launching speed. The numerical simulation analysis of the flow field in the bore provides a basis for the design and optimization of the two-stage light gas gun.

1. Introduction

The two-stage light gas gun is a launch device that uses compressed gas for high-speed impact. Compared with the traditional gun, the light gas gun has a higher initial velocity while receiving less stress. At present, the two-stage light gas gun has been widely used in aerospace, high-speed collision and other fields[1] .

According to the characteristics of gas guns, Zhao Junli et al.[2] gave a practical theory and technology of the internal ballistics of gas guns. Boganoff et al.[3] selected a new Godunov format to simulate the two-stage light gas gun, and compared with the measured results, and achieved good consistency. Huang Jie et al.[4] conducted a calculation and analysis on the gas flow and projectile piston models of the two-stage light gas gun, and optimized the firing performance of the two-stage light gas gun. Li Feng et al.[5] revised the classical equation of the internal ballistics of the first-stage gas gun, and pointed out that the volume of the gas chamber cannot be ignored. Xiong Hao et al.[6] analyzed the gas movement state in the inner ballistic bore and the influence of the valve on the inner ballistic. Su Hongxing et al.[7] analyzed the effects of three factors, namely the volume of the gas chamber, the initial pressure of the gas chamber, and the length of the launch tube.

At present, the numerical simulation related to the light gas gun is more for the first-level gas gun model, and the relevant simulation calculation for the second-level light gas gun is less. With the development of computational fluid dynamics, CFD-related software, dynamic mesh theory and 6DOF theory have achieved good results in ballistic calculation-related fields. In this paper, 6DOF and layered dynamic mesh technology are used to simulate the flow field in the two-stage light gas gun bore, and the influence of different factors on the projectile launch.

2. Numerical theoretical model

The gas situation in the chamber of the two-stage light gas gun is relatively complex. In order to facilitate the calculation, this paper makes the following assumptions for the research content:

1) The gas in the gas chamber and the pump tube is regarded as an ideal gas.

2) The air resistance before the bomb is ignored, and there is no air leakage during the movement of the piston.

3) Various energy losses in the process of piston motion are solved by introducing a virtual mass coefficient.

The gas flow in the chamber follows the following mathematical model[8] :

Mass conservation equation:

$$\frac{\partial\rho}{\partial t}+\frac{\partial(\rho u)}{\partial x}+\frac{\partial(\rho v)}{\partial y}=0 \tag{1}$$

t is time and density, u and v are components in x and y directions, respectively.

Momentum conservation equation:

$$\frac{\partial(\rho u_i)}{\partial t}+\frac{\partial(\rho u u_i)}{\partial x}+\frac{\partial(\rho v u_i)}{\partial y}=\frac{\partial}{\partial x}\left(\mu\frac{\partial u_i}{\partial x}\right)+\frac{\partial}{\partial y}\left(\mu\frac{\partial u_i}{\partial y}\right)-\frac{\partial p}{\partial x_i}+S_i \tag{2}$$

μ is the viscosity coefficient, p is the fluid pressure, u_i is the velocity in each direction of x and y, and s_i represents the direction of i, which are the x and y directions respectively.

$$\frac{\partial}{\partial t}(\rho E)+\frac{\partial}{\partial x_i}(u_i(\rho E+P))=\frac{\partial}{\partial x_i}\left(k\frac{\partial T}{\partial x_i}\right) \tag{3}$$

T is the temperature, k is the heat transfer coefficient of the fluid, and E is the total energy of the fluid. The Standard $k-\varepsilon$ Model is the most widely used model in the industry, the Realizable $k-\varepsilon$ model uses an improved method to calculate turbulent viscosity and provides better predictions for large pressure gradients, boundary layer properties in separation and recirculation flows, so Realizable $k-\varepsilon$ is chosen as the turbulence model[9] .

The turbulent energy transport equation is:

$$\frac{\partial}{\partial t}(\rho k)+\frac{\partial}{\partial x_j}(\rho k u_j)=\frac{\partial}{\partial x_j}\left[\left(\mu+\frac{\mu_t}{\sigma_k}\right)\frac{\partial k}{\partial x_j}\right]+G_k+G_b-\rho\varepsilon-Y_M+S_k \tag{4}$$

The turbulent energy dissipation transport equation is:

$$\frac{\partial}{\partial t}(\rho\varepsilon)+\frac{\partial}{\partial x_j}(\rho\varepsilon u_j)=\frac{\partial}{\partial x_j}\left[\left(\mu+\frac{\mu_t}{\sigma_\varepsilon}\right)\frac{\partial\varepsilon}{\partial x_j}\right]+\rho C_1 S\varepsilon-\rho C_2\frac{\varepsilon^2}{k+\sqrt{v\varepsilon}}+C_{1\varepsilon}\frac{\varepsilon}{k}C_{3\varepsilon}G_b+S_\varepsilon \tag{5}$$

$$C_1=\max\left[0.43,\frac{\eta}{\eta+5}\right] \tag{6}$$

$$\eta=S\frac{k}{\varepsilon} \tag{7}$$

$$S=\sqrt{2S_{ij}S_{ij}} \tag{8}$$

μ_t is the hydrodynamic viscosity, G_k is the turbulent kinetic energy due to the mean velocity gradient, G_b is the turbulent kinetic energy due to buoyancy, is the contribution of the pulsating expansion in compressible turbulence to the overall dissipation rate, and C_2 , $C_{1\varepsilon}$ are constants, and is the turbulent kinetic energy and the Prandtl number corresponding to the turbulent kinetic energy.

3. Initial conditions

The initial parameters of the air chamber of the two-stage light gas gun are shown in Table 1. According to the parameters in the table, the initial motion law of the piston in the pump tube is calculated according to the internal ballistic correlation equation in the literature[10] . The light gas in the pump tube is helium, the pressure is set to 0.12MPa through the patch, and the temperature is 300k. Due to the rapid change of the gas state in the light gas gun, the time step is selected as 10^{-6}s according to multiple calculations and analyses.

Table 1. Initial parameters.

Air chamber pressure (MPa)	Air chamber volume (m^3)	Piston mass (kg)	Pump tube diameter (mm)
10	0.12	20	180

Using the user-defined function (UDF), the above theoretical motion law is assigned to the piston through the DEFINE_CG_MOTION function. The time curve of piston movement speed is shown in Figure 1.

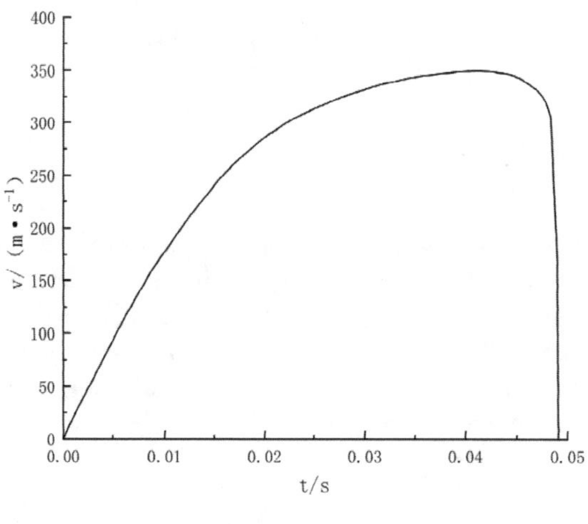

Figure 1. Velocity-time curve of piston.

The dynamic mesh in Fluent uses technologies such as mesh deformation and reconstruction to simulate the deformation and motion of the fluid boundary. The dynamic mesh area is the rightmost projectile motion area. In this paper, the layering algorithm is used, the splitting factor is selected as 0.4, and the merge factor is selected as 0.2. In order to simplify the model and improve the calculation efficiency, this paper adopts the two-dimensional model for simulation calculation.

The launch velocity of the projectile is determined by the gas pressure load it is subjected to, and the conditions of the internal fluid domain change with the movement of the projectile. The projectile mass is defined as 0.1kg, and its motion law is calculated by the 6DOF method, and the displacement of the projectile boundary is calculated at each time step to realize the solution of the projectile motion. The positions of the mesh nodes are also continuously updated as the motion of the mesh changes.

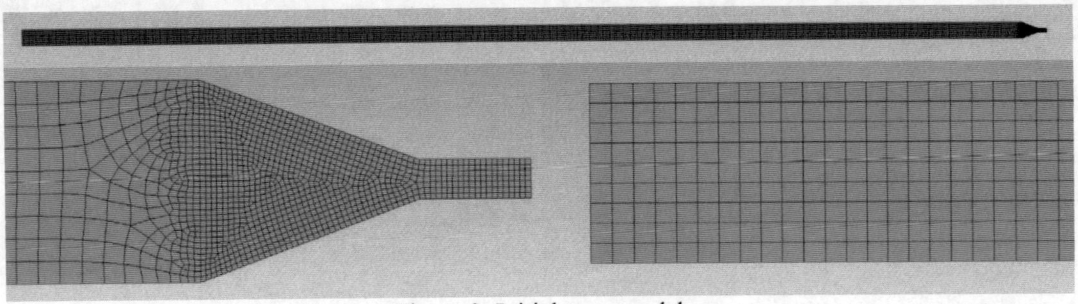

Figure 2. Initial state model.

4. Analysis of calculation results

4.1. Analysis of flow field in bore during launch

Figures 3 and 4 show the flow field in a 35mm two-stage light gas gun from the piston in the pump tube to the flow field in the launch tube bore. After the piston starts to move for 17.5ms, the airflow in the pump tube gathers at the bottom of the bomb and starts to return to mix with the gas at the bottom of the piston to form a complex state. During the process of compressing helium by the piston, the gas backflow in the pump tube constantly changes periodically, reaching a maximum pressure of 30.2MPa before the projectile is launched.

The gas density, pressure and velocity on the right side of the piston increase continuously with the increase of the piston speed. As the piston moves to the right, the helium gas in the pump tube is further squeezed at the cone section, and the pressure on the bottom of the bomb keeps rising. Therefore, the design of the sealing connection at this place should be considered to avoid damage to the cone segment structure due to excessive pressure.

(a) t=17.5ms

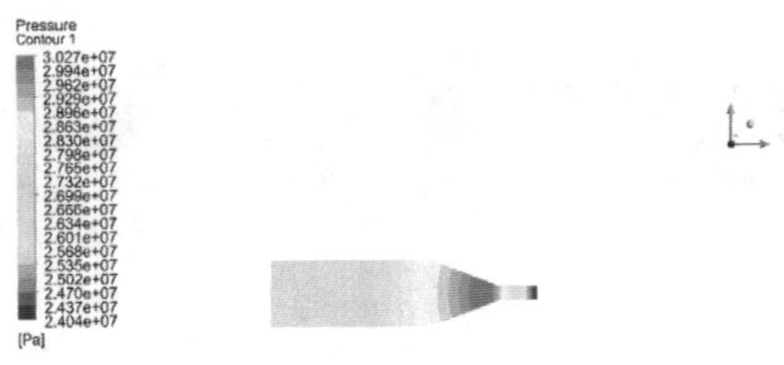

(b) t=45.1ms

(c) t=46.8ms

Figure 3. The pressure in the chamber at different times.

The low pressure of the projectile in the launch tube rises rapidly, and the projectile is pushed by the high-pressure helium gas and accelerates forward rapidly. It can be seen from Figure 4 that the gas velocity in the launch tube is basically linearly distributed.

Journal of Physics: Conference Series **2365** (2022) 012024 doi:10.1088/1742-6596/2365/1/012024

(a) t=17.5ms

(b) t=45.1ms

(c) t=46.8ms

Figure 4. Gas velocity in the chamber at different times.

The piston continuously squeezes the air column in the pump tube, the gas pressure in the pump tube increases rapidly, and at the same time, the gas density and flow rate are also continuously increased. The cloud diagrams at different times show that the gas pressure near the piston and the bottom of the bullet is larger relative to the middle part during the compression process, while the gas velocity is larger in the part near the piston.

4.2. Velocity of projectile

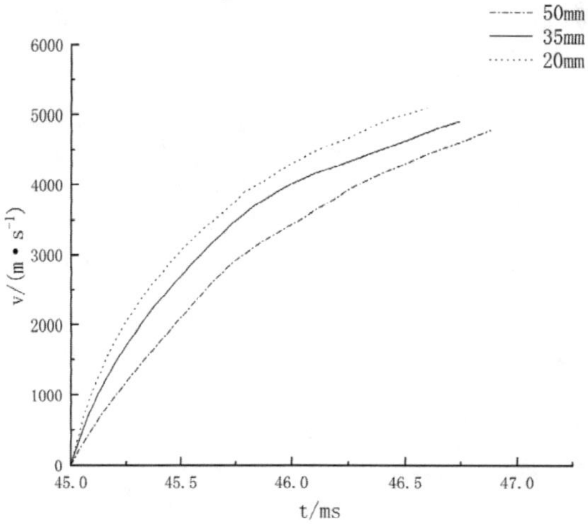

Figure 5. Projectile velocity curve of different caliber launch tubes.

The final projectile exit velocity time curve of the three kinds of launch tube calibers is shown in the figure, the launch tube is the 20mm caliber projectile with the smallest velocity, and the 50mm caliber projectile with the highest initial velocity. The speed increase trend in the three states is basically the same. The speed increases rapidly within 0.8ms after the launch, and then slows down until the projectile exits the chamber. It can be seen from the figure that the projectile exit speeds corresponding to the launch tube diameters of 20mm, 35mm, and 50mm are 5107m/s, 4906m/s, and 4776m/s, respectively.

4.3. Pressure on the piston

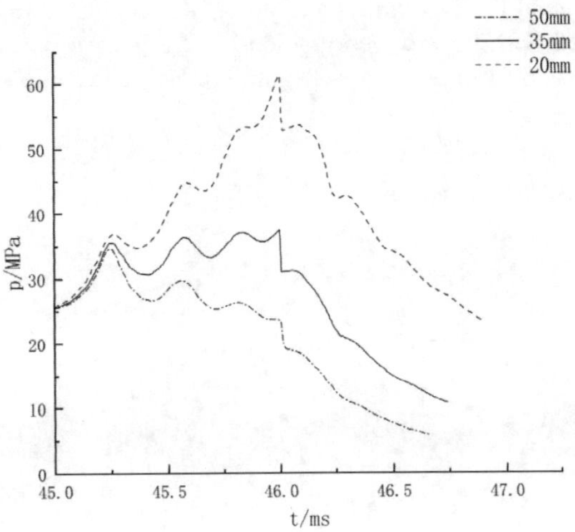

Figure 6. Force curve of piston of different launch tube.

After the projectile was launched, the pressure on the end face of the piston of the 20mm launch tube continued to rise. After the projectile moved, the pressure in the air chamber still appeared several peaks, and it began to slowly decrease after the pressure reached the maximum value of 61.3MPa. After the projectile is launched, the gas in the pump tube still exerts a large force on the piston after the reflection and superposition of the high-pressure cone segment. The acceleration effect of the 50mm launch tube gas on the projectile is more obvious.

5. Conclusions

In the process of being compressed by the piston, the gas in the pump tube will reciprocate to form periodic shock waves, and finally form a complex flow state. When the difference between the mass of the piston and the mass of the projectile is relatively large, appropriately increasing the diameter of the launch tube will help the gas in the gas chamber to act on the bottom of the projectile faster, thereby increasing the initial speed of the projectile.

References

[1] WANG D F, XIAO W K, PANG B J. A brief introduction on NASA's two stage light gas guns[J]. *Journal of Experiments in Fluid Mechanics*, 2014,28(04):99-104.

[2] ZHAO J L, GAO Y F. Study on the interior ballistics of the gas gun[J]. *Journal of Taiyuan University of Technology*, 2003, 34(3):3.

[3] Bogdanoff D W, Miller R J. New higher-order Godunov code for modelling performance of two-stage light gas guns[R]. 1995.

[4] HUANG J, LIANG S C, LI H Y, et al. Numerical research on interior ballistics of the launch process of two-stage light gas gun[J]. *Acta Aerodynamica Sinica*,2013,31(5): 657-661.

[5] LI F, BAI Y S, FENG X W. Modification and validation of interior ballistic equation for one-stage gas gun[J]. *Journal of Ballistics*, 2016, 28(1):5.

[6] XIONG H, WANG H Y, WANG H X. Research on internal Ballistics of single-stage gas gun based on fluent[J]. *Journal of Ordnance Equipment Engineering*, 2019, 040(003):77-81.

[7] SU H X, ZHAO J L, PENG S Z. Analysis of the influencing factors of the initial velocity of gas gun projectile[J]. *Journal of Projectiles, Rockets, Missiles and Guidance*, 2018, 38(1):5.
[8] WANG F J. Computational fluid dynamics analysis[M]. Tsinghua University Press, 2004.
[9] ANSYS Inc. ANSYS FLUENT: 14.0 theory guide[Z]. PA, Canonsburg,2011.
[10] WANG J G. Principle and technology of gas gun[M]. Beijing: National Defense Industry Press, 2001.

Dynamic modeling and simulation analysis of steel wire winding of reinforcement cage roll welder

Pengliang Huang [1,2], Dongmin Zhang[1,2], Yincang Xu [1,2]

[1]Shanghai Institute of Technology, College of Mechanical Engineering, Shanghai 201418, China

[2]School of Mechanical Engineering, Shanghai Engineering Research Center of Physical Vapor Deposition (PVD) Super Hard Coating and Equipment, Shanghai Institute of Technology, Shanghai 201418, China

charles_huang2020@foxmail.com

Abstract. In view of the problem that the spiral reinforcement material plate is out of sync due to the rotation speed during the work of the reinforcement cage roll welding machine, the spiral reinforcement is broken or loose. By discretizing the steel wire into several micro segments, the dynamic model of steel wire winding based on ADAMS virtual prototype technology is established. By analyzing the motion relationship between the material disc and the flower disc in the roll welding machine, the functional relationship between the material disc speed and time is fitted, so as to realize the dynamic simulation of steel wire winding. By analyzing the stress condition of steel wire micro segments, the accuracy of the material disc fitting speed is verified, At the same time, it provides a basis for the automation transformation and detection control scheme of roll welding machine.

1. Introduction

Prestressed concrete pipe piles are manufactured through factory production lines and widely used in residential, industrial buildings, road traffic, port terminals, super high-rise buildings, large-scale complex and other fields with reliable quality and reasonable cost. As the frame of pipe pile forming, the quality and efficiency of steel cage production and processing are particularly critical[1][2],At present, the enterprise steel bar rolling welder rotates through the disc, at the same time, the material disc rotates together to realize the spiral rolling weld of auxiliary reinforcement to the main reinforcement.

During the process of roll welding, experienced workers adjust the speed of disc in real time. With the speed of disc, the speed of disc is too fast or too slow, which results in wire breakage or looseness. It needs to be stopped for treatment, which directly affects the production efficiency and has potential safety hazards. The enterprise proposes to develop new reinforcing cage rolling welding equipment, which can realize the safe production of reinforcing cage rolling welding by automatically adjusting the rotating speed of disc. In this paper, a dynamic model of wire winding is established based on ADAMS software, which provides basis for new material disc speed control mode and equipment development[3][4].

Modeling and solving are the core problems of multi-body system dynamics. At present, there are three kinds of dynamic modeling methods for steel wire, which are based on MNF, Polyline and Bushing. Li Junwen and others[5] successfully established the coupling system model of impact drilling rig wire rope and impact drilling tool by using the macro command of ADAMS and Bushing connection method. Escalona et al.[6] proposed a discrete method for flexible multi-body modeling and established winch system by using the discrete program to simulate the span deformation of variable length wire rope. Du Zi et al.[7] In the process of bridge tower inspection and virtual prototype establishment, the wire rope system is connected and simulated by Bushing, which has certain reference meaning for the rigid-flexible coupling system with wire rope model.

Comparing the current methods of establishing dynamic model of steel wire, based on Bushing's discrete modeling method, the flexible connection between micro-segments is realized by separating steel wire into several micro-segments and adding Bushing between micro-segments, which can truly reflect the movement characteristics and winding process of steel wire. In this paper, the discrete modeling of steel wire is carried out, the dynamic model of wire winding is established, the dynamic simulation analysis is carried out on the process of wire disc collecting and disc loosening, the stress condition of steel wire during the process of rolling welding is analyzed, and the matching speed calculated theoretically is verified.

2. Establishment of dynamic model of wire winding

2.1. Ideas for establishing steel wire model

The core of dynamic model of wire winding in this paper is the establishment of wire model. By establishing the geometric models of material disc, disc and main reinforcement support disc in UG, the wire model is set up by array of micro-section cylinders, and then in ADAMS, adding Bushing between micro-sections of wire through macro command to realize the flexible connection between micro-sections of steel wire. During the winding process of steel wire micro-section, it is necessary to collide and friction with other components. The dynamic analysis of steel wire winding system requires the addition of contact force.

2.2. Establishment of geometric model of wire winding system

Bushing-based discretization method is used for simulation of wire winding, and a discrete wire model needs to be established in UG. The blank diameter of disc is 830mm and the diameter of wire is 6mm. The smaller the length of wire segment, the more accurate the simulation result is, but it will lead to long simulation time. Considering comprehensively, a circle of steel wire is discretized into 200 micro-segments, and the length of steel wire micro-segments is calculated as 26mm. At the same time, in order to improve the simulation efficiency, only the disc, disc and support are retained, and the number of winding rings of steel wire on the disc is reduced. The winding model of steel wire is shown in Figure 1.

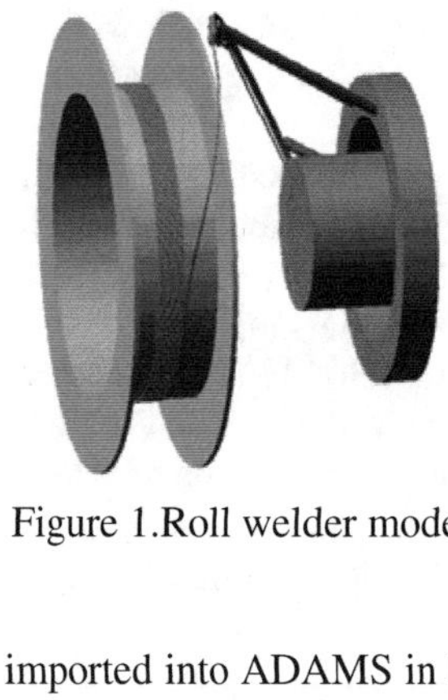

Figure 1.Roll welder model

2.3. Parameter setting in ADAMS

The completed model built in UG is imported into ADAMS in Parasolid format. First, gravity and the quality attributes of each component are added to the ADAMS.

2.3.1. Parameter solution and addition of bushing

The flexible connection between steel wire micro-segments is realized by Bushing, which is an interaction force applied to two members and is determined by parameters such as stiffness coefficient and damping coefficient. Bushing's stiffness coefficients can be calculated as in equation(1-4):

$$K_1 = \frac{EA}{L} \tag{1}$$

$$K_2 = K_3 = \frac{GA}{L} \tag{2}$$

$$K_4 = \frac{G\pi d^4}{32L} \tag{3}$$

$$K_5 = K_6 = \frac{EI}{L} \tag{4}$$

Formula: K_1 is the coefficient of tensile stiffness. K_2 and K_3 are shear stiffness coefficients. K_4 is the torsional stiffness coefficient. K_5 and K_6 are bending stiffness coefficients. E is the elastic modulus of the wire rope. A is the cross-sectional area of the wire rope; G is the shear modulus of the wire rope; L is the length of the slice of the wire rope; I is the moment of inertia of the wire rope micro section to the neutral axis; D is the diameter of the steel wire. The stiffness coefficients are calculated as shown in Table 1.

Table 1. Various stiffness coefficients of bushing.

Tensile stiffness coefficient	Shear stiffness coefficient	Torsional stiffness coefficient	Bending stiffness coefficient
2.175×10^5 N/mm	0.870×10^5 N/mm	0.682×10^5 N·mm/deg	0.086×10^5 N·mm/deg

In Bushing, there is no definite formula for various damping coefficients. Generally, the damping coefficient C is between 1 and 10. In order to reflect the true characteristics of steel wire, the tensile and shear damping coefficients are selected as 1 and the torsional and bending damping coefficients are 10. In ADAMS, Bushing[8] is added between adjacent steel wire segments by writing macro program.

2.3.2. Solving and adding parameters of contact force

The main parameters of contact force are stiffness contact coefficient K, damping coefficient C, non-linear impact index E and maximum penetration depth dmax. The stiffness coefficient is determined by the shock function method as in equation(5-6):

$$K = \frac{4}{3}\left(\frac{R_1 R_2}{R_1 + R_2}\right)^{\frac{1}{2}} E^* \tag{5}$$

$$\frac{1}{E^*} = \frac{1-\mu_1^2}{E_1} + \frac{1-\mu_2^2}{E_2} \tag{6}$$

Formula: R_1 is the radius of steel wire, R_2 is the radius of material tray, E_1 is the elastic modulus of steel wire, E_2 is the elastic modulus of material tray, μ_1 is the Poisson ratio of steel wire and μ_2 is the material tray Poisson ratio.

Calculated contact force stiffness coefficient K=1.626×105 N/mm1.5 between steel wire micro-section and disc.

The damping coefficient in the contact force is set to 10N.s/mm according to the empirical formula.

The non-linear impact index e is selected as e=1.5 by combining material properties. The maximum penetration depth is dmax=0.1mm.

The contact force parameters between the steel wire micro-section and the disc guide hole and the support are the same as those calculated above. Add contact force by writing macro commands.

2.3.3. Adding sports pairs

For the dynamic model of wire winding in the roll welder, the kinematic pair relationship between the components is added according to the actual motion as shown in Table 2.

Table 2. Kinematic pair relationship between components.

Component	Joint type	remarks
Flower Disk	Rotating pair	
Material tray	Rotating pair	With the ground
Main reinforcement support plate	Fixed pair	

2.3.4 Add motion

When the reinforcement cage roller welder works, the flower disc rotates at a constant angular speed to loosen the material, and the steel wire is wound on the main reinforcement of the reinforcement cage for spot welding; The steel wire tray rotates in the same direction for receiving,and its angular velocity changes with the change of steel wire diameter on the tray. In this paper Taking Φ500 specification reinforcement cage as an example, the quadratic curve is used for fitting, and the continuous functional relationship between the angular velocity of the material tray and time is obtained, as in equation(7).

$$\omega = -1.14\times10^{-5}t^2 - 2.55\times10^{-2}t + 378.00 \tag{7}$$

According to the establishment method of the above steel wire winding dynamic model, the models with disc diameters of 936mm, 971.2mm and 1096mm are established respectively, and the angular velocity is set in ADAMS software. In order to prevent the sudden change of steel wire stress, the step function is used to set the angular velocity of disc and flower disc respectively, as shown in Table 3.

Table 3. Angular speed setting of material tray and flower tray.

Component name	angular velocity (deg/s)
Model_A	STEP(time,0,0d,2,(-0.0000114× (time+24)^2-0.00255*(time+24)+378))
Model_B	STEP(time,0,0d,2,(-0.0000114× (time+811)^2-0.00255*(time+811)+378))
Model_C	STEP(time,0,0d,2,(-0.0000114× (time+1561)^2-0.00255*(time+1561)+378))
Flower Disk	STEP(time,0,0d,2,600d)

3. Dynamic simulation and result analysis of steel wire winding

Before the simulation calculation, the composition of the wire winding dynamic model,the degree of freedom and constraints of the system are tested[9]; Set the simulation time as 10s and the simulation step size as 10000, simulate three steel wire winding systems respectively, and observe the winding state of steel wire at the beginning, middle and end for comparative observation. As shown in Figures 2, 3 and 4, there is no loose falling off during the wire winding process, which verifies the accuracy of the fitting speed.

(a) 0 second (b) The 5th second (c) The 10th second

Figure 2. Simulation results of Model_A.

(d) 0 second (e) The 5th second (f) The 10th second

Figure 3. Simulation results of Model_B.

(g) 0 second (h) The 5th second (i) The 10th second

Figure 4. Simulation results of Model_C.

Select Model_C for post-processing analysis. The system discretizes 1000 steel wire segments, and select part10, part350, part700 and part1000 steel wire segments for stress analysis. As shown in Figure 5, the force variation trend of the steel wire micro segment conforms to the motion law, and the force is far less than the allowable tensile force of the steel wire,which verifies the accuracy of the fitting speed.

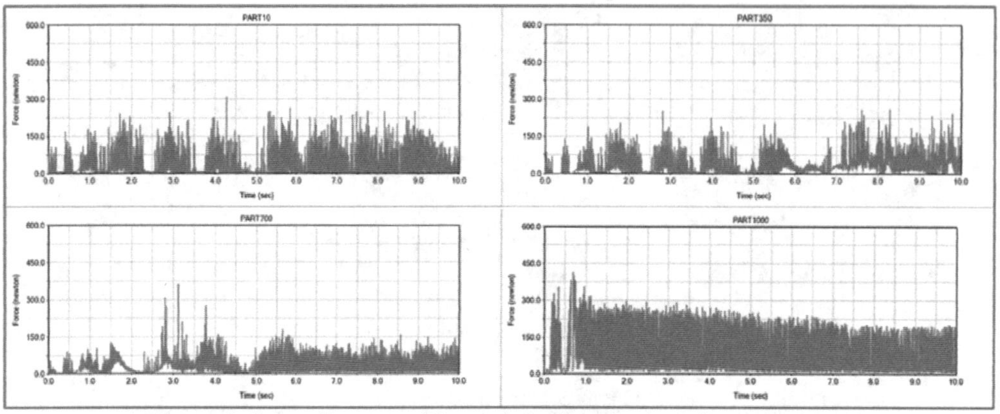
Figure 5. Stress diagram of four selected steel wire micro segments.

4. Conclusion

(1) Through the secondary development function of ADAMS, the dynamic model of steel wire winding is established, and the simulation and motion analysis of steel wire winding process are realized;

(2) Through the analysis of the motion relationship between the disc and the flower disc, the theoretical matching speed is calculated, and the functional relationship between the disc speed and time is obtained by quadratic curve fitting;

(3) The fitting speed of the material tray is simulated and verified in ADAMS, which verifies the accuracy of the fitting speed, and provides a strong basis for the automatic transformation and detection control scheme of the roll welding machine.

References

[1] Li Liang, Zhong Zhenhong, Zhou Zhicheng, Yang Baopeng Introduction to mechanical properties and application of ultra high performance concrete UHPC [J] Concrete world, 2018 (09): 56-62

[2] Zhao Ruijun Application of innovative technology in pipe pile roll welder [J] Concrete and cement products, 2014 (10): 76-78

[3] Wang Kun, Xing Haijun, Xu Mengchao, Zhang Linhao Simplified modeling and Simulation of multi rigid body dynamics based on ADAMS [J] Journal of graphics, 2019,40 (04): 733-738

[4] Yandong Wang,Guohua Cao,Wim T. van Horssen. Dynamic simulation of a multi-cable driven parallel suspension platform with slack cables[J]. Mechanism and Machine Theory,2018,126:

[5] Li Junwen, bu Changgen, Wang long Application of Adams macro command in virtual prototype modeling of wire rope percussion drill [J] Machine tools and hydraulics, 2011,39 (23): 150-153

[6] José L. Escalona,Grzegorz Orzechowski,Aki M. Mikkola. Flexible multibody modeling of reeving systems including transverse vibrations[J]. Multibody System Dynamics,2018,44(2):

[7] Du Zixue, Liu Mingdong Modeling of steel wire rope system of bridge tower detector based on ADAMS macro program [J] Mechanical design and research, 2013,29 (06): 140-143

[8] Lefan Wang. Modeling and Analysis of Steel Wire Rope Based on ADAMS Macro Command[J]. International Core Journal of Engineering,2019,5(11):

[9] Li Aiping, Jiang Chaoping, Liu Xuemei Simulation analysis of dynamic tension of steel wire rope based on ADAMS [J] Modern manufacturing engineering, 2010 (01): 43-46

Assembly and dynamic operation simulation research of tooth-locked-output harmonic gear reducer

1stBin Ju[1,a], 2stYuanxun Fan[1,*]

[1]School of Mechanical Engineering,Nanjing University of Science and Technology, Nanjing, Jiangsu, China

E-mail: [a]1767718676@qq.com; [*]fanyx@njust.edu.cn

Abstract. Aiming at the special structure of the tooth-locked-output harmonic gear reducer, a simulation analysis method of ring shape flexspline based on ABAQUS is proposed in this paper.In the process of simulating assembly, this paper proposes a new method of “interference assembly first and then interference elimination”;In the process of simulating dynamic operation, this paper uses the “Continuum Distributing” coupling function to solve the difficult problem of setting the boundary conditions of the ring shape flexspline, and completes the data transfer between the two analysis processes.The research shows that the method proposed in this paper effectively reduces the difficulty of the ring shape flexspline simulation analysis, and reduces the cost and time for subsequent research.

1. Introduction

Harmonic gear reducer is a new type of high-precision transmission device that meshes with rigid gears through the controllable elastic deformation of flexible gears. Because of its outstanding advantages such as high precision, small size, high efficiency, large transmission ratio, strong carrying capacity, and the ability to transmit motion and power to confined spaces, it is widely used in industrial robots, communication facilities, new energy, aerospace and other fields.

In recent years, with the advancement of computer technology and the maturity of finite element analysis theory, the research of harmonic reducer based on finite element analysis has developed in many aspects and in depth[1-4].In this paper, the research on the ring shape flexspline harmonic reducer with less research but obvious advantages is carried out, and a relatively complete analysis process is established to solve the problems of pre-simulation processing and solution setting caused by the special structure of the ring shape flexspline. An example is demonstrated, and the rationality and correctness of the simulation analysis process in this paper are verified by experiments.

2. Finite element modeling and solution

2.1 Geometric model and parameters

The tooth-locked-output is the most commonly used one among all the connection methods of the harmonic reducer[5]. Its structure is shown in the figure 1, consisting of a wave generator (1), a fixed circular spline (2),an output circular spline(3) and a ring shape flexspline (4).Under the action of the wave generator, there is a relative motion with a transmission ratio of $i=1:1$ between the output circular spline and the ring shape flexspline.The ring shape flexspline is a thin-walled external gear ring, and its structure is shown in the figure 2.where δ is the wall thickness of cylinder, d is the inner diameter, d_a is the addendum circle diameter, and d_f is the dedendum circle diameter.

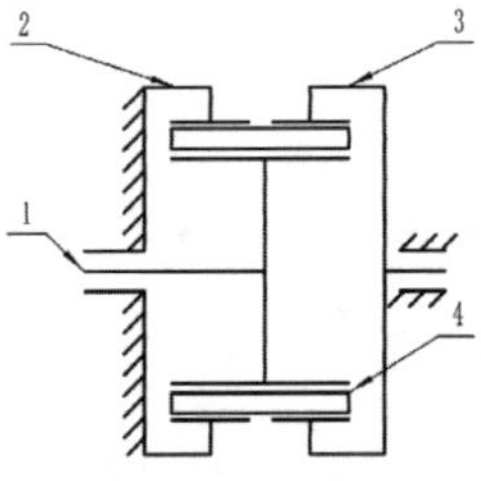

Figure 1. Schematic diagram of tooth-locked-output

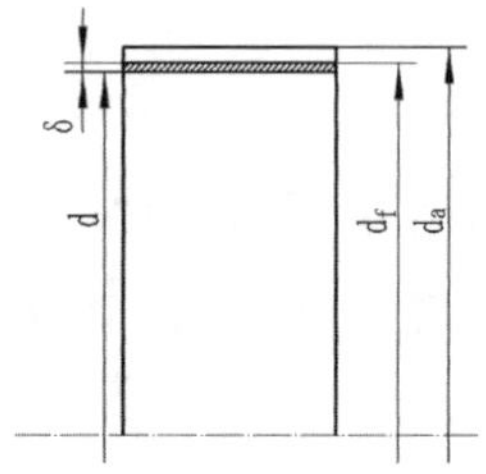

Figure 2. Schematic diagram of ring shape flexspline

The cam in the wave generator adopts a standard elliptical type, with a long axis radius of 40.47mm and a short axis radius of 39.52mm.Its material is 45 steel, the density is $7.89\times10^{-9}t/mm^3$, the elastic modulus is $2.10\times10^5 MPa$, and the Poisson's ratio is 0.269.The parameters and properties of other parts are shown in the table 1.

Table 1.Profile parameters and material properties

	Addendum circle radius/mm	Dedendum circle radius /mm	Facewidth /mm	Density/ t/mm^3	Elastic modulus /MPa	Poisson ratio
Flexspline	41.43	40.55	24	7.87×10^{-9}	2.09×10^5	0.295
Fixed circular spline	41.97	41.09	15	7.82×10^{-9}	2.11×10^5	0.277
Output circular spline	41.97	41.09	10	7.82×10^{-9}	2.11×10^5	0.277

2.2 Model simplification and meshing

Because the rigidity of the wave generator is larger than that of the flexspline, it is simplified as an elliptical ring rigid body, and its model and mesh division are shown in figure 3(a). As an important part of our concern, the flexspline needs to be divided into precise meshes to output its calculation results as accurately as possible. The model and mesh division are shown in figure 3(b).

For the two circular splines, we need to refine the mesh at the gear teeth for the same reason as the flexspline. At the same time, we need to reduce the mesh density on the outer ring of the circular spline to reduce the computational cost.For this reason, this paper divides the circular spline into two parts with the dedendum circle as the boundary through the command of "Partition".The gear teeth part adopts the "Medial axis" to divide the mesh more uniformly; the outer ring part adopts the "Advancing front" to achieve the transition from dense to sparse mesh density.After inspection, this method effectively reduces the total number of meshes and can ensure good mesh quality.The final result is shown in figure 3(c) and 3(d).

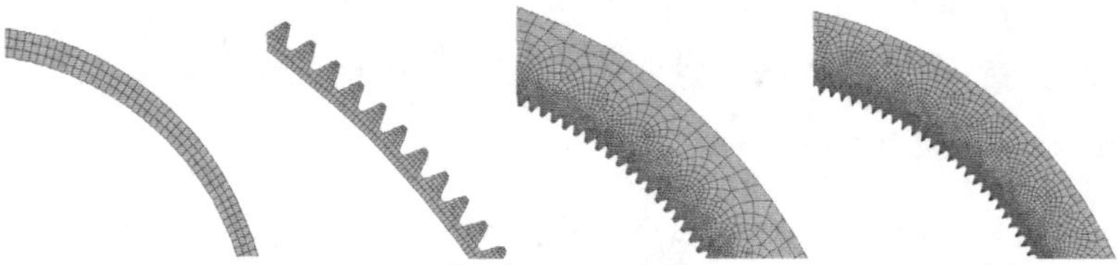

(a) (b) (c) (d)

Figure 3. Meshing of parts

2.3 Simulation Analysis Process

In this paper, a detailed analysis process is formulated for the ring shape flexspline, as shown in the figure 4.In the pre-processing stage, first modeling, then assigning materials, and finally dividing the mesh.This not only prevents loss of material properties, but also prevents boundary conditions from failing.In addition, assembly simulation will be elaborated in Chapter 2, and dynamic operation simulation will be elaborated in Chapter 3.

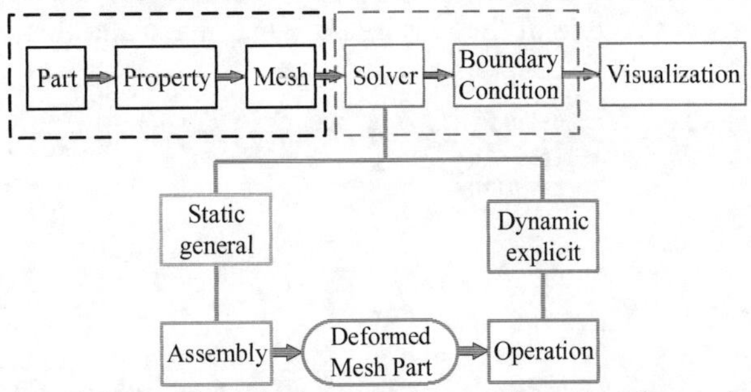

Figure 4.Flowchart of the tooth-locked-output harmonic gear reducer simulation analysis

3. Simulation analysis of assembly process

The assembly process mainly refers to the process that the ring shape flexspline changes from a perfect circle to an ellipse under the action of the elliptical cam wave generator.The simulation of this process is helpful to observe the stress distribution law of the ring shape flexspline, and to verify the radial deflection of the ring shape flexspline and the meshing and separation of the rigid and flexible gear teeth.

3.1 Assembly process solution settings

The process uses the "static, general" analysis step.In the contact setting, "surface-to-surface contact" is selected between the outer surface of the wave generator and the inner wall surface of the flexspline, and the complex contact between the rigid-flexible teeth can be handled by "General Contact", which can achieve better results after inspection.

In the process of simulating assembly, scholars usually divide the simplified wave generator into two ways to avoid the initial interference. This paper proposes a new method of "interference assembly first and then interference elimination".This method refers to importing into ABAQUS on the premise that the major axis of the elliptical cam and the flexspline interfere with the assembly;Then, in the "surface-to-surface contact" option, through the "Interference Fit", the interference is gradually eliminated to simulate the process of the flexspline being stretched into an ellipse.This method ensures the integrity of the wave generator, and the model is simpler;the contact surface is complete, and the setting of contact properties is more convenient;"Interference Fit" is to adjust the slave surface in the contact pair, the contact has less influence on the convergence, and the calculation cost and time cost are low.

3.2 Stress distribution law

The stress and distribution of the ring shape flexspline after assembly are shown in the figure 5. The position of the flexspline with large stress is concentrated in the major axis position and the minor axis position, and the maximum stress value is about 47.84MPa.

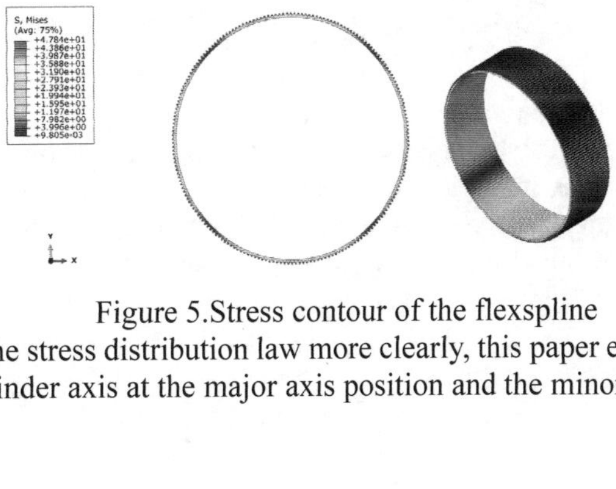

Figure 5.Stress contour of the flexspline

In order to express the stress distribution law more clearly, this paper extracts the stress values of the flexspline along the cylinder axis at the major axis position and the minor axis position respectively, as shown in figure 6.It can be seen that the stress fluctuation of the ring shape flexspline along the axial direction is very small.This is due to the ring shape structure of the flexspline, there is no cup opening angle problem of the cup-shaped or hat-shaped flexspline, and the stress value will not fluctuate greatly in the axial direction.This is also one of the reasons why the ring shape flexspline can have a longer fatigue life.

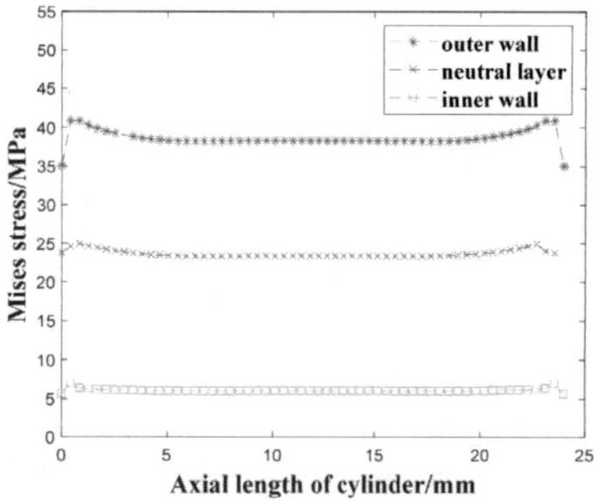

(a)The major axis position

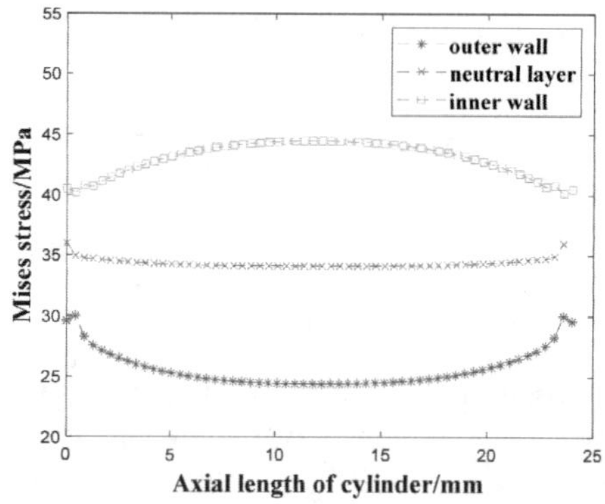

(b)The minor axis position

Figure 6. Stress distribution law

3.3 Radial deflection and meshing

The radial deflection refers to the deformation generated in the radial direction at any point on the neutral layer of flexspline's toothed ring under the action of the wave generator.Its calculation formula is $\omega = \omega^* m$, where ω^* is the radial deflection factor, which is taken as 1.156 in this paper; m is the modulus, which is taken as 0.41mm in this paper.

Extract the radial deflection of the ring shape flexspline at the the major axis and the minor axis position, as shown in figure 7.The radial deflection at the major axis is about 0.4735mm, which is similar to the theoretical calculation result; the radial deflection at the minor axis is slightly larger, about 0.4753mm.This result is basically consistent with the data of the major and minor axes of the ellipse.

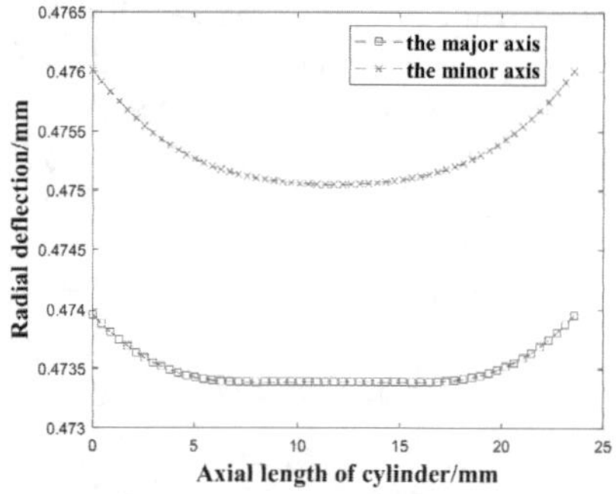

Figure 7. Radial deflection

After the assembly process is completed, the inner wall of the flexspline fits the outer contour of the wave generator, and the flexspline is stretched to become an ellipse, as shown in figure 8.At the same time, the meshing condition of the gear teeth can be observed: the meshing can be completed in the direction of the major axis position, and the meshing can be completely separated in the direction of the minor axis.In the transitional position of the two, there is a relatively uniform transition from meshing to separation, and no interference phenomenon is found.

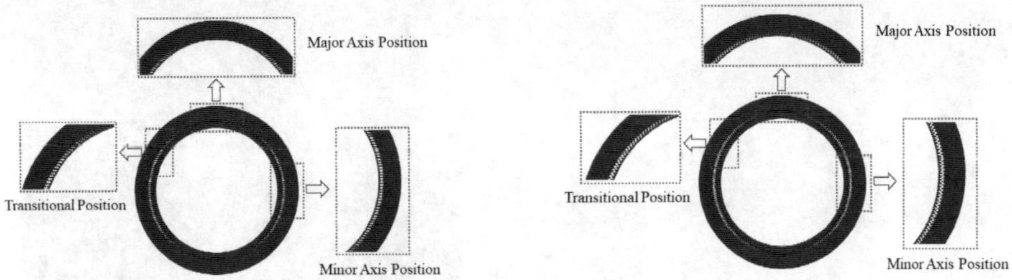

(a)Fixed circular spline side (b)Output circular spline side

Figure 8.Meshing after assembly

4. Simulation analysis of dynamic operation process

Avoiding gear teeth interference is a rule that must be followed in the design of harmonic reducers, and transmission accuracy is an important evaluation index, both of which need to be detected during dynamic operation.Therefore, dynamic operation analysis is necessary.

4.1 Dynamic operation solution settings

Because the contact is a strong nonlinear factor, it is very prone to non-convergence to continue the analysis using the "Static, General" analysis step. The "Dynamic, Explicit" analysis step, which is more suitable for contact analysis, cannot be used after the above analysis step.In order to solve this problem, this paper imports the deformed flexspline as a deformed mesh component, and continues the analysis in the new model, completing the data transfer from implicit to explicit analysis.

Unlike the flexspline with its own output section, the boundary conditions of the ring shape flexspline have nowhere to apply, which is the difficulty of simulating it.This paper addresses this dilemma with the "Continuum Distributing" coupling function in ABAQUS. A "Coupling" constraint constrains a constraint control point to the motion of a face.Among them, "Continuum Distributing" coupling can not only control the three degrees of freedom of the ring shape flexspline, UR1, UR2 and UR3, which is enough to realize the rotary motion of it around the axis; it can also allow the deformation of the coupling surface, which is consistent with the actual use of the ring shape flexspline.

4.2 Result analysis and experimental verification

The level of transmission accuracy is measured by the size of the transmission error. The smaller the transmission error, the higher the precision.The transmission error refers to the difference between the actual rotation angle and the theoretical rotation angle of the output shaft when the input shaft of the harmonic reducer rotates in one direction under the working state.Its calculation formula is $\Delta\theta=\theta_a-\theta_t$.

In the formula, θ_a is the actual output angle; θ_t is the theoretical output angle, $\theta_t=\frac{\theta_I}{i}$ and θ_I is the input angle, and the transmission ratio i=100.

The transmission accuracy of the harmonic reducer is generally measured by the maximum value $\Delta\theta_{max}$ of the transmission error within the range of one rotation of the output shaft.In this paper, ABAQUS is used to simulate the one-way rotation of the harmonic reducer to measure the one-way transmission error.Extract the information of the rotational displacement UR3 of the reference point of the output circular spline, as shown in figure 9.

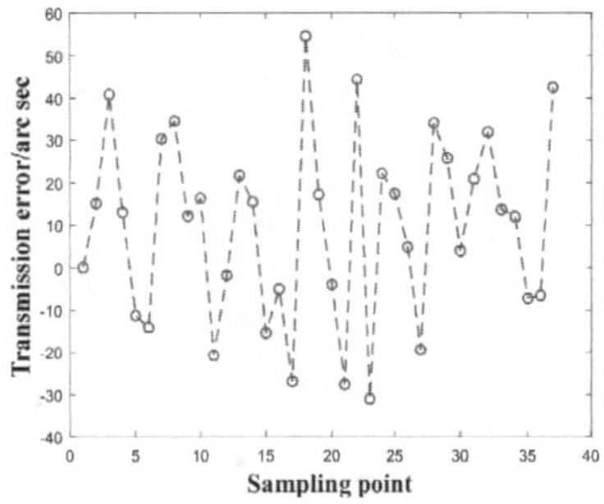

Figure 9 One-way transmission error simulation results

Figure 10 Transmission accuracy test bench

In order to verify the reliability of the simulation results, the transmission accuracy test of the locked-output harmonic gear reducer prototype under no-load conditions is carried out in this paper, as shown in figure 10. The experimental results are shown in Figure 11. The maximum transmission error is 60 arc seconds, which is close to the simulation result of 54.6 arc seconds.In order to more intuitively judge the closeness of the simulation results and the experimental results, the transmission error distribution diagram as shown in Figure 12 is drawn. The transmission error results of both simulation and experiment are concentrated below 30 arc seconds, and there are few results higher than 45 arc seconds.It can be seen that the distribution laws of the two are relatively close, and there is no tooth profile interference in the dynamic operation process animation, indicating that the simulation method and results in this paper are reliable to a certain extent.

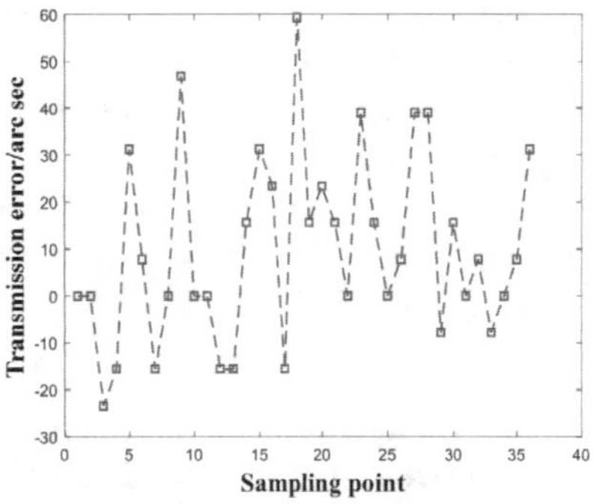

Figure 11 One-way transmission error experimental results

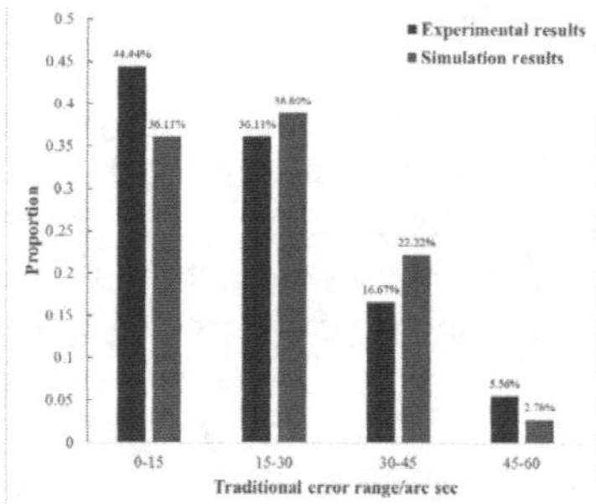

Figure 12 Comparison of transmission error distribution

5. Conclusion

This paper establishes the basic analysis flow of the finite element analysis of ring shape flexspline assembly and dynamic operation based on ABAQUS, and conducts a reasonable model simplification. Through the comparison of the finite element simulation results and experimental data, it is proved that the simulation analysis process of the ring shape flexspline in this paper is reasonable. Therefore, the verification of some performance parameters of the harmonic reducer can be completed by using simulation instead of experiment within a certain error range, which can reduce costs and improve efficiency.

References

[1] Li FJ, Li XP,Guo YJ,Shang DY,Analysis of Contact Mechanical Characteristics of Flexible Parts in Harmonic Gear Reducer, Shock and Vibration, vol. 2021, 17 pages, 2021.

[2] Zhu JX; Wang JX,;Zhou QH; Zhou BL,Finite element analysis on influence of surface treatments on mechanics of flexible gear material Journal of Central South University (Science and Technology), v 49, n 3, p 592-599, March 26, 2018.

[3] Li XZ, Song CS, Yang Y,Optimal design of wave generator profile for harmonic gear drive using support function,Mechanism and Machine Theory,Volume 152,2020.
[4] Yao, Y., Chen, X., Xing, J. et al. Tooth Position and Deformation of Flexspline Assembled with Cam in Harmonic Drive Based on Force Analysis. Chin. J. Mech. Eng. 34, 104 (2021).
[5] Fan YG, Harmonic Gear Transmission Technical Manual, National Defense Industry Press,BeiJing.
[6] Ji XJ. Tooth profile design and meshing analysis of CTC tooth profile harmonic drive [D].Nanjing University of Science and Technology,2013.
[7] Sha XC. Theoretical research and test of the transmission accuracy of the new tooth-shaped harmonic gear[D]. Nanjing University of Science and Technology,2014.
[8] GB/T 30819-2014, Harmonic gear reducer for robot [S]. China Standard Press, 2014.
[9] Zhao ZX,Wang FC,Chen C,Liu EL,Niu JH,Huang L. Research on load-carrying characteristics of annular precision harmonic drive flexible pulley [J]. Mechanical strength,2015,37(03):472-478.
[10] Shen YW, Ye QT. Theory and Design of Harmonic Gear Transmission [M]. BeiJing:,Machinery Industry Press 1985: 2-33.

Particle image recognition in the measurement of workpiece cleanliness

Zhiqiang long[1], jiangang wang[1*], fengqin He[2], mochuan sun[1], fengqi lian[3], jiancheng zhong[3], qianqian shangguan[3], qiuhong zhang[3]

[1]Shanghai Institute of Technology, College of Mechanical Engineering, Shanghai 201418, China

[2]the college of information, mechanical and electrical engineering, Shanghai normal university, Shanghai, 200234, China

[3]Shanghai Suying Cleanliness Technology Development Co., Ltd.

*wjg@sit.edu.cn

Abstract: Monitoring the cleaning degree of workpiece surface cleaning is a key step in high-end equipment manufacturing .The use of image recognition technology to identify the cleaning degree of workpiece surface can effectively improve the quality and efficiency of workpiece cleaning. In this paper,a particle image recognition technology based on opencv was proposed, which uses a series of methods such as grayscale, binarization, morphological processing, line contour detection, contour extraction to statistically calculate the number and average size of particles. So that the surface cleanliness of the workpiece can be determined.Judging the surface cleanliness classification according to the particle size and quantity is a precise quantitative cleaning method.

1. Introduction

Accuracy and surface roughness have a certain impact on the service life of the bearing cage[1],Cleanliness affects surface quality and the initial accuracy of the workpiece. Instruments cleaned in the field of medical and other high-precision instruments must meet high cleanliness standards[2].The surface of the workpiece is sufficiently clean to ensure the quality of the workpiece and to ensure the stability and functionality of the part, and the attachment particles in the process of mechanical workpieces will affect the stability of the machinery and the accuracy of the surface quality of the workpiece[3].

Cleanliness measurement methods includes: visual inspection method, contact angle method, fluorescent luminescence method, particle size quantity method, weight method. Particle size quantity method is a precision cleaning quantitative cleaning degree detection method. In recent years, with the development of technology, the industry has stricter requirements on the surface quality of parts. In the past, technicists mainly relied on their own vision to recognize particles on the surface of workpieces, while image recognition techniques can effectively replace visual inspection and make up for the shortcomings of human vision. Image recognition can evaluate the cleaning quality of the surface of the cleaned parts, which is conducive to shortening the cleaning time and improving the cleaning quality.

In this work, the image recognition technique using Python-OpenCV image library was utilized to quickly, accurately and quantitatively determined surface cleanliness of workpieses based on particle size quantity method.Opencv is an open source image processing library that uses Python-opencv combinations to recognize faces, gestures, soccer balls, and other objects that can be captured by cameras[4, 5].Grayscale and binarization is used to improve the efficiency of image recognition processing, and fine particles can be accurately identified through morphological processing and edge detection. The above series of operations can achieve the identification of the number of particles and the average area on the surface of the workpiece.

2. Basic principles of particle identification

OpenCV is an open source library, composed of a series of C languages such as C++, with interfaces such as python and MATLAB, It mainly includes three aspects: image processing, computer vision and machine learning[6, 7].It contains a wealth of image processing algorithms can complete the image recognition and processing and other tasks, Due to its powerful functions, it is widely used in industrial product testing, traffic monitoring, marine traffic and other fields[8, 9].Digital image processing is a method and technology that removes noise, enhances, restores, segments, extracts features and other processing methods and techniques for images by computers[10].The image taken by the camera this time contains a lot of noise. The key difficulty in this particle identification is to extract the contour and calculate the average area.

3. Construction of particle identification system

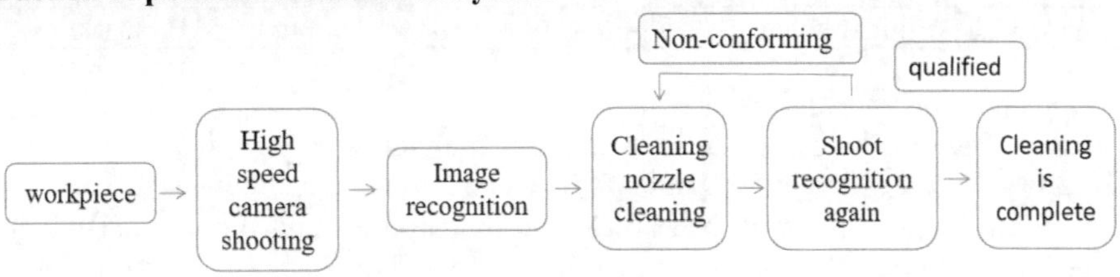

Figure 1. System identification process

The flow of particle identification system in this paper is shown in Figure 1. The particle size on the surface of the workpiece is micron level. Image quality has a significant impact on the accuracy of particle recognition,but the quality of the captured image is closely related to the on-site light source. high-speed camera light source is mainly LED light source,LED light sources are divided into the following four types according to shape: ring light source, strip light source, coaxial light source, and backlight source. The backlight can be mixed with different colors to meet the multi-color requirements of different analytes. The backlights used in this article are divided into parallel backlights, side backlights, and bottom backlights. The positions of different light sources are shown in Figure 2.

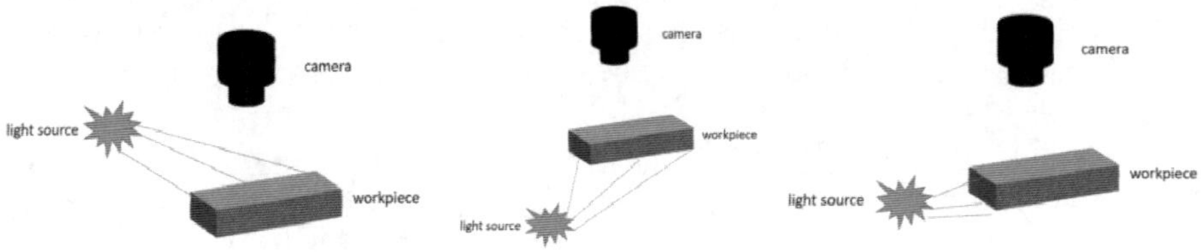

Figure 2. Light source position adjustment

4. Image pre-processing

The picture collected by the camera is generally a color image, the color image is larger than the grayscale image pixel value and contains a lot of noise, if it is not pre-processed, it will have a certain impact on the running speed and recognition accuracy of image recognition. The use of open-cv image processing library to preprocess the original image of the surface of the workpiece acquired.

4.1. Grayscale

In image processing, you need to put the pixels in the pixel matrix, with RGB three channels to represent the red, green, blue primary colors, R component, G component, B component of the value range is 0 ~ 255, when the value of the RGB three channels of a pixel is 225, the performance is the brightest white; when the RGB three-channel value of a pixel is 0, it shows the darkest black. Color image information is too large to affect the processing speed, grayscale can turn three channels into a single channel in order to improve the image processing speed, usually use image grayscale processing[11].The original image is shown in Figure 3.The grayscaled image is shown in Figure 4.

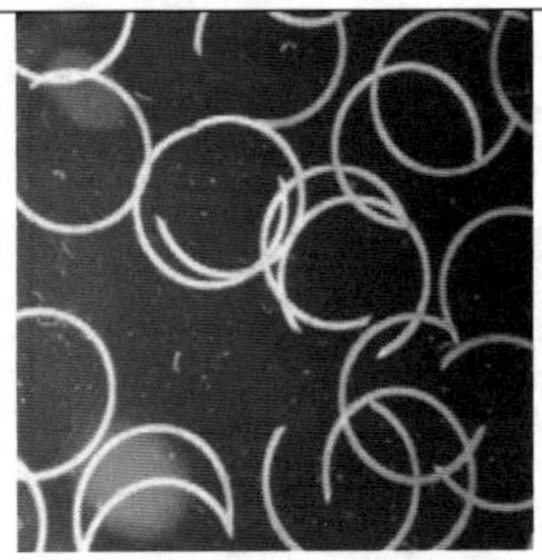

Figure 3.The original image

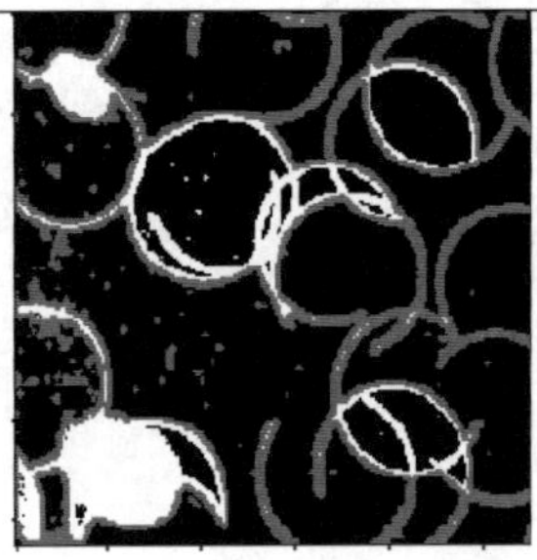

Figure 4. Grayscale

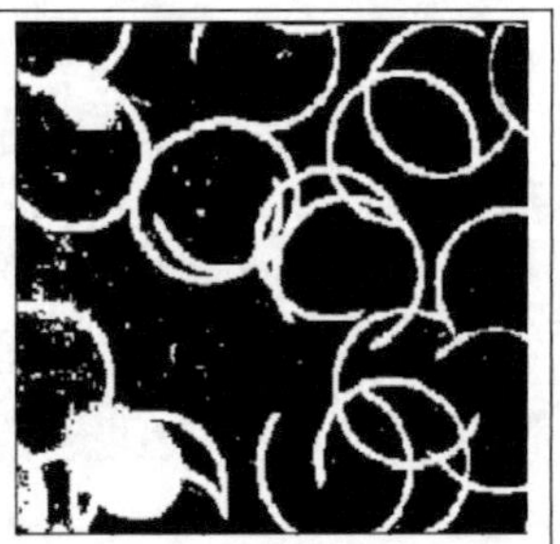

Figure 5. Binarization

4.2. Binarization

After binarization, the gray value of the particles is generally close to 250, and the gray value of the workpiece surface is generally lower than 120.Therefore, binarizing the image can enhance the difference between the gray value of the particle and the surface, and further enhance the particle recognition effect. Different thresholds have different effects on image processing, and the most important thing about image binarization is to find a suitable threshold [12,13]. In this paper, image binarization using the global Otsu method can quickly and efficiently find the inter-class segmentation threshold. The binarized image is shown in Figure 5.

4.3. Morphological denoising

After the image is processed by grayscale and binarization, many details on the image are enhanced. However, there are still a small amount of noise, such as the blurred edge of the particles leading to the inability to identify fine particles, adjacent particles misidentified and other errors.Using morphology to process particle images to further optimize image detail information can improve the accuracy of particle recognition[14, 15]。 Morphological denoising is a common and effective way to remove noise in image processing. Morphological denoising includes corrosion, expansion, open operation, closed operation, top hat, black hat, morphological gradient. Morphological processing related operation type functions are shown in Table 1.

Table 1. Morphological processing

The type of operation	function
Open operations	cv2.morphologyEx(img,cv2.MORPH_OPEN,kernel)
Closed operations	cv2.morphologyEx(img, cv2.MORPH_CLOSE, kernel)
Top hat	cv2.morphologyEx(src, cv2.MORPH_TOPHAT, kernel)
Black cap	cv2.morphologyEx(src, cv2.MORPH_BLACKHAT, kernel)
Morphological gradients	v2.morphologyEx(thresholdImg,cv.MORPH_GRADIENT, kernel)
corrode	cv2.erode(img,kernel,iterations)
swell	cv2.dilate(img,kernel,iterations)

The opening operation is to erode and then dilate, which can eliminate smaller bright spots in the image. Closed operations expand and then corrode, eliminating small black holes in the image. In this paper, the on and closed operations are used to reduce the noise around the particles, which greatly improves the image quality and highlights the particle contour characteristics. After the opening operation is shown in Figure 6, the closed operation is shown in Figure 7.

		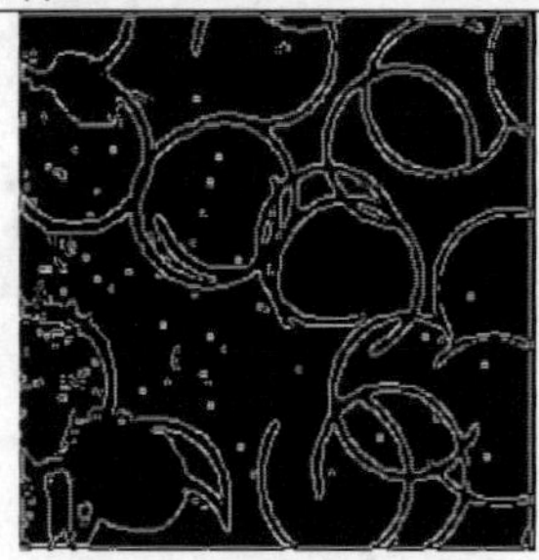
Figure 6. opening operation	Figure 7. closed operation	Figure 8. Edge detection image

4.4. Edge detection

After the morphological processing of the image, the grains of the image are detected at the edges, the particle contour boundaries are recognized, and then the contours are extracted for marking and counting. Since too small may be particles or noise that is not eliminated, this study sets a numerical value to default to background noise below the value.There are three types of edge detection: laplacian operators, Sobel and Canny functions, and the canny function is found to work best after experimentation. Edge detection is shown in Figure 8, and the edge detection procedure is as follows:

```
edge_output = cv2.Canny(out_img_GRAY, 10, 300)
contours,hierarchy=cv2.findContours(dst,cv2.RETR_EXTERNAL,cv2.CHAIN_APPROX_SIMPLE)
cv2.drawContours(dst,contours,-1,(120,0,0),2)
```

5. Image processing

After image pre-processing, the particle profile is extracted using an algorithm, and the method of filtering small particles with an area of less than 1 and workpieces greater than 250 is adopted, filtering smaller particles and oversized workpieces is the least error-prone approach. The average area of the identified particles is 1, and the particle counting effect in this article is shown in Figure 9, except for the tiny particles, all particles are basically identified.

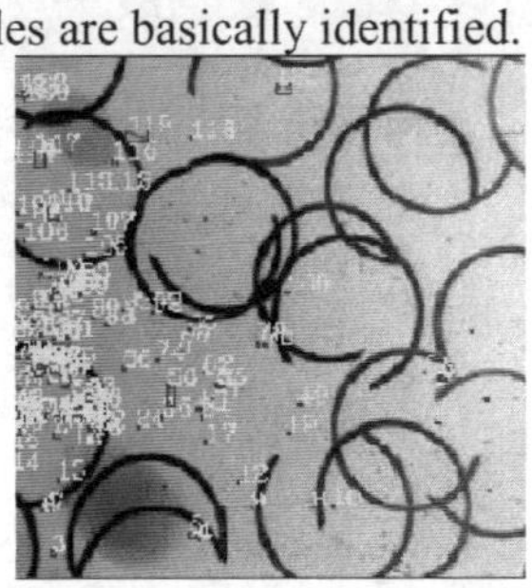

Figure 9. Particle count plot

6. Conclusion

opencv, high-speed camera and particle size quantity method was proposed to achieve rapid determination of workpiece surface cleanliness. Firstly, the initial image was grayscaled, and the gray value was reduced to improve the operation speed.The image is initially noise-reduced and then binarized to enhance the contrast between the background and foreground and improve the accuracy of particle recognition. Then perform the opening and closing operations,after grayscale and binary image noise is basically eliminated, but there is still a small amount of noise at the edge of the particle in the image, through the on operation, closed operation two morphological processing, the particle edge noise can be eliminated, further enhance the details in the image. Finally, edge detection is performed to basically identify the particle outline, and then particle counting and area calculation are performed. Finally, the size and quantity of particles are basically counted, and the particle size quantity method can be combined to determine whether the surface cleanliness of the workpiece is qualified.

In this paper, the particle size is filtered by area, and the accuracy of filtering small particles has a certain error. But with the advancement of technology, these disadvantages will be liminated. These drawbacks will be eliminated, and this paper is mainly aiming to provide a research direction for the automatic determination of workpiece cleanliness in the future.

Acknowledgement
This work was sponsored by Natural Science Foundation of Shanghai (Grant No.19ZR1437400).

References

[1] Balc, N., et al., Surface roughness obtained at stamping of bearing cages. MATEC Web of Conferences, 2017. 137.
[2] Mitschele, M. and D. Bartels, High Standards of Cleanliness for Specialist Parts. IST International Surface Technology, 2019. 12(1): p. 40-41.
[3] Lohse, A., Process-oriented approach to assuring cleanliness of parts. IST International Surface Technology, 2015. 8(3): p. 40-41.
[4] Adusumalli, H., et al., Face Mask Detection Using OpenCV, in 2021 Third International Conference on Intelligent Communication Technologies and Virtual Mobile Networks (ICICV). 2021. p. 1304-1309.
[5] Ismail, A.P., et al., Hand gesture recognition on python and opencv. IOP Conference Series: Materials Science and Engineering, 2021. 1045(1).
[6] Qashlim, A., et al., Smartphone Technology Applications for Milkfish Image Segmentation Using OpenCV Library. International Journal of Interactive Mobile Technologies (iJIM), 2020. 14(08).
[7] Wu, W.H., Y.J. Zhao, and Y.F. Che, Research and Implementation of Face Detection Based on OpenCV. Advanced Materials Research, 2014. 971-973: p. 1710-1713.
[8] Wang, W.J. and M. Gao, Vehicle Detection and Counting in Traffic Video Based on OpenCV. Applied Mechanics and Materials, 2013. 361-363: p. 2232-2235.
[9] Chen, D., et al., Research on Marine Radar Image Collection Technology Based on OpenCV. Advanced Materials Research, 2013. 798-799: p. 578-581.
[10] Ma, Y.F. and S.G. Liu, Wavelet De-Noising Algorithm Based on OpenCV for Images Edge Detection. Advanced Materials Research, 2014. 989-994: p. 3973-3976.
[11] Vishal, R.M., et al., Object Detection: Automatic License Plate Detection using Deep Learning and OpenCV. International Journal of Engineering and Advanced Technology, 2019. 9(1): p. 6022-6028.
[12] Marengoni, M. and D. Stringhini, High Level Computer Vision Using OpenCV, in 2011 24th SIBGRAPI Conference on Graphics, Patterns, and Images Tutorials. 2011. p. 11-24.
[13] He, Y.M. and J.Q. Du, Segment of Tongue and Tongue's Venation Based on OpenCV. Advanced Materials Research, 2013. 718-720: p. 2276-2280.
[14] Zhang, D.P., et al., The Vision Orientation System Based on OpenCV of the Glass Curtain Wall Cleaning Robot. Applied Mechanics and Materials, 2013. 336-338: p. 1063-1067.
[15] Xie, G. and W. Lu, Image Edge Detection Based On Opencv. International Journal of Electronics and Electrical Engineering, 2013. 1(2): p. 104-106.

Journal of Physics: Conference Series **2365** (2022) 012028 doi:10.1088/1742-6596/2365/1/012028

Research on braking energy recovery technology of electric vehicle in underground coal mine based on road condition recognition technology

Fan Zhang[1,*], Yuan Zhao[2] and Zhijun Hao[2]

[1]Shanxi Tiandi Coal Mining Machinery Co., Ltd. Taiyuan, China

[2]China Coal Technology & Engineering Group Taiyuan Research Institute Co., Ltd. Taiyuan, China

* zhangfan@tyccri.com

Abstract: Electric explosion-proof trackless rubber-wheeled vehicles in coal mines usually use braking energy recovery technology to improve cruising range. In order to improve the efficiency of vehicle braking energy recovery and improve vehicle braking safety performance, this paper proposes a braking energy recovery control strategy for electric vehicles in coal mines based on road condition recognition. This strategy is based on the coal mine underground positioning technology, and adopts different braking energy recovery control strategies for different road conditions, while improving the braking energy recovery ratio of the vehicle, it ensures a good feeling of the brake pedal and ensures the safety performance of the vehicle braking. Finally, the feasibility of the control strategy is verified by simulation.

1. Introduction

Explosion-proof trackless rubber-wheeled vehicles in coal mines have the advantages of flexible maneuvering and high transportation efficiency. They have become important auxiliary transportation tools in coal mines and are widely used in major mining areas. With the development of electric vehicle technology, all kinds of new pure electric explosion-proof trackless rubber-wheeled vehicles with explosion-proof batteries as power sources have emerged, and gradually began to replace the traditional diesel engine explosion-proof trackless rubber-tired vehicles. However, the high curb weight of pure electric explosion-proof trackless rubber-tired vehicles generally leads to a short cruising range, which is difficult to meet the needs of continuous operation in coal mines. Therefore, how to improve its cruising range has become an urgent problem to be solved in the development of electric explosion-proof trackless rubber-tired vehicles.

After calculation, the braking energy of electric vehicles accounts for 30% to 50% of the total energy consumption of the vehicle[1] [2] [3]. Therefore, the reasonable recovery of the braking energy of the vehicle is an effective means to improve the cruising range of the vehicle. However, the underground roads in coal mines are complex and harsh, and there are many continuous slopes. The traditional brake recovery system will lead to changes in the brake pedal feel of the vehicle, and at the same time increase the complexity of the braking system, which makes the electric explosion-proof rubber-wheeled vehicles used in coal mines, driving brings safety hazards.Therefore, in order to improve the cruising range and braking safety performance of electric explosion-proof rubber-wheeled vehicles in coal mines, based on the actual road conditions in coal mines and combined with zigbee vehicle positioning technology, this paper proposes a brake recovery control strategy for electric vehicles in coal mines based on road condition recognition.

2. Research on key technologies of braking energy recovery

The underground road environment of coal mines is harsh and there are many slopes. When rubber-wheeled vehicles are driving underground, in order to meet the demand for long-distance continuous braking, explosion-proof wet brakes are often used. The wet brake adopts a closed structure, and the key components such as the friction plate and the piston are sealed in the brake housing, which prevents the impact of the harsh external environment on the brake and has a good lubrication and heat dissipation effect. In new electric vehicles, a regenerative braking system is introduced to form a braking energy recovery system.

The braking energy recovery system includes two subsystems: motor regenerative braking and friction braking. During the deceleration and braking process of the vehicle, the wheel drive motor is reversed, the motor is in the state of generating electricity, and the kinetic energy of the vehicle is recovered and stored in the battery. At this time, the feedback braking torque provided by the motor acts on the front wheel of the vehicle through the transmission system.At the same time, the hydraulic control system modulates the hydraulic braking force of the friction braking system, and the hydraulic pressure acts on the brake wheel cylinder through the transmission of the hydraulic medium to generate the friction braking torque. Finally, the regenerative braking torque and the friction braking torque are dynamically coupled at the wheels to form a deceleration braking effect on the vehicle[4] [5] [6].

2.1. Research on Brake Force Distribution Technology

When the vehicle is braking in a coal mine, the front and rear wheels are locked at the same time. At this time, the distribution curve of the braking force F_1 and F_2 of the front and rear wheels is the I curve (Figure.1). At this time, the braking force relationship between the front and rear wheels is as follows:

$$F_2 = \frac{1}{2}\left[\frac{G}{h_g}\sqrt{b^2 + \frac{4h_g L F_1}{G}} - \left(\frac{Gb}{h_g} + 2F_1\right)\right] \tag{1}$$

where F_1 is the braking force of the front axle; F_2 is the braking force of the rear axle; L is the wheelbase of the vehicle; G is the weight of the vehicle; b is the distance from the center of mass of the vehicle to the rear axle; and hg is the height of the center of mass of the vehicle.

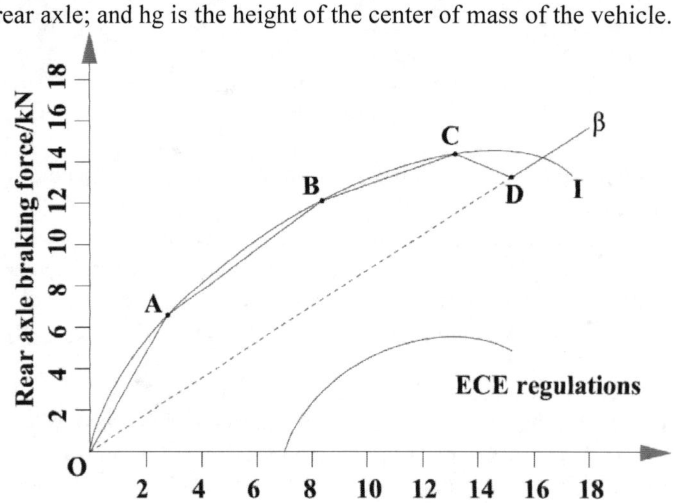

Figure 1. Front and rear axle braking force distribution curve

At the same time, in order to ensure braking efficiency and braking stability, Europe has formulated ECE regulations to clearly stipulate the distribution of braking force between the front and rear axles of automobiles, as follows[7] :

$$(F_1+F_2)^2+G\left(0.07+\frac{b}{h_g}\right)(F_1+F_2)-\frac{0.85GLF_1}{h_g}+\frac{0.07G^2b}{h_g}=0 \tag{2}$$

For the braking intensity Z, it is defined as:

$$Z=\frac{T}{GR} \tag{3}$$

where Z is the braking strength; T is the braking torque; R is the tire radius. The point on the I curve with the known braking intensity Z satisfies the following formula:

$$\begin{cases} F_1=\dfrac{GZ\left(b+Zh_g\right)}{L} \\ F_2=\dfrac{GZ\left(a-Zh_g\right)}{L} \end{cases} \tag{4}$$

Points A, B, C, and D in Figure X represent the points of braking intensity Z=0.2, Z=0.4, Z=0.6, and Z=0.8 respectively, of which points A, B and C are located on the I curve, and point D is located on β On the curve; when Z ≥ 0.8, the vehicle is considered to be in emergency braking condition, and the front and rear braking force ratio of the vehicle is distributed according to the β curve, as follows:

$$\begin{cases} F_1=GZ\beta \\ F_2=GZ-F_1 \end{cases} \tag{5}$$

2.2. Research on underground positioning technology of coal mine

The positioning of vehicles in coal mines is one of the prerequisites for realizing intelligent coal mines. Currently, the main technologies used are UWB, Zigbee, Wi-Fi technology, etc. Among them, Zigbee positioning technology is widely used in many coal mines due to its advantages of low energy consumption and high reliability. application.

The Zigbee wireless communication positioning module in the coal mine is mainly composed of: vehicle-mounted mobile node, fixed node and coordinator node. The on-board node is carried by the vehicle to be located, and transmits signals carrying vehicle information to the surroundings at any time. The fixed node is arranged in the coal mine roadway, and the position coordinates of the node are known. When the vehicle passes the reference node, the reference node receives the signal transmitted by the unknown node and transmits it to the coordinator. When the coordinator node receives information about the same unknown node from multiple reference nodes, it uses the ranging algorithm to accurately locate the vehicle position by measuring the distance between the vehicle mobile node and multiple different nodes. In this paper, the RSSI-based ranging algorithm is used. In the RSSI wireless wireless propagation model, the shadow shielding propagation model is used, which can be written as:

$$PL(d)=PL(d_0)-10n\lg\left(\frac{d}{d_0}\right)-X \tag{6}$$

In the above formula, PL(d) refers to the RSSI strength value of the signal at the propagation distance d; $PL(d_0)$ refers to the signal strength value of the signal at the reference distance d_0; n is used to describe the environmental parameters of signal attenuation with distance; X refers to the pair of environmental parameters. The effect of the signal has a mean of 0.

After only determining the distance between the vehicle-mounted mobile node and the fixed node, it is still impossible to determine the exact position of the vehicle. It is necessary to use the calculated distance to calculate the coordinates of the unknown point, that is, to realize the positioning of the underground trackless rubber-tired vehicle. In this paper, the triangular centroid algorithm is used to calculate the location of the underground vehicle (Figure 2)

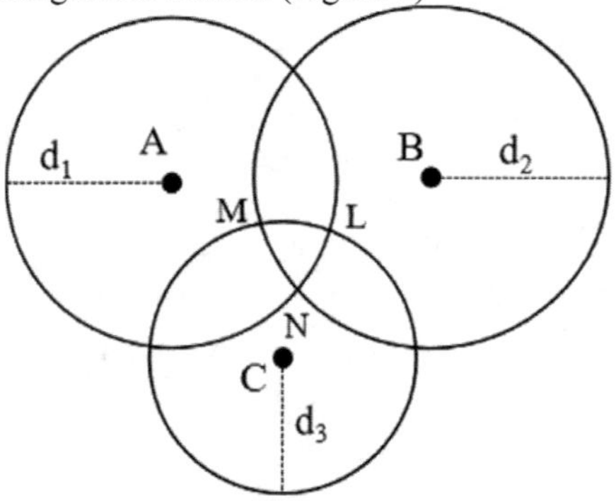

Figure 2. Triangle centroid positioning diagram

As shown in the figure above, the mobile node (position is uncertain) is denoted as point O, and the fixed node (position is determined) is three points A, B, and C, respectively. Through the RSSI ranging algorithm, it can be obtained that the distances from point O to points A, B, and C are d_1, d_2, and d_3, respectively. Then point O must be in a circle with points A, B, and C as the center and d_1, d_2, and d_3 as the radius. superior. Due to the loss of signal propagation in space, the three circles do not intersect at a point, but in an area, so point O is on the overlapping area of the three circles.

Note that the coordinates of the detected nodes are $A(x_a,y_a)$, $B(x_b,y_b)$, $C(x_c,y_c)$ respectively. Suppose the coordinates of the unknown node O are (x, y), and the intersection coordinates of the three circles are M respectively. (x_m, y_m), N (x_n, y_n), L (x_l, y_l). Then it can be expressed as:

$$\left.\begin{array}{l}\sqrt{(x_m-x_a)^2+(y_m-y_a)^2}\le d_1\\ \sqrt{(x_m-x_b)^2+(y_m-y_b)^2}=d_2\\ \sqrt{(x_m-x_c)^2+(y_m-y_c)^2}=d_3\end{array}\right\} \tag{7}$$

Then the coordinates of M point (x_m, y_m) can be calculated. Similarly, the coordinates of points N and L can be obtained as (x_n, y_n) and (x_l, y_l) respectively. Finally, the unknown node coordinates (x, y) are calculated according to the following formula, That is, the location of the unknown point is realized.

$$x=\frac{\dfrac{x_m}{d_1+d_2}+\dfrac{x_n}{d_2+d_3}+\dfrac{x_l}{d_3+d_1}}{\dfrac{1}{d_1+d_2}+\dfrac{1}{d_2+d_3}+\dfrac{1}{d_1+d_3}},\quad y=\frac{\dfrac{y_m}{d_1+d_2}+\dfrac{y_n}{d_2+d_3}+\dfrac{y_l}{d_3+d_1}}{\dfrac{1}{d_1+d_2}+\dfrac{1}{d_2+d_3}+\dfrac{1}{d_1+d_3}} \tag{8}$$

3. Vehicle brake recovery control strategy

3.1. Structure of regenerative coupling braking system

The overall structure of the motor feedback braking and friction braking coupling braking system is shown in Figure 3. A drive motor is arranged in the middle of the front axle. After the output torque of the motor is amplified by the deceleration and torque increase of the gearbox, it is transmitted through the half shafts, universal joints and other components, and finally acts on the wheels. The central controller receives the positioning signal sent by the Zigbee positioning module in real time, and judges the position of the vehicle in real time. Switch the braking amount recovery mode; the hydraulic braking force is adjusted by the hydraulic control unit, and through the hydraulic braking pipeline, it directly acts on the friction brake on the side of the wheel, thereby generating the friction braking force. In the regenerative braking process of the vehicle, the regenerative braking torque from the motor drive system and the hydraulic braking torque from the friction braking system are coupled at the wheels, and the two work together to complete the vehicle braking operation.

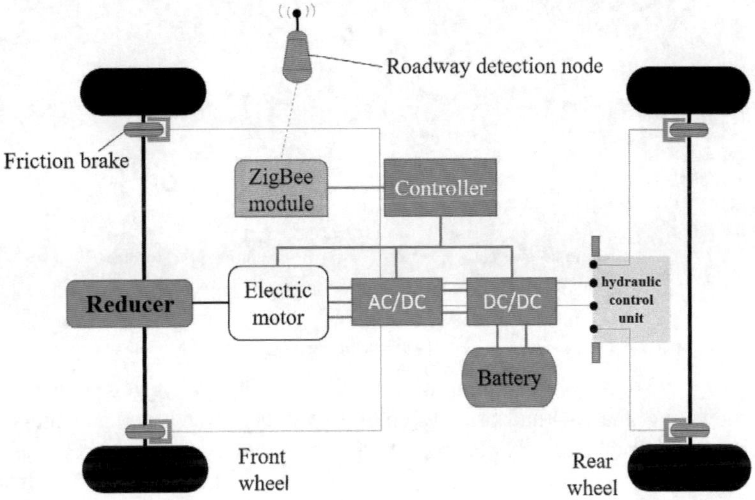

Figure 3. Schematic diagram of the regenerative braking system model

3.2. Braking energy recovery management strategy

When the electric explosion-proof rubber-wheeled vehicle is driving on complex and harsh roads in coal mines, limited by environmental restrictions, the vehicle starts and brakes frequently, resulting in the vehicle braking system frequently under regenerative braking, friction braking and coupling braking. switching, which makes the braking energy recovery efficiency not only limited by the state of the electric drive system including the motor, battery and other components, but also affected by the dynamic adjustment of the friction braking force, and the frequent switching of the friction braking system leads to hydraulic pressure The transient change of the brake wheel cylinder pressure causes frequent fluctuations in the brake pedal pressure, which seriously affects the handling performance of the vehicle and reduces the braking safety performance of the vehicle[8] [9].

Therefore, it is necessary to combine the vehicle real-time positioning system to determine the road conditions of the vehicle at that time, and formulate different braking energy recovery strategies to improve the braking energy recovery efficiency and improve the braking safety performance of the vehicle. For different roads in coal mines, this paper formulates four braking energy recovery management strategies, as shown in the following figure:

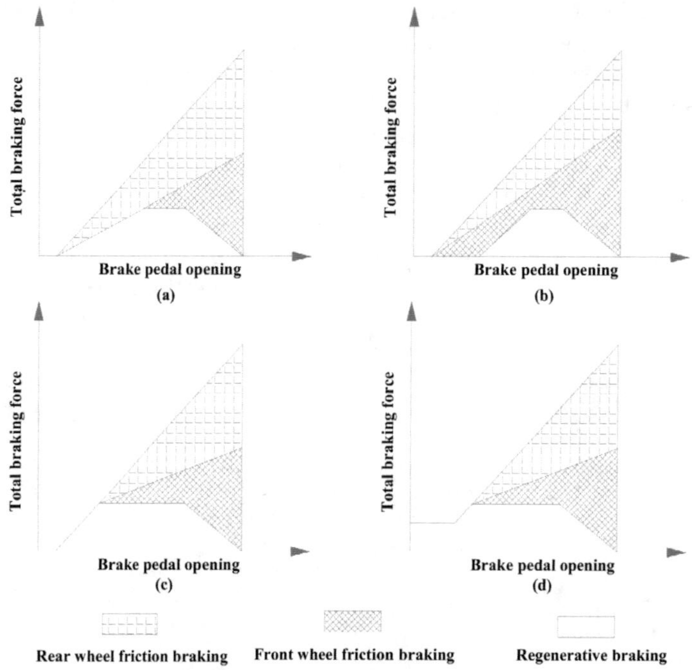

Figure 4. Schematic diagram of braking energy recovery strategy

Figure 7(a) shows the general road braking energy recovery strategy. When the vehicle is braked on a generally hardened and relatively flat road in a coal mine, the driving wheel (front wheel) first uses the motor to perform regenerative braking, and the rear wheel follows the The front and rear axle braking force distribution requires the application of the corresponding friction braking force. When the feedback braking force of the front wheel motor is insufficient, the friction braking force is supplemented. This strategy can effectively recover the braking energy, but when the front wheel friction braking intervenes, it will cause braking The sudden change of braking force caused by the pressure around the wheel cylinder has an impact on the safety during emergency braking, so it is suitable for general hardened and smooth roads with a low accident rate;

Figure 7(b) shows the strategy of good braking safety performance. This strategy is a control strategy to ensure good braking safety. When braking, the front and rear wheels are braked with friction braking force. When the initial braking force is established, Reduce the brake wheel cylinder pressure, and at the same time the motor feedback braking force begins to intervene. This strategy can ensure good braking safety and avoid sudden changes in brake pedal feel caused by sudden changes in brake wheel cylinder pressure during the braking process[10]. Part of the braking energy will be wasted in the initial stage, and this strategy is suitable for the use of the vehicle in the case of emergency braking in bad road conditions.

Figure 7(c) shows a good road energy recovery strategy, which is a strategy with maximum braking energy recovery efficiency. When the vehicle is driving on a wide transportation roadway with good road conditions and sufficient lighting, the vehicle speed is stable, and the occurrence of Braking behavior is less. At this time, when the vehicle is braking, only the motor feedback braking force is used for braking at the initial stage. When the motor feedback braking force is insufficient, the friction braking force is supplemented. This strategy can recover the braking energy to the greatest extent, but it will lead to braking The initial instantaneous braking force is low and the braking distance is significantly increased, so it is only suitable for non-emergency braking situations on good roads.

Figure 7(d) shows the braking energy recovery strategy for the continuous downhill road section. When the vehicle goes downhill continuously for a long distance under the coal mine, frequent braking is required to control the speed of the vehicle. Therefore, the control logic of this strategy is:

when When going downhill for a long distance, keep a certain value of the motor feedback braking force to control the speed of the vehicle. The braking force value is determined by the current slope. When the vehicle needs to brake on the slope, depress the brake pedal and the motor feedback braking force When the motor feedback braking force is insufficient, the friction braking force will be supplemented. This control strategy can effectively recover the high potential energy of the vehicle when going downhill, reduce the braking frequency when going downhill, and improve the braking safety performance.

4. Modeling and simulation

In order to verify the feasibility and effect of the braking energy recovery control strategy proposed in this paper, the Simulink environment is used to build a control strategy model simulation. Simplified, as the working condition used in the simulation model, under this working condition, the electric explosion-proof rubber-wheeled vehicle takes a total of 20 minutes to travel from the underground to the working face, and a round trip is 40 minutes. The speed of the uphill section is 8km/h. When driving under this working condition, the curve of the gradient with time is as follows:

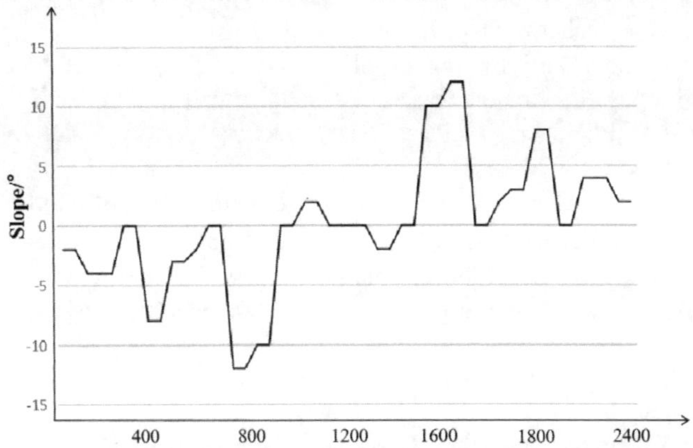

Figure 5. Simulated pavement slope curve

After building a simulation environment according to the actual transmission mode of electric explosion-proof vehicles in coal mines, the initial SOC value of the battery is set to 90%, and the total battery capacity is about 100 kWh. Compared with the fixed-proportional energy recovery strategy and the non-braking energy recovery strategy, the simulation results of the fixed-proportional control strategy are shown in Table 1 below:

Table 1. Simulation results of underground road conditions in coal mine

parameter	Energy recovery strategy in this paper	Fixed proportion energy recovery strategy	No energy recovery
Recover energy/kJ	10325.10	6854.57	/
SOC /%	56.85	55.74	52.85

It can be seen from the simulation results data that the braking energy recovery strategy based on road condition recognition proposed in this paper has certain advantages compared with the traditional fixed-proportion braking energy recovery strategy. 1.11% energy can be saved, which verifies the effectiveness of the control strategy proposed in this paper.

5. Conclusion

Aiming at the complex and harsh road environment in coal mines, this paper proposes a braking energy recovery strategy based on road condition recognition. This strategy uses the vehicle real-time positioning technology to determine the road environment where the vehicle is located in real time, and switches different braking energy recovery control strategies according to road conditions. In order to meet the braking performance required by the vehicle under different road conditions; the control strategy is used to simulate the actual road environment in the simulated coal mine. the effectiveness of the strategy.

Acknowledgments
This paper is independently supported by the fund project: Key projects of Tiandi Technology Co., Ltd.(2022-2-TD-ZD012)

References

[1] Lv Chen. Dynamic coupling control of regenerative braking force and friction braking force for electric drive vehicles [D]. Tsinghua University, 2015.

[2] DAI Y，TIAN H. Research on the development status of energy recovery control strategy for electric vehicles［J］. Journal of Science，Technology and Economics，2019，27（32）：91.

[3] ZHOU S，SUN Y，HUANG J. Research on influencing factors ofelectric vehicle braking energy recovery ［J］. Mechatronics，2019（5）.

[4] Zhang Junzhi, Lv Chen, Gou Jinfang, et al. Cooperative control of regenerative braking and hydraulic braking of an electrified passenger car[J]. Proceedings of the Institution ofMechanical Engineers, Part D: Journal of Automobile Engineering, 2012,226(10):1289-1302.

[5] PENG W，LI Z H，CHEN Q P，et al. Multi-objective optimization of braking stability and braking energy recovery of bev based on EMB［J］. Automobile Technology，2021（3）.

[6] Li L, Zhang Y, Yang C, et al. Model predictive control-based efficient energy recovery control strategy for regenerative braking system of hybrid electric bus[J].Energy Conversion and Management, 2016, 111:299–314.

[7] Farhat A, Hagen K, Cheok K C, et al. Neuro-fuzzy-based electronic brake system modeling using real time vehicle data[J]. EPiC Series in Computing, 2019,58:444–453.

[8] Dyar L R, Akita Y, Paul S, et al. Development of Advanced Braking System for Hybrid Sports Cars[J]. SAE International Journal of Passenger Cars-MechanicalSystems, 2016, 9(2016-01-1923):1151–1156.

[9] Zhao J, Chen Z, Zhu B, et al. Precise Active Brake-Pressure Control for a Novel Electro-Booster Brake System[J]. IEEE Transactions on Industrial Electronics, 2019.

[10] Bayar K, Biasini R, Onori S, et al. Modelling and control of a brake system for an extended range electric vehicle equipped with axle motors[J]. International journal of vehicle design, 2012, 58(2-4):399–426.

Design and analysis of a climbing robot for weld detection

Zhongqin Li[1,a], Jiaqing Fan[1,b*]

[1]Shanghai Power Equipment Research Institute Co., Ltd. Shanghai 200240, China

[a]lizhongqin@speri.com.cn; [b*]fanjiaqing@speri.com.cn

Abstract: The climbing robot is gradually applied to the safety weld detection on large metal walls to avoid manual operation. However, most robots are difficult to perform efficient operations for the poor performance on obstacle overcoming, flexible turning and motion stability. This paper presents a weld detection climbing robot, which integrates contact permanent magnet adsorption units and crawler chain mobile mechanism to obtain continuous, flexible and stable motion performance on large metal walls. By establishing the robot mechanical model, the minimum adsorption force can be calculated considering the instability conditions such as sliding, longitudinal overturning and lateral overturning. Combined with the Ansys Maxwell simulation, the magnet structure was optimized to obtain the structural size of a single permanent magnet adsorption unit. Experimental results show the robot can achieve stable motion and accurate detection on large metal walls with welds, and the maximum load of the robot can reach to 15.6 kg, which meets the engineering application requirements of weld detection on large metal walls.

1. Introduction

The large metal wall is widely used in petrochemical storage tanks and other industrial buildings, and it is generally welded by multiple metal plates [1]. As the service time increases, a series of defects including corrosion, surface peeling, cracks and fatigue fracture will inevitably appear at the wall weld in the harsh environment [2], which seriously endangers the safety of life and property. Therefore, conducting regular detection of large metal wall welds is of great significance.

Manual detection was used in traditional weld detection mostly [3]. However, due to uncontrollable environment and human factors, this method has large data deviation and poor consistency, and has low detection accuracy and efficiency. To solve the problems of manual detection, developing a climbing robot system for weld detection on large metal walls has become a hot spot in the field of robots [4-5].

Based on the theory and key technologies of the climb weld detection robot, domestic and foreign scholars have carried out extensive and in-depth research. For example, Longo et al. developed a two-wheeled wall mobile robot that uses an exhaust fan to generate negative pressure so that it can adsorb on the wall. It has a simple structure, but the flatness of the wall is high [6-7]. Woosub et al designed a reconfigurable multi-module climbing robot with permanent magnet adsorption, and the additional arm was used as a connection mechanism for reconfiguration. Obviously, it can achieve stability and high mobility, but hardly to achieve flexible movement [8]. Zhao et al studied the rust removal of oil tanks and designed an electromagnetic adsorption unit to form an adsorption track structure with large load. However, the robot is not suitable for the global detection of large metal walls due to its poor mobility [9]. Zhu et al. developed a climbing robot with permanent magnet inflatable wheels, and established the relationship between the adsorption force, tire pressure and obstacle-crossing ability to ensure obstacle-crossing performance. However, the lift gap of the permanent magnet was large, and the load was constrained [10].

According to the above studies, the existing weld detection robots generally have poor performance on obstacle overcoming, flexible turning and motion stability. Therefore, we present a weld detection climbing robot, which integrates contact permanent magnet adsorption units and crawler chain mobile mechanism to obtain continuous, flexible and stable motion performance on large metal walls. Through establishing the mechanical model of the robot, the minimum adsorption force is calculated considering the instability conditions such as sliding, longitudinal overturning and lateral overturning. Combined with the Ansys Maxwell simulation, the magnet structure was optimized to obtain the structural size of a single permanent magnet adsorption unit. To verify the performances of the robot, we built an experimental platform and conducted a series of experiments.

2. Mechanical Design

2.1. Design indexes

To ensure the stability of the climbing robot movement, it is required that the robot can be stably adsorbed on the wall and can move flexibly. Meanwhile, the robot should have a compact structure, lightweight and simple control under the condition of industrial applications. Considering the requirements of large metal walls weld detection and site operation environment, the design indexes of the robot are presented, as shown in table 1.

Table 1. Design indexes of the robot.

Technical requirement	Design parameters
Robot weight	$\leq 20\,kg$
Overall dimension	$\leq 600\,mm \times 450\,mm \times 200\,mm$
Obstacle ability	$\leq 20\,mm$
Creeping height	$20\,m$

2.2. Robot design

In this paper, a climbing robot is designed based on the idea of modularization and parameterization. To ensure the adsorption reliability and motion flexibility of the robot on large metal walls, the contact permanent magnet adsorption is determined. Meanwhile, the crawler chain walking mechanism is adopted to increase the contact area and ensure the stability of the moving process. According to the design indexes, the overall design size of the climbing robot is $490\times 372\times 172\,mm$.

Based on the analysis of the large metal wall working environment and weld detection process, the overall structure of the climbing robot is shown in figure 1, and it mainly includes the vehicle body frame, weld detection module, flexible movement module and driving module. The flexible wall mobile module of the robot adopts the permanent magnet adsorption unit and crawler sprocket type mobile mechanism, which is mainly composed of two modular magnetic adsorption crawler sprockets. The detection module is installed on the vehicle body frame.

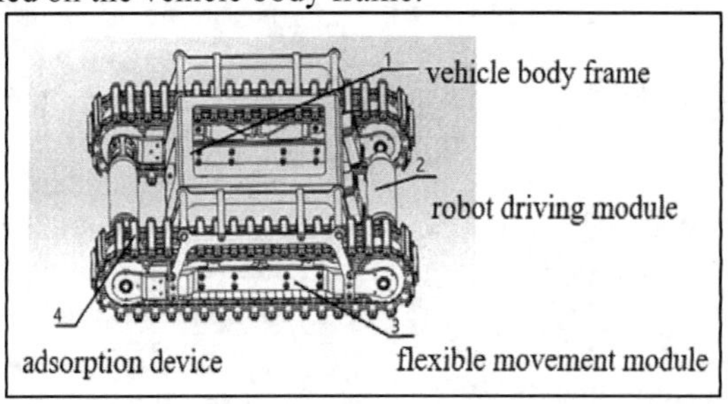

Figure 1. The basic module of the climbing robot.

The magnetic adsorption crawler sprocket is symmetrically installed on both sides of the vehicle body frame, which is mainly composed of the permanent magnet adsorption unit, crawler chain, sprocket. Each chain is composed of two chains, which are connected by a fixed plate with a permanent magnet.

The permanent magnet adsorption unit is composed of the permanent magnet, fixing plate, bolts and installation plate, which are installed on the chain through the fixing plate and are in direct contact with the metal wall to ensure safe and stable movement of the robot. There are 40 permanent magnet adsorption units on each track, and each permanent magnet adsorption unit is equipped with a permanent magnet. These permanent magnet adsorption units are connected through a chain, which can be adjusted through the chain tightening device during assembly.

The driving module adopts two DC motors, which are respectively installed on the magnetic adsorption crawler sprocket. Through the speed difference of the motor, the robot can turn flexibly on the large metal wall surface. The two DC motors are installed in front and rear of the body, making the overall structure of the robot more compact. The overall structure of the climbing robot is shown in figure 2. The prototype of the climbing robot is shown in figure 3.

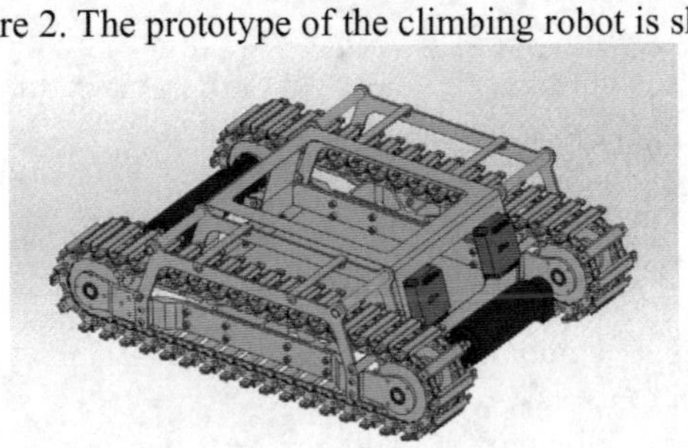

Figure 2. The overall structure of the climbing robot.

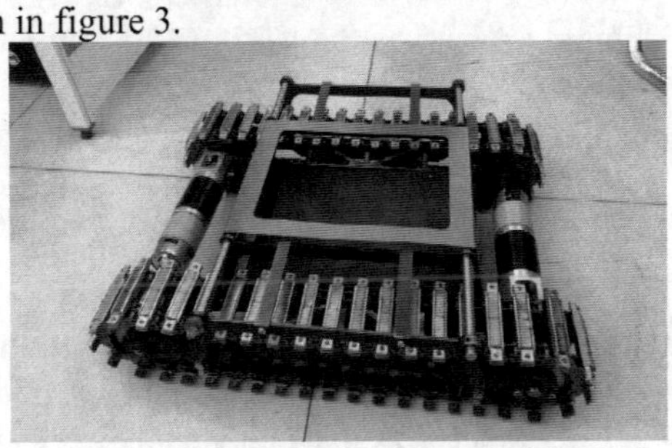

Figure 3. Prototype of the climbing robot.

3. Mechanical Analysis

There are three main forms of instability when the climbing robot is statically adsorbed on a large metal wall [11]. (1) The robot has a downward trend along the wall. (2) The robot tends to overturn longitudinally. (3) The robot tends to overturn laterally. According to the above three situations, the mechanical model of the robot is established, and the corresponding limit adsorption force in the case of instability is calculated.

3.1. Slide analysis

Assuming that the robot is stationary on the metal wall and the adsorption force of the magnet is uniform, the static force analysis of the robot is shown in figure 4 (the force generated by the contact between the weld detection module and metal wall can be ignored).

For mechanical analysis of the robot in the X direction, the force balance equation can be listed, as shown in equation (1):

$$\Sigma X = 0 \quad (F_{N1} + \cdots + F_{N17}) - (F_{k1} + \cdots + F_{k17}) = 0 \tag{1}$$

For mechanical analysis of the robot in the Y direction, the force balance equation can be listed, as shown in equation (2):

$$\Sigma Y = 0 \quad \frac{G}{2} - F_f = 0 \tag{2}$$

$$F_f \leq \mu(F_{N1} + F_{N2} + \cdots + F_{N17}) \tag{3}$$

Where, μ is the static friction coefficient; F_f is the magnet absorbs the static friction provided by the wall; $F_{N1}, F_{N2}, \cdots F_{N17}$ are support force of the large metal wall to the robot; $F_{k1}, F_{k2} \cdots F_{k17}$ are magnetic adsorption force between the robot and the metal wall; G is the total gravity of robot.

Assuming that the adsorption force provided by each permanent magnet is the same:

$$F_{k1} = F_{k2} = \cdots = F_{k17} = F_k \tag{4}$$

From equations (1), (2), (3) and (4), it can be obtained:

$$17\mu F_k \geq \frac{G}{2} \tag{5}$$

The results show that the adsorption force provided by the permanent magnet is:

$$F_k \geq \frac{G}{34\mu} \tag{6}$$

3.2. Longitudinal overturning analysis

The climbing robot has the tendency of longitudinal overturning at the separation point P due to its gravity, as shown in figure 4. The resistance to this force is the adsorption force of the permanent magnet adsorption unit on the wall. The robot does not have longitudinal overturning is: $\sum M_p \geq 0$.

$$\sum M_P = (F_{k1} - F_{N1})L - \frac{GH}{2} = 0 \tag{7}$$

Where, P is the vertical overturning limit point of the climbing robot; H is the gravity offset axis distance; L is the distance from the first adsorption unit to the last adsorption unit.

When the first adsorption unit is in the critical state of falling off the wall, $F_{N1} = 0$.

The adsorption force of each adsorption unit on the track shall meet the following condition:

$$F_k \geq \frac{GH}{2L} \tag{8}$$

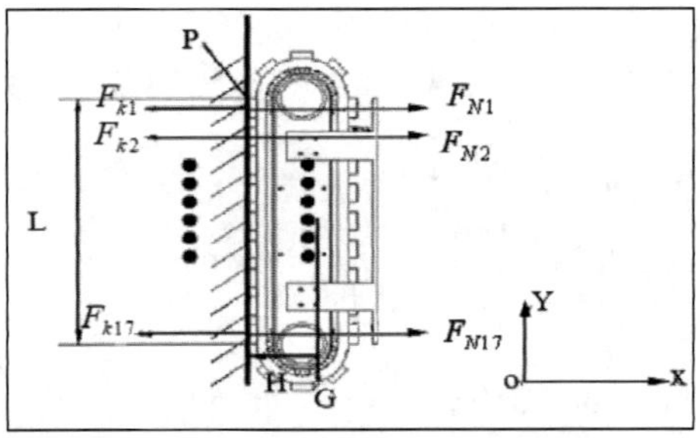

Figure 4. Force analysis of climbing robot

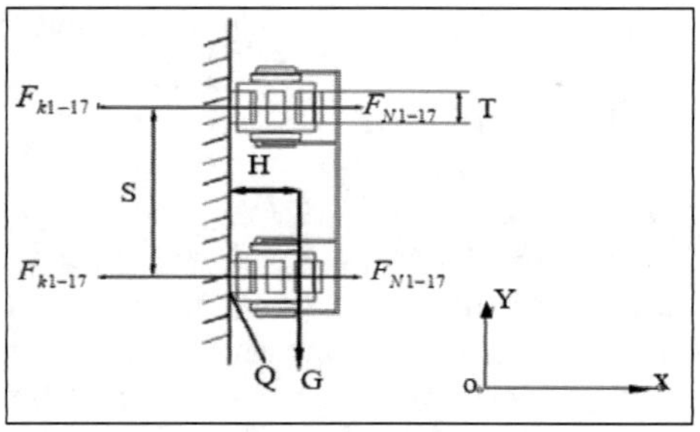

Figure 5. Force analysis of lateral overturning.

3.3. Lateral overturning analysis

The robot is horizontally adsorbed to the wall. At the separation point Q, the robot has the tendency of lateral overturning due to the torque generated by gravity, as shown in figure 5, for point Q:

$$n \times F_k \times 0 + n \times F_k \times S - G \times H \geq 0 \tag{9}$$

Where, Q is the lateral overturning limit point of the climbing robot; S is the distance between adsorption centers of two tracks; n is the number of permanent magnets on a single track.

Simplify equation (9), we can get:

$$F_k \geq \frac{GH}{nS} \tag{10}$$

3.4. Minimum adsorption force

According to the above common instability states of the climbing robot, the minimum adsorption force required by the climbing robot is:

$$F_k \geq K \bullet \max\left\{\frac{G}{34\mu}, \frac{GH}{2L}, \frac{GH}{nS}\right\} \tag{11}$$

Where, K represents safety factor. According to design parameters, $G = 500N$, $H = 75mm$, $L = 420mm$, $\mu = 0.4$, $K = 1.5$, and $S = 290mm$.

Substituting various parameters into equation (11), we can get:

$$F_k \geq 66.96N$$

At this point, the minimum adsorption force required to maintain stable adsorption between the robot and the wall can be calculated, and the force is at least 66.96N

4. Magnetic Circuit Optimization

4.1. Magnetic circuit design

The experimental research shows that [12-13], the adsorption unit including two permanent magnets and a yoke can produce the largest adsorption force in the permanent magnet adsorption unit. Therefore, the arrangement of the permanent magnet adsorption unit is shown in figure 6.

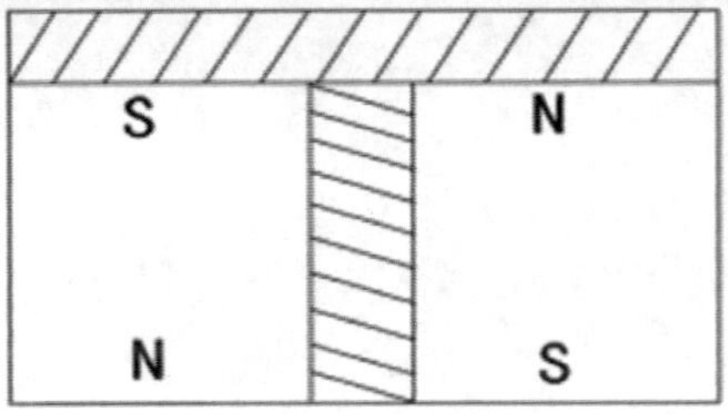

Figure 6. Structure of adsorption unit.

4.2. Materials selection

The disadvantages of NdFeB permanent magnet material are low Curie temperature, poor temperature stability and poor chemical stability, but these disadvantages can be overcome by adjusting the chemical composition and taking other measures. NdFeB permanent magnet materials have broad application and development prospects due to their excellent performance, high coercivity and magnetic energy product, its concrete performance parameters are shown in table 2[14-15].

Table 2. Performance parameters of NdFeB N38.

Performance index	Parameter
Brand	N38
Remanence induction intensity/ mT	1200-1260
Coercivity/($KA \cdot m^{-1}$)	860-915
Intrinsic coercivity/($KA \cdot m^{-1}$)	≥955
Maximum magnetic energy product/($kJ \cdot m^{-3}$)	287-303
Maximum operating temperature/ ^0C	80

4.3. Simulation optimization design

The adsorption force provided by the adsorption unit is: [16]

$$F = \frac{\mu_0 (BH) V}{K_1 K_2 L} \tag{12}$$

Where, (BH) is the magnetic energy product, V is the volume of the permanent magnet, K_1 is the magnetic leakage coefficient, K_2 is the magnetic resistance coefficient, μ_0 is the vacuum permeability, and L is the length of the air gap.

For specific magnet materials, (BH), K_1, K_2, μ_0 are constant value, the magnetic adsorption force is mainly related to V and L. The relationship between the size of the magnet unit and the adsorption force F provided by the adsorption unit can be obtained by Ansys Maxwell, as shown in figure 7, figure 8 and figure 9.

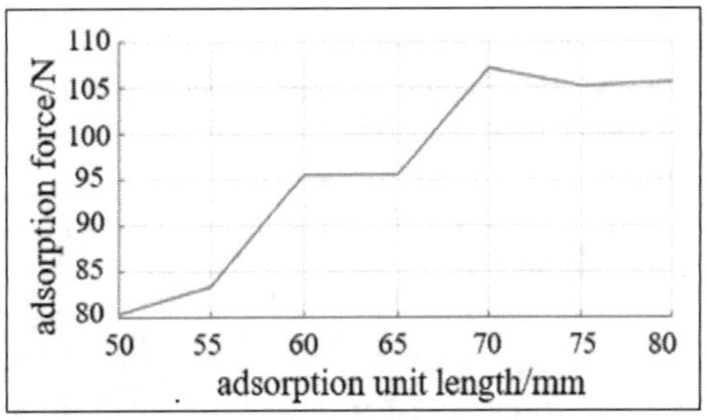

Figure 7. Relationship between adsorption unit length and adsorption force.

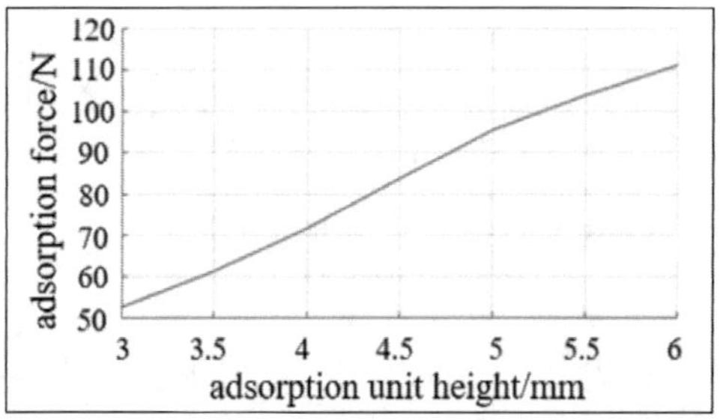

Figure 8. Relationship between adsorption unit height and adsorption force.

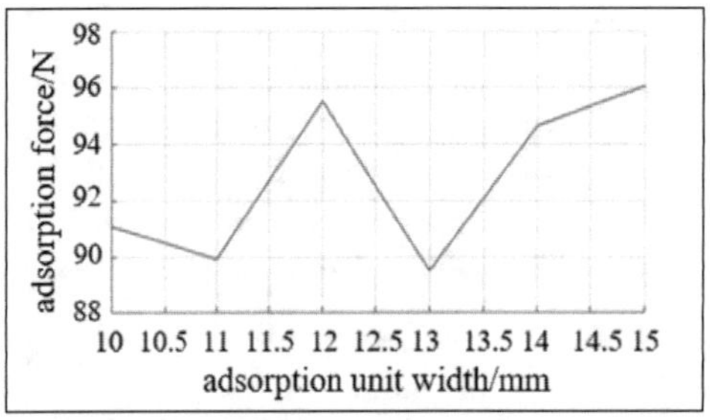

Figure 9. Relationship between adsorption unit width and adsorption force.

From figure 7, figure 8 and figure 9, we can see the relationship between the length, width and height of the magnet adsorption unit and the adsorption force. Since the adsorption force required by the design of the permanent magnet is 69.96 N, it can be obtained from the above figure that when the size of the adsorption unit is 70 mm, 12 mm and 5 mm, the adsorption force is approximately 82 N, which meets the application requirements. The permanent magnet adsorption unit, installed on the track, directly contact with the wall, and the influence of air gap width on the adsorption force can be ignored. Through the structural size optimization of magnetic force, the structural size of single permanent magnet adsorption unit can be determined. The magnetic induction line of the magnet adsorption unit is shown in figure 10.

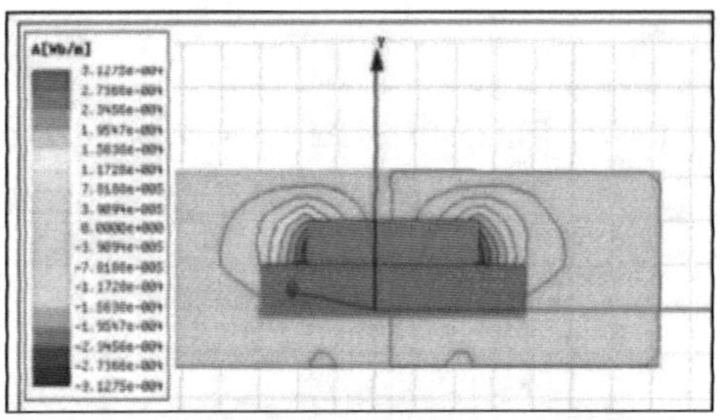

Figure 10. Magnetic induction line of adsorption unit.

5. Experiments

To verify the performance of the robot, a series of experiments were conducted to simulate the field environment on a $6\,mm$ thick metal wall made of Q235. The experimental platform of the climbing robot is shown in figure 11.

Figure 11. Experimental platform.

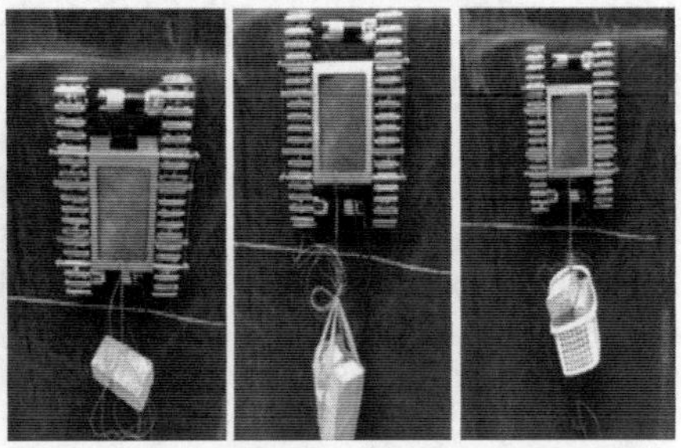
Figure 12. Robot load experiment.

Table3. Load capacity record.

Load/kg	Static state	Dynamic state
2.6	stable	stable
13	stable	stable
15.6	stable	occasionally carton
18.2	slip	easy to fall off

The climbing robot needs to carry detection equipment and cables, which undoubtedly increases the load when it moves on the wall. Therefore, it is necessary to test the load-carrying capacity of the climbing robot. The load counterweight was replaced with $2.6\,kg$ bricks, gradually increase the load of the robot in the static state and the motion state successively, and observe its adsorption state, as shown in figure 12. The load capacity record is shown in table 3.

From the experimental data, the robot can stably adsorb on the metal wall in the static state, when the load is less than $15.6\,kg$. when the load increases to $18.2\,kg$, the robot slips. In the dynamic state, the robot can steadily crawl on the metal wall when the load is less than $13\,kg$. When the load increases by $15.6\,kg$, occasional cartons occur in the crawling process. When the load increases to $18.2\,kg$, the robot will fall from the metal wall. Therefore, it can be obtained from the load experiment of the robot that the maximum load is $15.6\,kg$ and the adsorption force can meet the use requirements when the load increases.

When the robot crawls to $15\,mm$ weld, it is shown in figure 13. Experiments show that the robot can safely and stably cross the welds on the metal wall, which meets the requirements of the large metal wall for the movement performance of the climbing robot. Therefore, the rationality of robot structure design is verified

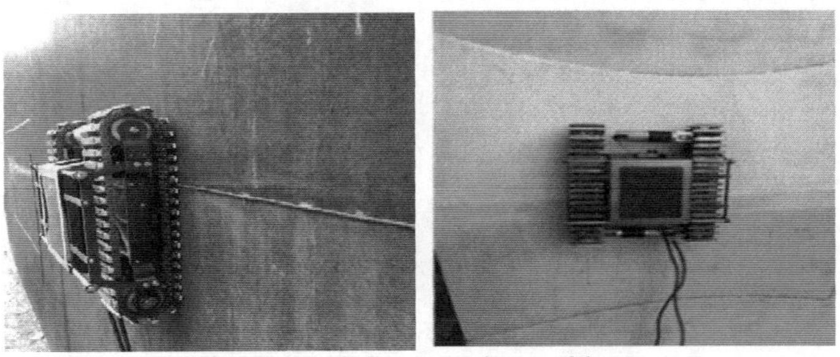
Figure 13. Robot crossing weld test.

6. Conclusion

In this paper, a robot was presented to achieve efficient detection of the large metal walls. The robot integrates contact permanent magnet adsorption units and a crawler chain mobile mechanism to obtain continuous, flexible and stable motion performance on large metal walls. Based on the mechanical model of the robot, the minimum adsorption force is calculated considering the instability conditions such as sliding, longitudinal overturning and lateral overturning. To obtain the structural size of a single permanent magnet adsorption unit, the magnetic structure was optimized by Ansys Maxwell. A series of experiments are tailored to verify the performance of the robot. Results show that the robot can achieve stable motion and accurate detection on large metal walls with welds, and the maximum load of the robot is 15.6 kg , which meets the engineering application requirements of large metal wall weld detection. In future work, the parameters of robot structure can be further optimized to improve the safety, accuracy and efficiency.

References

[1] Chen, Z.W. (2022) Application of nondestructive testing method for steel structure welds. Metallurgy and Materials, 14(01): 179-180.
[2] Zhang, W.P. (2005) Common defects detection and repair methods of the oil storage tank. Journal of Petroleum and Natural Gas, (S4): 186-188.
[3] Yan, L., Hu, W.L., Li, J.P. (2005) Research into the methods for ultrasonic testing of welds for steel-structure truss with connected lines. Journal of Nanchang Aviation University (Natural Science Edition), 19(2): 95-99.
[4] Wang, J., Mai, Z.H., Fei, Y.N. (2020) Research current status and development trends of welding seam inspection robot. Sensors and Microsystems, 39(2).
[5] Wang, Y., Bian, A.G., Wang, L. (2020) Design of saddle-type welding gun posture adaptive actuating mechanism. In: 2020 International Conference on Internet of Things, Artificial Intelligence and Mechanical Automation. Hangzhou. p. 12007.
[6] Longo, D., Muscato, G. (2006) The Alicia3 climbing robot. IEEE Robotics & Automation Magazine, 13(1): 42-50.
[7] Longo, D., Muscato, G. (2004) A modular approach for the design of the Alicia3 climbing robot for industrial inspection. Industrial Robot, 5(2): 10-18.
[8] Woosub, L., Masahiro, H. Shigeo, H. (2013) Gunryu III: reconfigurable magnetic wall-climbing robot for decommissioning of nuclear reactor. Advanced Robotics, 27(14): 1099-1111.
[9] Zhao, J.Y., Zhang, Y.N., Bi, X.D., Yan, X.C., Dong, Y.F. (2020) Optimization design of magnetic adsorption structure and machine performance test for sand blasting and rust-removing wall-climbing robot. Journal of China University of Petroleum (Natural Science Edition), 44(4): 94-99.
[10] Zhu, S.Q., Gao, Z.F., Song, W., Du, Z.T., Li, C.J., Zheng, M.J. (2020) Dynamics modelling of process of surmounting weld seam by wall-climbing robot. Transactions of the Chinese Society of Agricultural Machinery, 51(12): 377-383.
[11] Hou, J.R. (2018) Analysis of adsorption structure for wall climbing robots in wind power tower. Electromechanical Engineering, 35(9): 40-43.
[12] Gui, Z.C., Chen, Q., Sun, Z.G., Zhang, W.Z., Liu, K. (2006) Optimization of permanent-magnetic adhesion device for wall-climbing robot. Journal of Electrotechnics, 21(11): 2-5.
[13] Li, F.Q. (2002) Numerical calculation of electromagnetic fields and design of electromagnets. Tsinghua University Press, Beijing.
[14] Song, W., Jang, H.J., Wang, T., Gao, Z.F., Du, Z.T., Zhu, S.Q. (2018) Optimization design and experimental research on magnetic components for wall-climbing robot. Journal of Zhejiang University, 52(10): 3-4.
[15] Yi, Z.Y., Gong, Y.J., Wang, Z.W., Wang, X.R., Zhang, Z.M. (2010) Large wall climbing robots for boarding ship rust removal cleaner. Robotic, 32(04): 560-567.
[16] Liu, J.X. (2016) Research on weld-radiographic inspection robot. Southeast University.

Journal of Physics: Conference Series **2365** (2022) 012030 doi:10.1088/1742-6596/2365/1/012030

Design and test of a tree-climbing and pruning robot

Jianping Wu[1], Xi Yang[1], Chong Gao[1] and Wenhui Li[1*]

[1] Jiangmen Electric Power Bureau of Guangdong Power Grid, Jiangmen, Guangdong, 529085, China

*Corresponding author's e-mail: liwenhui721@163.com

Abstract. In view of the problems of manual pruning, such as time-consuming, labor-intensive, low efficiency and low degree of automation, the structure and working principle of the robot for tree-climbing and pruning were studied. Finding the problems in the process of product design, analyzing and defining them, transforming them into the standard mode of solving problems, and using TRIZ theory according to the essence of contradictions to solve the contradictions one by one.The virtual prototype was created to carry out the interference analysis and kinematics simulation analysis of the whole machine. The robot for tree-climbing and pruning was designed by modular design method, which can automatically climb to a certain height to realize the function of pruning and cutting the top of the tree trunk. The duty cycle of the motor output is gradually adjusted by PID algorithm to make multiple motors work synchronously and keep the robot stable in the up-tree pruning process and down-tree process. The robot was made and tested. The experimental results show that the diameter of climbing trunk is 100 mm～300 mm, and the climbing speed is 1 m/min. It can be used for tree-climbing, pruning lateral branches and cutting main trunks of high-straight trunk trees. It can be self-adaptively held, fall-resistant, obstacle-crossing, and manual operate to improve efficiency.

1. Introduction

In view of the high-straight trunk trees in the south, especially the tree under the high voltage transmission line, it will have a hidden danger to the high voltage transmission. At present, there are many research results on tree-climbing and pruning robots in China and abroad, mainly realized climbing and pruning to replace manual pruning for different occasions. Advaligno PATAS was developed by German Jordan family business[1]. The pruning robot with the energy-saving chain saw was developed by Qifu University in Japan. Coconut tree climbing and harvesting robot was developed in India[2]. The spiral lifting vertical pruning machine was developed by Beijing Forestry University. The tree-climbing and pruning robot for fast-growing forest was developed by Shandong Agricultural University[3]. The tree-climbing and pruning robot is developed to trim the lateral branches and cut the main trunk, so as to alleviate the increasingly prominent contradiction between tree lines and provide reference for relevant technical personnel.

2. TRIZ conflict analysis

TRIZ theory was proposed by the Soviet military technician Genrich S. and his colleagues in 1946, which is a systematic and practical theory and method system to solve the problem of invention. In TRIZ theory, the problem is described by engineering parameters, and the contradiction between parameters is completely overcome as the standard to solve the problem. At the same time, the corresponding invention solution principle is obtained by using the contradiction matrix table in TRIZ theory. Combined with practical problems, select the principle of invention in line with product design, get the

Journal of Physics: Conference Series **2365** (2022) 012030 doi:10.1088/1742-6596/2365/1/012030

best product design scheme. The situation that is encountered in solving problems is that when improving some performance of products, it may have an impact on relevant parts or have a negative impact on other aspects of performance, which is the technical conflict described by TRIZ theory[4].

The adaptability and versatility of the robot are improved, and the complexity of the device will increase. In order to solve this technical contradiction, the adaptability and versatility are used as the improvement parameters, and the complexity of the device is used as the deterioration parameter. By searching the contradictory conflict matrix, multiple invention principles to solve the contradiction are obtained : 15 (dynamic), 28 (substitution of mechanical system), 29 (pneumatic and hydraulic mechanism), 37 (thermal expansion). Create a conflict matrix as shown in Table 1.

Table 1. Contradiction matrix.

Improving parameters	Deterioration parameters	Recommended invention principles
Adaptability and versatility	Complexity of the device	15, 28, 29, 37

Linking theory with practice : Principle 15 is inspired. The electric push rod and plastic spring are used to clamp the tree trunk through inflatable tires and crawl along the tree trunk at a certain spiral angle, so that the climbing and clamping mechanism can move synchronously, and have a certain obstacle-crossing ability. When the tree is bumped or the tree stubble is encountered, the spring contraction is used to cross the obstacle and avoid sticking. Simple mechanism can reduce the design complexity of the whole machine, reduce the volume and weight of the whole machine, reduce the cost, and can meet the required functions.

3. Structure design of a tree-climbing and pruning robot

According to the functional characteristics, the tree-climbing and pruning robot can be divided into tree climbing module including clamping mechanism, pruning module and control module. The tree climbing module climbs to a certain height to assist the pruning module to complete the pruning work. Set a camera on the pruning module to ensure that the staff under the tree can observe in real time . In order to prune branches with different diameters, each wheel keeps the same angle with the cross section of the trunk, so that the device climbs around the trunk in a spiral manner. The movement is transmitted by the friction between the tires and the trunk, and the rising pitch is determined by the spiral angle and the diameter of the trunk[5].

As shown in Figure 1, assume that the angle between the wheel axis and the horizontal plane is θ; the displacement of the device for tree-climbing and pruning along the dip direction is v; v displacement along the horizontal direction is x ; v displacement along the vertical direction is y; the diameter of trunk is d ; the rising height of the device is h.

The rising height of the device around the tree is related to the inclination angle θ and the trunk diameter d. The higher the inclination angle θ and the larger the trunk diameter d, the higher the rising height of the machine around the tree after a week of rotation. Take θ= 75°, when the trunk diameter d is 300 mm, rotation a week rise height of 252.5 mm; when the trunk diameter d is 100mm, a week rise height is 84.2mm.

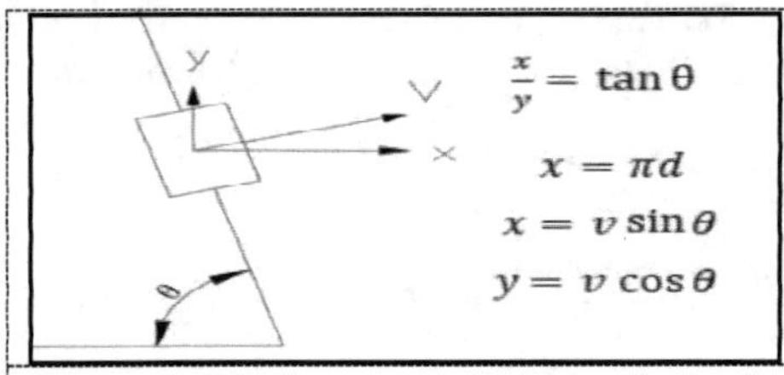

Figure 1. Displacement analysis of wheels

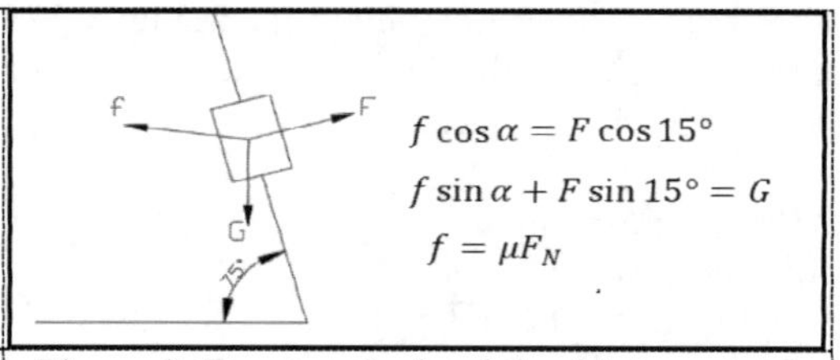

Figure 2. Force analysis of the driving wheel

The driving wheel is mainly subjected to four forces in the working process, namely gravity G, the friction force f between the tree trunk and the wheel, the axial driving force F of the wheel and the radial thrust F_N of tree trunk to wheel. As shown in Figure 2, it is assumed that the angle between f and horizontal direction is α, F_N = 2000N, μ= 0.4, M = 30 kg.

When the tree-climbing and pruning robot works, they are in danger of sliding downward and overturning. In order to avoid the instability of the tree-climbing and pruning robot, it is necessary to ensure that the static friction force generated by the active wheel and clamping mechanism is greater than the gravity force of the robot along the tree trunk, and the anti-tilting torque of the clamping mechanism is greater than the tilting torque of the robot under the action of gravity[6].

The overall framework of the device is semi-circular and semi-clasp structure. Four active rubber wheels are arranged in the middle of the overall frame, driven by four DC worm gear reducer motors, and four groups of driven rubber wheels are arranged up and down, with the wheel inclination angle of about 15°. The electric push rod on both sides controls the opening and closing angle of the locking mechanism, which can adapt to the change of trunk diameter from 300 mm to 100 mm. According to the actual working requirements, the engineering design of the tree climbing pruning intelligent device is carried out to determine the standard parts and main structural parameters.The overall structure of the device is shown in Figure 3. 1 - hold the electric push rod; 2 - subject framework; 3 - shock absorber; 4 - rubber wheel components; 5 - DC worm gear reducer motor; 6 - pruning side branch electric push rod; 7 - guide rail sleeves and electric saw seats; 8 - cutting side chain saw; 9 - active rubber wheel components; 10 - cutting the main electric push rod; 11 - cut main chain saw; 12 - lever; 13 - electric saw seat; 14 - follow-up framework. There are 12 controlled objects: 4 climbing motors, 2 holding electric push rods, 2 electric push rods of electric saw, 2 steering gear of camera platform and 2 electric saw. Operators need to use special handheld devices to control these 12 controlled objects to complete the tasks of climbing, pruning lateral branches and cutting main trunks.The three-dimensional model of the device is shown in Figure 4.

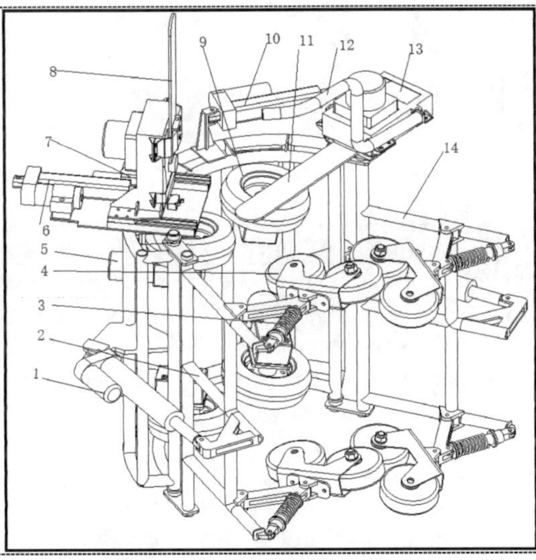

Figure 3. General structure

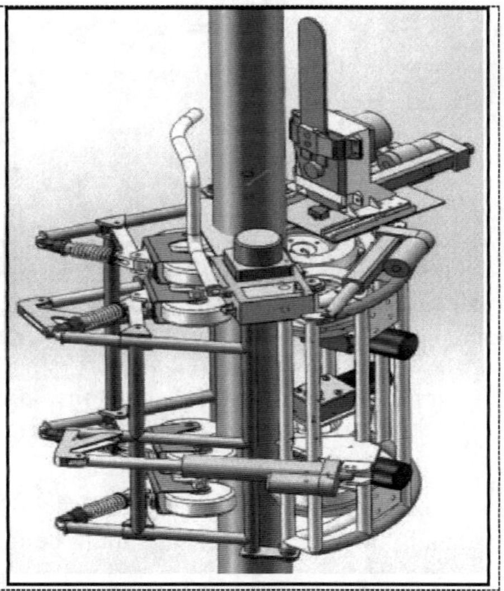

Figure 4. Pruning branches and trunks

The device is divided into up-tree pruning process and down-tree process.

(1) The process of up-tree pruning: The main frame of the device is manually set at the bottom of the tree trunk, and the wireless remote control is used to start the device. The electric push rod is held tightly to start working, and four driven rubber wheels are continuously pressed against the tree trunk until the tree trunk is held tightly. Four DC motors drive four active rubber wheels forward to realize the overall spiral rise of the device. The diameter of the tree trunk changes during the climbing process. The sensors feedback to the system in time to control the operation of the electric push rod to ensure that the wheels always hold the trunk during the climbing process. The chain saw trims lateral branches. When it rises to the required height, the chain saw cuts main trunk, and the rod pushes the trunk to one side at the same time.

(2) The process of down-tree: When the device completes the cutting of the trunk, the device reverses to achieve spiral decline. When it reaches the bottom of the tree, it stops working.

4. Control module design of a tree-climbing and pruning robot

The mechanical structure part needs to be combined with the controller, power supply and remote control to complete the tree climbing and pruning function. The handheld device is the upper computer, and the control unit of the controlled object is the lower computer. The upper computer controls the lower computer by remote control. The design of the overall control scheme is shown in Figure 5.

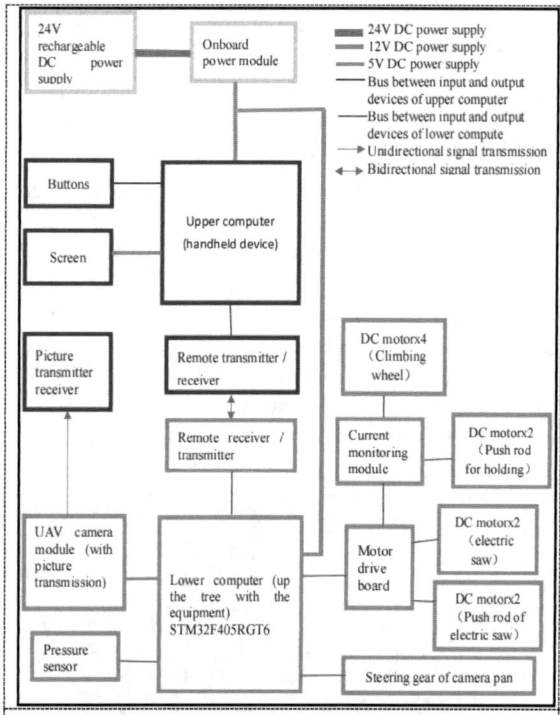

Figure 5. Overall controlling design

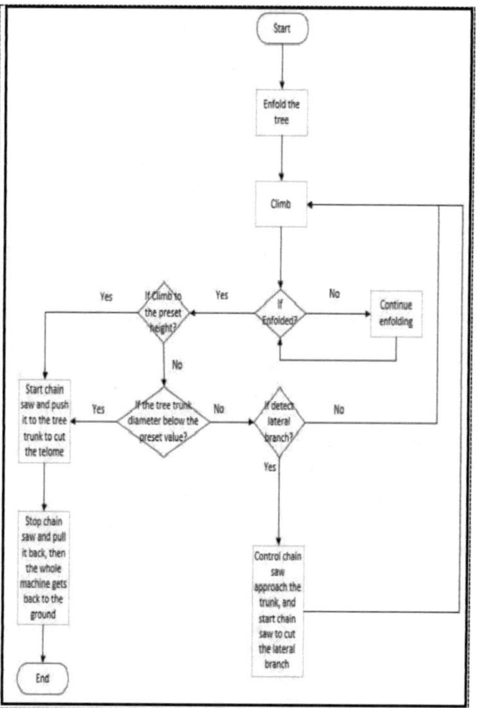

Figure 6. Automatic workflow flowchart

This device is divided into automatic and manual two ways, automatic working process is shown in Figure 6. Each action is controlled manually by the switch on the remote control. The main control chip reads the real-time return value of the motor encoder, calculates the data, and gradually adjusts the duty cycle of the motor output through the PID algorithm to make multiple motors work synchronous, so that the robot remains stable in the up-tree pruning process and down-tree process.

5. Kinematics simulation analysis of the whole device

The quality of each component should be ignored in the kinematic analysis of the whole device. The kinematic model is analyzed by using the rigid body dynamics software ADAMS, and the kinematic size of the components is extracted. The actions such as crawling around the tree trunk and holding the pruning are verified to ensure the smooth crawling action and no interference in the pruning action.

When the tree trunk is held tightly by the device, the electric push rod is combined with the chain saw to prune lateral branches or tree trunk. At this time, the frame and wheels are fixed. The electric push rod moves the cutting main trunk chain saw to the designated position to cut the trunk. After the calculation, the velocity and acceleration curves of the end point of the chain saw are derived, as shown in Figure 7 and Figure 8.

In Figure 7, the velocity curve in the Z-axis direction has an obvious inflection point. Here, the velocity curve in the Z-axis direction is Fourier transformed to obtain its spectrum, as shown in Figure 9. The Z-axis velocity spectrum shows that there are high amplitude low frequency components at each time node, and the low frequency value does not affect the normal operation of the device.

There are many inflection points in the acceleration curves of Y and Z directions in Figure 8, which proves that there are more shocks in Y and Z directions. Among them, the Z-axis acceleration spectrum is shown in Figure 10. The high amplitude is concentrated in the time from 0 s to 5 s, which proves that the electric opportunity connected with the chain saw is subjected to a certain impact force in the process of the chain saw moving to the specified position. In order to improve the safety and reliability of the chain saw, the motor should be started at a low speed, and the motor should be ensured to operate smoothly in the sawing process to avoid the damage of the motor caused by excessive impact force.

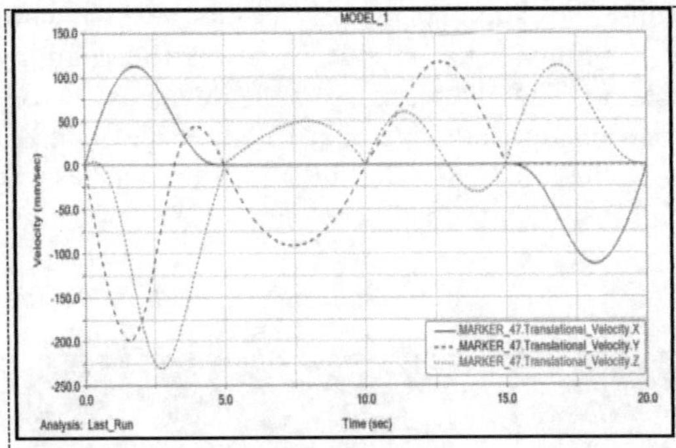

Figure 7. Velocity curve of the end point of the saw

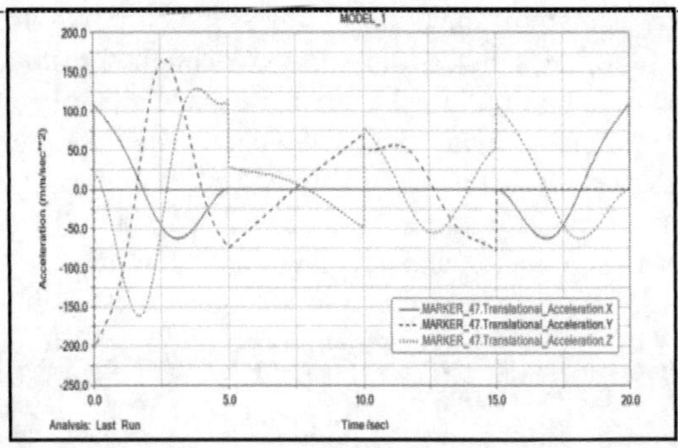

Figure 8.Acceleration curve of the end point of the saw

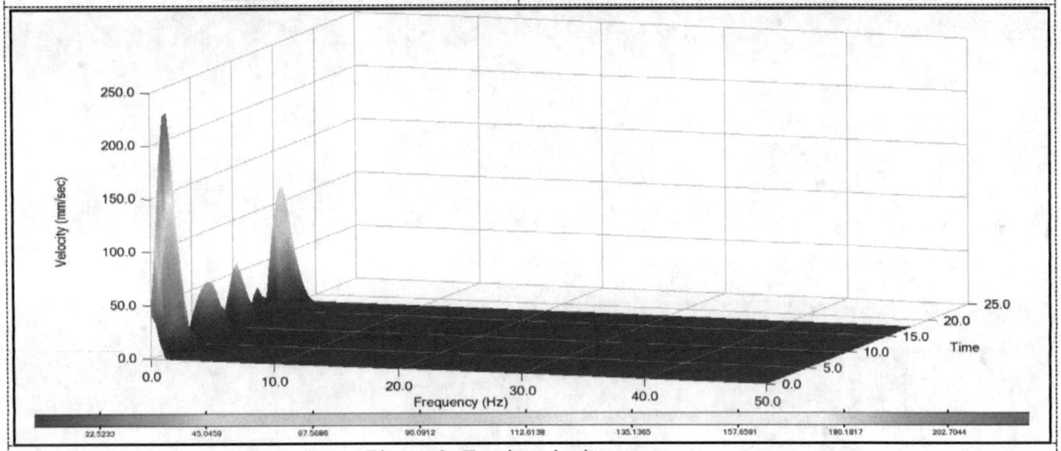

Figure 9. Z axis velocity spectrum

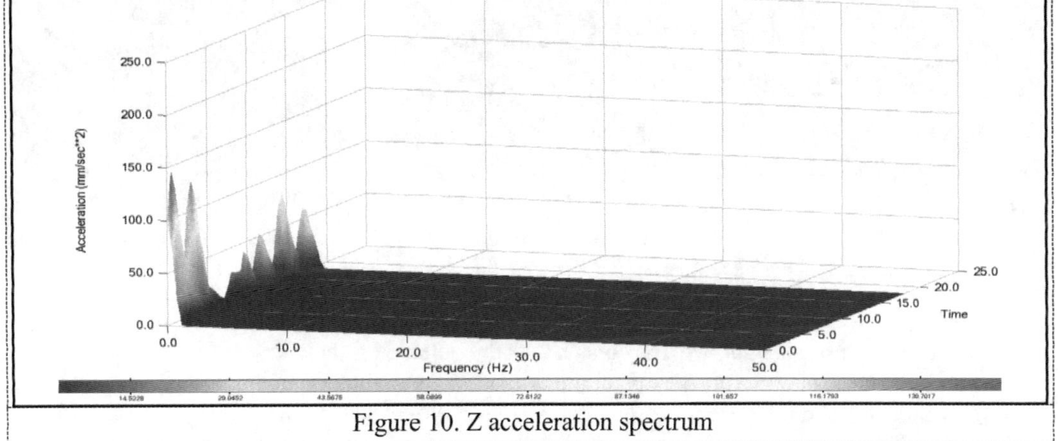

Figure 10. Z acceleration spectrum

6. Prototype production and experiment

The prototype production of tree-climbing and pruning robot was completed in March 2022. Limited by the experimental conditions, some performance tests were carried out on the device, as shown in Figure 11.

The results show that the device has been verified that the diameter range of climbing trunk is 80 mm～260 mm, and the climbing speed is about 1 m / min. The experimental test saws the side branches with 30 mm diameter, and the experimental test saws the trunks with 95 mm diameter. The climbing height is up to 6 m, which can cross the side branch stump. The overall mass of the device is 28 KG, which can be divided into three parts. The maximum part mass is not more than 15 KG, which realizes light weight and modularization. The device can achieve the functions of pruning lateral branches and cutting trunks, and can also be self-adaptively held, fall-resistant, obstacle-crossing, measuring height and tree diameter, and manual remote control.

A large number of tests were carried out on the prototype of the tree-climbing and pruning device, and the problems in design and operation were found to prepare for the improvement of the prototype.

(a)Prototype device | (b)pruning lateral branches | (c)After pruning lateral branches

(d)Cutting trunk and pushing down | (e) After cutting main trunk | (f)Cutting trunk's diameter was 95 mm

Figure 11. Tests of the tree-climbing and pruning robot.

7. Conclusion

Based on the results presented above, the conclusions are obtained as below:

(1) It is shown that the structural design of the device for tree-climbing and pruning uses Solidworks software to carry out the interference test on the virtual prototype. ADAMS software is used to carry out the simulation analysis of the kinematic model. The actions of spiral clamping around the tree trunk, such as crawling and pruning, are verified to ensure the smooth crawling action and no interference in

pruning action.

(2) The application experiment was carried out on the prototype device. According to the test results, the device for tree climbing and pruning is designed reasonably, stable and reliable. It is suitable for climbing and pruning lateral branches and cutting trunk of high straight trunk trees, and its market prospect is extremely broad.

Acknowledgments
This article is one of the phased achievements of China Southern Power Grid's science and technology project "Design and Development of Intelligent Device for Pruning High-Straight Trees" (GDKJXM20200479). I gratefully acknowledge the financial support of China Southern Power Grid Corporation.

References

[1] Wang J C, Zheng H, Zhang C, Meng C Y. 2021. Analysis and Research on Tree-climbing and Pruning Robots. In: the 2nd International Conference on Mechatronics Technology and Intelligent Manufacturing, ICMTIM 2021.Hangzhou, China. p 012062-012068.
[2] Cui S H, Jiang S H, Geng Z B. 2019. Research progress in tree pruning machinery for plantation. World Forestry Research, 32(5) : p 47-51.
[3] Gui P F, Tang L Q, Mukhopadhyay S. 2018. A novel robotic tree climbing mechanism with antifalling functionality for tree pruning. Journal of Mechanisms and Robotics, 10(1) : p 014502.
[4] Li W H, Yang J J, Lian X. 2021. Structural Design and Experiment of a Vertical Bicycle Parking Device Based on TRIZ Theory. In: the 4th World Conference on Mechanical Engineering and Intelligent Manufacturing, WCMEIM 2021.Shanghai, China. p 539-542.
[5] Fu G H, Liu X M, Chen Y F. 2015. Fast-growing forest pruning robot structure design and climbing control. Advances in Manufacturing, 3(2) : p 166-172.
[6] Yang X, Wang J C, Yan M X, Fan S. 2021. Structure Design of Surrounding Tree-climbing and Pruning Robot. In: the 2nd International Conference on Mechatronics Technology and Intelligent Manufacturing, ICMTIM 2021.Hangzhou, China.p 012052-012058.

Finite Element Analysis and optimization design of sprinkler Truss Structure

Changhu Wang[1a], Weiyun Meng[1b*], Jianxiang Wang[1c], Xiangmei Min[2d*]

[1]College of Mechanical and Electronic Engineering,Shandong Agriculture and Engineering University, Jinan,Shandong, 250100, China.

[2]Jining Industrial technician College, Jining,Shandong, 272100, China.

[a]z2013168@sdaeu.edu.cn,[b*]z2017019@sdaeu.edu.cn, [c]3381584317@qq.com, [d*]852754779@qq.com.

Abstract: This paper includes three sections. Truss structures have been widely applied in bridges and buildings due to their light weight, reasonable load-bearing, and high material utilization rate. For truss structures serving as load-bearing carriers, their structural strength should be guaranteed, and moreover, their economic performance should be considered. First, the finite element theory was introduced, and the computer-aided simulation analysis was performed by combining numerical analysis and finite element theory. Second, the truss structure of a sprinkler was subjected to a mechanical analysis, and its stability was explored with its maximum displacement and maximum stress as constraints. Finally, the truss structure was optimally designed until its stability, strength, and stiffness met relevant requirements.

1. Introduction

The finite element method (FEM) initially aimed to study structural mechanics and was widely applied to such fields as fluid mechanics and heat transfer science later on. In this study, the target truss structure was analyzed via the finite element theory. The reliable truss operation is of vital importance to the sprinkler and is associated with the normal operation of sprinkling and fertilization. The mechanical strength of truss structures is required to be very high. Beams—load-carrying structures in the truss—need to bear very great acting forces. Hence, the truss structure should be subjected to a force analysis and optimized in order to improve the mechanical properties of the whole structure and ensure the reliability and safety of truss operation. In this study, the mechanical properties of the truss structure were mainly analyzed through finite element software. On this basis, the truss structure was optimized and practically applied in a test base, during which good application effects were achieved.

2. Overview of Finite Element Method (FEM) Theory

When it comes to the finite element method (FEM), a practical problem is solved after being simplified into a mathematical model via a simplified model. And thus, the practical problem is regarded as a mathematical problem solved in a computer-aided fashion according to the numerical analysis and the FEM theory [1]. Then, the boundary value problem of differential equations is solved, and the physical model obtained on the basis of a mathematical theory is subjected to the discretization simulation analysis[2]. The FEM theory exhibits excellent processing effect and discretization accuracy when handling practical problems like stress concentration, especially in the case of complex geometrical shape and boundary conditions. Moreover, the FEM theory has comparative advantages in analyzing intricately shaped structures and processing complex boundary conditions[3]. In addition, the accuracy problem is solved through formulas like high-order interpolation functions so that the obtained conclusions can meet the practicability and accuracy requirements.

Furthermore, the FEM theory can be used to process various materials like anisotropic and orthogonal isotropic materials and analyze complex composite structures consisting of multiple materials[4] Besides, the FEM theory mainly aims to explore discrete structures[5], namely, the analysis object is simplified into discrete objects consisting of materials of the same kind and divided into all kinds of small units[6]. After a physical entity is discretized, the force is transferred among units, and the force acting upon units is equivalently substituted by nodal forces, followed by simulation analysis and investigation.

Three processes are mainly involved in the simulation analysis via ANSYS: First, a finite element model of the study object is established according to its physical mode. Then, the load is applied and solved according to the engineering situation. Finally, the results are checked and analyzed. Mesh generation is done after the finite element modeling. In the event of too refined mesh generation, the calculation will be more time-consuming, and unnecessary computer resources will be occupied, and moreover, the final calculation result may not always be appropriate[7]. If the meshes are generated too sparsely, the obtained result will be inaccurate despite the simple calculation process. Hence, the mesh refinement should be appropriate.

When a 3D geometrical model was established via software, secondary factors were neglected in order not to affect the overall calculation results. Moreover, it was necessary to perform mesh generation for the geometrical model. Specifically, meshes were automatically generated by selecting a type of mesh element. Then, the model was solved through a computational analysis module. Subsequently, the calculation results were exploratively analyzed. The truss structure, which was made of Q235 steel, was 33,000 mm in full length. In this prototype structure, a layer of support was chosen from this prototype model as the analysis model. Two types of elements—Shell and Solid—were used in the finite element modeling process. A structural failure would be identified when the ultimate stress of a component was approximate to the yield stress. Given this, the changes in the ultimate strength of the truss structure under different circumstances were verified.

3. Study on Mechanical Analysis of sprinkler Truss Structure

The steel pipes in the truss structure of a sprinkler exert the effects of transporting water and fertilizers and bearing the axial pressure and bending moment. The web member support arranged in the rack is made of profile steel, with the main functions of connecting water pipes and bearing pressures. The truss structure, which aims to ensure the overall stability of the sprinkler, bears the tensile force applied by the entire structure as well as its own gravity, wind power, and the water load in water pipes. However, the stability of the truss structure will be affected in the case of its too high center of gravity.

The test base is located in Yuncheng County, Shandong Province, which is subjected to small wind power at ordinary times. Therefore, the effect of wind power was neglected. The mass of the truss structure was made to be the lightest, in which the cross-sectional dimensions of each member bar served as design variables and the maximum displacement and maximum stress of the truss structure as constraint conditions. On this basis, the stability and economical efficiency of the truss structure were explored and discussed, thus improving its mechanical properties and optimization design to obtain the optimal dimensions.

3.1 Mechanical properties of truss structure

In the finite element software, both water pipes and profile steels were tension and compression components, and the supports at both ends were relatively fixed with the truss structure, accompanied by the constraints applied. The forces borne by the truss structure mainly included the water gravity, the gravity of the truss structure itself, and the wind load. The truss structure was bilaterally symmetrical to such forces, which belonged to uniformly distributed loads. In the force analysis, the dead weight of the truss structure and the water gravity were mainly considered, where the latter was regarded as a uniformly distributed load, which was applied according to the product between the gravitational acceleration and the density of water pipes.

Multi-nozzle truss-type sprinklers have the advantages of low energy consumption and evener sprinkling over single-nozzle sprinklers. However, truss-type ones are characterized by a large span and poor operating environment, making them difficult to run freely and smoothly. Moreover, they tend to slide and even turn over along steep slopes, thus affecting normal operation. Therefore, it is necessary to investigate the stress conditions of the truss structure under harsh environments, and exploratively analyze the height of its center of mass and the wheel track for the sake of optimization design and enhancing its stability and reliability.

The truss design has a great influence on fertilization efficiency. The work efficiency can be improved by strengthening the truss strength and relieving the mass of sprinkler equipment, and meanwhile, the economical efficiency should be taken into account. In the finite element analysis, the stresses borne by the truss structure were analyzed mainly considering the bending deformation and axial deformation of member bars as beam elements.

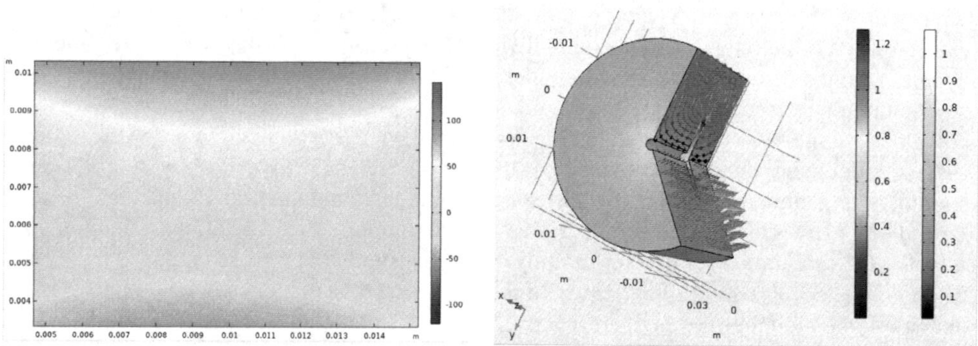

(a) (b)

Fig. 1 Stress analysis diagram of truss structure

According to the material properties and stress conditions of the truss structure, the simulation analysis was performed at the fertilizer bucket volume of 90 L and 140 L, thereby obtaining the stress nephograms of the truss structure. The strain firstly grew and then declined with the increase in the stroke close to the middle position.It could be observed from the stress nephograms that the maximum stress in the middle of the truss structure was 190 and 346 MPa, respectively, both of which were greater than the permissible stress of 180 MPa. Hence, the fertilizer bucket was too large, failing to meet requirements, and thus its volume should be reduced.

As reflected by the stress, deformation, and displacement simulation analysis of the truss structure, the middle position was subjected to the maximum stress (exceeding the permissible stress) and the maximum deformation under the joint action of uniformly distributed load and concentrated force, namely, the middle position was susceptible to the fatigue failure. Therefore, this position must be checked during the truss strength test, and the dimensions should be optimized to ensure the reliability and economical efficiency of the truss structure.

3.2 Optimization design and parameter determination of truss

From the stress analysis chart obtained through the simulation analysis of the truss structure, it could be known that the external force borne by the truss structure should be changed and the volume of the fertilizer bucket should be reduced. For the fertilizer bucket in the middle position of the truss structure, a concentrated load was applied at the support node as the boundary condition, and finally, gravity was applied to the whole system.

According to the finite element simulation and calculation, one part of the truss structure was intercepted when the volume of the fertilizer bucket was 55 L, and the maximum concentrated force was obtained as 550 KN. As revealed by the numerical calculation and finite element analysis, satisfactory results could be achieved if the pipe radius was above 11 mm and the thickness was greater than 1 mm.The scheme harvested the most evident optimization effect at the volume of 55 L. Hence, the truss structure could be transformed according to this scheme, thereby enhancing its overall performance. In addition, the pipe requirements of the truss structure in the aspects of portability and load-bearing of fertilizer buckets could be relatively satisfied using stainless steel pipes.The reliability of this truss structure was strengthened through its structural optimization and improvement, thus laying a good foundation for agricultural production.

The truss structure was subjected to finite element simulation with the optimization objective of minimizing the truss mass. On the premise of meeting the dimensions and permissible stress, the requirements for reliability and economical efficiency should be considered, and efforts were made to minimize the truss mass. Next, an objective function was established based on the reliability of truss mass. Moreover, a mathematical model was constructed by using the node coordinates and cross-sectional area as design variables of shape and dimensions and the strength and deformation amount of member bars as constraint conditions.

Table.1 Parameters of sprinkler truss

name of index	Specification
length（m）	33
Axis of highly（m）	0.37
Axial spacing（m）	0.8
Spacing between ventral rod joints（m）	0.60
Length of winding rod（m）	3.0
Length of lower chord（m）	3.0
Length of straight belly rod（m）	0.48
Length of diagonal belly rod（m）	0.49
Pull rod length（m）	0.46
Steel pipe diameter（mm）	26~30

Dimensions should conform to the requirements for economical efficiency, small size, and light weight. Meanwhile, strength and stiffness are also needed to meet relevant requirements. By reference to the dimensions of the existing member bars, the maximum deformation occurred in the middle of the truss structure, the stiffness, slenderness ratio and strength of each member bar should meet requirements, and thus the permissible ranges of design parameters for the dimensions of each member bar were provided.

Favorable stability, extensibility, and light weight of member bars are required in order to operate on the rugged pavement between paddy fields and reduce the bend loss of water. As seen in Table 1, the truss structure was designed on the precondition of its guaranteed stability, strength, and stiffness. In addition, the pipe diameter and the height of the axial center were designed, enabling the truss structure to meet the stability and lightweight requirements.

4. Conclusion

The truss structure in a sprinkler mainly bears water weight and dead weight, and its structural reliability, stiffness, and strength should be considered. The maximum deformation generally tends to occur at the middle position of the truss structure. In this study, the mechanical properties of members were investigated through finite element analysis, and the truss structure was optimized, thereby guaranteeing better sprinkling and fertilization of fertilization devices.

(1) Three volume values were designed and finite element analysis was implemented according to the actual situation. In the end, the third scheme was selected. When the volume of the fertilizer bucket was set at 55 L, the maximum stress of the optimized truss structure conformed to the permissible stress requirement.

(2) The optimized scheme was applied to engineering test practice and good effects were achieved. Moreover, the reliability and stability of truss operation were significantly enhanced, thus contributing to better sprinkling and fertilization of fertilization devices.

Acknowledgments
This work was financially supported by the Science and Technology Poverty Alleviation Project of Shandong Agriculture and Engineering University (30220004) .

References

[1] Tang Yue，ZhuXiangyuan，Mei Xingxin，et al. Per-formance test of water turbine of JP50 reel sprinkler[J]. China rural water and hydropower，2014，(2):26—29.

[2] Wang Tao，Wang Chuan，Kong Fanyu，et al. Theoretical，experimental，and numerical study of special impeller used in turbine mode of centrifugal pump as turbine[J]. Energy，2017，130:473—485.

[3] Li Zhantai，Yang Jungang，Zou Guoyuan，et al. Current situation analysis of light and simplified production in greenhouse vegetable gardens in Beijing ［J］ . China Vegetables，2019(8) : 68-75.

[4] Liu Jinping, Sun Feifei, Wang Xia, et al. Effects of fertilizer reduction on growth and quality of Tomato [J]. Northern Horticulture,2021(8):52-56.

[5] Han Xue, Qu Mei, Li Yinkun, et al. Effects of different fertilization levels on greenhouse tomato growth, nitrogen uptake and Soil and Fertilizer Science in China,2021(2):162-169.

[6] Yue Huanfang，Li Chao，Li Youli，et al. Studies on irrigation decision in substrate culture of high quality tomato ［J］ . China Vegetables，2020(4) : 56-60.

[7] Pan Jing, Ge wen, Ye Yongwei. Integrated water Research on advantages, mechanism and significance of chemical technology [J]. New Agriculture, 2022 (3) : 16.

Calculation and Analysis of Deformation Characteristic and Energy Dissipation of Flexible Bearings in Harmonic Drive for Industrial Robots

Zhaoxiang Xu, Yazhen Wang*, Haifeng Wang

School of Mechatronic Engineering and Automation, Shanghai University, Shanghai, P.R. China.

*meyzwang@shu.edu.cn

Abstract. The flexible bearing is a key component of harmonic gear reducer for industrial robots. The wave generator cam as assembly part will significantly affect the flexible bearing transmission state. This paper derives the flexible bearing deformation coordination equation formula based on the harmonic drive theories, and the deformation characteristics of the flexible bearing formed by different wave generator cam shapes were studied. Combined the theory of clabellon with the load and deformation which are obtained, the deformation energy dissipation of the outer ring of the flexible bearing is discussed and quantitatively evaluated. The results showed that the wave generator cam shapes have a greater effect on the carrying capacity of the flexible bearing, and the dissipation of deformation energy of the flexible bearing can be effectively reduced by selecting appropriate wall thickness of the outer ring. Moreover, it was found that the two-roller wave generator deformation energy dissipation is the least at the same constraint.

1. Introduction

Harmonic gear reducer is widely used in the field of the industrial robots to adjust the movement speed and output torque of robot arms. It has the advantages of small return difference, large transmission ratio, high transmission precision and high transmission efficiency. The harmonic gear reducer is mainly composed of three components: a circular spline(CS), a flexspline(FS) and a wave generator(WG). The WG consists of a flexible bearing (FB) and a wave generator cam. When the harmonic gear reducer works, the outer ring of the FB will produce corresponding periodic elastic deformation by the wave generator cam rotates. Therefore, the structural parameters of the WG will greatly influence the reliability and efficiency of the harmonic gear reducer for industrial robots.

Many experts and scholars have made a large amount of research about the FBs. Bourdon A[1] solved the load and displacement of rolling bearings structures. Shen[2] calculated the load of the FB in wave generator based on the meshing force distribution of the harmonic drive theories. W. Ostapsk[3] proposed the calculation method of optimizing the size of wave generator by analyzing the fit of different specifications of cam and the FB. Li[4] calculated the load distribution of the FB under cosine cam assembly based on analyzing the deformation and stress of the FB. Yu[5] proposed a three-force ring calculation model and obtained the load distribution of the FB assembled with cosine cam. Lyuminarsky[6] proposed a method to improve the transmission efficiency of the FB by studying the contact deformation of the FB structure, Shao[7] established the expression of the relationship between interface energy dissipation and contact deformation, and studied the characteristics of energy dissipation caused by plastic deformation and interface friction. Shi[8] obtained that the friction dissipation caused by elastic deformation of the harmonic gear reducer accounts for 20% ~ 50% of the total transmission energy dissipation by introducing the energy dissipation relationship of elastic deformation.

The FB is one of the most important components for a harmonic gear reducer, so it is necessary to study the performance of the FB. In this paper, the load distribution of the FB assembled with different wave generator cams are calculated and analyzed. The effects of thin-walled part structures of the FB deformation energy dissipation are studied. It provides references for the optimization of the structural parameters and the improvement of transmission performances of the harmonic gear reducers.

2. Analysis of deformation characteristic of the flexible bearing

2.1. Calculation of elastic deformation of the outer ring

The FB inner and outer rings will be forcibly deformed after assembled with cams. The deformation of the inner ring is decided by the cam shape, and the outer ring produces periodic elastic deformation in working condition[9]. This paper analyzes the deformation characteristics of the FB in the form of the cosine-cam and roller-cam. The deformation shapes of the WG are shown in figure 1.

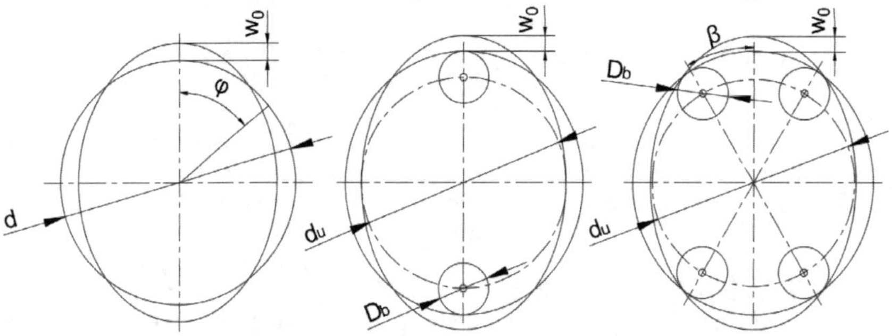

(a) cosine-cam WG (b) two-roller WG (c) four-roller WG

Figure 1. Three kinds of diagram of conventional mechanical WG

(1) Deformation function of the cosine-cam:

$$w(\varphi) = w_0 \cos 2\varphi \tag{1}$$

Where w_0 is the maximum pre-deformation amount.

(2) Deformation function of the roller-cam:

$$w(\varphi) = \frac{w_0}{\sum\limits_{n=2,4,6...} \frac{\cos n\beta}{(n^2-1)^2}} \sum_{n=2,4,6} \frac{\cos n\beta \cos n\varphi}{(n^2-1)^2} \tag{2}$$

Where the four-roller WG β value is 30°. two-roller WG can be regarded as a special case of four-roller WG when β = 0°.

In the harmonic gear reducer, the outer ring surface of the FB will be affected by the meshing reaction between the flexible gear and rigid gear. It is assumed that radial component of the meshing force can transfer to the FB without consumption. The radial load expressions are:

$$q_r = q_{t\max} \tan \tau \cos[(\varphi - \Phi_1)\pi / (2\Phi_2)] \tag{3}$$

$$q_{t\max} = \pi T / (2\Phi_2 d_r^2 b) \tag{4}$$

Where $q_{t\max}$ is the maximum of tangential load, τ is the tooth profile angle, d_r is the pitch circle of flexspline tooth, b is the width of the outer ring, Φ_1 =0 when partial load is not considered, angle of radial load $\Phi_2 = \pi/3$.

The radial deformation of the outer ring under this radial distributed load is[9]:

$$q_{rk} = 8\Phi_2 q_{t\max} \cos(k\Phi_2)\tan\tau(\pi^2 - 4k^2\Phi_2^2) \tag{5}$$

$$w_q = \frac{12(d+t/2)^4}{Et^3} \sum_{k=2,4,6...} q_{rk} / \left(k^2-1\right)^2 \cos[k(\varphi-\Phi_1)] \tag{6}$$

Where q_{rk} is the coefficient of the term of the series.

Combined with the cam deformation function, when the FB bears the load, the equation of the total deformation of the outer ring can be obtained as follows:

$$w_i = w_{\varphi,i} - w_{q,i} \tag{7}$$

2.2. *Calculation of load distribution of the flexible bearing*

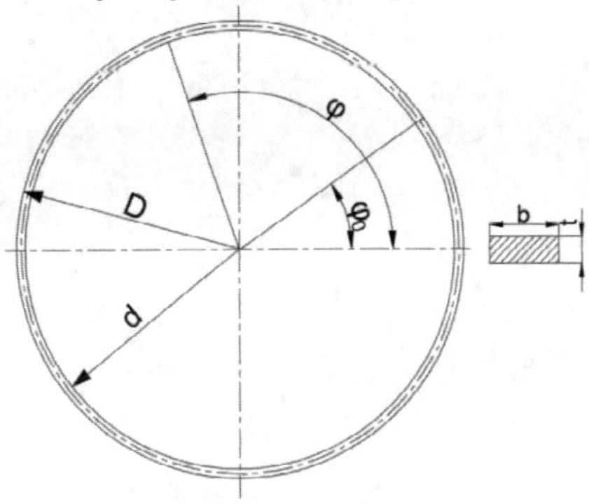

Figure 2. Diagram of thin-walled ring

Structure diagram of thin-walled ring as shown in figure 2. According to the thin-walled ring theory, the overall deflection deformation of the ring should be considered to determine the load distribution of the rolling element. The outer ring section is equivalent to a rectangle to simplify the calculation of the section moment of inertia. Therefore, the differential equation of radial displacement under bending moment[9] is transformed to:

$$\frac{d^2 w}{d\varphi^2} + w = -\frac{12M}{Ebt^3}(d+t/2)^2 \tag{8}$$

Where M is the connecting bending moment, w is the radial displacement, E is the modulus of elasticity, d is the inner radius of the outer ring, t is the thickness of the outer ring. φ is angular position on section of the ring.

The relationship between the ring load and radial displacement is deduced as follows[5]:

$$w_i = \begin{cases} \dfrac{6F(d+t/2)^3}{Ebt^3}\begin{bmatrix} \frac{1}{2}(\sin\varphi + h_2\varphi\sin\varphi - \varphi\cos\varphi) + \\ \frac{1}{2}(\varphi_0 - \tan\varphi_0 - h_3 - \pi h_1)\cos\varphi - h_1 \end{bmatrix} & 0 \le \varphi \le \varphi_0 \\ \dfrac{6F(d+t/2)^3}{Ebt^3}\begin{bmatrix} -\frac{1}{2}\varphi_0\tan\varphi_0\sin\varphi_0 - \frac{1}{2}(h_3 + \pi h_1)\cos\varphi_0 \\ +\frac{1}{2}h_3\varphi\sin\varphi - h_1 \end{bmatrix} & \varphi_0 \le \varphi \le \pi \end{cases} \tag{9}$$

Where φ_0 is the angle of adjacent balls, and define the function of $h_{1,2,3}$ as φ_0 [5]. If the radial displacement coefficient is H, the radial load expression of the ring can be obtained as follows:

$$F = w_i H(\varphi) \tag{10}$$

Table 1. Main structural parameters.

Parameter	Value
Outer radius of outer ring D /mm	24.5
Inner radius of outer ring d /mm	23.3
Thickness of outer ring t /mm	1.2
Number of balls N	23
Angle of adjacent balls φ_0 /°	15.7
Maximal pre-deformation w_0 /mm	0.3
Width of outer ring b /mm	7.2
Pitch diameter of flexspline d_r /mm	50.8
Profile angle τ /°	20

From the relevant calculation parameters are listed in table 1,the interal load distributions of the FB under working load are shown in figure 3.

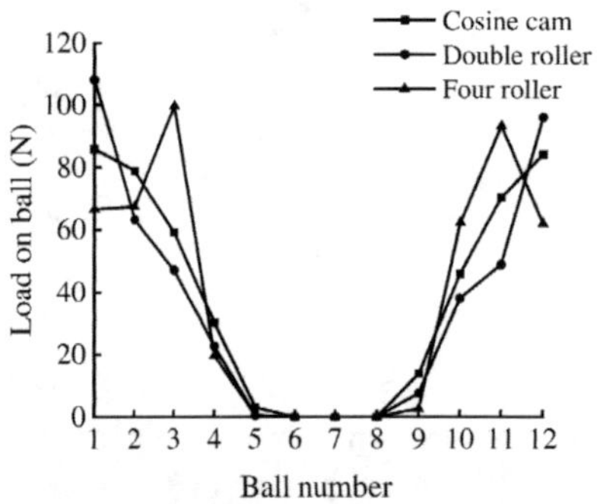

Figure 3. Ball load distribution of bearing

From the load distribution curve in the half region of the FB, It can be seen that the ball loads of the FB are symmetrically distributed under different WG cams. The FB in two-roller and cosine-cam WG has the maximum load at the longer axis, which are about 108.1N and 87.4N respectively. The maximum load of the FB in four-roller WG is at the rollers position instead of the longer axis. Which is about 101.7N.

The result shows that the cosine-cam WG has the minimum load range in three wave generators. The rollers position has a great influence on the load distribution of the FB. However, the positions of unload balls in three wave generators are same from the caculation results.

3. Calculation of deformation energy dissipation of the the flexible bearing

The elastic deformation of the outer ring of the FB will cause the blindingly high dissipation of transmission energy. The energy dissipation of flexspline deformation is equal to the outer ring[9]. Therefore, the elastic deformation energy dissipation have an important impact on the transmission efficiency of the harmonic gear reducer.

The elastic deformation energy in the process of deformation is a state quantity based on the theory of clabellon,which only depends on the final value of external force and displacement. Suppose that the displacement corresponding to each point increases in proportion α . If α increase to d_α , the corresponding increment of displacement is:

$$w_1 \mathrm{d}\alpha, w_2 \mathrm{d}\alpha, w_3 \mathrm{d}\alpha \cdots w_n \mathrm{d}\alpha \tag{11}$$

The force increment is:

$$F_1(\alpha+\mathrm{d}\alpha),F_2(\alpha+\mathrm{d}\alpha),F_3(\alpha+\mathrm{d}\alpha)\cdots F_n(\alpha+\mathrm{d}\alpha) \quad (12)$$

The elastic deformation energy generated on the corresponding displacement increment is:

$$\mathrm{d}W=\sum_{i=1}^{n}(\alpha+\mathrm{d}\alpha)F_i w_i \mathrm{d}\alpha=\sum_{i=1}^{n}\alpha F_i w_i \mathrm{d}\alpha \quad (13)$$

Integrate the above formula:

$$W=\int_0^1\sum_{i=1}^{n}F_i w_i \alpha \mathrm{d}\alpha=\frac{1}{2}(F_1w_1+F_2w_2+\cdots+F_nw_n)=\sum_{i=1}^{n}\frac{1}{2}F_iw_i \quad (14)$$

The elastic deformation energy calculation equation of the outer ring of the FB as follows:

$$W(t,T,N)=\frac{1}{2}\sum_{1}^{n}F_iw_i \quad (15)$$

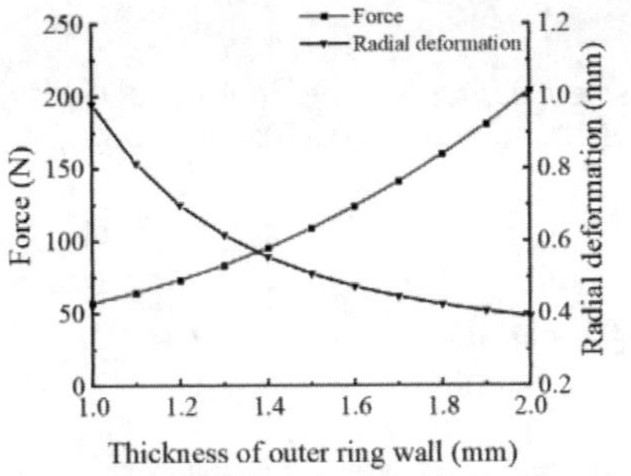

Figure 4. Diagram of load and deformation

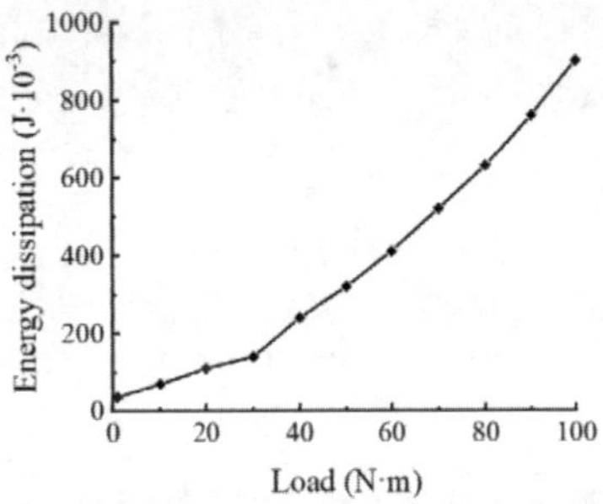

Figure 5. Diagram of load and energy dissipation

As is shown in figure 4, The changing trend of the ball load and radial deformation at the longer axis of the outer ring is opposite. The main argument is that increase the outer ring thickness leads to a decrease in corresponding deflection, and the force required to produce the same deformation is greater from the perspective of elastic deformation. The figure 5 shows that the deformation energy dissipation of the outer ring increases with the changing of the load torque. The main argument is that the radial displacement increased caused by the large load, the more deformation energy is dissipation.

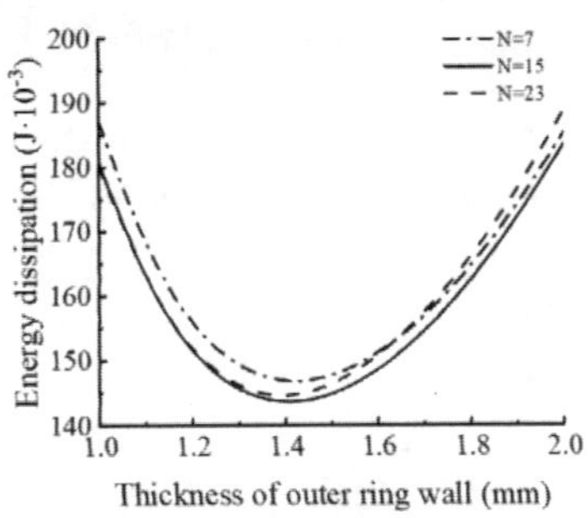

Figure 6. Energy dissipation about the outer ring thickness

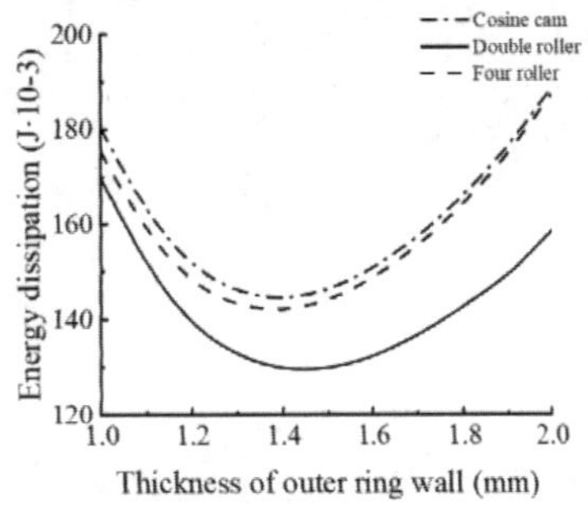

Figure 7. Energy dissipation about the wave generator cams

As is shown in figure 6, The energy dissipation caused by elastic deformation exist a minimum value with the changing of the thickness of the outer ring wall and the number of balls under working load. The minimum outer ring deformation energy dissipation corresponding to the wall thickness is about 1.4mm. The number of bearing balls has limited impact on the outer ring deformation energy dissipation.

However, The friction dissipation inside the bearing increases with the the number of balls raised, which has a great impact on the transmission efficiency of the FB. The trend of deformation energy dissipation of the different wave generators are similarity as shown in figure 7. The outer ring has the best wall thickness to minimize the energy dissipation of the FB. Morever, the deformation energy dissipation of the two-roller WG is obviously less than the other wave generators.

The above analyses show that selecting the appropriate outer ring wall thickness can effectively reduce the deformation energy dissipation of the FB. It provides a theoretical reference for the optimal design of the FB structural parameters.

4. Conclusions

(1) The caculation results show that the wave generator cam are an important parameter which has a greater effect on the deformation and carrying capacity of the FB. The maximum ball load of the FB assembled with cosine-cam appear at the longer axis, while the roller-cam are at the rollers position.

(2) The relationship between the deformation energy dissipation and the outer ring thickness of the FB is analyzed theoretically. It found that select appropriate outer ring thickness can effectively reduce the deformation energy dissipation of the FB. The deformation energy dissipation of the two-roller wave generator is the minimum at the same constraint.

References

[1] Bourdon A, Rigal J. F, Play D 1999 Static Rolling Bearing Models in a C.A.D. Environment for the Study of Complex Mechanisms: Part I-Rolling Bearing Model [J]. *Journal of Tribology,* 205-214.

[2] Shen YW, YE QT 1985 *Theory and design of harmonic gear transmission* [M]. (Beijing: China Machine Press) pp 1-20.

[3] Ostapski W 2010 Analysis of the stress state in the harmonic drive generator-flexspline system in relation to selected structural parameters and manufacturing deviations [J]. Bulletin of the Polish Academy of Sciences: *Technical Sciences*, **58(4)** : 683-698.

[4] Li L, You BD, Wen JM 2016 Load distribution calculation of flexible ball bearing with cosine cam wave generator [C]; *Materials and Structural Engineering*, 49-55.

[5] Yu Y, Zhu E, Chen XY 2021 Load analysis and deformation research of the FB based on a three-force ring superposition method [J]. *Mechanics & Industry*, **22** 34.

[6] Lyuminarsky S.E 2020 Features of determining the efficiency of the cam wave generator of the wave gear[J]. *Materials Science and Engineering*, **971(4)** 042009.

[7] Shao YM, Xiao HF 2011 Deformation and energy dissipation of single layered and discontinuous structural interface of a power system [J]. *Journal of Vibration and Shock*, **30(4)** 217-222.

[8] Shi Y, Yin HC, Li JY, Tang T 2022 Modeling and Parameter Identification of Friction Characteristics of Harmonic gear reducer[J]. *Journal of Northeastern University(Natural Science)*, 89-96.

[9] Ivanov M H 1981 *Harmonic Gear Drive* [M]. (Moscow: Visajas Kola Press).

Design and Kinematics Model of Autonomous Wheel-Claw Reconfigurable Amphibious Robot

1stBing Bai[1,a], Meili Song[1,b*], Wei Bao[1,c], Zhen Xiao[2,d]

[1]School of Mechanical Engineering, Nanjing University of Science and Technology, Nanjing, Jiangsu Province, Chinese

[2]School of Computer Science and Engineering, Nanjing University of Science and Technology, Nanjing, Jiangsu Province, Chinese

[a]Bai_0528@njust.edu.cn; [b*]384667579@qq.com; [c]Baow_0326@njust.edu.cn; [d]1437123029@qq.com

Abstract. In order to improve the adaptability of traditional wheeled robots to complex environments, a passive wheel-claw reconfigurable amphibious robot is proposed in this paper. This robot has strong mobility when moving on the ground. In complex scenes, if there are obstacles, it will automatically switch to claw mode. By studying the motion mechanism of the wheel-claw reconstruction process, the kinematics model of the robot under multiple working conditions is established by using the coordinate transformation method, and the wheel-claw transformation process is analysed in detail. Based on the virtual prototype of Adams, the dynamic simulation under the conditions of flat ground and obstacle crossing is carried out, the established dynamic mathematical model is verified. The simulation results provide theoretical data basis for the actual design of the robot. The experimental results of the prototype platform show that the designed passive wheel claw robot has the ability to work in complex scenes.

1. Introduction

Wheeled robot has the advantage of fast moving, but it cannot pass through obstacles larger than its wheel diameter. Claw robot has the advantage of high trafficability on rough ground, but it is impacted greatly when it travels. Therefore, it is very necessary to develop a robot with the function of wheel claw mode switching.

In 2006 and 2007, LRP (Laboratoire de roboticique de Paris) successively put forward two design ideas of scaling wheels [1], both of which use four-bar mechanism to design rigid mechanical devices to achieve the scaling of wheels. The lunar rover designed by BUAA can be deployed and contracted under the control of the on-board system, and the wheel diameter varies from 200 to 390mm [2]. Beijing Jiaotong University designed a wheel leg reconfigurable robot that deforms through the parallelogram structure, and carried out simulation experiments to verify that it can climb slopes and vertical walls [3]. Huo Jianhua and others designed a walking mechanism with reconfigurable wheel claws through the planetary gear train, which can significantly improve the obstacle climbing ability of the robot [4]. Rhex robot developed by the University of Michigan is a wheel legged robot with semi-circular arc leg wheels. Through gait planning, it can cross obstacles 1.7 times its own height and travel in wheel mode on a flat road, reaching a speed of 2.25m/s [5]. Long Bai et al Of Chongqing University designed a wheel leg variable robot (LDR) based on the spatial folding mechanism and the metamorphism principle, which can complete the mode switching by changing the position of the disc, and the LDR can pass through obstacles 2.8 times higher than the wheel diameter [6]. KIMY proposes a deformable wheel structure similar to LDR, and the obstacle clearance height is 3.25 times of the wheel structure[7]. CHENSC completes the switch of wheel leg mode by using two arc-shaped wheel legs, and realizes the function of wheel leg switch through driving internal and external motors respectively under different working conditions [8].

The wheel claw reconfigurable mechanisms proposed in the literature mostly rely on additional power sources and sensors to switch modes, which increases the design difficulty and design cost. The main contributions of this paper are:

A kind of wheel-claw mode switching mechanism without additional power source is designed by using plane scroll spring, and the kinematic model is established by the method of coordinate transformation, and the dynamic model of this structure under different working conditions is analyzed in detail. And through ADAMS to build a virtual prototype for simulation analysis. Finally, a prototype platform is built to verify the working effect of the structure under complex working conditions.

This paper is organized as follows. In Section 2, The proposed passive wheel-claw structure is introduced and designed Wheel-Claw Reconfigurable Amphibious Robot(WCRAR). In Section 3, the kinematic models of the robot under different working conditions are analysed in detail. The simulation and experimental results are shown in Section 4, which verify the feasibility and kinematics model of the proposed wheel-claw mechanism.

2. Mechanical design

Under the premise of no inertial force, the round wheel cannot pass through obstacles larger than its radius, but relying on the inertial force can easily cause damage to the parts and cannot guarantee the passage. Therefore, it is required to design a reconfigurable wheel that can not only be suitable for flat ground conditions, but also can pass obstacles larger than 10% of its radius; in order to meet the requirements of lightweight, the robot needs to be free from sensors and additional power sources. In order to meet the above requirements, the switching mechanism in this paper is realized by a plane four-bar mechanism with a plane scroll spring. The mechanism and structure of wheel-claw reconstruction are shown in Figure 1.

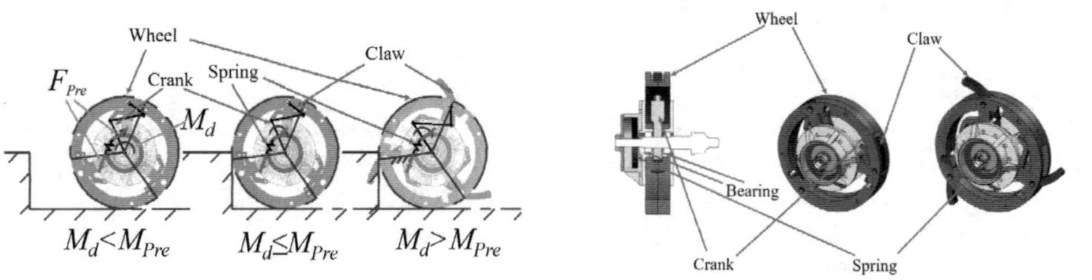

(a) The principle of refactoring (b) Structure

Figure1. Autonomous wheel-claw reconfigurable wheel

As shown in Figure 1(a), when the robot is running on level ground, the driving torque M_d of the motor is less than the preload torque $M_{\mathrm{Pre}}(F_{\mathrm{Pre}})$ of the spring. When the robot encounters an obstacle, the motor torque begins to increase due to maintaining the desired speed. After the preload of the spring, the pawls are reconfigured and deformed. The designed wheel claw reconstruction structure is shown in Figure1(b). In the structure involved, the crank and the axle are fixedly connected, and a bearing is added between the wheel and the axle to ensure the relative rotation of the wheel and the axle. A spring is installed between the axle and the wheel, and the purpose of self-reconfiguration can be achieved through preload.

3. Kinematics Analysis of WCRAR

The robot designed above can adapt to a variety of environments, but the kinematic model of the robot under different working conditions is also different. In order to analyse the motion mechanism of the robot under different working conditions in detail, a kinematic model is established by the coordinate transformation method, and assumptions are made to simplify the kinematic model:

① In the reconstruction process, the robot will not be pushed up by the claw;②Ignore the impact and vibration in the process of reconstruction and obstacle crossing; ③The changes of yaw (ψ_0) and roll (ϕ_0) angles are not considered during obstacle crossing; ④ In the process of obstacle surmounting, there is no relative sliding between the tip of the claw and the ground; ⑤The wheel and the ground roll longitudinally, and the center of mass of the body does not slip laterally.

3.1. kinematic model

The geodetic coordinate system $\{D\}$, the body coordinate system $\{B\}$ at the centroid of the robot body, and the rectangular coordinate system $\{L_i\}$ at the centroid of the *ith* wheel are established. As shown in figure 2.

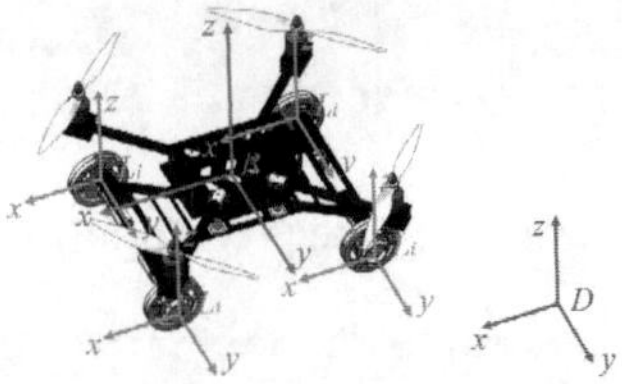

Figure 2. reference coordinate system.

The rotation mode from the body coordinate system to the geodetic coordinate system is to rotate ψ_0 around the z_B axis (yaw) ϕ_0, rotate θ_0 around the y_B' axis (pitch), and rotate ϕ_0 around the x_B'' axis (roll); The rotation mode from the wheel coordinate system $\{L_i\}$ to the body coordinate system $\{B\}$ is to rotate φ_i around the y_L axis. Their rotation matrices are

$$\begin{cases} {}_B^DR = \begin{bmatrix} C_{\psi_0}C_{\theta_0} & C_{\psi_0}S_{\phi_0}C_{\theta_0} - C_{\phi_0}S_{\psi_0} & S_{\phi_0}S_{\psi_0} + C_{\psi_0}C_{\phi_0}S_{\theta_0} \\ C_{\theta_0}S_{\psi_0} & C_{\psi_0}C_{\phi_0} + S_{\phi_0}S_{\psi_0}S_{\theta_0} & C_{\phi_0}S_{\psi_0}S_{\theta_0} - C_{\psi_0}S_{\phi_0} \\ -S_{\theta_0} & C_{\theta_0}S_{\phi_0} & C_{\phi_0}C_{\theta_0} \end{bmatrix} \\ {}_{L_i}^BR = \begin{bmatrix} C_{\varphi_i} & 0 & S_{\varphi_i} \\ 0 & 1 & 0 \\ -S_{\varphi_i} & 0 & C_{\varphi_i} \end{bmatrix} \end{cases} \quad (1)$$

Where $C_{\theta_0} = \cos\theta_0; S_{\theta_0} = \sin\theta_0$, the rest are the same.

Because the mass and moment of inertia of the claw and connecting rod are relatively small, in order to simplify the model, the claw and connecting rod are considered as a part of the wheel. The rotation angles of the *ith* wheel and crank are φ_{W_i} and φ_{C_i} respectively, and the angular velocities of each component in its reference coordinate system are respectively

$$\begin{cases} {}^{B}\boldsymbol{\omega}_0 = {}^{B}_{\omega_0}\Omega\begin{bmatrix}\dot{\phi}_0 & \dot{\theta}_0 & \dot{\psi}_0\end{bmatrix}^{\mathrm{T}} \\ {}^{L_i}\boldsymbol{\omega}_{W_i} = \begin{bmatrix}0 & \dot{\varphi}_{W_i} & 0\end{bmatrix}^{\mathrm{T}} + {}^{L_i}\boldsymbol{\omega}_B \\ {}^{L_i}\boldsymbol{\omega}_{C_i} = \begin{bmatrix}0 & \dot{\varphi}_{C_i} & 0\end{bmatrix}^{\mathrm{T}} + {}^{L_i}\boldsymbol{\omega}_B \end{cases} \tag{2}$$

Where ${}^{L_i}\boldsymbol{\omega}_B = {}^{L_i}_{B}R\,{}^{B}\boldsymbol{\omega}_B$; ${}^{B}_{\omega_0}\Omega$ is Euler kinematic equation

$${}^{B}_{\omega_0}\Omega = \begin{bmatrix} 1 & 0 & -S_{\theta_0} \\ 0 & C_{\phi_0} & S_{\phi_0}C_{\theta_0} \\ 0 & -S_{\phi_0} & C_{\phi_0}C_{\theta_0} \end{bmatrix} \tag{3}$$

The velocity of the centroid of each component in the $\{D\}$ coordinate system is:

$$\begin{cases} {}^{D}\boldsymbol{v}_B = \begin{bmatrix}\dot{x}_0 & \dot{y}_0 & \dot{z}_0\end{bmatrix}^{\mathrm{T}} \\ {}^{D}\boldsymbol{v}_{W_i} = {}^{D}\boldsymbol{v}_B + {}^{D}_{B}R\left({}^{B}\boldsymbol{\omega}_0 \times {}^{B}\boldsymbol{r}_{W_i}\right) = {}^{D}\boldsymbol{v}_B - {}^{D}_{B}R\left[{}^{B}\boldsymbol{r}_{W_i}\right]_{\mathrm{x}}{}^{B}\boldsymbol{\omega}_0 \\ {}^{D}\boldsymbol{v}_{C_i} = {}^{D}\boldsymbol{v}_B + {}^{D}_{B}R\left({}^{B}\boldsymbol{\omega}_0 \times {}^{B}\boldsymbol{r}_{C_i}\right) = {}^{D}\boldsymbol{v}_B - {}^{D}_{B}R\left[{}^{B}\boldsymbol{r}_{C_i}\right]_{\mathrm{x}}{}^{B}\boldsymbol{\omega}_0 \end{cases} \tag{4}$$

Where, ${}^{B}\boldsymbol{r}_{W_i}$ 、 ${}^{B}\boldsymbol{r}_{Q_i}$ are the vectors from the wheel centroid L_i and the crank centroid C_i to the body centroid B in the coordinate system $\{B\}$; $\left[{}^{B}\boldsymbol{r}_{L_i}\right]_{\mathrm{x}}$ is vector antisymmetric matrix.

3.2. Nonholonomic Constraint Analysis

Since WCRAR can work in a variety of working conditions, it is necessary to analyze the motion model under different working conditions. WCRAR is a non-holonomic system, and performs non-holonomic constraint analysis for various working conditions.

When the robot surmount obstacles, it can be divided into three stages: reconstruction stage, claw obstacle transition stage and obstacle crossing stage. The front wheel obstacle crossing can be divided into the reconstruction phase, the claw-obstacle transition phase and the obstacle crossing phase, and the rear wheel is the same, so the nonholonomic constraint analysis is carried out for the above phases.

3.2.1. Flat condition

According to the assumption, if the robot wheel rolls longitudinally and the body centroid does not slip laterally, there is

$$\begin{cases} {}^{B}\boldsymbol{\omega}_0 = \begin{bmatrix}0 & \omega_y & 0\end{bmatrix}^{\mathrm{T}} \\ {}^{B}\boldsymbol{v}_0 = \begin{bmatrix}v_x & 0 & 0\end{bmatrix}^{\mathrm{T}} \\ {}^{B}\boldsymbol{v}_{Ni} = {}^{B}_{D}R \cdot {}^{D}\boldsymbol{v}_i - {}^{B}_{L_i}R\,{}^{L_i}\left[\boldsymbol{r}_{Cn_i}\right]_{\mathrm{x}}\begin{bmatrix}0 & \dot{\varphi}_{W_i} & 0\end{bmatrix}^{\mathrm{T}} \end{cases} \tag{5}$$

Where, ${}^{L_i}\left[\boldsymbol{r}_{Cn_i}\right]_{\mathrm{x}}$ represents the antisymmetric matrix formed by the vector from the wheel centroid to the ground in the $\{L_i\}$ coordinate system.

Due to the longitudinal pure rolling of the robot, there is

$$\begin{cases} v_0 = \dfrac{v_L + v_R}{2} 、 \omega_0 = \dfrac{v_R - v_L}{2B} \\ \dot{\varphi}_{W_1} r = \dot{\varphi}_{W_4} r = v_L 、 \dot{\varphi}_{W_2} r = \dot{\varphi}_{W_3} r = v_R \end{cases} \tag{6}$$

3.2.2. Obstacle crossing condition

As mentioned above, there are three processes in front wheel obstacle crossing. In the reconstruction stage and the transition stage, because the robot does not travel in the claw mode, the wheel will idle under the obstruction of obstacles, and if the body does not move, there will be ${}^{D}\boldsymbol{v}_B = 0$、${}^{D}\omega_B = 0$.

As shown in Figure 3, when crossing obstacles in claw mode, there is

$$\begin{cases} r'S_{\varphi_{W_i}-\varphi_0} + e = l_0 S_{\theta_0} \\ x_{W_{1,4}} + x_{W_{2,3}} = r'C_{\varphi_{W_{1,4}}-\varphi_0} + l_0 C_{\theta_0} \end{cases} \tag{7}$$

Therefore, in the process of obstacle climbing, the pitch angle and angular velocity of the robot are

$$\begin{cases} \theta_0 = \arcsin\left(\dfrac{r'\sin\left(\varphi_{W_i} - \varphi_0\right) + e}{l_0}\right) \\ \dot{\theta}_0 = \dfrac{\dfrac{r'}{l_0}\dot{\varphi}_{W_i}\cos\left(\varphi_{W_i} - \varphi_0\right)}{\sqrt{1 - \left(\dfrac{r'\sin\left(\varphi_{W_i} - \varphi_0\right) + e}{l_0}\right)^2}} \end{cases} \tag{8}$$

The analysis methods of transition stage and rear wheel obstacle crossing stage are the same as those of flat ground driving and front wheel obstacle crossing respectively, it will not be repeated in the article. The kinematic equations under constraints can be obtained by combining equations 7, 8 and 9 with equations 2 and 5.

4. Simulation

In order to verify the correctness of the kinematic model, the robot is simulated and analyzed based on ADAMS. The kinematics results of the robot under multiple working conditions are obtained through the virtual prototype simulation, and the angular velocity of the wheel is substituted into the kinematics model built above. Finally, the numerical simulation results obtained from the theoretical model are compared with the simulation results of the virtual prototype.

The desired speed drive of the virtual prototype is a step function, specifically expressed as $STEP\left(time, t_0, const_0, t_1, const_1\right)$.This function represents that from t_0 to t_1, the motion pair moves from $const_0$ to $const_1$, and its state function conforms to $y = -at^3 + bt^2$. In order to simulate the function of the spring, a torsional spring damper is applied between the crank and the wheel, and its preload and stiffness parameters are set as the preload and equivalent stiffness of the spring. The distance from the obstacle to the ground is 60mm, and the radius of the wheel is 50mm. ADAMS virtual prototype is shown in Figure 4.

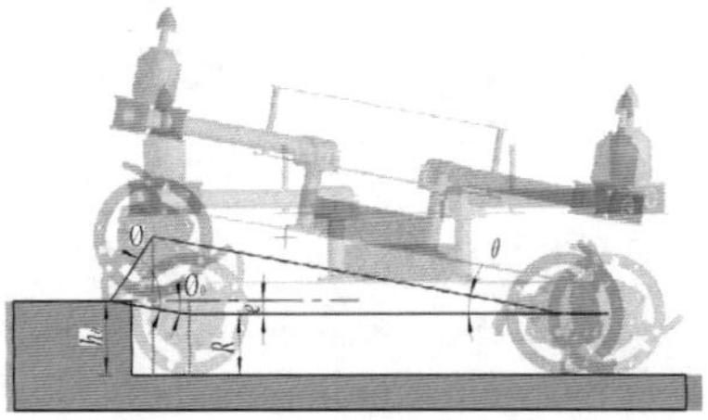

Figure 3.Obstacle crossing process of front wheel.

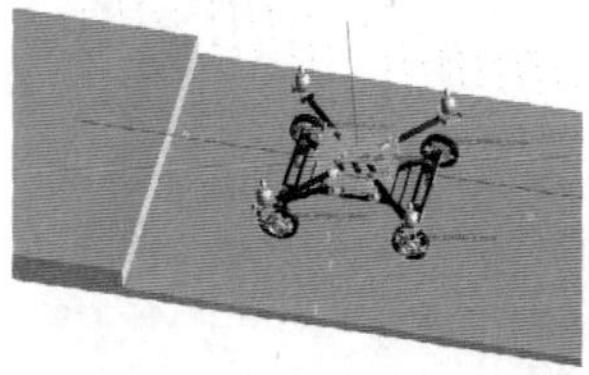

Figure 4. ADAMS virtual prototype

After the simulation in ADAMS virtual prototype, the speed of each wheel is extracted and fitted into the kinematic theoretical model. The fitting curve and simulation curve are shown in Figure 5. The simulation results of Adams are extracted, and the body centroid velocity, angular velocity and the velocity component along the X axis of the front wheel centroid in the solved kinematic theoretical model are drawn as curves. The comparison results are shown in Figures 6,7 and 8.

Figure 6 shows the simulation results and theoretical model calculation results of the body centroid along the X and Z axes, and figure 8 shows the speed of the front wheel centroid along the X axis. According to figures 6 and 8, the simulation results of the body centroid speed are almost consistent with the theoretical results. At 2s and 9s, due to the switching of the driving acceleration direction, the impact of friction and collision is introduced into the virtual prototype, while the theoretical model does not consider the impact of slip and impact, there is a slight difference in the curve. Figure 7 shows the curve of the angular velocity of the body. It can be seen from the figure that at 4.4s, the body starts to cross the obstacle, while at 5.2s, the driving torque of the motor is less than the pre tightening torque of the spring. The claw retracts, and there is a collision between the body and the obstacle, which will cause a sudden change in the angular velocity. However, the theoretical kinematic model does not consider the impact of the collision, so the curve is different.

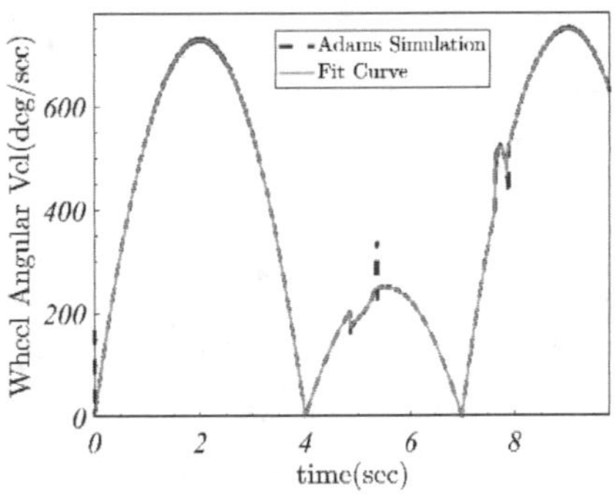

Figure5. Motor drive fitting curve

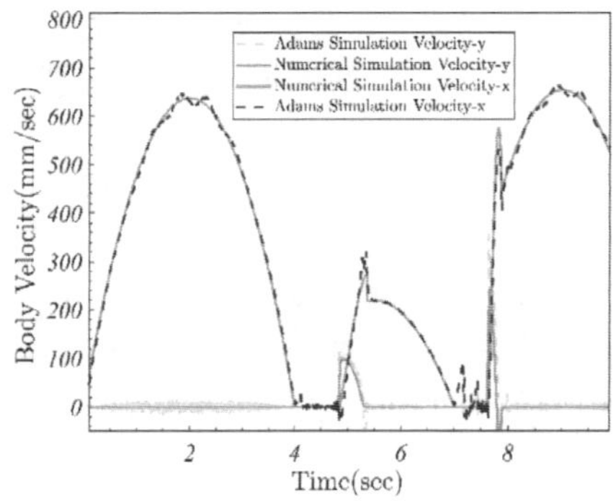

Figure 6. Body centroid velocity

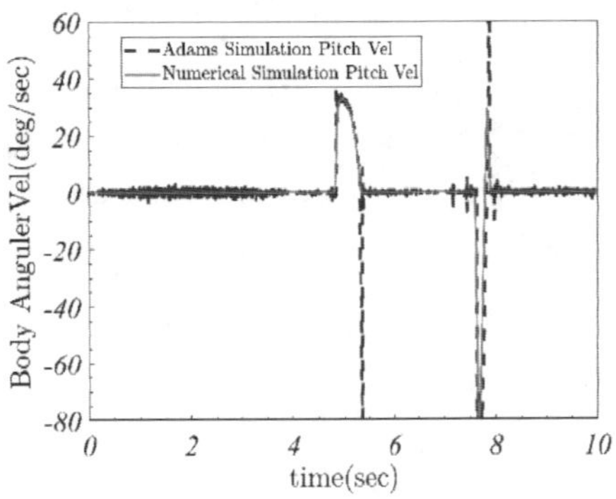

Figure 7. Body angular velocity curve

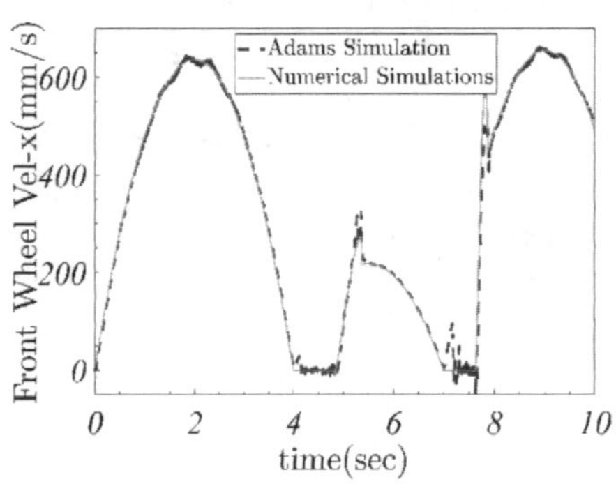

Figure 8. X-axis component of front wheel centroid speed

According to the simulation results of the virtual prototype, the dynamic analysis of WCRAR is carried out. The motor driving torque during the motion of the virtual prototype is shown in Figure 9. Combined with the virtual prototype working animation, it can be seen that in the 0-4s stage, the motor torque of WCRAR under the flat ground is related to friction and acceleration; In the 4-4.4s stage, the robot encounters obstacles, and the motor increases the output torque to maintain the desired speed; In the 4.4s-4.7s stage, the output torque of the motor is greater than the pre-tightening torque of the spring, and the reconfigurable mechanism of the wheel claw begins to reconstruct; In the 4.7-4.8s stage, because the claw did not touch the obstacle, the robot skids in the claw mode; In the 4.8-5.2s stage, the robot crosses the obstacle, and the motor torque increases accordingly; In the 5.2-7s stage, the robot transits from the obstacle crossing state of the front wheel to the obstacle crossing state of the rear wheel, and completes the obstacle crossing process in 7s-10s. The torque of the virtual prototype in the obstacle crossing process is shown in figure 9.

According to the simulation data in Figure 9, the driving torque of the robot in the non obstacle crossing process is less than 200 $N \cdot mm$, and the motor output torque is greater than 170 $N \cdot mm$ during obstacle crossing. In order to avoid the false triggering of the wheel claw, the normal output torque of the motor can be controlled to be less than 170 $N \cdot mm$ under normal working conditions when the prototype is developed, when the obstacle is crossed, the output torque of the motor increases, and the wheel mode is converted to the claw mode to realize the obstacle crossing. The height of the obstacle is 60mm, which meets the requirement of more than 10% of its outer diameter.

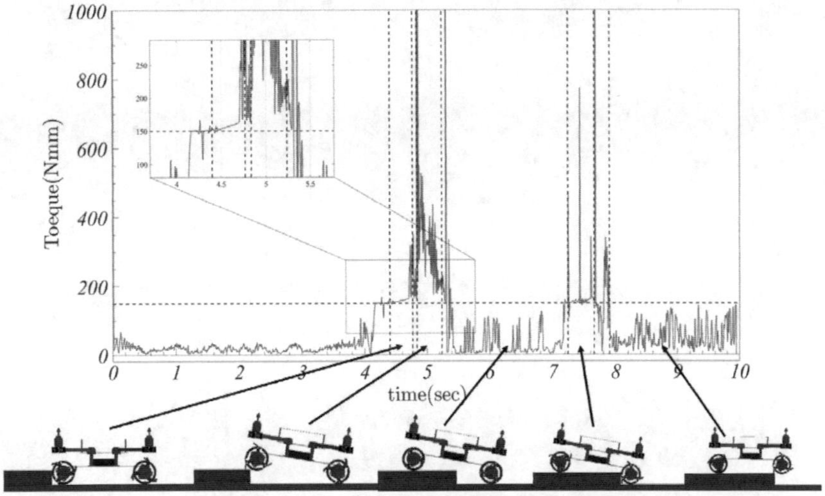

Figure 9. Torque analysis of virtual prototype.

5. Conclusion

In order to improve the ability of traditional wheeled robot to adapt to complex environment, this paper proposes a wheel claw reconfigurable land air amphibious robot, which realizes the autonomous switching between wheel mode and claw mode through the relative movement between crank, wheel and spring. The motion principle of the reconfigurable wheel claw mechanism of the robot is analyzed in detail, and its kinematics is theoretically modeled, and the nonholonomic constraints of multiple working conditions are analyzed; The virtual prototype simulation based on ADAMS and the numerical simulation of the theoretical model are carried out. The simulation results verify the correctness of the theoretical model and the virtual prototype model; The robot meets the requirement of passing the radius of more than 10% wheels; The driving torque values of multiple working conditions are obtained through virtual prototype, which provides a reference for the design of controller and the selection of spring and motor.

Reference

[1] B. Philippe,B. Faiz,P. Sebastien.(2008)An Expandable Mechanism for Deployment and ContactSurface Adaptation of Rover Wheels. Laboratoire de Robotique de Paris.1053- 1060.
[2] Sun G,Gao F,Sun P.(2008)Variable diameter wheel lunar rover and its obstaclesurmounting capability analysis. Machine Design,2008(05):21-23.

[3] Liu C, Tao X, Yao Y.(2022) Design and Analysis of a Novel Deformable Wheeled-Legged Robot. Journal of Mechanical Engineering,58(03):65-74.
[4] Huo J, Wu FX, Wei XH. Luo HD.(2021)Design and Research on walking mechanism of deformable wheel leg robot. Journal of Mechanical Transmission, 45(01):157-161.
[5] Wang D. (2016)Design and Research of Deformable Wheel Mobile Robot.Beijing: Beijing Jiaotong University.
[6] Long B , Jian G. (2018)An optional passive/active transformable wheel-legged mobility concept. Robotics & Autonomous Systems,107: 145-155.
[7] Kim Y, Jung G, Kim H.(2014)Wheel Transformer: A Wheel-Leg Hybrid Robot With Passive Transformable Wheels. Robotics IEEE Transactions on,30(6): 1487-1498.
[8] Chen SC, Huang KJ, Chen WH.(2014) Quattroped:A Leg-Wheel Transformable Robot. IEEE/ASME Transactions on Mechatronics, 19(2): 730-742.

Journal of Physics: Conference Series **2365** (2022) 012034 doi:10.1088/1742-6596/2365/1/012034

Matlab-based Study of the Dynamic Performance of Flcxible Jointed Robots

NanJun Huang[1,*]

[1] Jiangxi Vocational and Technical College of Industry, China

*Corresponding author e-mail: 68527828@qq.com

Abstract. In the development and research of industrial robots, due to the defects of traditional mechanical structures in the drive system and sensors, control units and other aspects, so in order to solve these problems and improve the efficiency of flexible work, the author proposes a Matlab based core motion characteristics analysis method, that is mathematical modelling method (GUI) to improve the elastic joint robot dynamics The relationship between performance index evaluation and prediction model parameters. At present, most of the research on elastically articulated robots at home and abroad is focused on the development of one or several functionalised components. In this paper, the kinematic model is developed and its performance characteristics are analysed based on the Matlab c language platform. The kinetic equations are discretized to obtain the velocity and acceleration curves with time for each degree of freedom, so that the non-linear contact angle matrix and the driving torque can be established.

1. Introduction

Robots are highly intelligent automatic control systems that can perform many tasks that are unreachable or impossible with traditional technology, and their performance is also very impressive. As the level of industrial automation continues to rise, the field of computer applications is expanding and the structure of the robot itself is becoming more powerful, making the requirements for power devices more stringent and complex. Robot kinematics and dynamics are important methods for studying the dynamic performance of robots. When investigating the optimisation of mechanical structures using kinetic performance, the spatial position, size and degrees of freedom of the joints are usually analysed before designing functional modules that are reasonable, economical, practical and easy to control and co-ordinate to fulfil the task requirements, taking into account the differences between different drive motor types and performance indicators, in order to meet the motion planning requirements of the robot under special operating conditions.

MATLAB is a very common computer programming tool. It can be used to handle complex systems. MATLAB has many powerful functions, such as data analysis, problem solving and modelling, as well as the advantages of strong fault tolerance and visibility, in addition to many types of file processing and graphical display operations, and can well reflect these advantages.

2. Kinematic position inversion of the robot

Inverse solution of the first chain position. Establish the kinematic equations.

$$(A_0)^{-1}T = {}^0A_1{}^1A_2{}^2A_3{}^3A_4{}^4A_5 \quad (1)$$

$$a = L_2 + \sqrt{L_3^2 - P_2^2} \quad (2)$$

$$b = P_1 - R + r \tag{3}$$

$$n = \frac{a^2 - b^2 - P_3^2 - L_1}{2L_1} \tag{4}$$

$$x = \frac{P_3 - \sqrt{P_3^2 + b^2 - n^2}}{n - b} \tag{5}$$

$$x = \tan\frac{\theta_{11}}{2} \tag{6}$$

Similarly, the position solution of the second chain and the third chain can be obtained, and then the derivative of time can be found to obtain the angular velocity and angular acceleration of each rod of each chain. The trajectory of the center end of the moving platform is shown in Figure 1.

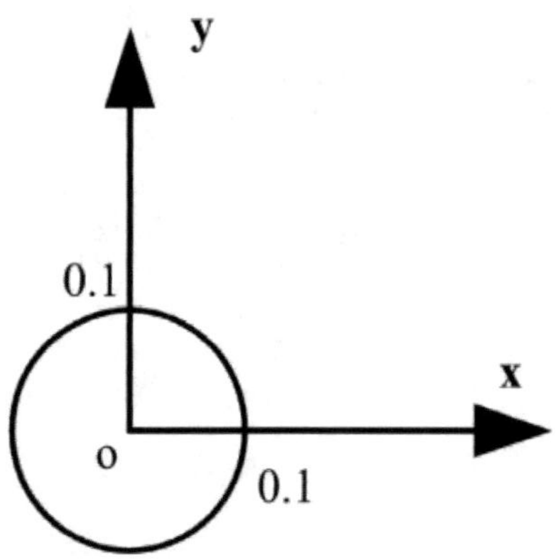

Fig. 1 Trajectory of the center of the moving platform

When considering the rigidity of the components, the ideal trajectory is a circle, the angle of the driving rod of each branch chain is shown in Fig. 2, the angular velocity is shown in Fig. 3 and the angular acceleration is shown in Fig. 4, all of which show the change of waveform without any special fluctuation or change. The amplitude and period of the changes in angle, angular velocity and angular acceleration of the three chains are more or less the same, but the phases are different.

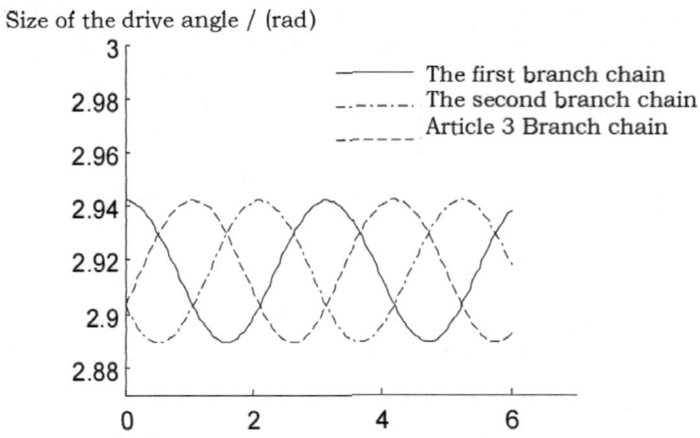

Fig. 2 Magnitude of the driving angle of each chain

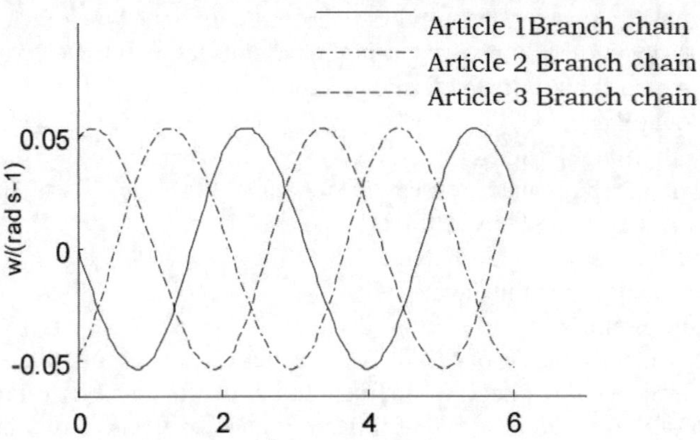

Fig. 3 Angular velocity of each chain

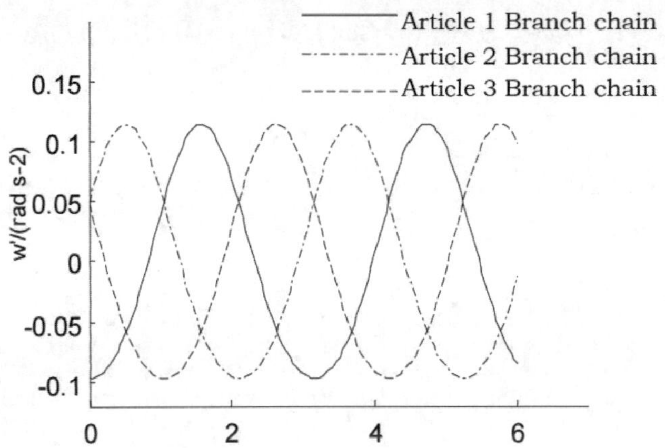

Fig. 4 Angular acceleration of each chain

3. The 3-RRRT parallel mechanism is used as an example for modelling the dynamics of a flexible jointed robot.

PMATLAB Sim u link is a general-purpose simulation software that provides all the inputs and outputs needed to run a program. Its main functions are to model, analyse and diagnose systems and provide solutions. In industrial design, many different types of products have been developed to meet the requirements of the production process and to simulate the contradictions between working conditions and performance in a complex environment. With the continuous development of computer technology and the improvement of manufacturing processes, the simulation data is becoming more and more diverse, which makes the calculation volume of the simulation much higher and the calculation speed faster.

The structural dimensions of the 3-RRRT parallel mechanism are R=0.4m for the outer radius of the static platform, r=0.1m for the outer radius of the dynamic platform, L1=0.4m, L2=0.1, L3=0.8m, L1 is the length of the active arm, L2, L3 is the length of the driven arm, the weight of rod L1 is 2kg, the weight of rod L2 is 0.5kg, the weight of rod L3 is 2kg, The kinematic modelling steps are as follows.

(1) Use MATLAB Sim u link modelling software to build a 3D model of the parts of the 3-RRRT mechanism, then carry out assembly and save the assembly file as Parasolid (*.x-t) format.

(2) Create a new file New Model in MATLAB, select MMKS(mm,kg,N,s,deg), import the Parasolid(*.x-t) format file saved in the previous step and define the type of each part as rigid body.

(3) Apply 6 rotating pairs and 3 universal joints constraints to the 3-RRRT parallel mechanism, apply fixed constraints to the static platform, and use Model verify in ADAMS to check the degrees of freedom as shown in the following data

VERIFY MODEL: MODEL 3RRRT

3 Gruebler Count lap proximate degree of freedom)11 Moving Parts (not including ground)9 Revolute Joints3 Universal Joints1 Fixed Joints

3 Degrees of Freedom for MODEL 3RRRT

There are no redundant constraint equations.

Model verified successfully

(4) Using the point drive function of MATLAB, a general motion is applied at the center of the circle of the moving platform, the three degrees of freedom of rotation are FREE unchanged, and the three degrees of freedom of movement are set to a trajectory curve that runs from the center of the circle to the edge of the circle, then runs a circle counterclockwise and then returns to the center of the circle. Enter the moving platform point drive function as follows.

Tra X=IF(time-5 : IF(time-1 : 100*time , 100 , 100*cos(90d*(time-1))) , 100 ,
100*(6-time))

Tra Y=IF(time-5 : IF(time-1 : 0 , 0 , 100*sin(90d*(time-1))) , 0 ,0)

Tra Z = -565.69

The trajectory function curves are shown in Figure 5.

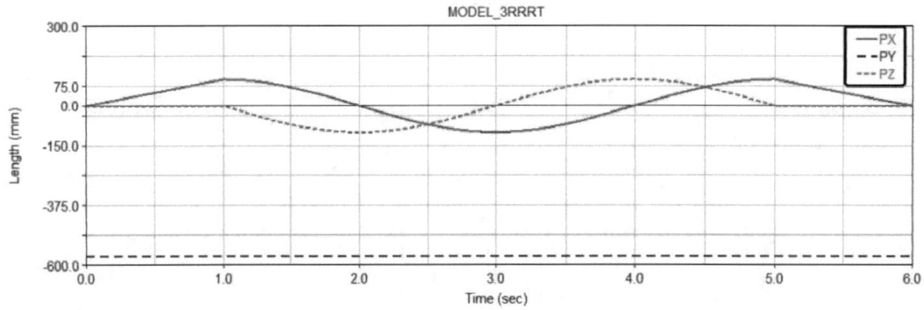

Fig. 5 Displacement of the center of the end-action platform of the mechanism

(5) Enter MATLAB for simulation, end time set to 6, Steps set to 3000, click on assembly and initial position, then click on start for simulation and use ADAMS/Post Processor post-processing module to measure the three active arm turning angles. The kinematic modelling procedure is as follows: set the gravitational acceleration property of the model to the negative direction of the Y-axis of the global coordinate system. With the kinematic modelling process above, a running simulation was performed to measure the incremental swing angles of the three active joints, using the tool Ax/Ay/Az Projected Rotation. as shown in Figure 6.

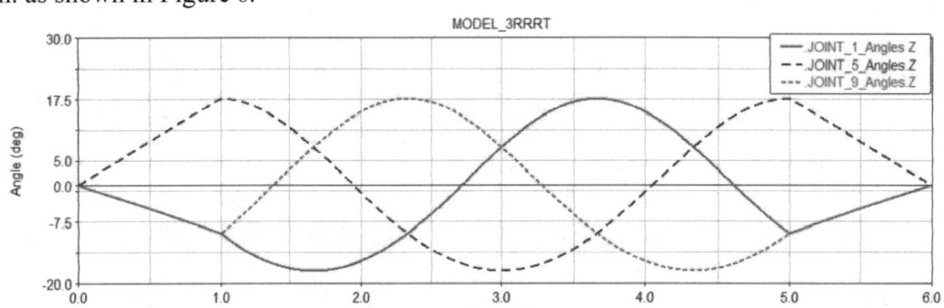

Figure 6 Angular displacement versus time curve

(6) Import the obtained drive angle swing angle curves into MATLAB, use the discretization tool option to regenerate the three spline curves SPLINE_1, SPLINE_2, SPLIE_3, add rotational drives to Joint1, Joint5 and Joint9, respectively, which are connected to the static platform and the drive arm. Define the material properties and mass-inertia tensor and gravity parameters for each component. Change the drive parameters of the drive motors MOTION_1, MOTION_2 and MOTION_3 and apply the following functions in turn.

CUBSPL(time,0,SPLINE_1,0)*1d,
CUBSPL(time,0,SPLINE_2,0)*1d,
CUBSPL(time,0,SPLINE_3,0)*1d,

At this point, the direction of the function should be set and the units are guaranteed to be uniform, so as to ensure the correctness of the simulation results, at which point the dynamics simulation model has been established. Use MATLAB to simulate its dynamics, end time is set to 6, steps is set to 3000, type is set to Dynamic. Use the function of creating trajectory tracking line to check whether the trajectory of the dynamic platform meets the requirements, if not, it is likely that the direction of the drive function is wrong or the unit is not uniform (for example, the unit is converted to angle or radian, add torque). If this is not the case, it is likely that the drive function is in the wrong direction or that the units are inconsistent (e.g. the units are all converted to angles or radians, and the direction is easily reversed when adding torque). Using the torque measurement tool, the forces and moments of the three active arms are measured to obtain the required data to complete the dynamics simulation. Using the post-processing module the drive moments for the drive joints can be obtained as shown in Figure 7.

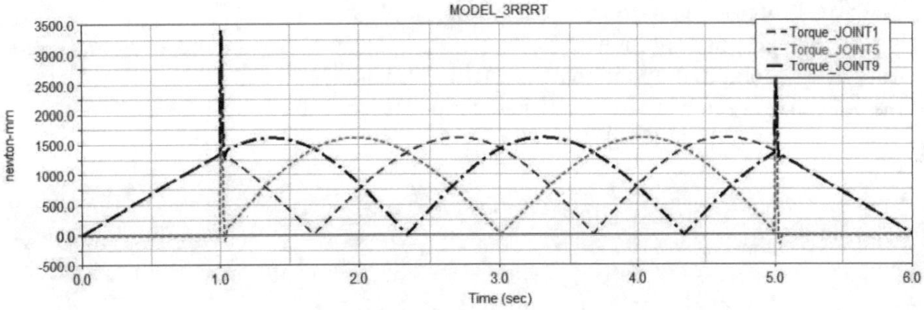

Figure 7 Drive moment diagram for the drive rod

Another factor that causes the sudden change in the moment at these two moments is that the trajectory of the end stage changes from a linear motion to a curved motion, with a very small radius of curvature at the transition point, which causes the moment to change. This means that when planning the motion of the end trajectory, you should try to ensure that the radius of curvature of the trajectory curve is relatively large, and try not to have sudden changes in speed, if there will be a great impact on the accuracy of the motion trajectory of the end trajectory. And affect the life of the mechanism reducer motor and other components.

4. Conclusion

In the field of robotics, there are many different types of robot, such as kinetics, dynamics and drive systems. All of these power sources are implemented mechanically. This paper focuses on the structural design of elastically jointed robots and the kinematic analysis methods used to build up a 3D virtual model for simulation calculations as required and to draw final conclusions. A brief description of the interrelationships between the various subsystems is then presented, and finally the results are translated into machine simulation using the MATLAB language on Matlab software, thus making the whole system work better and more reliably.

References

[1] Zuo Fuyong, Hu Xiaoping, Xie Ke, et al. Trajectory planning and simulation of SCARA robot based on MATLAB Robotics-r toolbox [J]. Journal of Hunan University of Science and Technology (Natural Science Edition), 2017, 27(2): 41-44.
[2] Wang Zixing, Fan Wenxin, Zhang Baocheng, et al. Kinematic analysis and simulation of industrial robots based on Matlab [J]. Electromechanical Engineering, 2018, 29(1): 34-37.
[3] Gan Minyao, Ma Junjian, Chen Yongxing, et al. Kinematic simulation of PUMA robots based on MATLAB [J]. Journal of Kunming University of Technology (Science and Technology Edition), 2018, 28(6): 5O-I 53.
[4] Cai Zixing. Robotics [M]. Beijing: Tsinghua University Press, 2019.
[5] Hu Yunfu, Zhu Yanhe, Wu Xiaoguang, et al. Simulation and analysis of a simple bipedal passive walking model[J]. Journal of Harbin Institute of Technology, 2020, 42(3): 490-499.
[6] Liu, Li-Mei, Tian, Yan-Tao, Li, Jian-Fei, et al. Passive walking robot switching control for variable road conditions[J]. Control and decision making, 2017, 26(8): 1203-1208.
[7] Fu, C.L., Huang, Y.L., Wang, J.M., et al. Quasi-open-loop control of a semi-passive bipedal robot[J]. Robotics, 2019, 31(2): 110-117.
[8] Zhang Y, Liu F, He GP. Dynamics and control of the support phase of a single-legged jumping robot[J]. Mechanical Design and Manufacturing, 2012 (12): 165-167.
[9] Xia Yanfeng. Design and experimental study of flexible joints for light robotic arms[D]. Harbin Institute of Technology, 2014.
[10] Junping S, Yi D. Research on Ship Target Auto-Recognition Technique for Low Resolution Radar[C]. International Conference on Radar, 2016: 1-4.
[11] Nelson G, Blankespoor K, Raibert M. Walking BigDog: Insights and challenges from legged robotics[J]. Journal of biomechanics, 2016, 39: S360.
[12] Collins S, Ruina A, Tedrake R, et al. Efficient bipedal robots based on passive-dynamic walkers[J]. Science, 2015, 307(5712): 1082-1085.

Non-singular terminal extended sliding mode control of permanent magnet synchronous motor based on variable exponential reaching law

Chuanxi Zheng[1,a], Yuandong Gu[2,b]

[1]Faculty of Microelectronics, Shanghai University, Shanghai, China

[2]Faculty of Microelectronics, Shanghai University, Shanghai, China

[a]1256832828@qq.com; [b]Alex.Gu@shu.edu.cn

Abstract. A non-singular terminal extended sliding mode control method based on variable exponential reaching law is presented because the permanent magnet synchronous motor speed control system is subject to parameter disturbance, external disturbance, and end effect. In order to accomplish global rapid convergence and get rid of singularity, a non-singular terminal extended sliding mode surface is first created. After the system reaches the sliding mode surface, a new extended state variable is constructed in accordance with the first-order angular velocity model, and a new better exponential reaching law is utilized to eliminate chattering. Finally, a disturbance observer is designed to detect the disturbance during motor operation in real-time and add it to the speed control loop as a compensation. The simulation findings demonstrate that, in comparison to the conventional sliding mode control strategy, the suggested variable-parameter non-singular terminal sliding mode control strategy lowers the effect of disturbance on the motor's real operation and has the benefits of rapid convergence speed, good resilience, minimal speed overshoot, and high control precision.

1. Introduction

With the advancement of precision motion control technology, the application of permanent magnet synchronous motors has gradually increased [1-2]. Due to its high stiffness, high thrust, and high precision, permanent magnet synchronous motors have rapidly gained popularity in the sectors of high-precision instrumentation, transportation technology, and semiconductor chip fabrication [3-5].

One of the most popular nonlinear control theories is sliding mode theory, and in [6] and [7], an extended state observer with load torque as an extended state variable was devised for PMSM motors. However, under load perturbations, the performance of PMSM drive systems with linear ESO is constrained. Researchers created a non-singular fast terminal sliding mode controller with adaptive ability to address the singular issue in the conventional synovial controller. This controller has great anti-disturbance ability but still exhibits some overshoot [8].

The non-singular terminal extended sliding mode speed regulation technique presented in this research is based on variable exponential reaching law and aims to improve PMSM speed control performance. The non-singularity of the non-singular terminal sliding mode approach must be satisfied before this method may add a new extended state. In order to reduce chattering once the system has reached the sliding mode surface, a new variable exponential reaching rule is built together with a disturbance observer as a feedforward compensation term. In comparison to the conventional sliding mode control method, the enhanced sliding mode speed control approach described in this study not only minimizes the impact of disturbance on the motor's real operation, but also offers the benefits of quick convergence, strong resilience, little speed overshoot, and high control precision.

2. Mathematical model of permanent magnet synchronous motor

When PMSM builds a mathematical model to streamline the analysis, the following presumptions are made: 1.It is assumed that the magnetic linearity is not saturated and the effect of higher harmonics is not considered; 2. It ignores the hysteresis and eddy current losses of the iron core on the damping winding of the rotor; 3. Permanent magnet synchronous motor stator windings are 120° apart from one another in space, have symmetric electrical angles, and have a sinusoidally distributed magnetic field. 4.The influence of external environmental factors on the motor winding resistance is ignored [1-3]. The surface mount permanent magnet synchronous motor mathematical model may be expressed as follows in the two-phase rotating coordinate system.

$$\left.\begin{aligned} u_d &= Ri_d + L_d\dot{i}_d - w_e L_q i_q \\ u_q &= Ri_q + L_q\dot{i}_q + w_e(L_d i_d + \psi_f) \\ J\dot{w}_m &= T_e - T_L - Bw_m \\ T_e &= 1.5 p_n i_q [i_d(L_d - L_q) + \psi_f] \end{aligned}\right\} \tag{1}$$

3. PMSM control system design

3.1. Non-singular terminal extended sliding mode surface

The typical non-singular fast terminal sliding mode surface is shown by the formula below

$$s = x_1 + \alpha x_1^{\gamma} + \beta x_2^{\lambda} \tag{2}$$

In (2): x_1, x_2 are system state variables; α, β are system state variable coefficients, and α, β >0; $\gamma = a/b, \lambda = c/d$, a, b, c and d are positive odd numbers, and $a/b > c/d$, $1 < c/d < 2$.

From (2), it is clear that when the system state variable x_1's absolute is large and $|x_1| > 1$, the convergence speed of the error is mainly affected by the system state. when $|x_1| < 1$, the higher-order term of the system state can be ignored at this time, which can be approximately regarded as a non-singular terminal sliding mode surface. Due to this, global rapid convergence can be achieved during the sliding stage using the non-singular fast terminal sliding mode control, and $1 < c/d < a/b, a/b > 1$, which prevents the negative exponential state after deriving the sliding mode surface, preventing the singularity problem from happening. When the x_1 state is fixed, the system state x_2 will become the main factor affecting the system convergence. In light of the aforementioned considerations, this research suggests the following enhanced non-singular fast terminal sliding mode surface

$$\begin{aligned} s &= x_1 + a|x_1|^{k_1}\operatorname{sgn}(x_1) + b|x_2|^{k_2}\operatorname{sgn}(x_2) \\ &+ c|x_3|^{k3}\operatorname{sgn}(x_3) \end{aligned} \tag{3}$$

In (3): x_3 is a new extended system variable associated to x_1 and x_2, sgn is a sign function; a, b, c are all positive values; and k_1, k_2, k_3 satisfy $k_1 > k_2 > k_3, 1 < k_3, k_1 < 2$.

In addition to adding a new system state that can adapt to more complicated systems, the enhanced sliding mode surface also makes it possible for a system to get there faster by setting the right parameters. The introduction of the sgn function avoids that negative fractional-order system states may produce sliding mode surfaces with complex solutions. Therefore, the range of possible values for k_1, k_2 and k_3 becomes larger, and more appropriate values can be selected according to the actual motor state.

3.2. Improved non-singular terminal extended sliding mode controller

The error cannot converge to zero in a finite period of time because classical sliding mode control employs linear sliding mode surfaces. Because of this, several researchers have created nonlinear sliding mode surfaces for terminal sliding mode controllers. The terminal sliding mode surface has a singular problem, hence a non-singular terminal sliding mode speed controller is developed in this study to guarantee that the system converges to zero in a finite amount of time while also avoiding the unique problem. In the actual operation process, the PMSM system is easily affected by external load disturbances. To counteract the impact of disturbances during the actual operation of the motor, disturbance variables are added to the speed loop controller developed in this study and supplied back into the speed loop.

Introducing the disturbance $f(t)$, we can get.

$$\dot{w}_m = \frac{3p_n\psi_f i_q}{2J} - \frac{B}{J}w_m + f(t) \tag{4}$$

The SPMSM's intermediate state variables are specified as follows to aid in the controller's design

$$\begin{cases} x_1 = w_r - w_m \\ x_2 = \int (w_r - w_m)dt \\ x_3 = \int (p_1x_1 + p_2x_2)dt \end{cases} \tag{5}$$

where the reference angular velocity is denoted by w_r. Taking the derivative of (5), we can get

$$\begin{cases} \dot{x}_1 = -\dot{w}_m \\ \dot{x}_2 = x_1 = w_r - w_m \\ \dot{x}_3 = p_1x_1 + p_2x_2 \end{cases} \tag{6}$$

The surface for the sliding mode is chosen to be Equation (3). The sliding mode surface is derived without a negative exponential power term, which prevents the singularity problem because of $k_1 > k_2 > k_3 > 1$. Taking the derivative of (3), we can get

$$\begin{aligned} \dot{s} = \dot{x}_1 + ak_1 |x_1|^{k_1-1}\dot{x}_1 + bk_2|x_2|^{k_2-1}\dot{x}_2 \\ +ck_3|x_3|^{k_3-1}\dot{x}_3 \end{aligned} \tag{7}$$

Using variable exponential reaching law

$$\begin{cases} \dot{s} = -\varepsilon f(s)sat(s) - qs \\ f(s) = \dfrac{1}{\dfrac{1}{m[\log_{10}(|s|+10)]^n} + \dfrac{1}{|s|+1}} \end{cases} \tag{8}$$

In (8): $\varepsilon > 0, q > 0, m > 0, n > 0$.

Substituting (4), (6) and (8) into (7), The speed controller's expression may be found as

$$\begin{aligned} i_q^* = \frac{1}{D}(\varepsilon f(s)sat(s) - qs - f(t) + \frac{Bw_m}{J} + \\ ak_1|x_1|^{k_1-1}\dot{x}_1 + bk_2|x_2|^{k_2-1}\dot{x}_2 + ck_3|x_3|^{k_3-1}\dot{x}_3) \end{aligned} \tag{9}$$

In (9), $D = 3P_n\psi_f / 2J$, i_q^* is the reference current.

The Lyapunov function may be described as follows when used to check the stability of the proposed speed controller

$$V = \frac{1}{2}s^2 \tag{10}$$

By substituting the reaching rule for the derivative of (10) we may get

$$\dot{V} = s\dot{s} = -\varepsilon s f(s) sat(s) - qs^2 \tag{11}$$

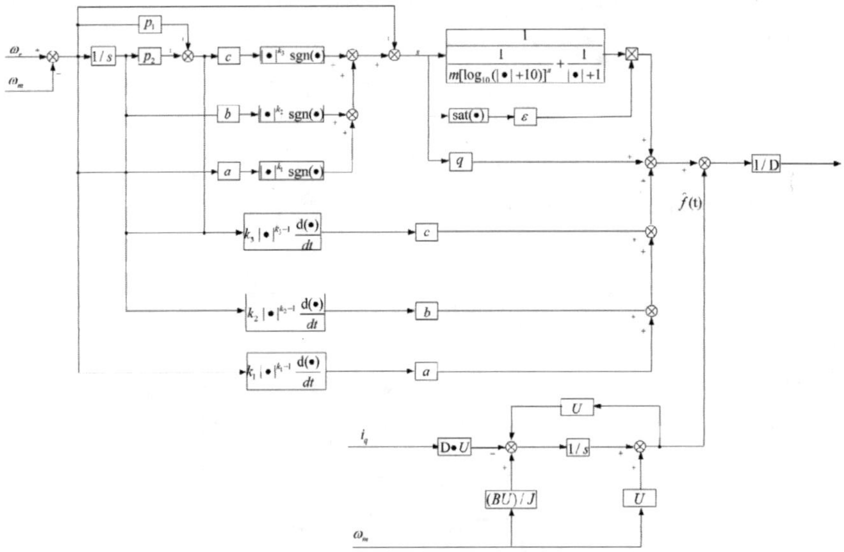

Figure 1. Schematic diagram of the improved control algorithm

Since $\varepsilon > 0, q > 0, m > 0, n > 0$, so $s \cdot sat(s) > 0$ and because of $|s| \geq 0$, then $f(s) > 0$, which proves $\dot{V} < 0$, as a result, the developed extended sliding mode controller for non-singular terminals can be considered stable. Figure 1 displays the speed loop control's schematic design.

3.3. Disturbance observer

In order to maintain the system's stable operation and potentially improve the system's anti-interference capability and control precision, the disturbance observer is designed to continuously monitor any disturbance in the system and make necessary corrections. The concept behind the disturbance observer is to adjust the estimated value of the disturbance using the difference between the estimated and real values. According to the prior knowledge, the change of the disturbance quantity is slow, according to (4), we can get

$$\begin{cases} \dot{w}_m = \dfrac{3p_n\psi_f i_q}{2J} - \dfrac{B}{J}w_m + f(t) \\ \dot{f}(t) = 0 \end{cases} \tag{12}$$

Then the error of the observed disturbance is:

$$\tilde{f}(t) = f(t) - \hat{f}(t) \tag{13}$$

Define $\dot{\hat{f}}(t) = U(f(t) - \hat{f}(t))$, where U is the gain of the disturbance observer. By including (12) into the derivative of (13) we obtain

$$\dot{\tilde{f}}(t) = -U\tilde{f}(t) \tag{14}$$

Therefore, the dynamic equation of the observation error is

$$\dot{\tilde{f}}(t) + U\tilde{f}(t) = 0 \tag{15}$$

The anticipated value of the disturbance may be exponentially approximated to the true value of the disturbance by changing the value of U, which means that the disturbance inaccuracy will approach zero. In the process of obtaining the derivative of the state variable velocity signal in (15), many noise signals will be amplified, so the derivative cannot be directly obtained. In the observation of disturbance variables, by introducing an intermediate quantity G, the amplification of the noise signal is avoided, and the intermediate variable is defined as $G = f(t) - Uw_m$, and the disturbance observer is obtained as

$$\begin{cases} \dot{G} = U(-\dfrac{3p_n\psi_f i_q}{2J} + \dfrac{B}{J}w_m - \hat{f}(t)) \\ \hat{f}(t) = G + Uw_m \end{cases} \tag{16}$$

Figure 1 shows the principle block diagram of the disturbance observer designed according to (16).

4. Simulation experiment and analysis

The PMSM control system's speed regulation block diagram is shown in Figure 2. A current and a speed loop make up the double closed-loop control system. The speed loop is managed using an enhanced sliding mode speed control approach, while the current loop is managed by the most basic PI control method. Following the construction of the Simulink simulation model of the PMSM speed control system, the simulation's results are contrasted with those of a typical controller based on the conventional exponential reaching law.

Table 1 displays the characteristics of the permanent magnet synchronous motor.

Table 1. Simulation parameter

Parameter	Value
Rated power (kw)	1.5
Stator inductance $L_s(mH)$	8.5
Viscous damping $B(N \cdot m \cdot s)$	0.008
Number of phases	3
Rated torque $(N \cdot m)$	10
Flux linkage $\psi_f(Wb)$	0.175
Pole pairs P_n	4
Stator resistance $R(\Omega)$	2.875
Inertia $J(kg \cdot m^2)$	0.003
Rated speed $(r/\min)$	1500

The following describe the control algorithm's parameters: the PI control current loop's parameter is $K_p = 45$; $K_i = 220$; the classic sliding mode controller's parameter is $c = 60$; $\varepsilon = 200$; $q = 300$; the parameter of the improved algorithm is $p_1 = 9$; $p_2 = 1$; $\varepsilon = 400$; $m = 12$; $n = 10$; $q = 600$; $a = 1$; $k_1 = 1.04$; $b = 0.7$; $k_2 = 1.03$; $c = 1.4$; $k_3 = 1.01$; $U = 800$ 。

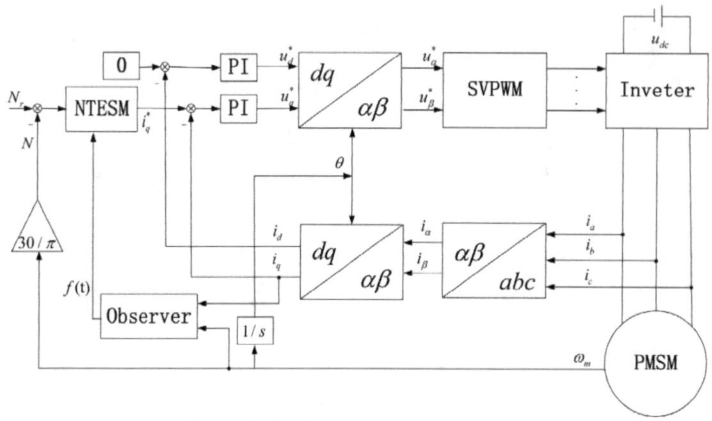

Figure 2 System structure diagram

4.1. Speed loop controller performance

The non-singular terminal extended sliding mode controller developed in this study is compared to the conventional sliding mode controller through simulation. The motor's rated speed is 1500 rpm, and at the instant of 0.2 seconds, a torque of 10 N•m addition is made. The motor's reaction speed curves for two distinct techniques are shown in Figure 3. Compared to the suggested approach, the typical sliding mode control system overshoots by 37.4% at launch, according to the comparison. This approach also provides a quicker reaction time. When subjected to external disturbances, the speed of the traditional sliding mode control system drops by around 36 r/min, but the speed of the improved sliding mode control is only about 17 r/min. It can effectively reduce the fluctuation brought by the disturbance to the system.

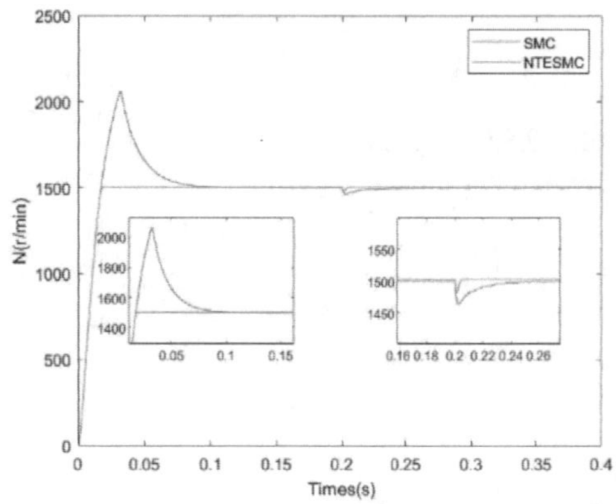

Figure 3 Speed response under

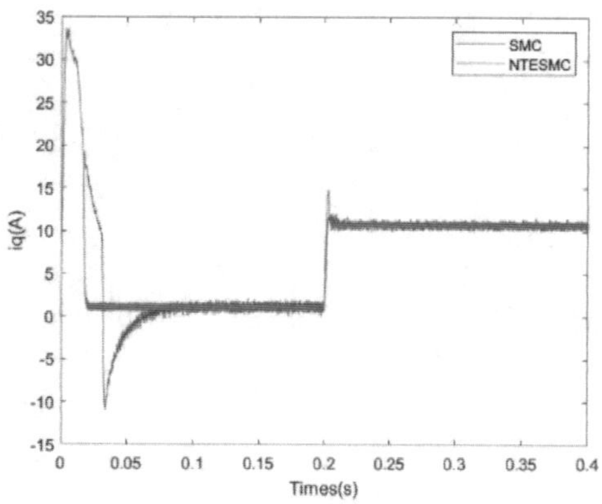

Figure 4 Q-axis current

The response curves for the q-axis current under two distinct approaches are shown in Figure 4. Figure 4 shows that, in contrast to the conventional sliding mode control approach, which may quickly meet the system's specified torque, when the system is subjected to external disturbances, the sliding mode control q-axis current is only marginally impacted by the load disturbance.

4.2. Disturbance observer performance

After simulation and verification, a load torque of 10 N•m was abruptly applied at 0.2 seconds to the specified disturbance observer model. Figure 5 depicts the relationship between the external disturbance's estimated and real values, while Figure 6 depicts the inaccuracy in the estimation of the disturbance. The disturbance observer developed in this research may be shown from the figure to be able to immediately follow the external disturbance that the system has encountered. The intended observer can immediately identify changes in the disturbance when the system is abruptly perturbed by an external load at 0.2 seconds, which lessens chattering and improves the system's capacity to resist disturbance.

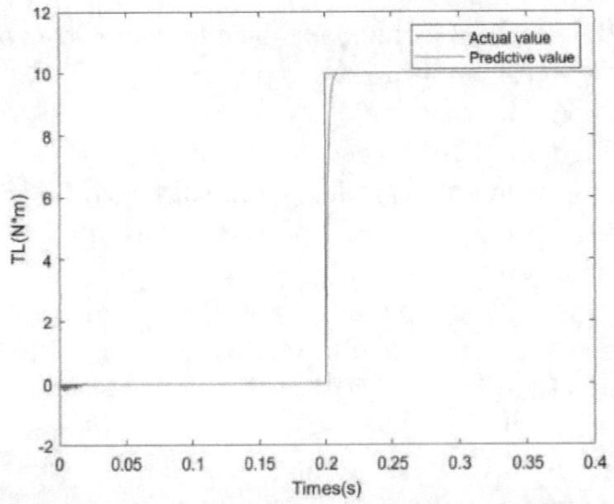

Figure 5 disturbance observations and actual values

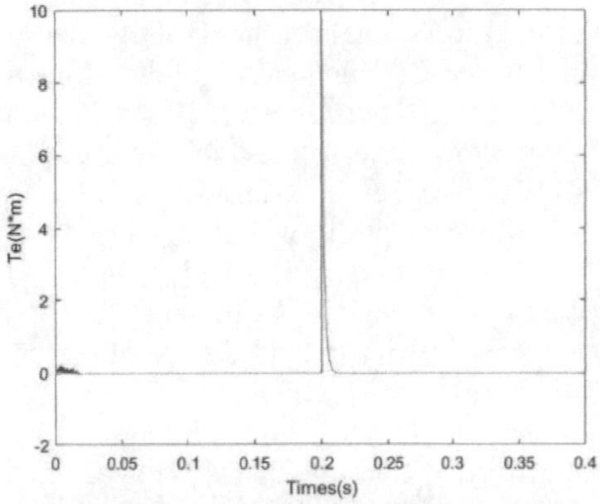

Figure 6 Estimation error of load disturbance

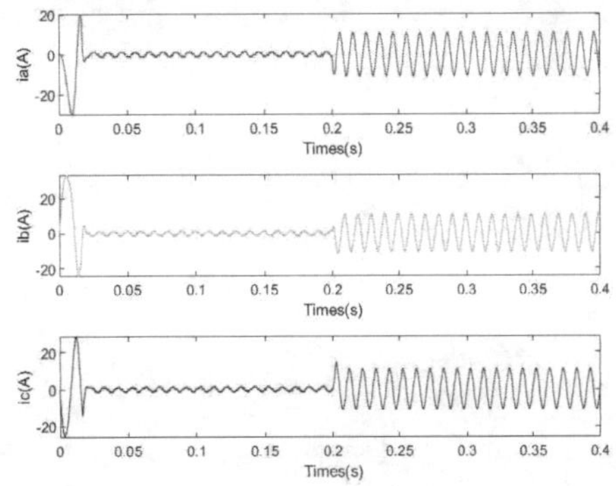

Figure 7 Three-phase current change curve

The three-phase current curve diagram in Figure 7 shows that each phase's current waveform is an ideal sine wave.

5. Conclusion

This work suggests a control strategy that combines a variable exponential reaching rule and a non-singular terminal extended sliding mode controller. The PMSM control system may converge quickly in a constrained amount of time thanks to the non-singular terminal extended sliding mode surface, and the variable exponential reaching law can lessen chattering after reaching the sliding mode surface. The system's built-in disturbance observer can precisely track disturbance information in real time and adjust the speed loop when it is subjected to external disturbance. This reduces speed fluctuation and increases the system's resilience. The simulation results shown that the control approach described in this study can successfully increase the system's control accuracy, eliminate chattering, and shorten the system's reaction time, proving the viability of the provided control method..

References

[1] L. Hou, J. Ma and W. Wang, "Sliding Mode Predictive Current Control of Permanent Magnet Synchronous Motor With Cascaded Variable Rate Sliding Mode Speed Controller," in IEEE Access, vol. 10, pp. 33992-34002, 2022.

[2] H. Alnami, C. Pang and Q. Wang, "A Novel Sliding Mode Control Method of Interior-Mounted PMSM," 2021 IEEE Texas Power and Energy Conference (TPEC), 2021, pp. 1-6.
[3] A. K. Junejo, W. Xu, C. Mu, M. M. Ismail and Y. Liu, "Adaptive Speed Control of PMSM Drive System Based a New Sliding-Mode Reaching Law," in IEEE Transactions on Power Electronics, vol. 35, no. 11, pp. 12110-12121, Nov. 2020.
[4] Y. -C. Liu, S. Laghrouche, D. Depernet, A. Djerdir and M. Cirrincione, "Disturbance-Observer-Based Complementary Sliding-Mode Speed Control for PMSM Drives: A Super-Twisting Sliding-Mode Observer-Based Approach," in IEEE Journal of Emerging and Selected Topics in Power Electronics, vol. 9, no. 5, pp. 5416-5428, Oct. 2021.
[5] Y. Wang, Y. Feng, X. Zhang and J. Liang, "A New Reaching Law for Antidisturbance Sliding-Mode Control of PMSM Speed Regulation System," in IEEE Transactions on Power Electronics, vol. 35, no. 4, pp. 4117-4126, April 2020.
[6] W. Xu, A. K. Junejo, Y. Liu and M. R. Islam, "Improved continuous fast terminal sliding mode control with extended state observer for speed regulation of PMSM drive system", IEEE Trans. Veh. Technol., vol. 68, no. 11, pp. 10465-104676, Nov. 2019.
[7] J. Yang, Q. Lv, B. Liu, L. Wang, Y. Li and Y. Xiao, "ESO-based Finite Set Model Predictive Current Control PMSM Delay Compensation," 2020 Chinese Automation Congress (CAC), 2020, pp. 3826-3831.
[8] B. Xu, L. Zhang and W. Ji, "Improved Non-Singular Fast Terminal Sliding Mode Control With Disturbance Observer for PMSM Drives," in IEEE Transactions on Transportation Electrification, vol. 7, no. 4, pp. 2753-2762, Dec. 2021.

Optimal Design of a Cable Driven Parallel Robot for Upper Limb Rehabilitations

Bin Liao[1,2], Jun Li[2], Shichao Deng[2], Jianmin Li[1] and Guohui Li[3]

[1] School of Mechanical Engineering, Tianjin University, Tianjin 300350, China

[2] School of Mechanical and Electrical Engineering, Guilin University of Electronic Technology, Guilin 541004, China

[3] Teda Orking Hi-Tech Co., Ltd., Tianjin 300072, China

Email: liao.bin@hotmail.com

Abstract. A novel cable driven parallel robot is introduced to assist upper limb in rehabilitation treatments in this paper, which has a lightweight structure. This robot is characterized by a novel and flexible end-effector that is easy to change assembly for different therapies. And it is designed for desktop usage, so its size is not too large. After presenting the structure, the kinematic model is presented for optimal design. Then, the minimal singular value is considered as the velocity transmission index. For upper limb rehabilitation, a new optimal design problem with the transmission index and the robot size is formulated. Last, the optimization presents a design with satisfying performances in velocity transmission capability and robot size.

1. Introduction

Upper limbs play significant roles in people's lives. Without them, our lives will become hard. A significant cause of impairments of the arms and the wrists is stroke. In general, rehabilitation trainings are required to tread or reduce stroke consequences. However, rehabilitation trainings usually are one-to-one programs between the patient and the therapist with rehabilitation tools. These programs are labor-intensive and time-consuming. For intensive, accurate and safe rehabilitation, rehabilitation robots are the best choice, which can also supply repetitive movements to patients' limbs [1, 2].

In general, according to rehabilitation robot structures, they split into two groups: end-effectors and exoskeletons [3]. End-effectors are characterized by their function to assist limbs with rehabilitation processes, which require trajectories with different shapes and sizes [4]. This kind of robot applies its movable handle to drive the patient's limb [5]. By comparison with end-effector robots, exoskeleton robot joint axes provide given trajectories [6]. And usually they are only feasible for one side limbs [7]. Thus, in this paper, the end-effector robots are focused on. To deal with upper limb rehabilitation tasks, one cable driven parallel robot is designed.

Cable driven parallel robots present the advantages of low moving mass, low cost, high speed/acceleration [8]. And since the ease of assembly and reconfiguration, they are applied to rehabilitation tasks well [9]. Instead of rigid links, their parallel architectures with end-effectors only use cables to connect to the fixed frame. The end-effectors can be linked to upper limbs. In previous literatures, cable driven architectures are designed for rehabilitations [10-12]. Most of proposed robots have bulky structures or asymmetric workspaces, which must be addressed. After type synthesis, usually, optimal design must be implemented. To improve the performances of parallel robots, many effective indices are employed, such as workspace [13], dexterity [14], and stiffness [15].

A novel cable driven parallel robot is optimized for upper limb rehabilitations in this paper. The remainder is organized as follows. Robot's structure and geometric model are introduced in next section. In section 3, the optimal design problem which focuses on upper limb rehabilitations and the simulation results are discussed. At last, conclusions are drawn.

2. Structure and modeling

Since forearm and upper arm revolve around elbow joint and shoulder joint respectively, the cable driven parallel robot should assist the rotational motion of a limb. However, the orientation of the end-effector is considered passive, if this rehabilitation robot moves around an arc. The robot has eight cables. And the fixed frame is a cuboid frame with eight pulleys at its eight vertices (see Fig. 1). In Fig. 2, the end-effector has a spherical joint which ensures there is no interference between upper limb and robot. The Velcro is applied to connect upper limb.

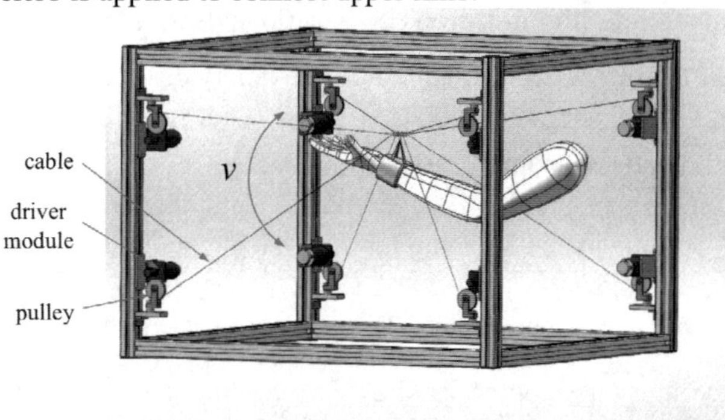

Figure 1. 3D model of the upper limb rehabilitation robot

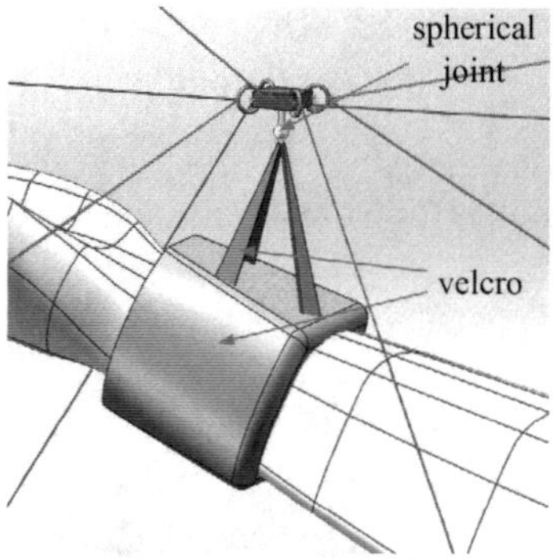

Figure 2. The end-effector

Rehabilitation processes are divided into two main categories: active therapy and passive therapy. In Fig. 1, this case is passive therapy, in which the patient could make no effort. Active therapy is applied to improve the joint range of motion. An example of active therapy is shown in Fig. 3, in which the assembly of the end-effector has a different orientation.

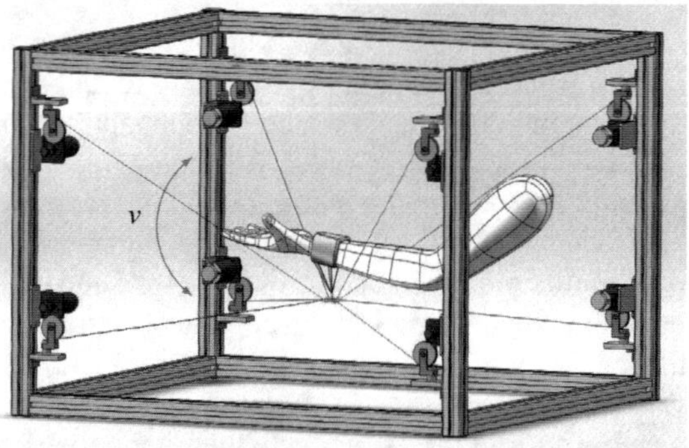

Figure 3. An example of active therapy

In Fig. 4, there is a coordinate frame *o-xyz*. The x axis and the y axis are on the bottom of the cuboid, and the origin o is in the center of this plane. The vertices are A_1 to A_8. Since the motion of the Velcro is passive and the moving platform is very small, the moving platform can be regarded as a mass point which can be controlled by cables. Suppose the mass point is B. The length between A_i and B is L_i, $i=1,\cdots,8$.

Suppose the coordinates of point B are (x, y, z) and the coordinates of point A_i are (x_i, y_i, z_i). Then, we have geometric equations as follows.

$$L_i^2 = \|\boldsymbol{B} - \boldsymbol{A}_i\|^2, \quad i=1,\cdots,8. \tag{1}$$

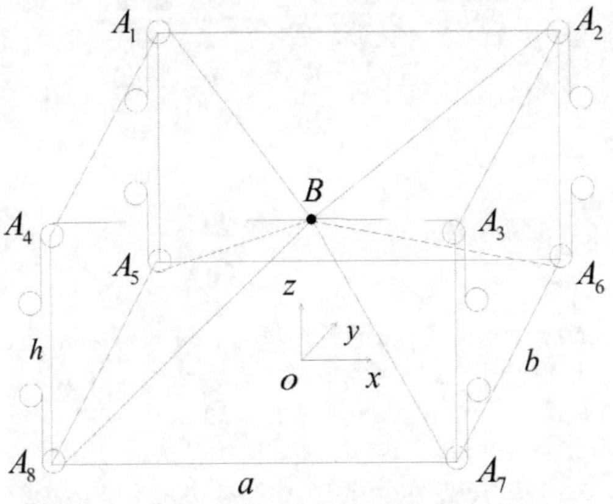

Figure 4. Notations and coordinate system

Differentiating these equations, the velocity relationships between the cables and the end-effector can be obtained. Then, the velocity equations can be expressed as matrix form.

$$\boldsymbol{J}_l \dot{\boldsymbol{L}} = \boldsymbol{J}_x \dot{\boldsymbol{X}}, \tag{2}$$

where $\dot{\boldsymbol{L}} = \left[\dot{L}_1, \cdots, \dot{L}_8\right]^{\mathrm{T}}$ is the velocity vector of the cables, $\dot{\boldsymbol{X}} = [\dot{x}, \dot{y}, \dot{z}]^{\mathrm{T}}$ is the velocity vector of the end-effector, $\boldsymbol{J}_l$ is a 8×8 diagonal matrix, whose diagonal elements are L_1 to L_8, and

$$\boldsymbol{J}_x = \begin{bmatrix} x+a/2 & x-a/2 & x-a/2 & x+a/2 & x+a/2 & x-a/2 & x-a/2 & x+a/2 \\ y-b/2 & y-b/2 & y+b/2 & y+b/2 & y-b/2 & y-b/2 & y+b/2 & y+b/2 \\ z-h & z-h & z-h & z-h & z & z & z & z \end{bmatrix}^{\mathrm{T}}. \tag{3}$$

The kinematics Jacobian matrix can be defined as

$$\boldsymbol{J} = \boldsymbol{J}_l^{-1} \boldsymbol{J}_x. \tag{4}$$

3. Optimal design

3.1. Problem formulation

This cable drive parallel robot is designed for upper limb rehabilitations. To improve efficiencies of the processes, the velocity transmission capability should be guaranteed. The kinematics Jacobian matrix's minimal singular value, $\sigma_{\min}(\boldsymbol{J})$, restricts the parallel robot's minimal velocity in the workspace. Thus, the transmission index can be defined as follows:

$$\eta = \frac{1}{V} \int_V \sigma_{\min} dV. \tag{5}$$

where, V is a regular workspace.

Considering patients can sit at any side of the robot, the bottom of the frame can be a square, that is $a = b$. Since the lengths of most people's forearms and upper arms are not more than 0.7 m and 0.3 m respectively, the regular workspace can be a cylinder, whose height is 0.7 m and cross section radius is 0.3 m. The symmetry axis of this cylinder is coincident with the fix frame of the robot. To avoid collision between upper limb and pulleys, the minimum length of the cables $l_{\min}$ must be limited. Because this robot will be use on the desk. The volume of the robot ($V_e = abh$) also should not be too large.

According to the above analysis, the design parameters are designated as $\boldsymbol{\alpha} = [a, h, l_{\min}]^{\mathrm{T}}$. Then, we can formulate the optimal design problem as follows.

$$\begin{aligned} &\max_{\alpha} \quad \eta \\ &\text{subject to} \quad 0.6\text{m} \le a \le 0.8\text{m} \\ &\qquad\qquad 0.8\text{m} \le h \le 1.0\text{m} \\ &\qquad\qquad 0.1\text{m} \le l_{\min} \le 0.2\text{m} \\ &\qquad\qquad V_e \le 0.5\text{m}^3 \end{aligned} \tag{6}$$

3.2. Results and analysis

This formulation is a nonlinear optimal problem, which can be solved by the genetic algorithm. The MATLAB GA Toolbox is employed. Table 1 shows the initial/optimal parameters and the corresponding performances. The initial parameters are chosen from the bounds, which are the mid-values. After optimal design, the side length of the bottom is increased by 12.04%, the height is decreased by 9.69%, and the minimum length of the cables is upped by 0.0216 m. The transmission index is improved obviously, which is increased by 39.32%. However, the volume of the robot reaches the allowable boundary.

Table 1. The initial/optimal parameters and corresponding initial/optimal performances

Parameters/performances	$a(\text{m})$	$h(\text{m})$	$l_{\min}(\text{m})$	η	$V_e(\text{m}^3)$
Initial	0.7000	0.9000	0.1500	1.4798	0.4410
Optimal	0.7843	0.8128	0.1716	1.5505	0.5000

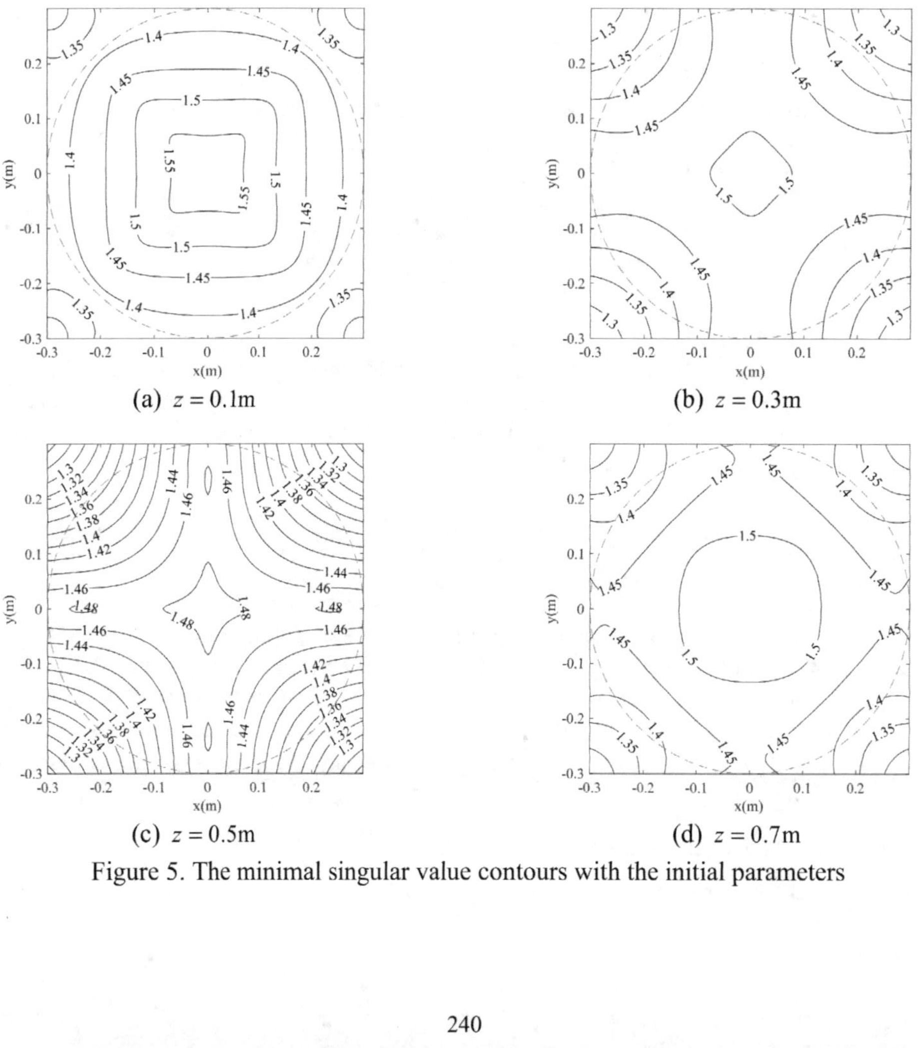

(a) $z = 0.1\text{m}$ (b) $z = 0.3\text{m}$

(c) $z = 0.5\text{m}$ (d) $z = 0.7\text{m}$

Figure 5. The minimal singular value contours with the initial parameters

Fig. 5 and 6 show the minimal singular value contours generated by initial and optimal parameters respectively. Chose the cross sections at $z =$ 0.1 m, 0.3 m, 0.5 m and 0.7 m to analyze. The improvement is remarkable. The minimal singular values are increased obviously in the regular workspace (in the red circles). Thus, the optimal design is effective.

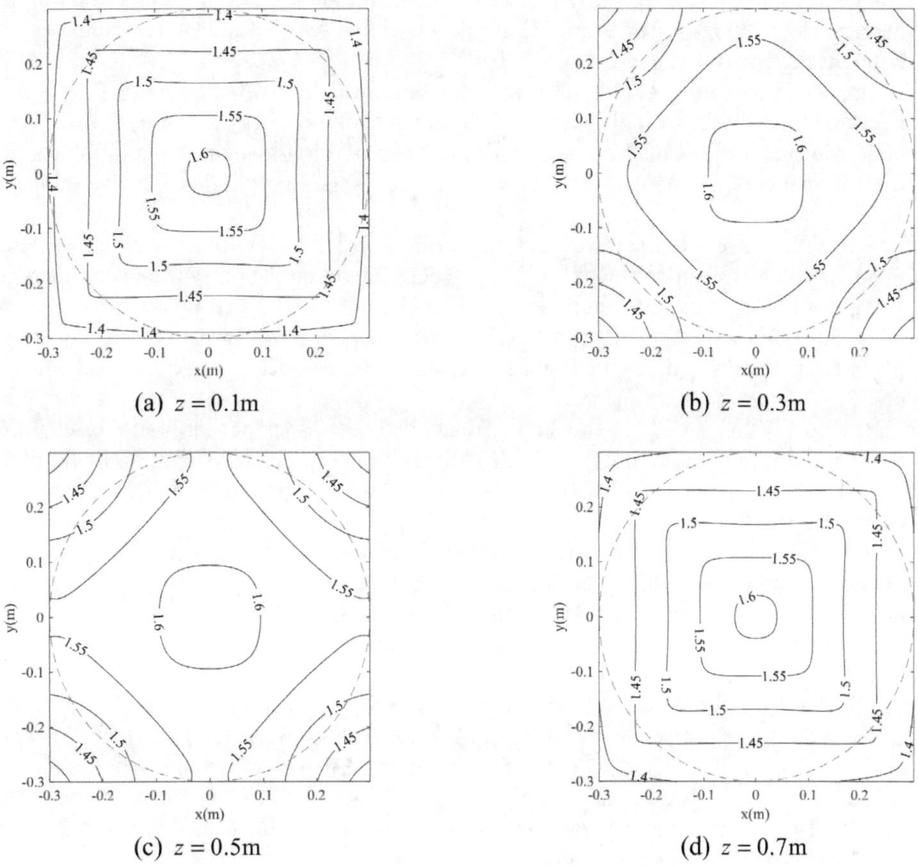

Figure 6. The minimal singular value contours with the optimal parameters

4. Conclusion

This paper introduces a novel cable driven parallel robot, which can be used for upper limb rehabilitations. This robot has a lightweight structure and a flexible end-effector, which can be employed for passive therapy and active therapy in different assembly types. The kinematic model is presented by algebraic derivation. Then, for rehabilitation processes, a new optimal design formulation is proposed. The simulation results show after optimisation the velocity transmission capability in the regular workspace is improved and the robot is not too large for desktop usage.

Acknowledgments

This work was supported in part by Guangxi Natural Science Foundation under Grant 2019GXNSFBA245024, in part by Guangxi Science and Technology Program under Grant AD19245045, in part by the National Key Research and Development Program of China under Grant 2018YFC2000701.

References

[1] Hu X L, Tong K, Song R, et al. (2009) A comparison between electromyography-driven robot and passive motion device on wrist rehabilitation for chronic stroke. Neurorehabilitation and neural repair, 23(8): 837-846.
[2] Sebastian G, Li Z, Tan Y, et al. (2019) Force Observer for an Upper Limb Rehabilitation Robotic Device using Iterative Learning Control. In: 12th Asian Control Conference (ASCC). Kitakyushu. pp. 1607-1612.
[3] Nef T, Quinter G, Müller R, et al. (2009) Effects of arm training with the robotic device ARMin I in chronic stroke: three single cases[J]. Neurodegenerative diseases, 6(5-6): 240-251.
[4] Bertani R, Melegari C, De Cola M C, et al. (2017) Effects of robot-assisted upper limb rehabilitation in stroke patients: a systematic review with meta-analysis. Neurological Sciences, 38(9): 1561-1569.
[5] Islam M R, Spiewak C, Rahman M H, et al. (2017) A brief review on robotic exoskeletons for upper extremity rehabilitation to find the gap between research porotype and commercial type. Advances in Robotics and Automation, 6(3): 177.
[6] Głowiński S, Błażejewski A. (2019) An exoskeleton arm optimal configuration determination using inverse kinematics and genetic algorithm. Acta of bioengineering and biomechanics, 21(1): 45-53.
[7] Sheng B, Zhang Y, Meng W, et al. (2016) Bilateral robots for upper-limb stroke rehabilitation: State of the art and future prospects[J]. Medical engineering & physics, 38(7): 587-606.
[8] Qian S, Zi B, Shang W W, et al. (2018) A review on cable-driven parallel robots[J]. Chinese Journal of Mechanical Engineering, 31: 66.
[9] Merlet J P. (2013) Wire-driven parallel robot: open issues. In: Romansy 19–Robot Design, Dynamics and Control. Springer, Vienna, pp. 3-10.
[10] Surdilovic D, Zhang J, Bernhardt R. (2007) STRING-MAN: Wire-robot technology for safe, flexible and human-friendly gait rehabilitation. In: IEEE 10th International Conference on Rehabilitation Robotics. Noordwijk. pp. 446-453.
[11] Ceccarelli M. (2014) Problems and experiences on cable-based service robots for physiotherapy applications. In: New trends in medical and service robots. Springer, Cham, pp. 27-42.
[12] Rosati G, Masiero S, Rossi A. (2017) On the use of cable-driven robots in early inpatient stroke rehabilitation. In: Advances in Italian Mechanism Science. Springer, Cham, pp. 551-558.
[13] Merlet J P. (1997) Designing a parallel manipulator for a specific workspace. the International Journal of Robotics research, 16(4): 545-556.
[14] Wu G, Bai S. (2019) Design and kinematic analysis of a 3-RRR spherical parallel manipulator reconfigured with four–bar linkages. Robotics and Computer-Integrated Manufacturing, 56: 55-65.
[15] Sun T, Lian B. (2018) Stiffness and mass optimization of parallel kinematic machine. Mechanism and Machine Theory, 120: 73-88.

Research on synchronization of cooperative trajectory of dual industrial robots based on fuzzy adaptive PID control strategy

Qiang Xiong[a], Jianqun Wang[*]

Beijing Institute of Technology, Beijing, China, 100081

[a]3220200348@bit.edu.cn, *wangjq@bit.edu.cn

Abstract. Through the research on the synchronization of the cooperative trajectory of dual industrial robots, this paper analyses the characteristics of master-slave control and distributed control, and puts forward a fuzzy adaptive PID control method suitable for the cooperative scene of dual industrial robots, which can dynamically adjust the parameters of PID control online and adapt the dynamic system of industrial robots in real time. Taking two same 6-DOF ABB industrial robots as examples, common PID controller and fuzzy PID controller are compared and simulated under master-slave control mode and distributed control mode respectively. The simulation results verified that the fuzzy adaptive PID control method is effective, and the distributed control mode has more applicability and expansibility than the master-slave control mode.

1. Introduction

The multi-industrial robot cooperation system breaks through the bottleneck of the single industrial robot system, has stronger operating capacity and load capacity, more flexible system structure and organization, and can quickly and effectively complete complex and difficult tasks. As a typical representative of multi-industrial robot cooperation, the trajectory synchronization of dual industrial robot cooperation is the basis of high-quality collaborative operation of dual industrial robots, and the control mode and control method are the key factors to improve the trajectory synchronization of dual industrial robot cooperation.

On control mode, one is master-slave control mode, Mansor N N et al[1] simulated and verified the effectiveness of the master-slave control mode of dual industrial robots in the V-rep environment by using three controllers: P, PD and PID. The other is the distributed model, Azar A T et al[2] designed a distributed robust controller of industrial cooperative robot considering system disturbance and other factors. In the experiment, the manipulator in several industrial robot work units is controlled synchronously, which can effectively reduce the tracking error of the robot. He Y H et al[3] implemented a distributed optimization algorithm for cooperative robot control on the dobby robot platform, which greatly enhances the internal and external control capabilities of multi-robot systems in dynamic environments.

In terms of control method, using intelligent control algorithm to improve and optimize the traditional control method. Ouyang F et al[4] used genetic algorithm to optimize PD damping control method for force coordination control in the process of dual industrial robot collaborative transportation, which reduces the force error of robotic system. Wang J Y et al[5] used neural network algorithm to adapt PID parameters of PID feedforward control in ADMAS simulation system. Tao Y et al[6] applied a fuzzy PID controller to the deburring environment of the industrial robot to reduce the error of end trajectory.

In view of the current industrial robots mainly focus on the control optimization of a single industrial robot and the shortcomings of common control modes and control methods, this paper takes the cooperation of dual industrial robots as the research background, designs a fuzzy adaptive PID control method, and studies the trajectory synchronization optimization of dual industrial robots based on this control method in master-slave mode and distributed mode.

2. Synchronous control mode of cooperative trajectory of dual industrial robots

The trajectory synchronization in the dual robot cooperation synchronization plays an important role in ensuring that the two industrial robots complete the task with high precision. Master-slave mode and distributed mode are the main modes of trajectory synchronization control of two-robot cooperation.

The master-slave mode means that one industrial robot is the master industrial robot, and the other industrial robot is the slave industrial robot. Its control mode structure is shown in figure 1. It only assigns relevant instructions to the master industrial robot, and the slave industrial robot acts with the actual relevant parameters in the motion process of the master industrial robot as the target parameters.

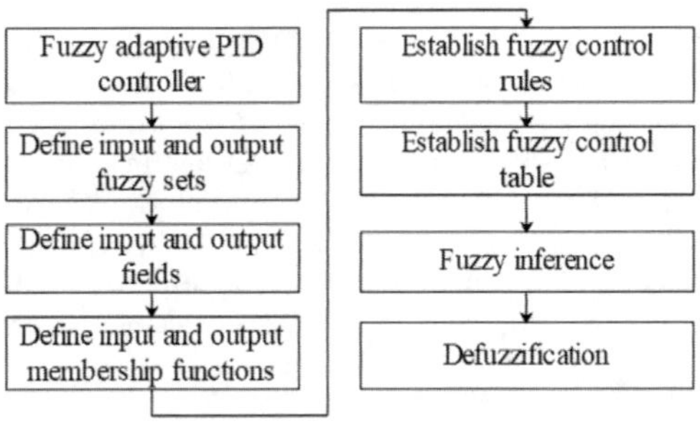

Figure 1. Master-slave control mode.

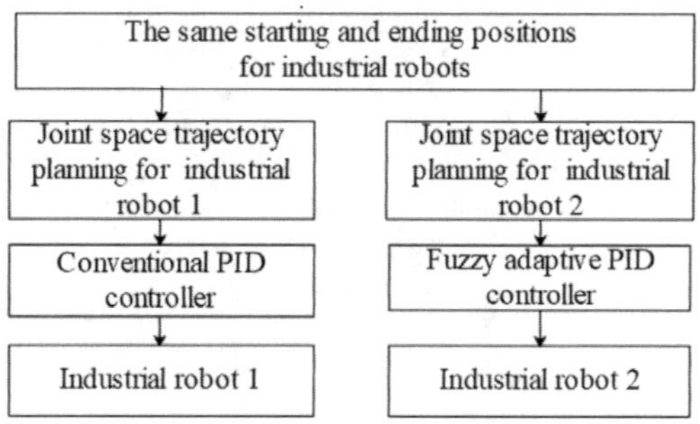

Figure 2. Distributed control mode.

The distributed control mode means that the two industrial robots establish task instructions separately. In the actual movement process, the dependence between the two robots on parameters is no longer carried out. Its control mode structure is shown in figure 2. It can be extended to the collaborative operation between multiple industrial robots.

3. Joint space trajectory planning

The control of industrial robots requires model of robots, and precise control of the planned trajectory of industrial robots is achieved through kinematics and dynamics-related model.

3.1 6-DOF industrial robot model

In this paper, as shown in figure 3-5, a 6-DOF ABB industrial robot (IRB1600) is taken as an example to establish a three-dimensional simulation model through physical objects.

Figure 3. 3D simulation model-view 1.

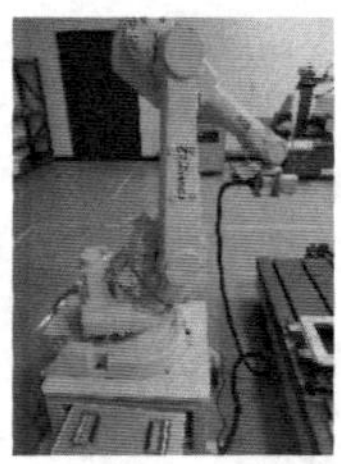

Figure 4. Robot- IRB1600.

Figure 5. 3D simulation model-view 2.

As shown in table 1 below, the improved DH parameters of IRB1600 industrial robot are obtained through relevant literature query and parameter measurement.

Table 1 Improved DH parameters of IRB1600 industrial robot.

	Theta(degree)	Alpha(degree)	A(mm)	D(mm)	Offset(degree)	Range (degree)
Joint 1	0	90	135	481.5	0	-180 ~ +180
Joint 2	0	0	475	0	90	-063 ~ +136
Joint 3	0	90	0	0	0	-235 ~ +055
Joint 4	0	90	0	600	180	-200 ~ +200
Joint 5	0	90	0	0	180	-115 ~ +115
Joint 6	0	0	0	72.5	0	-400 ~ +400

Through the above improved DH parameters, the robot simulation model based on irb1600 is established. In addition, corresponding modules are written according to the principles of forward kinematics and inverse kinematics. The corresponding angle values of each joint space of the 6-DOF industrial robot are obtained by specifying the pose, which provides an input source for the subsequent joint space trajectory planning.

3.2 Joint space trajectory planning by quintic polynomial

This paper mainly discusses the influence of different control methods on the trajectory synchronization of industrial robot cooperation. The joint space is directly related to the controller, so this paper chooses trajectory planning in joint space. In this paper, the joint space trajectory of the 6-DOF ABB industrial robot is carried out through the quintic polynomial. The quintic polynomial can solve the problems such as the impact caused by the discontinuous acceleration at the beginning and end of the low order polynomial. The expression of quintic polynomial is:

$$\theta(t) = a_0 + a_1 t + a_2 t^2 + a_3 t^3 + a_4 t^4 + a_5 t^5 \tag{1}$$

The constraints are:

$$\theta_0 = a_0 , \theta_f = a_0 + a_1 t_f + a_2 t_f^2 + a_3 t_f^3 + a_4 t_f^4 + a_5 t_f^5 \tag{2}$$

$$\dot{\theta}_0 = a_1 , \dot{\theta}_f = a_1 + 2a_2 t_f + 3a_3 t_f^2 + 4a_4 t_f^3 + 5a_5 t_f^4 \tag{3}$$

$$\ddot{\theta}_0 = 2a_2 , \ddot{\theta}_f = 2a_2 + 6a_3 t_f + 12a_4 t_f^2 + 20a_5 t_f^3 \tag{4}$$

The coefficients of the quintic polynomial obtained from the solution of the linear equations composed of the above constraints are:

$$a_0 = \theta_0 , a_3 = (10\theta_f - 10\theta_0 - (4\dot{\theta}_f + 6\dot{\theta}_0)t_f + (\ddot{\theta}_f / 2 - 3\ddot{\theta}_0 / 2)\, t_f^2) / t_f^3 \tag{5}$$

$$a_1 = \dot{\theta}_0 , a_4 = (-15\theta_f + 15\theta_0 + (7\dot{\theta}_f + 8\dot{\theta}_0)\, t_f - (\ddot{\theta}_f - 3\ddot{\theta}_0 / 2)\, t_f^2) / t_f^4 \tag{6}$$

$$a_2 = \ddot{\theta}_0 / 2 , a_5 = (6\theta_f - 6\theta_0 - (3\dot{\theta}_f + 3\dot{\theta}_0)t_f + (\ddot{\theta}_f / 2 - \ddot{\theta}_0 / 2)\, t_f^2) / t_f^5 \tag{7}$$

The six joint angles corresponding to the specified initial and end pose are obtained through the inverse kinematics module, and then the six initial and end joint angles are interpolated through the quintic polynomial to achieve trajectory planning, and the target trajectory of the subsequent control joint is obtained. The end pose of 6-DOF ABB industrial robot is [x y z R P Y], the initial state of the robot (each joint angle is 0) is set as the initial pose: [807.5 0 956.5 0 90 180], and the end pose is [610 0 -46 30 60 -90]. As shown in figure 6-9, Trajectories of various indicators are planned according to the quintic polynomial.

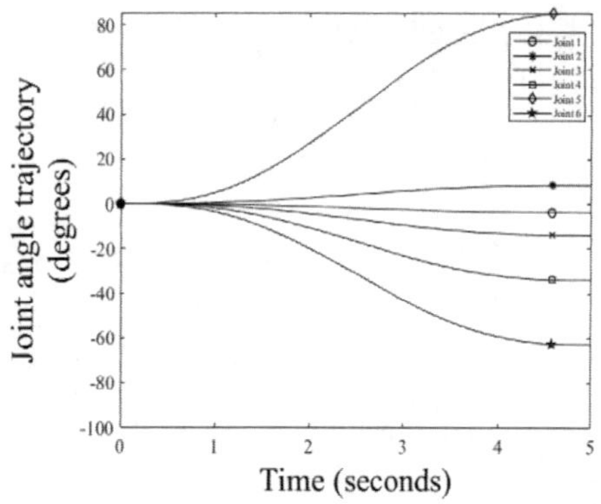

Figure 6. Angle curve.

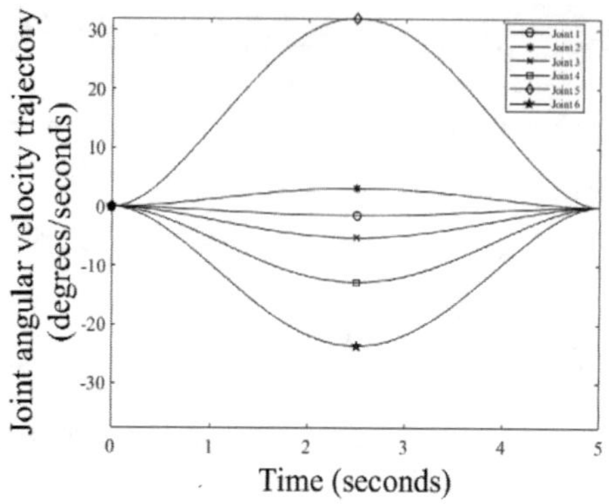

Figure 7. Angular velocity curve.

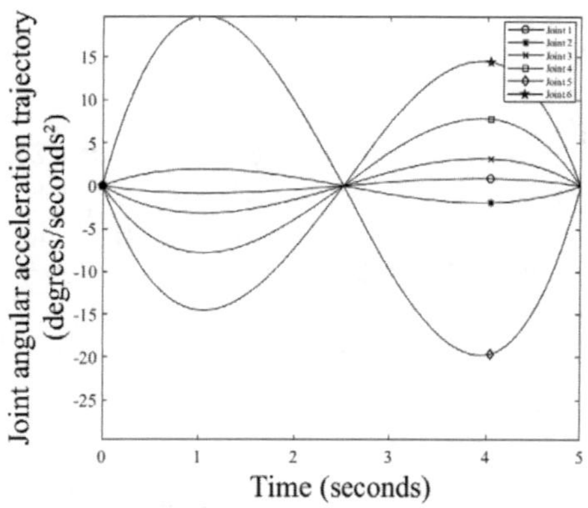

Figure 8. Angular acceleration curve.

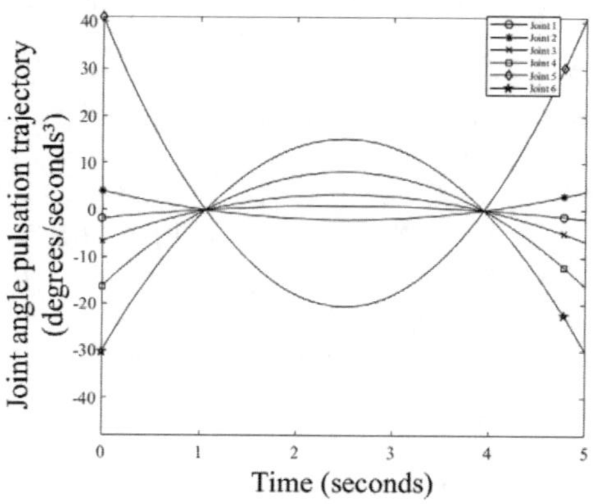

Figure 9. Angle pulsation curve.

4. Fuzzy adaptive PID control strategy

4.1 Common PID control

Because PID control method does not require complex system dynamic model, this method is much used in the field of industrial robots. It is relatively simple to control and has certain control stability compared with other control methods.

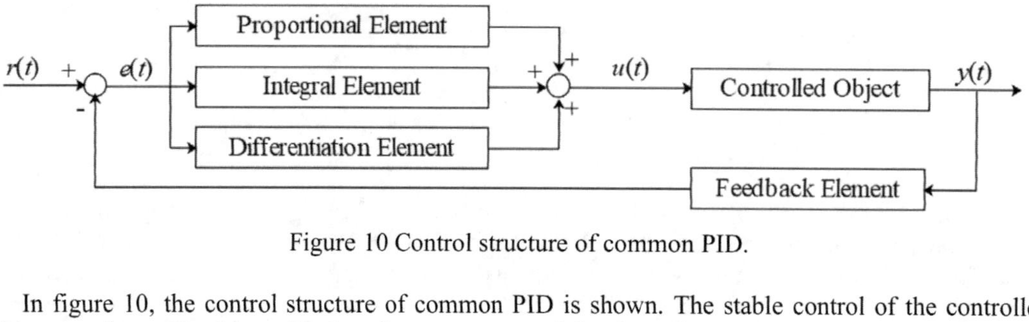

Figure 10 Control structure of common PID.

In figure 10, the control structure of common PID is shown. The stable control of the controlled object can be realized by adjusting the parameters of the three links of proportion, integration and differentiation. The corresponding common PID control law is:

$$u(t) = K_p(e(t) + \frac{1}{T_i}\int_0^t e(t)dt + T_d\frac{de(t)}{dt}) \tag{8}$$

order $K_i = K_p / T_i, K_i = K_p \cdot T_d$, the above formula can be simplified to:

$$u(t) = K_p e(t) + K_i \int_0^t e(t)dt + K_d \frac{de(t)}{dt} \quad (9)$$

$e(t) = r(t) - y(t)$: control input, $u(t)$: control output, K_p: proportional coefficient, K_i: integral coefficient, K_d: differential coefficient.

4.2 Fuzzy adaptive PID control

Common PID control cannot meet the requirements of high precision, strong anti-interference and high robustness due to the reduced applicability of the determined parameters such as proportion, integral and differential. Fuzzy control can enhance the disturbance and robustness of the control because it does not require the model accuracy of the controlled object. Therefore, Adding fuzzy control theory and method to PID dynamic control, fuzzy control can dynamically adjust PID parameters(K_p 、 K_i 、 K_d) on-line in real time, so that the three parameters of PID can adapt to the system change process, and improve the control accuracy and anti-interference.

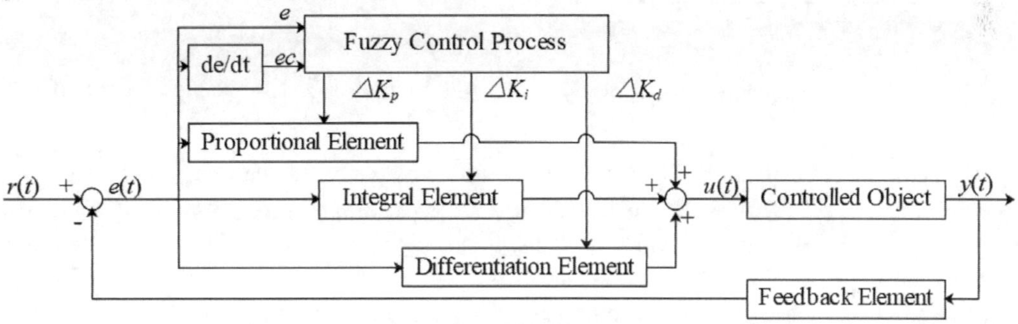

Figure 11 Control structure of fuzzy adaptive PID.

In figure 11, control structure of fuzzy adaptive PID is shown. On the basis of the common PID control, the fuzzy control process is added. e and ec act on fuzzy control inputs. Through the fuzzy control process, the change parameters (K_p 、 K_i 、 K_d) are obtained, which are respectively superimposed into the corresponding adjustment parameters of the PID control to form new fuzzy PID adjustment parameters, so as to achieve self-tuning of fuzzy PID parameters. Corresponding fuzzy adaptive PID control law is:

$$u(t) = (K_p + \Delta K_p)e(t) + (K_i + \Delta K_i)\int_0^t e(t)dt + (K_d + \Delta K_d)\frac{de(t)}{dt} \quad (10)$$

K_p , K_i and K_d are the parameters of common PID, ΔK_p, ΔK_i and ΔK_d are the variation of three parameters for on-line adaptive adjustment of fuzzy control process.

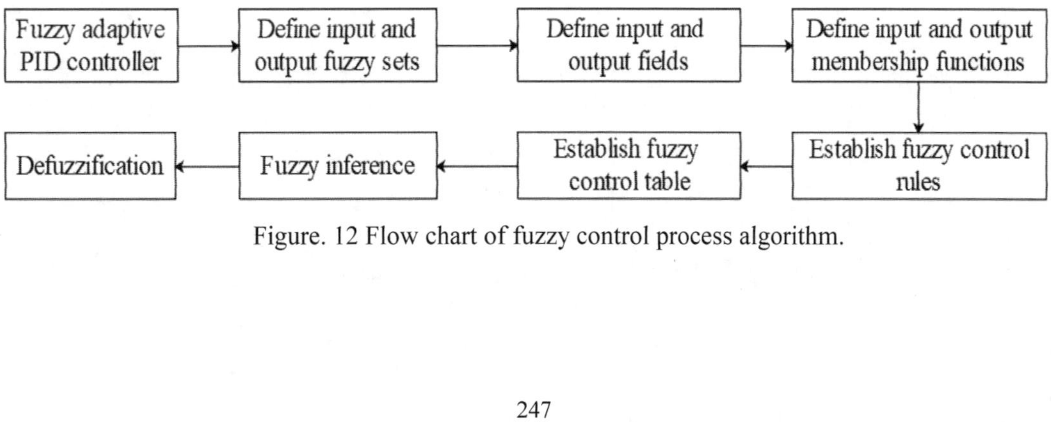

Figure. 12 Flow chart of fuzzy control process algorithm.

Fuzzy control process is the process of using methods and theories of fuzzy mathematics to implement control on input and output data, including fuzzification, fuzzy inference and defuzzification. The algorithm flow chart of fuzzy control process is shown in figure 12.

A multivariable two-dimensional fuzzy control structure (MIMO) with two inputs and three outputs is decoupled into three two inputs and one output single variable two-dimensional fuzzy control structures (MISO). Because the actual error range obtained by PID control of industrial robot is small, the inputs (e , ec) domain range can be set to be [-1, 1], and the outputs (ΔK_p, ΔK_i , ΔK_d) domain range are [-6, 6]. as shown in figure 13-14, the input and output membership function adopts the combination form of trigonometric function and Gaussian function.

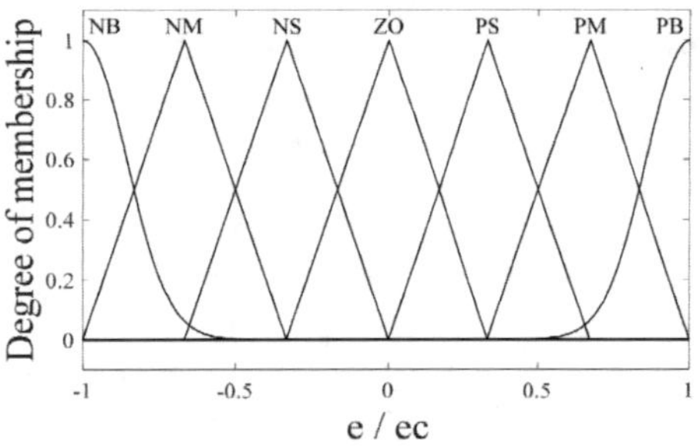

Figure 13. Membership function of e / ec.

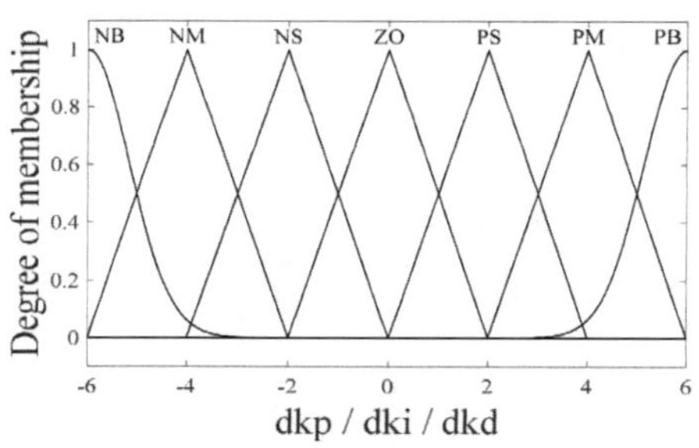

Figure 14. Membership function of ΔK_p / ΔK_i / ΔK_d.

Fuzzy control rules is to establish the adaptive adjustment relationship between inputs (e, ec) and outputs (ΔK_p, ΔK_i , ΔK_d), which will further affect the performance of the system including the corresponding speed, overshoot, oscillation degree, and steady-state error. Through the actual PID parameter adjustment experience and related laws, the fuzzy control table is obtained from the fuzzy rules. The process of fuzzy reasoning is a process of self-adaptive adjustment of ΔK_p, ΔK_i and ΔK_d parameters according to fuzzy control tables. Defuzzification is the process of converting fuzzy data into accurate data. In this paper, the defuzzification adopts the area centroid method.

5. Experiment and simulation analysis

5.1 Comparative experimental of two control methods

Through the previous theoretical analysis and design, taking 6-DOF ABB industrial robot (irb1600) as an example, the industrial robot simulation system is built. Common PID control and fuzzy adaptive PID control are used for comparative experiments. The industrial robot joint target angle trajectory and the actual angle trajectory form a trajectory error. The six joint trajectories and corresponding trajectory errors obtained by the two control methods are shown in figure 15-26.

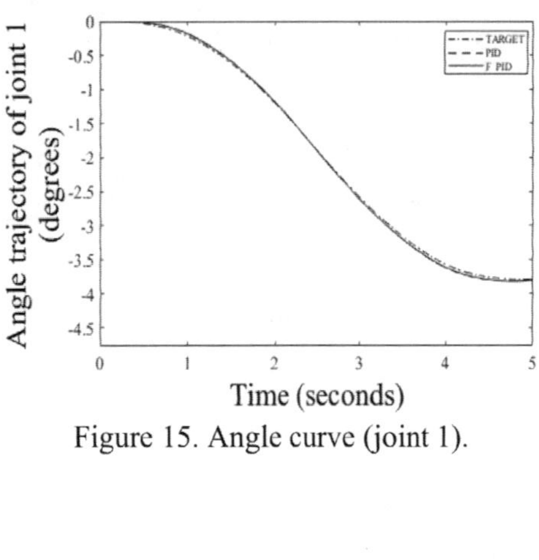

Figure 15. Angle curve (joint 1).

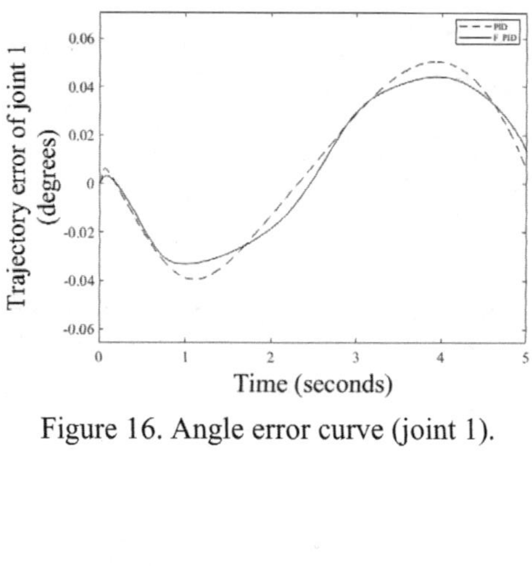

Figure 16. Angle error curve (joint 1).

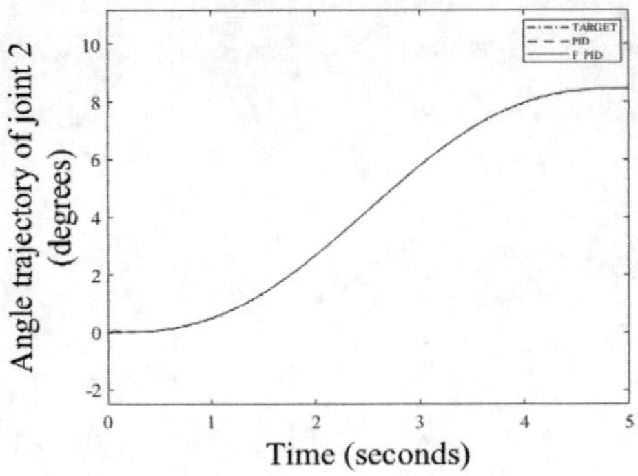

Figure 17. Angle curve (joint 2).

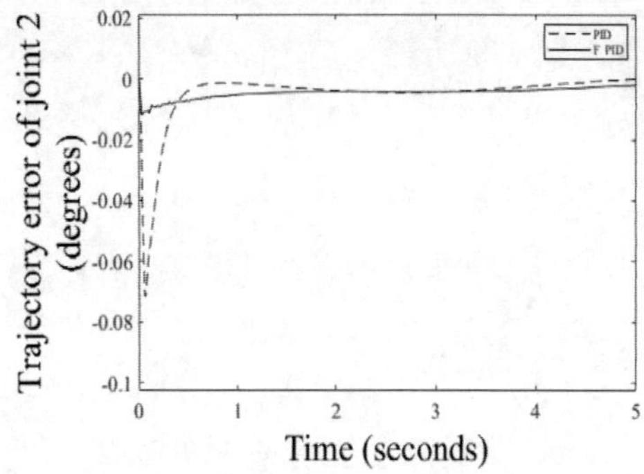

Figure 18. Angle error curve (joint 2).

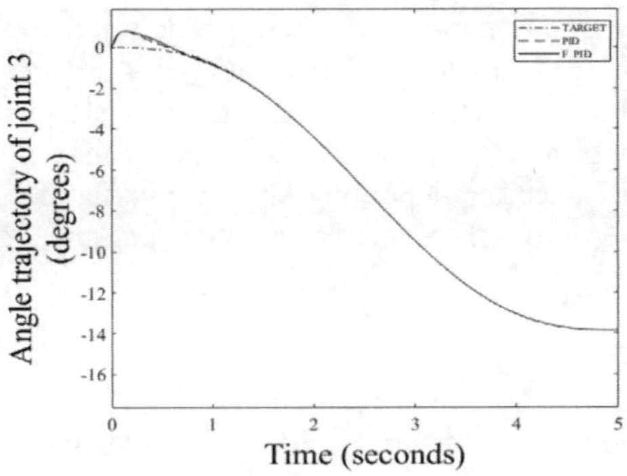

Figure 19. Angle curve (joint 3).

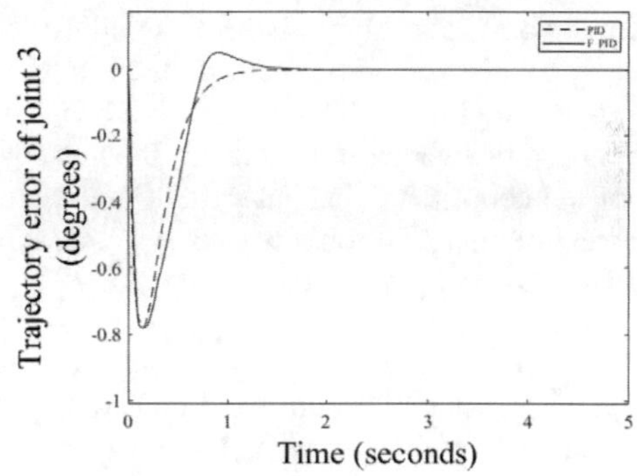

Figure 20. Angle error curve (joint 3).

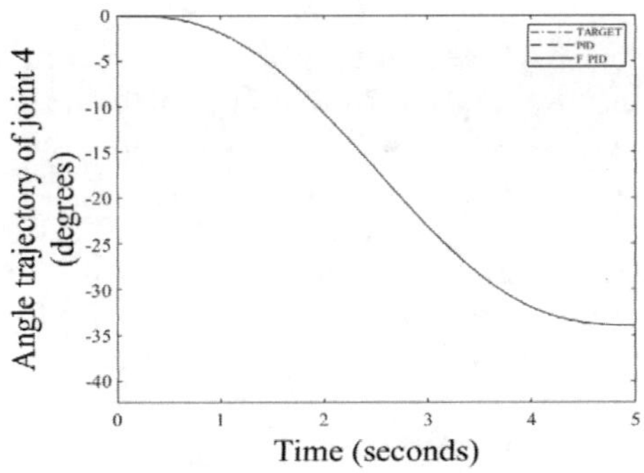

Figure 21. Angle curve (joint 4).

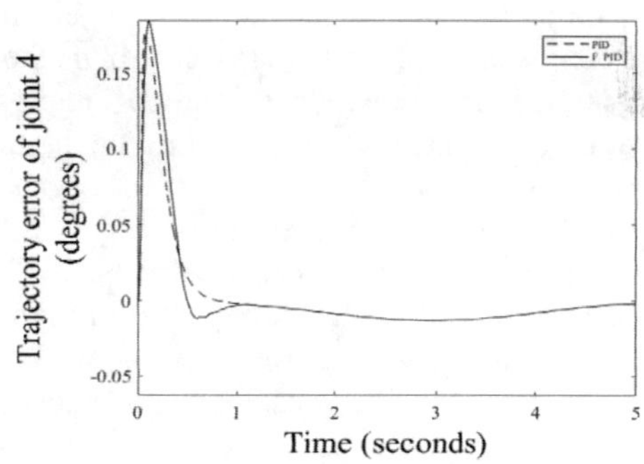

Figure 22. Angle error curve (joint 4).

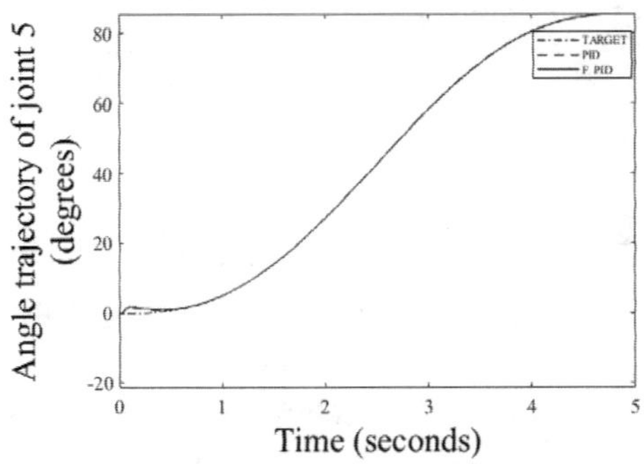

Figure 23. Angle curve (joint 5).

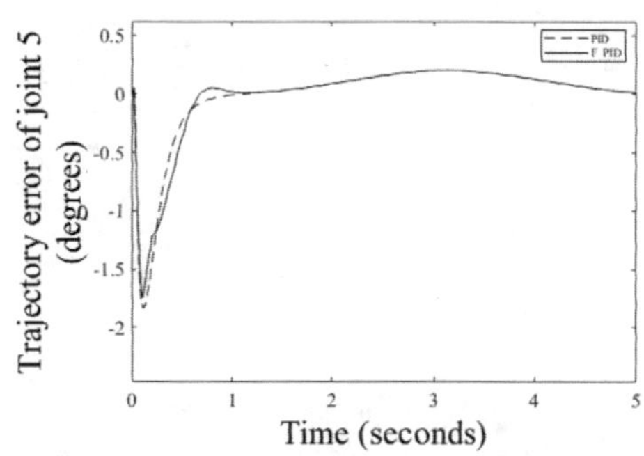

Figure 24. Angle error curve (joint 5).

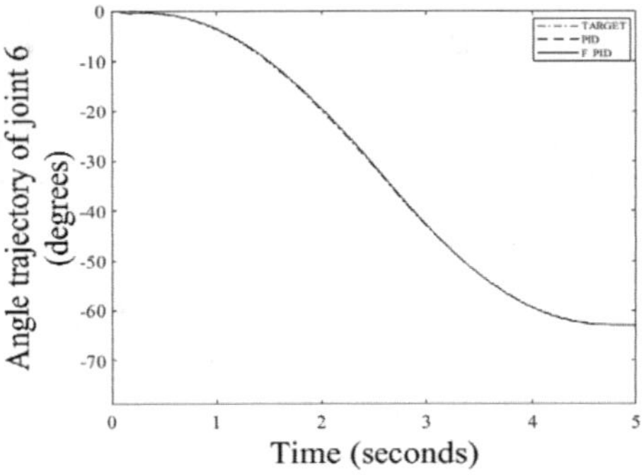

Figure 25. Angle curve (joint 6).

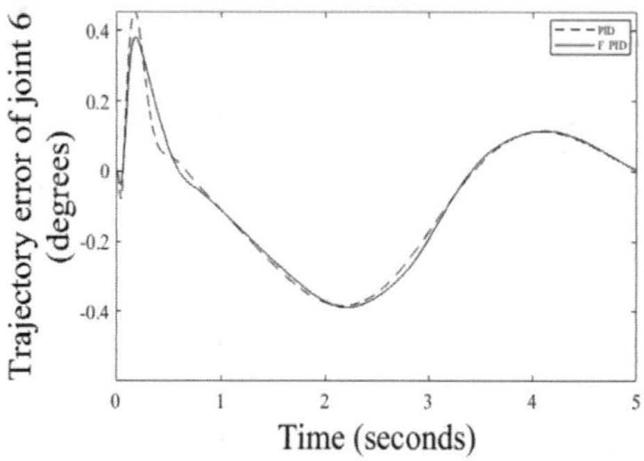

Figure 26. Angle error curve (joint 6).

Simulation results show that for the adjusted experimental parameters, the maximum error of the trajectory of the two control methods is controlled within ±1° and the stability error is controlled within ±0.1°. Compared with PID control, F_PID control has better effect. Although the effect on each joint is different, on the whole, it has the advantages of reducing the mutation peak and reducing the stability time of the controller. This has certain reference significance for improving the anti-interference and robustness of industrial robot control systems, and further shows the effectiveness of the designed fuzzy adaptive PID control method.

5.2 Comparative experiment of cooperative control modes of dual robots

The fuzzy adaptive PID control and PID control are applied to the cooperation of dual industrial robots. Comparative experiments were carried out on two identical industrial robots in master-slave and distributed modes.

The master-slave industrial robot adopts common PID control. To better track the trajectory of the master robot, the slave robot adopts fuzzy adaptive PID control with better control effect. In order to form a comparative effect, the same control method is adopted for the distributed mode, one robot adopts PID control, and the other industrial robot adopts fuzzy adaptive PID control. As shown in figure 27-32, simulation experiment results show the angular trajectory errors of six joints of two industrial robots in two modes.

From the figures, when the control method is that one industrial robot is under common PID control and one industrial robot is under fuzzy adaptive PID control, in terms of control performance, distributed control is better than master-slave control. The overall error range is reduced from [-2° 0.25°] to [-0.25° 0.1°], which can reduce the error of the cooperative trajectory of the two industrial robots and improve the synchronization of the cooperative trajectory.

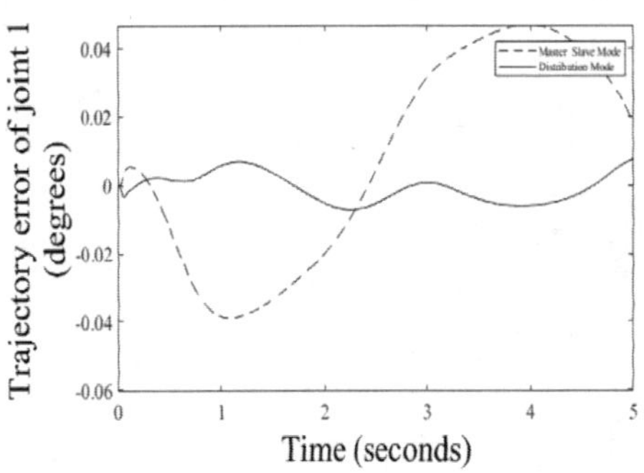

Figure 27. Angle error curve (joint 1).

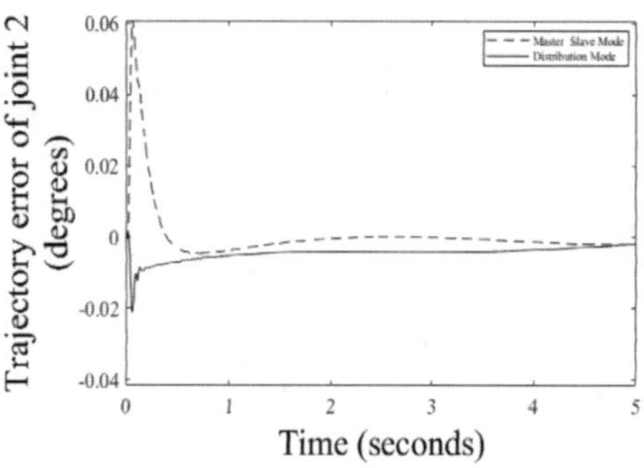

Figure 28. Angle error curve (joint 2).

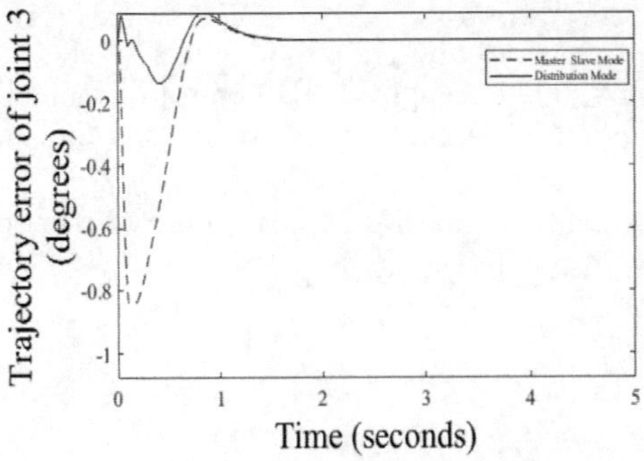

Figure 29. Angle error curve (joint 3).

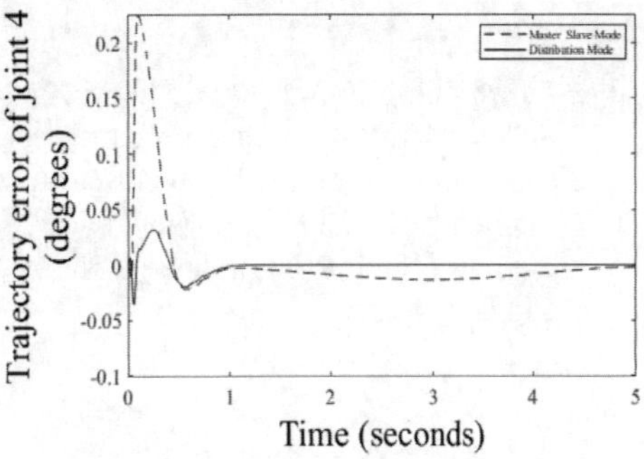

Figure 30. Angle error curve (joint 4).

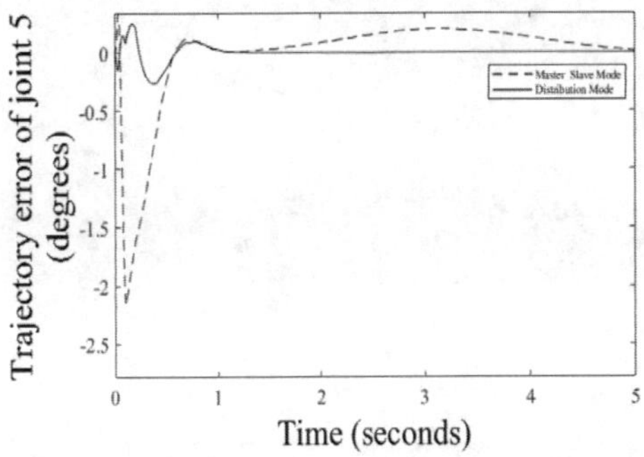

Figure 31. Angle error curve (joint 5).

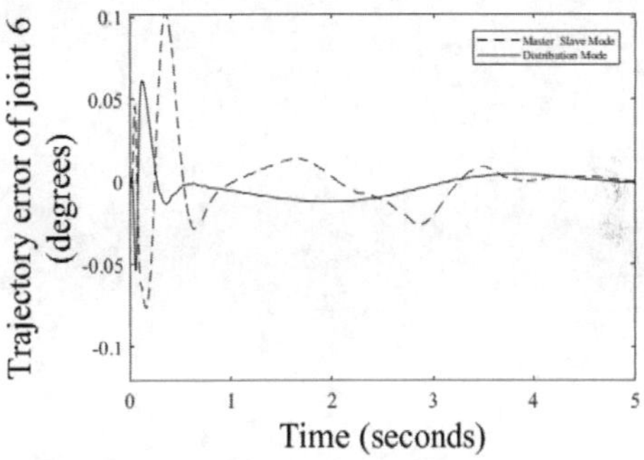

Figure 32. Angle error curve (joint 6).

6. Conclusion

In this paper, according to the actual performance requirements of industrial robots, a fuzzy adaptive PID control method is proposed. The fuzzy control is added to the conventional PID control, and the PID parameters are dynamically adjusted online to reduce the sudden peak value and reduce the stabilization time of the controller.

Compared with the master-slave control mode, the fuzzy adaptive PID control in the distributed control mode has better performance in obtaining the parameter index. It can reduce the trajectory error of two industrial robots' cooperation and improve the synchronization accuracy of the cooperation trajectory. This has a certain reference significance for the multi-industrial robot cooperation system to improve the trajectory synchronization, anti-interference and robustness.

References

[1] Mansor N N, Jamaluddin M H, Shukor A Z and Lok C C. 2018. A study of accuracy and time delay for bilateral master-slave industrial robotic arm manipulator system. In: Malaysian Technical Universities Conference on Engineering and Technology. Kuala Lumpur. v 150, p1015.

[2] Azar A T, Serrano F E, Hameed I A, Kamal N A and Vaidyanathan S. 2019. Robust H-infinity decentralized control for industrial cooperative robots. In: 5th International Conference on Advanced Intelligent Systems and Informatics. Cairo. v 1058, p 254-265.

[3] He Y H, Wu M and Liu S. 2022. A distributed optimal control framework for multi-robot cooperative manipulation in dynamic environments. Journal of Intelligent and Robotic Systems: Theory and Applications. v 105, n 1, p 1-18.

[4] Ouyang F, Zhang T and Chen Y. 2016. Genetic algorithm-optimized pd damping control of dual-robot coordinated transportation. Journal of Shanghai Jiaotong University. v 50, n 6, p 879-883.

[5] Wang J Y, Zhu Y G, Qi R L, Zheng X G and Li W. 2020. Adaptive PID control of multi-DOF industrial robot based on neural network. Journal of Ambient Intelligence and Humanized Computing. v 11, n 12, p 6249-6260.

[6] Tao Y, Zheng J Q, Lin Y C, Wang T M, Xiong H G, He G T and Xu D. 2015. Fuzzy PID control method of deburring industrial robots. Journal of Intelligent and Fuzzy Systems. v 29, n 6, p 2447-2455.

Path Planning of Manipulator Based on Improved RRT* Algorithm

Mengmeng Ye[1], Xiang Dong[1*], Jianjun Zhao[1], Zexia Huang[1]

[1]School of Electrical Engineering and Automation, Anhui University, Hefei, 230000, Anhui Province, China

*xdong@ahu.edu.cn.

Abstract. The traditional RRT* algorithm lacks directionality in sampling, and often generates a large number of invalid nodes, resulting in low utilization of later nodes. A large number of calculations between nodes cause the time spent by the manipulator to plan a path greatly increased and the path is not smooth. For the shortcomings of the above RRT* algorithm, an improved RRT* algorithm is proposed in this paper. First, a random point is generated with variable probability bias. Second, the sampling method is improved and an adaptive step size strategy is introduced. Finally, for the initial path, a path point connection screening strategy is introduced, and the screened path points are optimized again using a cubic b-spline curve. The experimental results show that the improved RRT* algorithm can quickly find a collision-free path, the path is more optimized, and the planning time is greatly reduced.

1. Introduction

Path planning technology was proposed in the 1970s and is still one of the important research directions in the field of robotics. Path planning means that given the starting point and the target point, the robot can avoid obstacles and plan a collision-free path by itself. Commonly used path planning algorithms include artificial potential field method, ant colony algorithm, and A* algorithm [1-3]. The working space of the manipulator is a three-dimensional space, when the above algorithm is applied to the three-dimensional space, the computational complexity is high and time-consuming, and it needs to occupy large memory space. Besides, the artificial potential field method easily falls into a local minimum while finding a path. The RRT algorithm is a discrete sampling path planning algorithm, which continuously extends the tree shape from the starting point and finally reaches the target point. The RRT algorithm not only finds a path fast, but also has probabilistic completeness, and is easy to combine with other algorithms to solve path planning in complex environments [4]. The disadvantage is that the path planned by the algorithm is not smooth enough, usually far from the optimal path, and the calculation of the later nodes is large.

Among the many derived RRT algorithms, the RRT* algorithm uses an asymptotically optimal way to find a path and is widely used in the obstacle avoidance process of multi-degree-of-freedom manipulators [5]. When the number of iterations is large enough, the length of the path planned by the RRT* algorithm gradually shortens, but it takes a long time to plan the path. Chen L[6] proposed a novel approach to RRT* in collaboration with a double-tree structure to separate the extension and optimization procedure. F. Islam [7] proposed RRT*-Smart, they used path optimization and intelligent sampling to improve the efficiency of RRT*-Smart.

To solve the problem of low sampling efficiency and long planning time of the RRT* algorithm, this paper uses a variable probability method to speed up the expansion of the random tree to the target point during sampling and introduces adaptive step size. Besides, controlling the sampling range to avoid the generation of a large number of invalid nodes, improves the efficiency of path planning. In addition, the initial path is optimized again, which reduces the length of the path and increases its smoothness.

2. RRT* algorithm

The RRT* algorithm performs two layers of optimization based on the RRT algorithm. The first layer of optimization is reselecting the parent node. The specific process is shown in Figure 1.

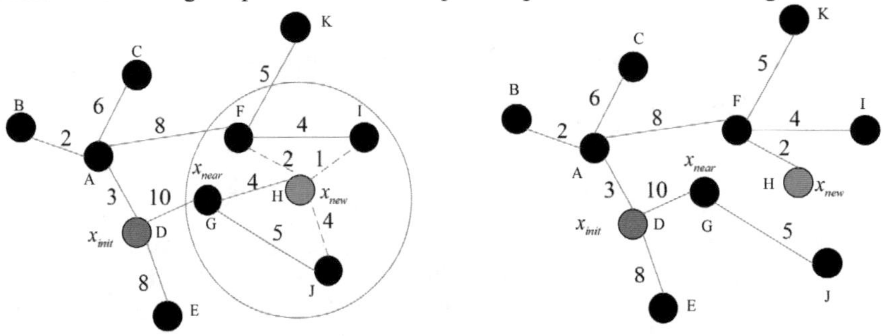

Figure 1. Reselect parent node.

Taking the node x_{new} as the center and the given radius to get the circular sampling area. The candidate nodes of the x_{new} are G, J, F, and I. The initial path planned by the RRT* algorithm is D-G-H, and the path cost is 14. Connect the candidate node to the x_{new}, the paths are D-A-F-H, D-A-F-I-H, D-G-J-H, and the path costs are 13, 16, and 19. The cost of the path from D-A-F-H is less than the initial path, so the parent node of the x_{new} is replaced by G with F.

The second layer optimization is random tree rewiring. The specific process is shown in Figure 2.

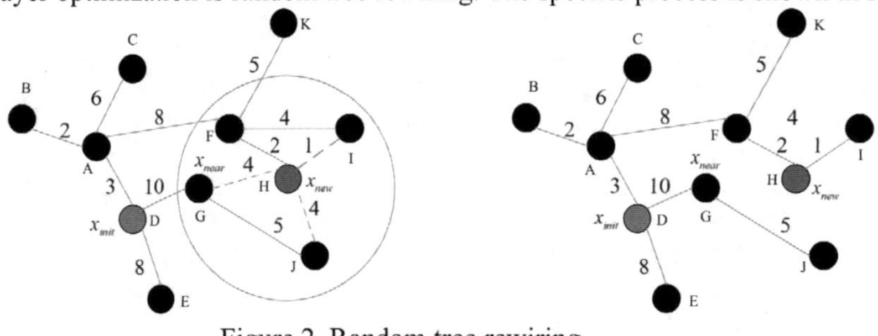

Figure 2. Random tree rewiring.

After re-selecting the parent node, the planned path cost is greatly reduced. The random tree rerouting process is to replace the parent node of the candidate neighbor point with the x_{new}. If the path cost after the replacement is less than the path cost before the change, the parent node of the neighbor point is replaced with the x_{new}. As shown in the figure, G, J, and I are the neighbors of the x_{new}. Before changes, the parent node of G is D, and the path cost is 10. The parent node of J is G, and the path cost is 15. The parent node of I is F and the path cost is 15. After replacing its parent node with x_{new}, the paths become D-A-F-H-G, D-A-F-H-J, D-A-F-H-I, and the path costs become 17, 17, and 14. Therefore, the rewiring path is D-A-F-H-I. The flow of the RRT* algorithm is as follows:

1) Randomly generate a sampling point in the workspace of the manipulator and denote it as x_{rand}.

2) Traverse the entire tree T, find the closest node to x_{rand} and denote it as x_{near}.

3) Generate a new node x_{new} with a fixed step size along the line connecting x_{near} and x_{rand}.

4) Connect x_{new} and x_{near} to perform collision detection. If there is no collision, add node x_{new} to tree, if there is a collision, return to step 1.

5) Reselect the parent node for x_{new}.

6) Prune the tree T.

7) Determine whether the Euclidean distance between node x_{new} and target point x_{goal} is less than a given threshold. If it is less than the given threshold, the two nodes are connected, and the random tree expansion ends. Otherwise, go back to step 1.

The algorithm chart is shown as in Figure3.

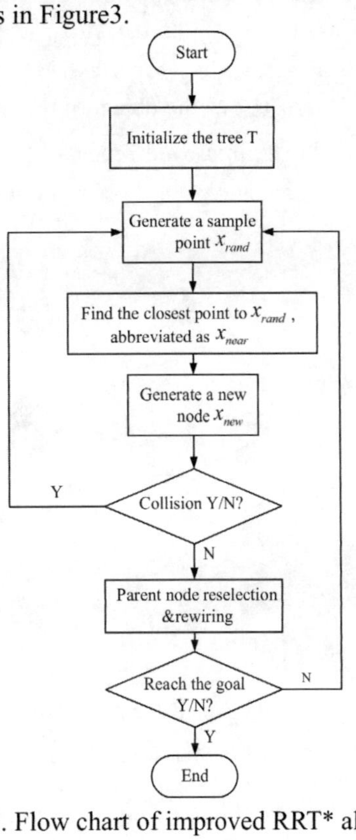

Figure 3. Flow chart of improved RRT* algorithm.

3. Improved RRT* algorithm

Given of the blindness of random point generation by the RRT* algorithm, random points are generated by variable probability. First, the random point is biased to the target point with probability p, and the probability q is uniformly sampled in the working space of the manipulator. When the line between the generated random points x_{rand} and x_{near} collides with an obstacle, the value of q is increased and the value of p is decreased in the next sampling, and the relationship between p and q always satisfies p +q=1.

The RRT* algorithm uses a fixed step size λ_{init} to expand. During the expansion process, if the step size λ_{init} is set too large, it is easy to collide with obstacles in places with dense obstacles. If the step size λ_{init} is set small, the number of random tree expansions will increase, the search efficiency will be low, and it will take a long time to find the path. To solve this problem, this paper proposes an adaptive step size adjustment strategy. The specific scheme is to give an initial step size λ_{init} and a minimum step size $\lambda_{\min}$, and the initial step size is n times the minimum step size. First, a new node x_{new} is obtained by expansion with a fixed step size, and it is judged whether the connection line between x_{near} and x_{new} collides with the obstacle. If no collision occurs, add x_{new} to the random tree. If there is a collision, divide the connection line into n equal parts with $\lambda_{\min}$ as the smallest unit. Connect with x_{near} in order from the rightmost sub-node. If a collision occurs, the next node is checked for collision, if there is no collision, the current node will be added to the random tree as the x_{new}. When all nodes do not meet the collision detection, then resample in the manipulator workspace.

The RRT* algorithm takes x_{new} as the center of the circle and reselects the parent node in the circular area. To further reduce the path cost, this paper expands the scope of re-selection of parent nodes. Take the ancestor node (x_1 , x_2) of x_{new} as the object of the re-selection of parent node and adds it to the x_{near} set, as shown in Figure 4. The improved process of the re-selection of parent nodes can optimize the random tree to a greater extent and get a shorter initial path length.

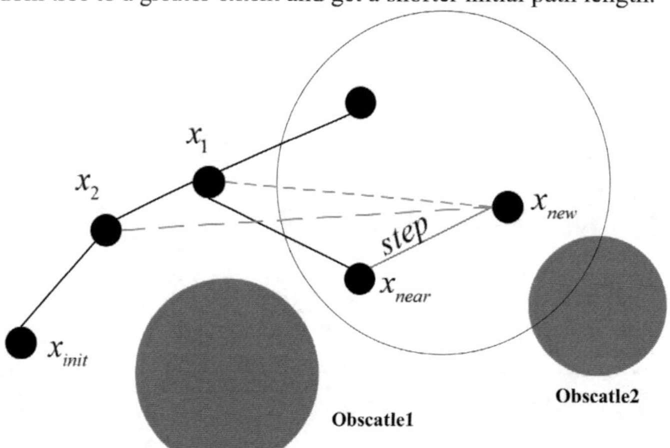

Figure 4. The ancestor node of x_{new} .

After the initial path is obtained through the above scheme, there are still some redundant nodes in the initial path. This paper adopts the path point connection screening strategy. Denote the initial path point as $[R_1, R_2, ..., R_n]$, take R_n as the base point, and trace upwards in turn. If the line connecting $[R_i, R_m]$ does not collide, all path points between the two points are eliminated. If a collision occurs, the collision path point is used as a new starting point, and the above steps are repeated until the entire path point sequence is traversed. The optimized path still has a turning point, when the manipulator passes through the turning point, it will cause wear to the manipulator and reduce its service life of the manipulator. To obtain a smooth path, this paper uses a cubic b-spline curve to smooth the path. The mathematical description of the B-spline curve is:

$$M(u) = \sum_{j=0}^{n} M_j N_{j,k(u)} \tag{1}$$

The base functions of cubic b-spline:

$$\begin{cases} N_{0,3}(u)=\frac{1}{6}\left(-u^3+3u^2-3u+1\right) \\ N_{1,3}(u)=\frac{1}{6}\left(3u^3-6u^2+4\right) \\ N_{2,3}(u)=\frac{1}{6}\left(-3u^3+3u^2+3u+1\right) \\ N_{3,3}(u)=\frac{1}{6}u^3 \end{cases} \quad u\in[0,1] \qquad (2)$$

Further mathematical description of the cubic b-spline curve can be obtained:

$$M_{0,3}(u)=M_0\times N_{0,3}(u)+M_1\times N_{1,3}(u)+M_2\times N_{2,3}(u)+M_3\times N_{3,3}(u) \quad u\in[0,1] \quad (3)$$

Given the control point M_j, the curve point that meets the requirements can be obtained by using the above formula.

4. Experiments and results

To verify the performance of the improved RRT* algorithm, conducting a simulation experiment to verify the improved RRT* algorithm on the MATLAB R2018a. The experimental hardware configuration is Lenovo Xiao Xin Pro14, the system is Win11.

Two-dimensional simulation environment, the starting point is (30, 30) and the target point is (400, 450). The initial step size is 40, the minimum step size is 2, the initial bias is 0.3, the threshold is 30, and the maximum number of iterations is 2000. What's more, reselect the parent node with a radius of 60, and the value of the ancestor node is set to 2.

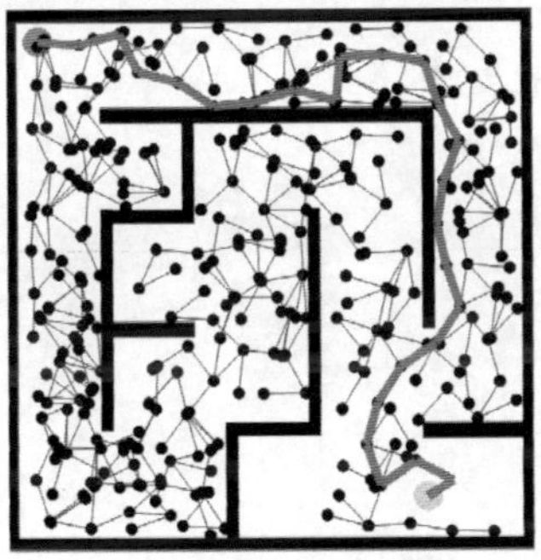

Figure 5. RRT-2D.

Figure 6. RRT*-2D.

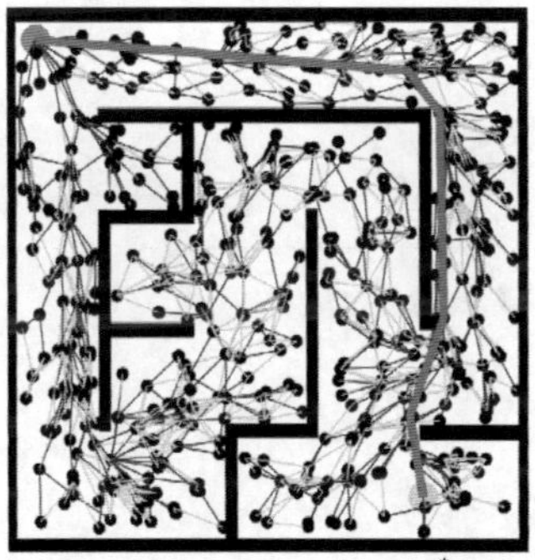

Figure 7. Improved RRT*-2D.

We experiment with the three algorithms 30 times, and the average performance parameters of the algorithm are shown in Table 1. The cost of the RRT* algorithm is significantly decreased compared with the RRT algorithm, but the time-consuming is also significantly increased. The time consumption of the improved RRT* algorithm is not much different from that of the RRT algorithm, and the path cost is also the smallest among the three algorithms.

Table 1. 2D environment simulation data.

Algorithm	RRT	RRT*	Improved RRT*
Time(s)	5.87	25.19	6.51
Path length(m)	975.66	801.29	768.38

Three-dimensional simulation environment, the starting point is (0, 0, 0), the target point is (100, 100, 1000), the initial step size is 3, the minimum step size is 0.3, the target point threshold is 2, and the maximum number of iterations is 2000. In addition, reselect the parent node with a radius of 20, and the value of the ancestor node is set to 2.

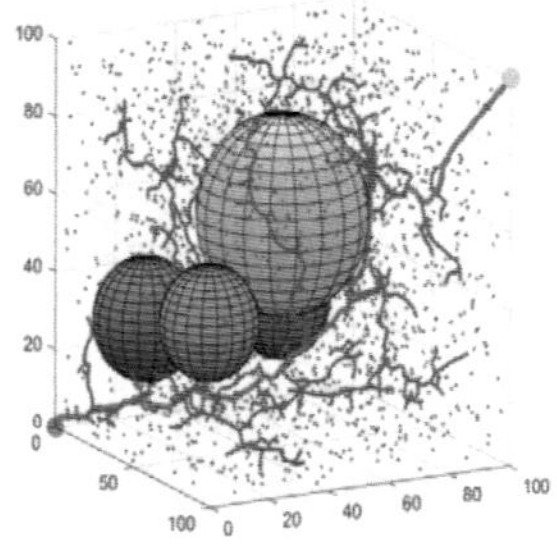

Figure 8. RRT-3D.

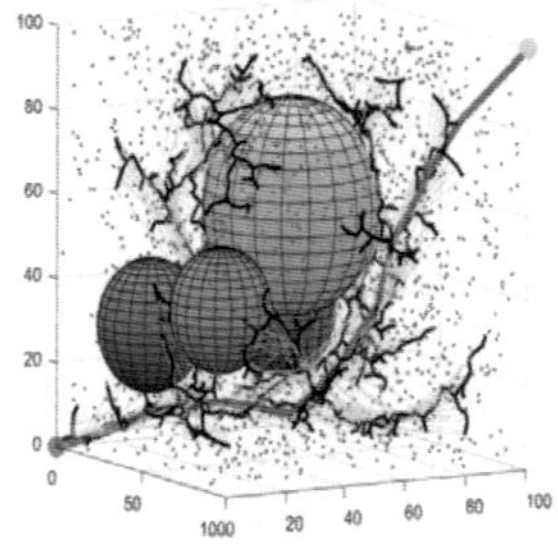

Figure 9. RRT*-3D.

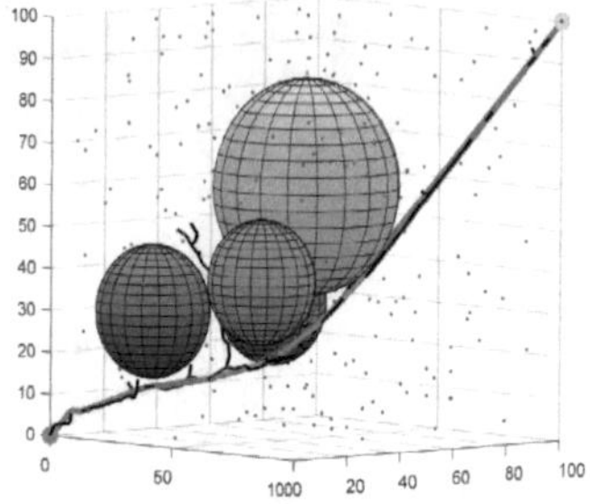

Figure 10. Improved RRT*-3D.

The same as the two-dimensional environment, 30 simulation experiments were performed, and the data obtained are shown in Table 2, the improved RRT* algorithm performs better than the other two algorithms in time and path cost.

Table 2. 3D environment simulation data.

Algorithm	RRT	RRT*	Improved RRT*
Time(s)	31.76	54.26	20.67
Path length(m)	252.01	207.67	189.71

The gluon manipulator model is established in this paper, and the improved RRT* algorithm is used for obstacle avoidance of the gluon manipulator. As shown in Figure 11. It can be seen from the motion process of the manipulator that the path planned by the improved RRT* algorithm is smooth and the path cost is small. The manipulator can stably complete the entire obstacle avoidance process.

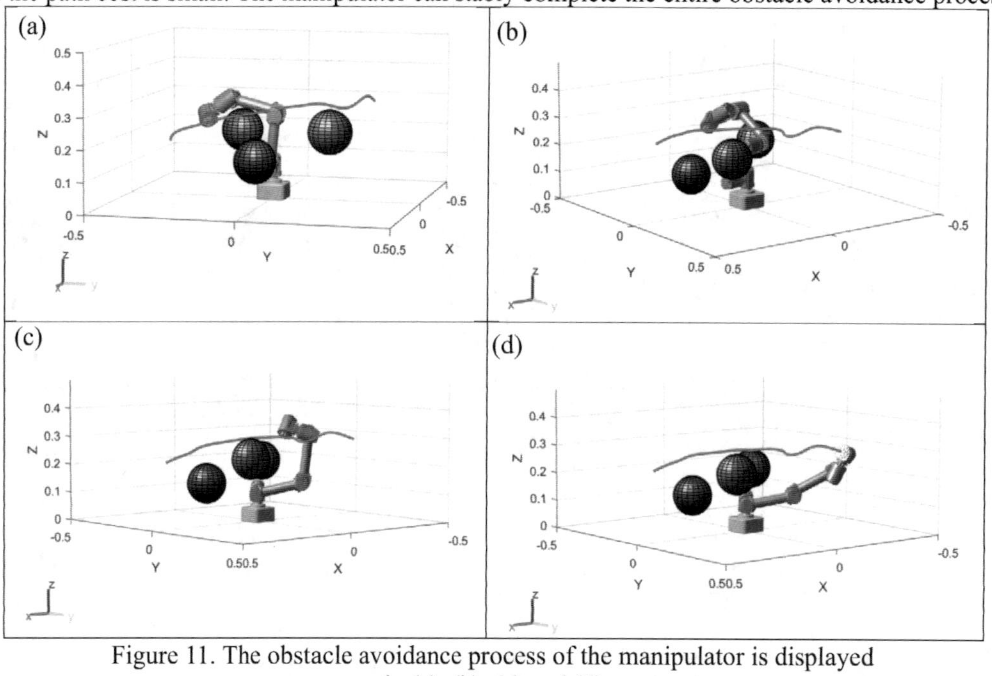

Figure 11. The obstacle avoidance process of the manipulator is displayed in (a), (b), (c) and (d).

5. Conclusion

Aiming at the problem of manipulator path planning, this paper proposes an improved RRT* algorithm. First, random points are generated by the variable probability target bias method, which reduces the amount of computation between nodes in the later stage. Second, the sampling range is increased and the path cost of the initial path is reduced. Finally, the node screening strategy and cubic b-spline curve are used to further reduce the path cost and smooth the turning points of the path. The simulation results show that the improved RRT* algorithm has certain advantages over the RRT algorithm and the RRT* algorithm.

Acknowledgment
This work was partially supported by the National Key Research and Development Project Monitoring and Prevention of Major Natural Disasters Special Program under Grant No. 2020YFC1512202; partly supported by the Anhui University Cooperative Innovation Project under Grant GXXT-2019-003.

References

[1] Hwang J , Lee J , Park C . Collision avoidance control for formation flying of multiple spacecraft using artificial potential field[J]. Advances in Space Research, 2022, 69(5):2197-2209.
[2] Miao C , Chen G , Yan C , et al. Path planning optimization of indoor mobile robot based on adaptive ant colony algorithm[J]. Computers & Industrial Engineering, 2021:107230.
[3] Hein B , H Wörn. Fast Hierarchical A* Path Planning for Industrial Robots Based on Efficient Use of Distance Computations[J]. 2006.
[4] Wei K , Ren B . A Method on Dynamic Path Planning for Robotic Manipulator Autonomous Obstacle Avoidance Based on an Improved RRT Algorithm[J]. Sensors, 2018, 18(2):571.
[5] Li Z , Gao Q , Song Y , et al. Motion Planning of Redundant Dual-arm Based on Improved RRTstar[C]// 2020 Chinese Automation Congress (CAC). 2020.
[6] Chen L , Shan Y , Tian W , et al. a fast and efficient double-tree rrt * -like sampling-based planner applying on mobile robotic systems.
[7] Islam F , Nasir J , Malik U , et al. RRT-Smart: Rapid convergence implementation of RRT towards optimal solution[C]// Mechatronics and Automation (ICMA), 2012 International Conference on. IEEE, 2012.

AHSM-Net: Unsupervised Stereo Matching Algorithm Based on Attention Mechanism and Hybrid Dilated Convolution

Xiaoge Li[1], Xingfang Zhao[2] and Long Yan[3,4]

[1] Department of Information and Electronic Engineering, Shandong Technology and Business University, Binhai Street, Yantai, 264005, Shandong, China.

[2] Testing and Technology Center for Industrial Products of Shenzhen Customs District, No.149, 7th Industry Road, District Nanshan, 518067, Shenzhen, China.

[3] Department of Management Science and Engineering, Shandong Technology and Business University, Binhai Street, Yantai, 264005, Shandong, China.

[4] Author to whom any correspondence should be addressed

yanlong@sdtbu.edu.cn

Abstract. Too many parameters often accompany the stereo matching algorithm of the convolutional neural network. The dense matching makes the constructed cost volume increase with the resolution, resulting in high occupied memory and processing time. To perform depth estimation more efficiently, this paper proposes a stereo matching algorithm AHSM-Net based on attention mechanism and hybrid dilated convolution (HDC), which uses an unsupervised method to train the stereo matching network end-to-end. This method obviously enhances the consistency of objects. AHSM-Net is trained on the SceneFlow dataset, and a large number of experiments and analyses are carried out on the KITTI2012 and KITTI2015 datasets, which verifies that the method proposed in this paper has great matching accuracy and speed under the condition of unsupervised.

1. Introduction

Stereo matching is an important task in three-dimensional reconstruction, and it is a very critical technology in the field of robotics. It has been widely used in unmanned aerial vehicles and autonomous driving. Binocular vision is an important dependency of stereo matching. It can be described as a binocular camera that simulates human left and right eyes to capture images so that objects are imaged from two viewpoint directions respectively. The pixel points of the images are matched, and the corresponding disparity is calculated to estimate the depth through the principle of triangulation.

For stereo matching, deep learning has been able to successfully learn features from original data. Zbontar et al. [1] and Kendall et al. [2] generated four-dimensional cost volumes by connecting left and right images to generate unary features. The cost of this strategy is higher memory requirements and runtime. Followed by [3], using one-dimensional association to reduce the feature dimension to generate a three-dimensional cost volume improves efficiency. However, when the disparity range is large, the calculation time of the matching cost is also greatly increased.

In this paper, we propose an unsupervised disparity regression network to learn pixel-to-pixel correspondence, adding a parallax-attention mechanism module to enable the network to learn a wide range of important association features at low resolution. In order to make the matching results more reliable, this paper is different from the previous work in calculating the feature difference, using a combination of the Manhattan distance metric and Euclidean distance to calculate feature differences and similarities. To perform fast stereo matching, in the subsequent steps, we only match candidate pixels with a certain disparity range, estimate residual disparity, continuously correct the disparity map estimated in the previous step, and use the refinement network to refine the initial disparity.

2. Related Work

The depth estimation of stereo images has been greatly developed, and the development upsurge has gradually transitioned from traditional stereo matching algorithms to learning-based algorithms, which benefit from the emergence of many standard datasets.

2.1. Attention Mechanism

To make up for the limitations of convolution operations and enable the network to learn the correlation between different features, the attention mechanism can well reflect the contribution between features. Zhang et al. [4] used the self-attention mechanism to enable the network-generated pictures to learn the regions that should be paid attention to. Wang et al. [5] captured the similarity of stereo figures by building a parallel-attention mechanism module.

2.2. Hybrid Dilated Convolution

Dilated convolution is more popular in the field of semantic segmentation, and it is used to expand the receptive field under the characteristics of avoiding pooling operation. Chen et al. [6] proposed to mix dilated convolution and dense upsampling convolution to effectively improve the accuracy of semantic segmentation.

2.3. Unsupervised Learning

Despite the successful results of supervised deep networks, such methods rely heavily on the richness and diversity of precisely annotated data, and unsupervised learning methods [7][8] simplify the process.

3. Network Architecture

The first part extracts features through an hourglass network and builds a parallax-attention module to regress the initial disparity map at low resolution. At high resolution, the disparity range is specified for the low-level feature maps corresponding to the left and right, and the Manhattan distance metric method is used to construct the cost volume regression to obtain the residual disparity, which is used to iteratively correct the disparity of the previous stage. In the second part, the disparity map is refined using HDC. We present the network in Figure 1.

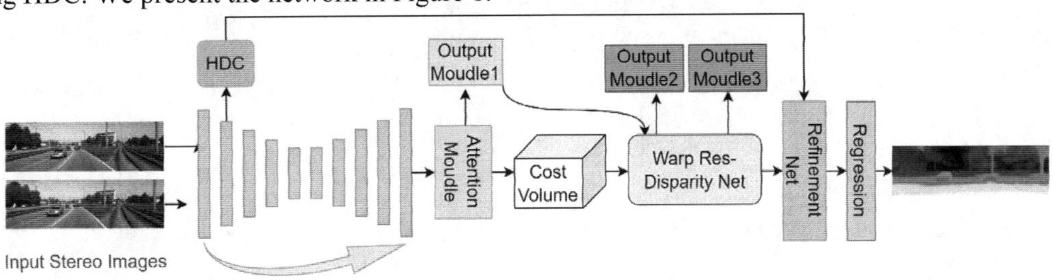

Figure 1. Architecture overview of proposed AHSM-Net.

3.1. Feature Extraction

In the encoder part, we use small convolutions for each layer to reduce the input size, and follow batch normalization and activation function operations to obtain a deeper feature map of size 1/32 of the input image and the number of channels of the feature map is 60 times that of the input image. In particular, we build a branch to apply a HDC to the feature map at 1/4 resolution, which is composed of three dilated convolutions with dilation rates [1, 2, 7] respectively. Without the use of pooling operations, keeping the internal data structure and increasing the size of the receptive field usually leads to higher accuracy, which we use to fuse context information and subsequently use it for the disparity refinement sub-network. In the decoder part, convolution is used to gradually restore the resolution, and the feature map in the encoding process is skipped to the corresponding part. Finally, the size of the left and right feature maps output by the network is 1/4 of the input, and the features are compressed to fewer channels.

3.2. Parallax-Attention Mechanism

Stereo matching aims to find the correspondence between the left and right images. Different regions in the image contribute greatly to the corresponding points to be matched. To make it easier for our network to learn important associated features for all regions, including edge regions, we add a parallax-attention mechanism.

The specific description is as follows: input a pair of left and right images, the feature map after the feature extraction of the hourglass network is $F \in R^{H\times W\times C}$, and its resolution is 1/16 of the input, and then sent to the new convolution layer, using 3×3 convolution to enhance features, followed by batch normalization and *Relu*, and then sent to 1×1 convolution to obtain new left and right feature maps, denoted as $K, P \in R^{H\times W\times C}$. Then, transpose P to get $N \in R^{H\times C\times W}$, perform matrix multiplication on K and N ,then perform softmax operation to get the attention map. The entire attention module can be described as:

$$A=\sigma\left(\left(f^{1\times 1}\left(f^{3\times 3}[F_L]\right)\right)\left(f^{1\times 1}\left(f^{3\times 3}[F_R]\right)\right)^T\right) \tag{1}$$

where σ represents the softmax operation, $f^{1\times 1}$ represents the convolution operation with a filter size of 1×1, and $f^{3\times 3}$ represents the convolution operation with a filter size of 3×3.

Among them, the element a_{ij} in the attention map represents the similarity between the points on the right image along the epipolar line and the left image (i, j), which can be described as:

$$a_{ij}=\frac{\exp\left(M_i\times N_j\right)}{\sum_{i=1}^{D}\exp\left(M_i\times N_j\right)} \tag{2}$$

At this point, each weight of the attention map is in the interval [0,1].

In addition, the dot product is used to calculate the similarity, and the matching cost , $Cost_{att} \in R^{H\times W\times W}$ is also included in the process of building the attention map.

3.3. Residual disparity

In the above steps, the initial disparity d_1 of our left image is obtained. In this part, we have to regress the residual disparity to correct the disparity map $\widehat{D}_1$. Here we refer to the idea of Pang et al. [9] and make improvements.

Use d_1 to warp the right image F_R to the left image F_L', resulting in a composite left image, described as follows:

$$F_L'(x,y)=F_R\left(x+d_1(x,y),y\right) \tag{3}$$

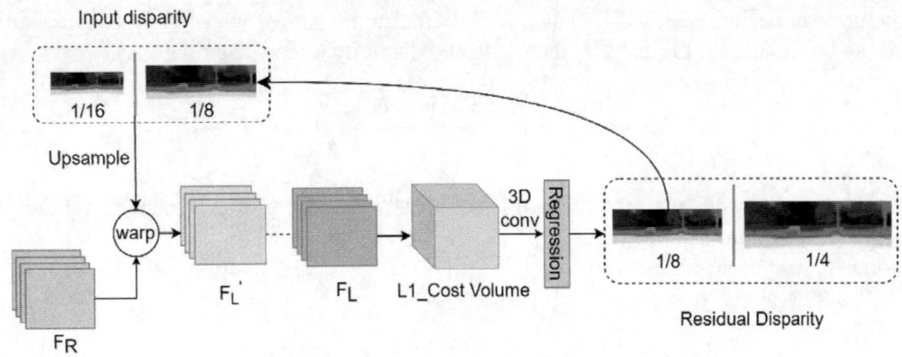

Figure 2. Residual disparity regression.

Then F_L, F_R, and $F_L^{'}$ are used to construct a cost volume using the distance metric, whose scale is $H \times W \times M$, where M is the maximum disparity, which is used to regress our residual disparity $\widehat{D}_{res}$,Then the new disparity map is described as $\widehat{D}_2(x,y) = \widehat{D}_{res1}(x,y) + \widehat{D}_1(x,y)$,and similarly $\widehat{D}_3(x,y) = \widehat{D}_{res2}(x,y) + \widehat{D}_2(x,y)$. Unlike [36], in the step of recovering the resolution as the decoder, we iterate this process twice, limiting the maximum disparity to specify $M = 12$ and $M = 3$, such as the disparity offset range of $M = 3$ as $(-1,0,1)$, we use the error $E = | F_L(x,y) - F_L^{'}(x,y) |$ for unsupervised end-to-end training. In addition, we feed the cost volume to the convolutional layers for refinement with 3×3 kernels, each layer is accompanied by batch normalization and *Relu*.

3.4. Disparity Refinement

In order not to lose the resolution and still expand the receptive field, we use a dilated convolution with a mixed dilation rate of [1, 2, 7] in the encoder stage to generate a feature map with a resolution of 1/4 of the original resolution, and then Fusion. This will not make the effective information in the receptive field too sparse, thus truncating the local connection between pixels is cut off.

Under the condition that the original parameters remain unchanged, the 3×3 convolution can obtain a larger receptive field, which is beneficial to improve the accuracy of larger objects.

After the HDC operation, the resulting sub-feature map has the same height and width as the input feature map. These output feature maps are connected with the input to obtain the feature map, which does not need to introduce additional parameters and can capture multi-scale context information, retains the internal data structure, which is beneficial to improve the accuracy of disparity estimation.

3.5. Disparity Regression

We get the initial disparity with a resolution of 1/16 of the original resolution and the residual disparity with a resolution of 1/8 and 1/4 of the original resolution, respectively. At low resolution or when a smaller disparity range is specified, the image noise may affect the cost we calculated and produces interference. We use the disparity calculation method proposed by Kendall et al. [2] to calculate the weighted average of the cost, which is described as follows:

$$\hat{D}_{ij} = \sum_{k=0}^{M} k \times \frac{\exp\left(-C_{ijk}\right)}{\sum_{k^{'}=0}^{M} \exp\left(-C_{ijk^{'}}\right)} \qquad (4)$$

4. Loss

Usually, the reconstructed image may have great distortion, and it is not enough to compare the reconstructed image with the original image. We use the combination of a mean absolute error (MAE) and a structural similarity index (SSIM) as our photometric image reconstruction loss, using the input image and the left image synthesized by the estimated disparity to construct the loss.

$$L_{ls} = \frac{1}{N}\sum_{i,j} \alpha \frac{1-\mathrm{SSIM}\left(I_{ij}^{L}, \tilde{I}_{ij}^{L}\right)}{2} + (1-\alpha)\left\| I_{ij}^{L} - \tilde{I}_{ij}^{L} \right\| \quad (5)$$

Among them, is the effective pixel, which is set to 0.85 according to experience, and the similarity error occupies a larger proportion.

Additionally, we use an edge-aware cue edge gradient for local smoothing due to the discontinuity of depth at the edge of the image.

$$L_{sm} = \frac{1}{N}\sum_{i,j} \left|\partial_x d_{ij}^{L}\right| e^{-\left\|\partial_x I_{ij}^{L}\right\|} + \left|\partial_y d_{ij}^{L}\right| e^{-\left\|\partial_y I_{ij}^{L}\right\|} \quad (6)$$

In summary, the total loss is defined as:

$$Loss = L_{ls} + \lambda L_{sm} \quad (7)$$

5. Experiments

5.1. Experimental datasets
We use SceneFlow dataset and KITTI dataset for training and testing. The SceneFlow dataset is a large synthetic dataset. KITTI is an improved scene traffic estimation dataset and benchmark. We first pre-train on SceneFlow dataset, finetune on KITTI dataset and get test results. We compare the results with previous work on SceneFlow and KITTI datasets. The sources of the comparison results in Section 5.3 are all from official data and references are cited.

5.2. Experimental details
Our network is implemented on PyTorch and all training and testing are performed on a GeForce RTX 3070. The network uses Adam (β1=0.9 , β2=0.999) when training with SceneFlow the initial learning rate is 0.001, the learning rate is updated every 5 epochs, and the batch size is set to 6 and the adjustment factor is 0.1. When fine tuning using the KITTI dataset, the initial learning rate is 0.0001, and the learning rate is updated every 6 epochs for 100 iterations.

5.3. Analysis of disparity results
In this paper, the results are evaluated on KITTI online platform, where ground truth and evaluation areas are all pixels. The percentage of background pixels, foreground pixels, and the constant value of stereoscopic disparity of all pixels in the first frame (D1) is used as the evaluation indicators, and the pixel value with an error of less than 3 pixels is considered to be correct.

The comparison results of the algorithm proposed in this paper and other methods on the KITTI2015 test are shown in Figure 3. It can be seen that the estimated disparity at the edges of objects (cars, billboards), shadow areas and image edges is more accurate and smoother; the disparity inside the object is relatively uniform and the outline is clear. In the occluded part of the image edge, the predicted disparity is more accurate and the transition is smooth.

As can be seen from the data in Table 1, the average outlier value of pixels in the foreground area of the method in this paper is the lowest relative to other methods. Compared with the OASM-Net method, the average outlier value in the foreground area is in the non-occlusion area region and all pixel regions have a 15% drop, respectively, and in all pixel regions, the combined background and foreground errors also have a good performance, with a drop of 2%. In addition, the single image inference speed of this method on KITTI2015 dataset is 0.39s, OASM-Net needs 0.73s. It can be seen that our inference speed is faster, and almost 2 times faster than OASM-Net, our network performs better in speed and accuracy on the whole.

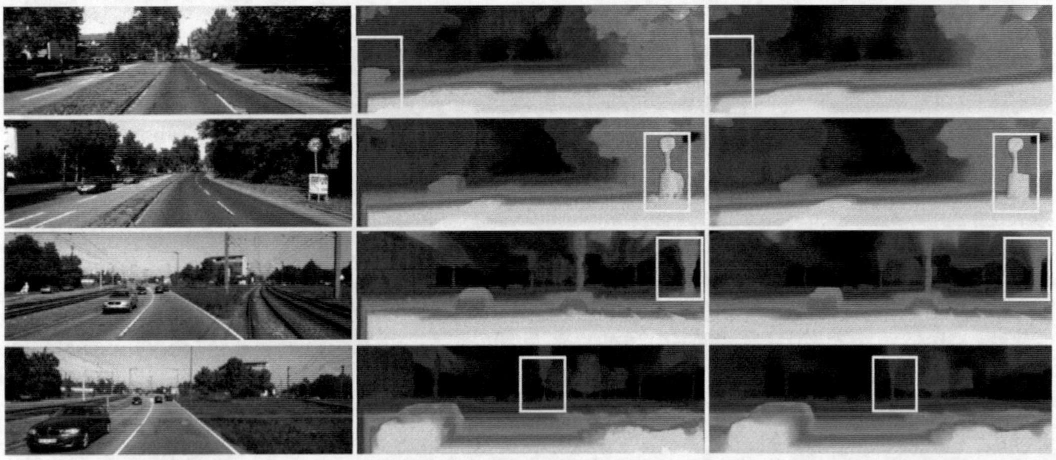

Figure 3. Comparison of results on the KITTI2015 dataset. The first column is the input image, the second column is the disparity map predicted by OASM-Net, the third column is the disparity map predicted in this paper.

Table 1. Performance comparison on KITTI2015.

Method	Noc			All		
	D1-bg	D1-fg	D1-all	D1-bg	D1-fg	D1-all
Zhou[7]	-	-	8.61	-	-	9.91
SegStereo[8]	-	-	7.70	-	-	8.79
REAF[10]	7.77	17.01	9.29	8.43	18.51	10.11
SGBM[11]	7.62	18.81	9.47	8.92	20.59	10.86
OASM-Net[12]	5.44	17.30	7.39	6.89	19.42	8.98
Ours	6.43	14.79	7.81	7.25	16.46	8.78

As can be seen from Table 2, the method in this paper has a lower error rate when the error is greater than 3, 4, and 5 pixels, respectively, and shows the best effect compared to OASM-Net and the traditional method SGBM. The method in this paper can better predict the internal disparity of the foreground and generate a more accurate disparity map.

Table 2. Performance comparison on KITTI2012.

Method	>3px		>4px		>5px	
	Noc	All	Noc	All	Noc	All
OASM-Net[12]	6.39	8.60	5.11	7.31	4.32	6.50
Ours	6.86	8.18	5.12	6.25	4.10	5.06

This paper uses part of the KITTI2015 data as a validation set to conduct ablation experiments to verify the effectiveness of HDC on accuracy. Fully dense extraction of information does not result in the best performance in disparity estimation. As can be seen from the Table 3, the use of mixed dilated convolution for multi-scale feature fusion makes the error smaller, and the use of mixed dilated convolution (dilation rate [1, 2, 7]) is better than that without dilation convolution. The endpoint error is reduced by 3%, the error rate is reduced by 2% for errors greater than 2 pixels, and the error rate is reduced by 3% for errors greater than 5 pixels.

Table 3. Comparison of validation results on some KITTI2015 datasets.

Method	EPE	>2 pixel	>5 pixel
No HDC	1.50	12.05	4.30
HDC-125	1.46	11.98	4.20
HDC-127	1.45	11.85	4.18

6. Conclusion

In this paper, we propose an unsupervised and efficient stereo matching network, which uses two distance metrics to calculate the similarity between pixels, and adds a parallax-attention mechanism. Since the process of obtaining disparity from stereo matching is a regression problem, so we use the predicted left image for unsupervised training. Our network achieves more reliable results in foreground regions, edges and contour regions, but needs to be improved in background regions. At the same time, due to the limitation of the disparity calculation range in the regression step, it also shows a very high rate in the prediction time of the disparity map. In stereo matching networks, the networks trained in an unsupervised approach are less than supervised ones, and our network shows more advanced performance in terms of speed and accuracy.

References

[1] Zbontar J and LeCun Y 2016 Stereo matching by training a convolutional neural network to compare image patches. *J. Mach. Learn. Res.* **17(1)** 2287-2318.

[2] Kendall A, Martirosyan H, Dasgupta S, Henry P, Kennedy R, Bachrach A and Bry A 2017 End-to-end learning of geometry and context for deep stereo regression. In: *IEEE international conference on computer vision* Venice, Italy pp. 66-75.

[3] Mayer N, Ilg E, Hausser P, Fischer P, Cremers D, Dosovitskiy A, and Brox T 2016 A large dataset to train convolutional networks for disparity, optical flow, and scene flow estimation. In: *IEEE conference on computer vision and pattern recognition* Las Vegas, NV, USA pp. 4040-48.

[4] Zhang H, Goodfellow I, Metaxas D, Odena A 2019 Self-attention generative adversarial networks. In: *International Conference on Machine Learning* Faridabad, India pp. 7354–73.

[5] Wang L, Guo Y, Wang Y, Liang Z, Lin Z, Yang J, An W 2020 Parallax attention for unsupervised stereo correspondence learning. *IEEE transactions on pattern analysis and machine intelligence* **44** pp. 2108 – 25

[6] Wang P, Chen P, Yuan Y, Liu D, Huang Z, Hou X, Cottrell G 2018 Understanding convolution for semantic segmentation. In: *IEEE Winter Conference on Applications of Computer Vision* Lake Tahoe, NV, USA pp. 1451–60

[7] Zhou C, Zhang H, Shen X, Jia J 2017 Unsupervised learning of stereo matching. In: *IEEE International Conference on Computer Vision* Venice, Italy pp. 1567–75

[8] Yang G, Zhao H, Shi J, Deng Z, Jia J 2018 Segstereo: Exploiting semantic information for disparity estimation. In: *European Conference on Computer Vision* Munich, Germany pp. 636–651

[9] Pang J, Sun W, Ren J S, Yang C Yan 2017 Cascade residual learning: A two-stage convolutional neural network for stereo matching. In: *IEEE International Conference on Computer Vision Workshops* Venice, Italy pp. 887–895

[10] Cigla C 2015 Recursive edge-aware filters for stereo matching. In: *IEEE Conference on Computer Vision and Pattern Recognition Workshops* Boston, MA, USA pp. 27–34

[11] Hirschmuller H 2008 Stereo processing by semiglobal matching and mutual information. *IEEE Transactions on pattern analysis and machine intelligence* **30(2)** 328–341
[12] Li A, Yuan Z 2018 Occlusion aware stereo matching via cooperative unsupervised learning. In: *Asian Conference on Computer Vision* Perth, *Australia* pp 197–213.

A Method of Connecting Traditional Energy-Consuming Equipment to the Internet of Things

Bingchen Li [1], Lei Xiong[2], Zhongwen Jin, Jie Wang[2], Haoyu Tian[1]

[1]Zhejiang Energy Digital Technology Co., Ltd., Hangzhou 311121, China

[2]Department of Electronics and Information Engineering, Hangzhou Dianzi University, Hangzhou 310018, China

Corresponding author: Bingchen Li (li_bc@foxmail.com).

Abstract. The concept of the Internet of things affects the way that people control electrical appliances. Moreover, due to the increasing importance of energy saving, the transformation of electrical power and energy system (EPES) is urgently needed. However, massive waste of energy caused by the inappropriate use of power-consuming equipment and most of the electrical devices controlled in traditional methods. This paper illustrates an Internet of Things (IoT) and edge computing (EC) technique to intellectualize standard energy-consuming equipment via LoRa wireless technology. The proposed system is low power and supports ad-hoc networks. Each node connects to the Internet via a LoRa based gateway. We mainly focus on controlling regular air conditioning and lighting system in office buildings via web pages, apps, etc. To simplify the process of installing the purposed node, several techniques were used to take over the control of traditional power-consuming equipment via user-side. The EC node can receive the short control command and operate itself according to the environment sensing results without control intelligence from cloud server. The experimental results indicate that our intellectualized network can work reliably and save energy-consuming equipment from wasting power.

1. INTRODUCTION

In recent years, the demand for power-consuming equipment has increased, and effective energy management is still a key issue. The massive waste of energy caused by the inappropriate use of power-consuming equipment, such as the incorrect configurations and unnecessary usage of the air conditioning system and lighting system. Most of the electrical devices controlled in traditional methods, such as infrared remoter or central control panels. It is difficult to achieve intelligent automatic central control for all the equipment remotely. Therefore, a simple and intelligent method to take over control of the high-energy-consuming equipment in buildings is urgently needed. To solve the problem, equipping the control system of power-consuming equipment with microcontrollers and the wireless sensor network (WSN) is a promising solution. Using WSN to control energy-consuming equipment is flexible in deployment and can save time for end-user of managing the equipment and cut energy costs [1]. However, modifying the interior of high-energy-consuming equipment will not only lead to high installation costs but also cause the overall system to become unstable. Unlike other solutions of intellectualizing traditional equipment by planting a control module into the machine, the designed WSN nodes control the equipment via refitting user-side control system. They do not change the structure of the equipment. Therefore, it is a cheap and friendly solution to control traditional energy-consuming equipment by simulating the old method of controlling. Furthermore, after the comparison of different wireless technologies available for IoT applicants, the designed nodes equipped with LoRa wireless technology that can provide the most comprehensive communication range and highest receiver sensitivity with relatively low power consumption. In addition, just connecting electrical devices to the cloud is not enough for a smart building. To acknowledge the real-time circumstances of every equipment, an impressive amount of data is unavoidable [2]. Considering the data transportation pressure and privacy protection, EC is a perfect solution for the purposed method. The algorithm deployed in the node can process simple control information sent from the cloud. EC not only reduces the pressure on the communication link but also increases the flexibility of control. Besides, with the EC algorithm and ability of environmental awareness, the nodes can evaluate the control effectiveness through the feedback of the perception environment, and finally achieve a more efficient configuration of electrical devices and a better user experience.

In this paper, we present a method that focuses on connecting existing energy-consuming equipment into the IoT at a low cost and realize intelligent control of traditional equipment via WSN, EC and the Internet. By installing the designed nodes, administrators and users of the building can control the traditional power-consuming equipment via web pages and app. Each node will be equipped with one microcontroller unit (MCU), one temperature sensor, several programmable input/output ports (I/O) and one LoRa Module are equipped on each node to realize intelligent control of traditional panels to save energy. The complete network structure is shown in Figure 1. The network provides an effective solution to the intellectualizing application. Furthermore, we will test the affection of the building infrastructures to the sensitivity of LoRa, the stability of LoRa communication, the performance of EC nodes in the actual use, including the low power consumption performance and the effectiveness[3] [4].

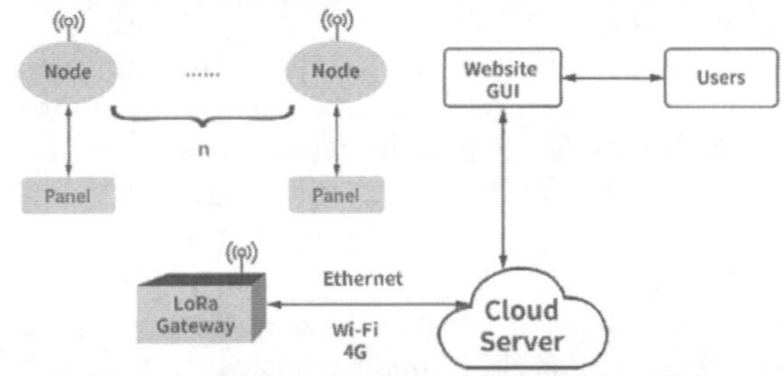

FIGURE 1. Network diagram of the proposed system

The remaining parts of this paper organized as follows. Section II outlines the system implementation and software algorithm of the nodes. In Section III, we discuss the performance of LoRa in office building and the effectiveness of EC nodes. Finally, this paper concludes in Section IV.

2. SYSTEM DESIGN

This study introduced a method of connecting traditional electrical devices to IoT with an embedded system and LoRa wireless technology to realize intelligent control and energy saving. To achieve that goal, power consumption, control method, and communication stability of the embedded system and communication module should be carefully considered. The entire system comprised two parts, which were "data collection implementation side" and "communication and data storage system". In this study, we illustrate the system with descriptions of hardware implementation and software design. Table 1 provides a summary of the components used in the node. Detailed information on each part is provided in the following sections.

TABLE 1. MAIN COMPONENTS USED IN EC NODES

Components	Model	Power Consumption
DC-DC converter	TPS54331	Shut down: 1 μA
LDO	XC6206P332	1 μA (Battery)
RF Module	LoRa (SX1268)	Sleep: 3 μA
MCU	STM32F103C8T6	Sleep < 10 μA

Temperature	DS18B20	5 μA
MOSFET array	AO3401	N/A

2.1. HARDWARE IMPLEMENTATION

The hardware of the purposed node consisted of five modules, which were power management module, microcontroller module, wireless communication module, sensing module and control module, as shown in Figure 2. An image of the hardware is shown in Figure 3. Detailed descriptions of the modules are presented in the following sections.

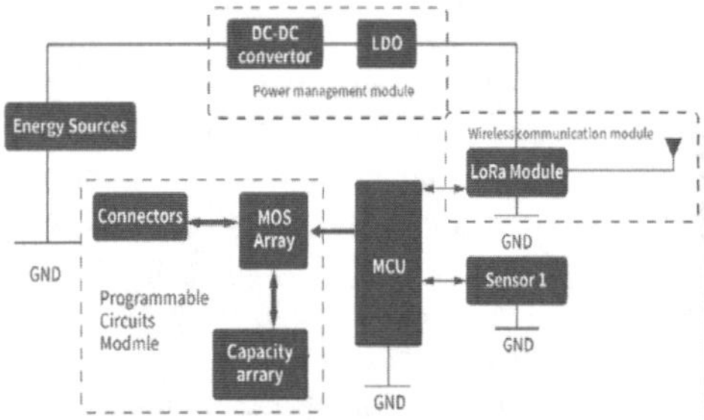

FIGURE 2. Schematic diagram of the node.

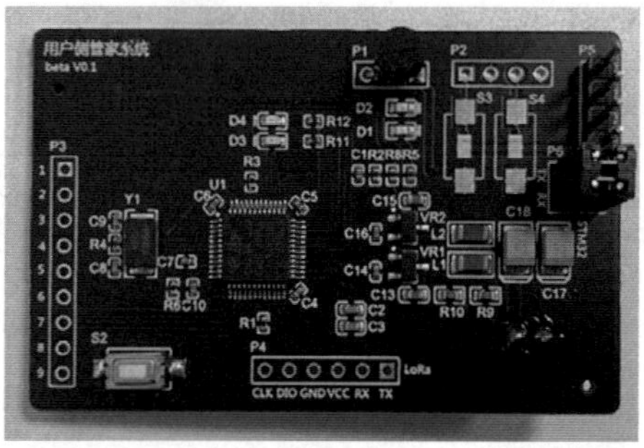

FIGURE 3. Sensing node picture.

2.1.1Wireless Communication Module

The wireless communication module built-in supports LoRa remote modem. In this paper, an SX1268 module is used to establish stable communication for the node. It mainly uses for long-distance spread spectrum communication, which has strong anti-interference and can minimize current consumption. SX1268 has a high sensitivity of over -148-dBm, a power output of +20-dBm, a long transmission distance, and high reliability [5]. Simultaneously, compared to traditional modulation technology, LoRa modulation technology also has obvious advantages in terms of anti-blocking, transmission distance, and low power consumption, which solves the problem of distance, anti-interference, and power consumption that traditional design solutions cannot consider simultaneously.

SX1268's operating voltage is 2.4-V to 3.7-V. The sleep mode current is only 2-μA. The receive-current is very low, which is only 12.15-mA. The LoRa chip communicates with MCU via the Serial Peripheral Interface bus (SPI). The module is LoRa Ra-01 from Ai-Thinker. More detailed performance of this module will be tested, and results will be shown in the next section.

2.1.2 control module

Controlling traditional energy-consuming equipment always require a solid connection between users and products. The designed nodes control the equipment by user-side without changing this connection. In that case, it is difficult for the nodes to acknowledge the situation of control. Therefore, the control must be successful with no feedback. In this work, the nodes are used to control a capacitance touch panel of the central air conditioning system. Nodes are equipped with capacitance convertible circuits to simulate the finger touch of the users. The principle of detecting finger touch is using the variation of capacitance and most familiar chip is AD7147. We made use of the detecting skill that applied in the chip and designed a unique method to simulate the finger touch without mechanical movement.

AD7147 can set sensitivity of detecting sensor contact to suit the size of finger. However, either small finger or large finger contacts produces the difference of electric charge on surface of the sensor [6]. Therefore, to simulate the finger touch, a variation of electric charge is needed.

With the principle of detecting, the designed capacitance variation circuit consists of a MOSFET and a capacitor. It creates a difference in capacity to simulate the process of finger touch. It connects directly to MCU. It operates by pulling down the grid electrode of MOSFET to connect a 10-μF capacitor into the circuit. Furthermore, the control pin of MCU should be set to the open-drain output mode. For other types

Journal of Physics: Conference Series **2365** (2022) 012040 doi:10.1088/1742-6596/2365/1/012040

of control panels, the designed control circuits can be programmed for different functions, such as infrared control and capacitance touch control, as illustrated in Figure 4.

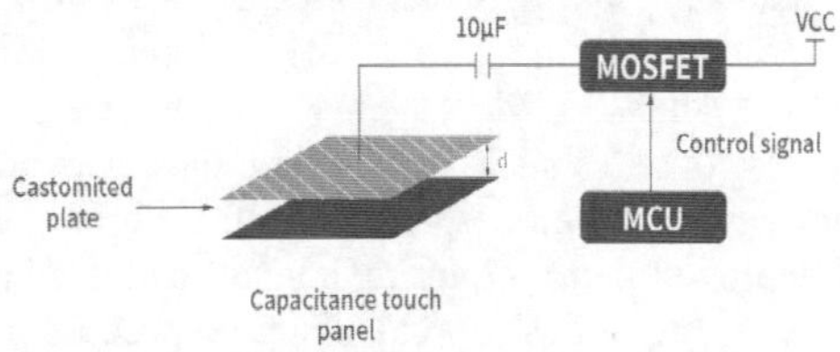

FIGURE 4. Control system setup.

2.1.3 System integration module

In this study, the system integration module read and organized data from the wireless communication module and sensor module and operated the command via a control module. It also controls the power consumption of the node. STM32F103 is used as MCU in this work. The MCU can work in sleep mode when there is no command so that it can save energy with a working current of 2-μA. It has 64KB flash, 12-bits A/D converters, and 37 pins, which can provide different functions. The nominal voltage for the MCU is 3.3-V, while the clock speed is 8MHz [7].

Through the system integration module, the algorithm of EC can be deployed on the nodes. It can produce data with accuracy and stability, process communication protocol, and realize real-time intelligent control of targeted devices.

2.2 SOFTWARE AND ALGORITHM DESIGN

The software and algorithm design for the purposed system consists of two parts, which are EC node and cloud server.

2.2.1 Basic Algorithm of Nodes

The nodes are programmed to wake up half a minute to track the temperature of the environment, upload the situation of the current operation, and receive commands from the cloud server. The node is equipped with two keys, which realize the functions of shutdown and extended use time, respectively. With basic algorithm, nodes are able to send heartbeat message to the cloud server every half a minute, receive command and translate communication protocol. The software algorithm for the node is presented in Figure 5.

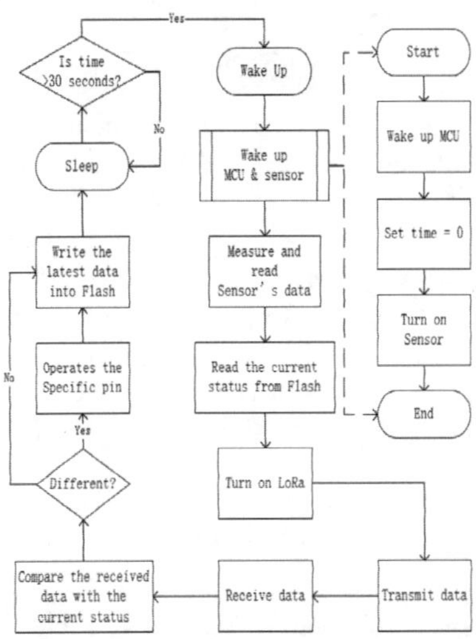

FIGURE 5. EC node's basic software algorithm.

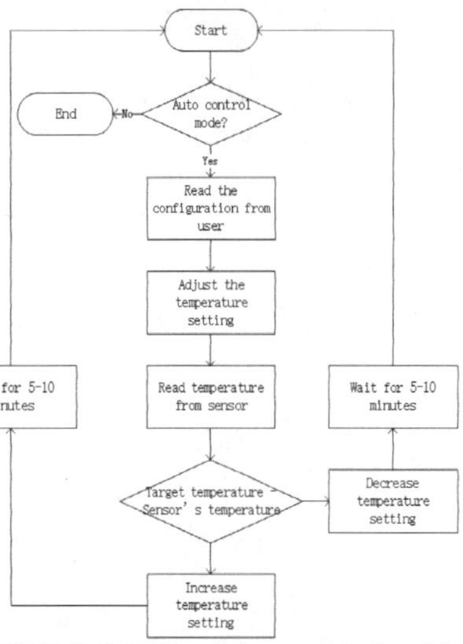

FIGURE 6. EC node's control algorithm.

2.2.2 Edge Computing

EC reduces the load of the communication link between the nodes to the Internet. Unlike the interior controller, user-side control is one-way control which does not have feedback result of the command. Therefore, the current situation and the expected result of the equipment that was controlled must be saved both in the nodes and on the cloud database. Moreover, for different electrical equipment, the required instruction form is also diverse. To transfer such a huge data packet is not affordable for a low-power communication system. In this paper, the nodes can save the current configuration of the targeted equipment. When a command deployed from the users, a small data package of expected configuration will be sent to the nodes. After that, the nodes can calculate the process needed locally and operate the circuits to realize the control.

Furthermore, the node needs to estimate the status saved in the flash. Every time the node receives the package from the cloud server, the node can realize a process of self-estimate. This process can reduce the possibility of missing the status of panels and compensate the error.

2.2.3 Environment-Awareness-Based Control

In order to provide a better user experience, the node has an environment-awareness-based control algorithm. After the user sets the temperature value, the node will save the setting value in the node, then use the setting value minus environment temperature value which is read by temperature sensor. When the result greater than 0, this node will actually increase the set temperature value, otherwise decrease. The delay would be 5-10 minutes. However, nodes will upload the final configuration to the cloud server to make sure the status is correctly recorded. If the node receives a manual control command, the node will automatically exit the auto control mode. The control algorithm is shown in Figure 6.

2.2.4 Cloud Server

To connect the nodes to the Internet, the gateway that connects to the cloud server via Ethernet receives the data from both the nodes and the server in JavaScript Object Notation (JSON). Every node has an address and a unique User Identification (UID). The command consists of the address of the target node and specific command for the node. The networking topology is a star-type network. The node transmits the data and receives the command from the gateway directly without network hopping to reduce communication delay.

The website page displays the data in a Graphical User Interface (GUI), records the operations of users, and stores the data collects both from the nodes and users' commands into a local MySQL database.

3. EXPERIMENTAL RESULTS

3.1 Control Accuracy and Stability

Because the node uploads heartbeat message to the cloud server every half a minute. The communication protocol stipulates that the command of successful communication will get the feedback information of the node. According to the protocol, if the node feedback information is not received, the lost command will be sent again to ensure that the node will receive every command. Therefore, the experimental design tests the control accuracy of the nodes by simultaneously receiving the heartbeat message of 5 nodes in different locations. When the heartbeat message is uploaded to the cloud, the cloud will send an instruction to indicate that the heartbeat message has been received, and the node will send a feedback message to the cloud after receiving the instruction. In the test, the node was tested continuously for 12 hours. After that, the node will restart the test lasting 24 hours. so that the results can find out if the stability will be affected by the working time.

In the first test that lasts for 12 hours, the average control accuracy of the nodes is 97.72% (1407.2/1440). For all the nodes, the minimum of recall rate is 95.63% (1377/1440) while the maximum of the rate is 99.24% (1429/1440).

In the second test that lasts for 24 hours, the average control accuracy of the nodes is 97.89% (2819.2/2880). The minimum of the rate is 96.60% (2782/2880) while the maximum is 98.98%

(2850/2880). In addition, every command that transmitted to the node successfully was operated by the node. No command was missed during the 12 hours test and 24 hours test.

None of the nodes crashed or disconnected from the communication link during the test. In conclusion, the algorithm of nodes and gateway is strong enough to overcome the effects from long-term working. Even if the heartbeat message was missed temporary, the algorithm can fix the error in due time. Detail results of the test are shown in Table 2.

TABLE 2 CONTROL ACCURACY AND STABILITY OF LORA NODES

Node Address	Proposed Number of Heartbeat	Actual Number of Heartbeat	Percentage
1	1377	1440	95.63%
2	1385	1440	96.19%
3	1424	1440	98.86%
4	1429	1440	99.24%
5	1421	1440	98.67%
1	2818	2880	97.84%
2	2782	2880	96.60%
3	2803	2880	97.33%
4	2850	2880	98.98%
5	2843	2880	98.73%

3.2 Ability of Crossing Walls

One of the advantages of LoRa wireless technology is the ability of crossing the wall. In the building, it is of vital importance that nodes can be set in the different room of the gateway. The better ability of crossing the wall reduces the total number of the gateway that needed. In this test, the nodes are set at different distances to the gateway. They are divided into three groups and set in three independent rooms blocked by walls. The test lasted for 24 hours; we use the spectrum analyzer to analyze the signal strength. Figure 7 illustrates the test environment.

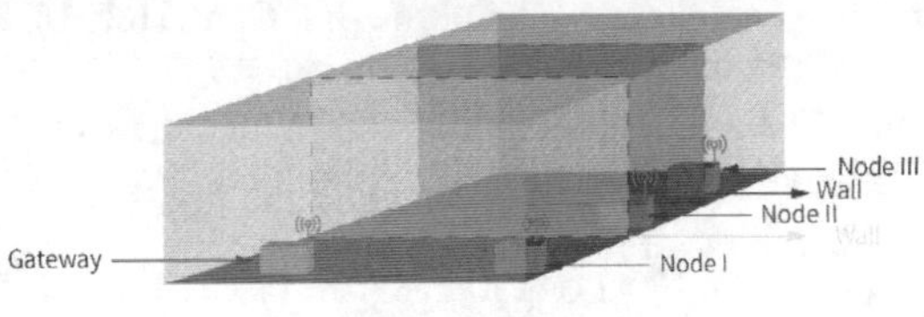

FIGURE 7. The environment of communication stability testing.

The test shows that the Received Signal Strength Indication (RSSI) dropped when the distance increased. The nodes set in the same room of gateway receive the effective command with the average RSSI of -57.21-dBm while the nodes in the next room decreased to -74.413-dBm. The nodes in the most distant room also have the average RSSI of -97.40-dBm. No commands are sent to the wrong node.

Moreover, the control accuracy stayed roughly the same while the RSSI decreased within -120-dBm. The accuracy is 97.6% for the closest node, 97.1% for the node next room, and 97.4% for the furthest node. The variation of RSSI is recorded as shown in Figure 8.

In conclusion, compared to the method of Zigbee, Wi-Fi, and BLE, the control accuracy and stability is much better via LoRa. Moreover, with the software algorithm, the missed command will be transmitted again if the nodes do not receive the command.

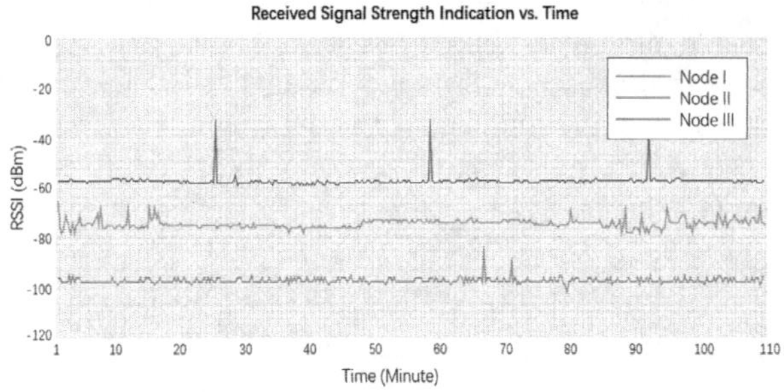

FIGURE 8. The variation of RSSI of the nodes.

4. CONCLUSION

This paper has presented a method of connecting traditional energy-consuming equipment to the IoT and realize intelligent control of these equipment to save energy from overuse. Current work in this paper primarily focuses on hardware developments. Some sample data is collected by the node and sent to the cloud server via LoRa network. The aim of this project is to provide a solution for the traditional power consuming devices to save energy and become more intelligent. Such an IoT platform will enlarge the usage of some old air conditioners or other energy-consuming products. Our future work will focus on more equipment that requires to be connected to the IoT platform, more experimental results, and performances of LoRa in other sceneries.

REFERENCE

[1] T. Torfs, T. Sterken, S. Brebels, J. Santana, R. van den Hoven, V. Spiering, N. Bertsch, D. Trapani, and D. Zonta, 2012. "Low Power Wireless Sensor Network for Building Monitoring," IEEE Sensors Journal, vol. 13, no. 3, pp. 1–1.

[2] Weisong Shi, Jie Cao, Quan Zheng, Youhuizi Li, and Lanyu Xu, 2016. "Edge Computing: Vision and Challenges," IEEE Internet of Things Journal vol. 3, no. 5, pp. 637-646.

[3] J. I. Lee, C. S. Choi, W. K. Park, J. S. Han, and I.W. Lee, 2011. "A study on the use cases of the smart grid home energy management system," in Proc. Int. Conf. (ICTC), Seoul, South Korea, pp. 46-50.

[4] A. Mahmood, A Ahmad, HT Javed, Z Mehmood, and N Javaid, 2015. "A survey of 'user comfort' home energy management systems in smart grid," in Proc. WAINA, Gwangiu, South Korea, pp. 36-43.

[5] Semtech, 2018. "SX1268 LoRa Core™ Long Range Low Power LoRa® Transceiver +22dBm, China frequency bands", [Online]. Available: https://www.semtech.com/products/wireless-rf/lora-transceivers/sx1268#datasheet.

[6] Analog Devices, 2015. "CapTouch™ Programmable Controller for Single-Electrode Capacitance Sensors [Datasheet]", [Online]. Available: https://www.analog.com/media/en/technical-documentation/data-sheets/AD7147.pdf.

[7] STMicroelectronics, 2015. "Low-density performance line, ARM-based 32-bit MCU with 16 or 32 KB Flash, USB, CAN, 6 timers, 2 ADCs, 6 com. interfaces", [Online]. Available: https://www.st.com/resource/en/datasheet/stm32f103c4.pdf

Journal of Physics: Conference Series **2365** (2022) 012041 doi:10.1088/1742-6596/2365/1/012041

Design and implementation of IoT gas meter based on security chip

Xiaoqing Xu[1*], Xiaowei Liang[2] and Cheng Shen [3]

[1] School of Information Engineering, Eurasia University, Xi'an, Shaanxi, 710065, China

[2] Innochip Technology company, Nanjing, Jiangsu, 211500, China

[3] School of Information Engineering, Eurasia University, Xi'an, Shaanxi, 710065, China

[*]Corresponding author's e-mail: xuxiaoqing@eurasia.edu

Abstract. IoT gas meter is facing the challenge of information security. This paper designs and implements a scheme of gas meter based on security chip. The scheme uses stm32l073 as MCU to collect gas metering data, cooperates with Hua Hong's security chip, and uses NB_IoT as the communication module. The business data is pushed to the gas meter business management system through China Telecom Tianyi IoT platform CTWing, so as to realize the remote safety management of gas meters.

1. Introduction

The diversification and enhancement of the application of the Internet of Things technology makes the information security problem in the Internet of Things era very serious[1]. The IoT gas meter architecture consists of the meter end, data acquisition and processing system, and business system[2]. Various computing keys and sensitive data of the table, such as configuration parameter information, key business data, and metering parameters, are stored in a designated area in the security chip. The platform layer is faced with the needs of massive intelligent terminal data collection and processing, and faced with security challenges such as preventing illegal internal supply. It is necessary to establish a reliable authentication mechanism, reliable key management, ciphertext query and confidential data mining and other security mechanisms[3]. In response to the above problems, this paper proposes a systematic solution. The security of data collection and processing and background business is realized. It uses digital certificates and key encryption and other cryptographic techniques for data security protection. The data collection system uses The various keys of the encryption machine are stored in the specific security area of the encryption machine[4]. The business system is the control core of the security platform, providing application management and encryption information collection services, and providing application activation for the security chip.

2. Gas meter safety scheme

The built-in security chip is used at the meter end to strengthen the information security management efficiency and realize the smart gas meter. The functional block diagram of the IoT gas meter is shown in Figure 1 and Figure 2.

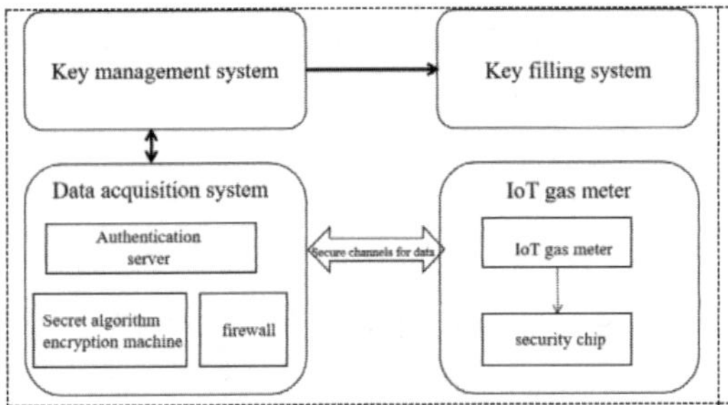

Figure 1. IoT gas meter system structure diagram.

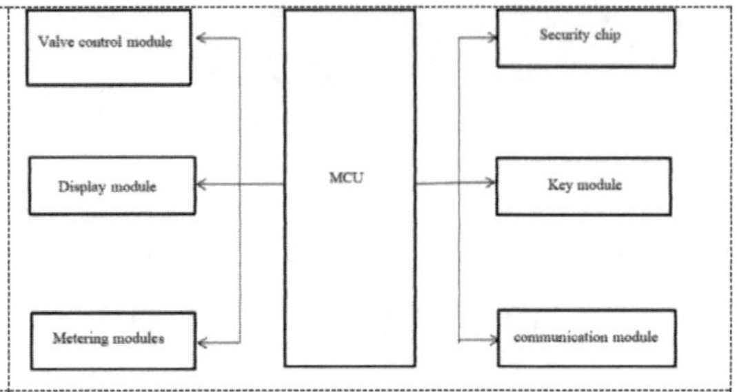

Figure 2. IoT gas meter function module.

2.1. Security Chip Hardware Solution

The connection between ESAM and MCU is shown in Figure 3.

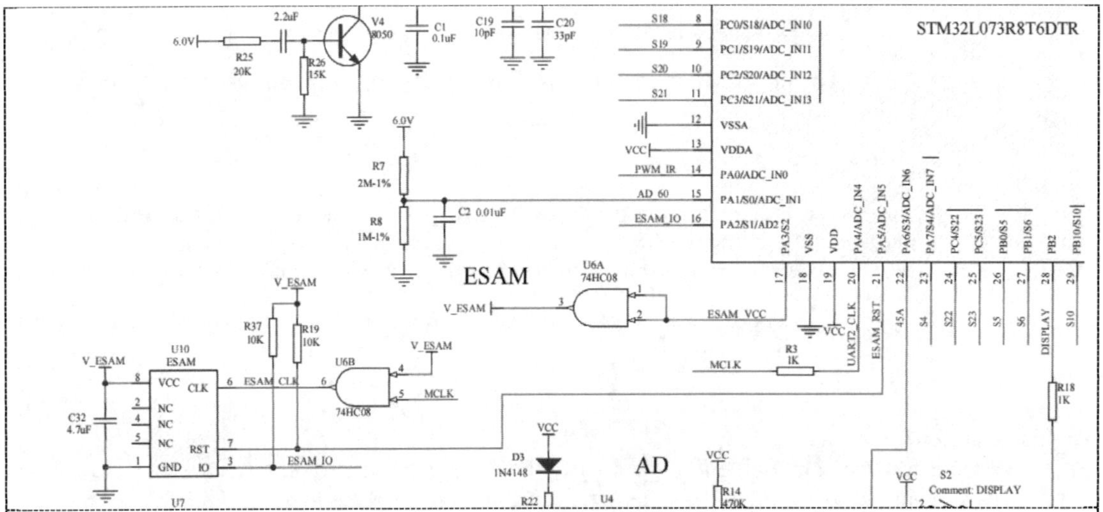

Figure 3. Schematic diagram of MCU and ESAM hardware interface.

2.2. Software implementation

The software of the gas meter terminal of the Internet of Things can be divided into three parts: the main program, the interrupt service program and the function module subprogram according to the hierarchical structure[5]. Interrupt operations include button interruption, reed switch measurement interruption, and timer interruption. The interrupt service routine only does some simple conditional judgments and flags the occurrence of interrupt events. Function module subroutines mainly include timer operation module, gas metering module, buzzer reminder module, data storage module, LCD display module, motor valve control module, infrared process communication module, NB network communication module, voltage detection and power-down processing modules, security chip modules, etc. The software is developed in C language, and the development environment is IAR. For the security module part, it mainly completes the design and implementation of the security chip driver.

2.2.1 Registration authentication interaction

First, the gas meter initiates a registration handshake authentication message, and the data acquisition system responds to complete the registration handshake authentication[6]. The registration uplink message and the registration downlink message carry MAC verification data to ensure data integrity, and at the same time complete the authentication of the identities of both parties. After the registration and authentication are completed, the application root key ID of the subsequent interaction is determined through negotiation, and the two parties use the random number of the message in the protocol interaction process to generate a working key for the communication process to complete the key negotiation. The registration interaction process is shown in Figure 4.

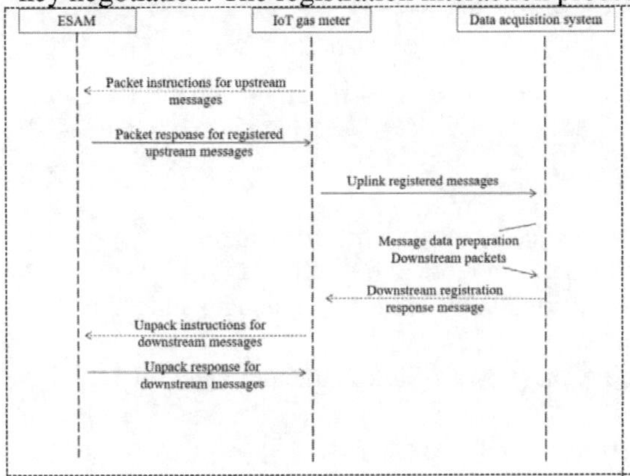

Figure 4. Registration authentication interaction.

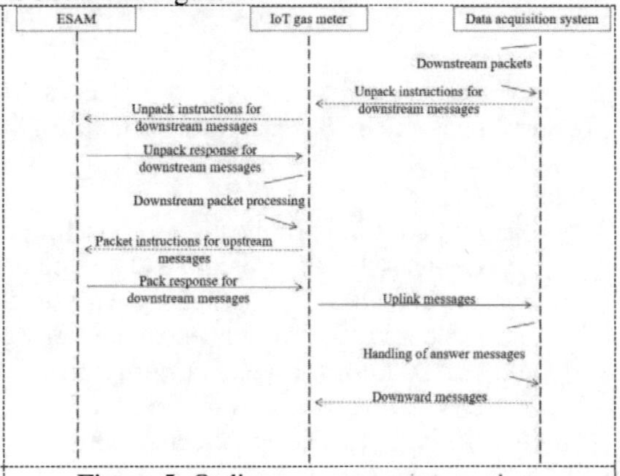

Figure 5. Ordinary message interaction.

2.2.2 Ordinary message interaction

After the registration and authentication process is completed, the subsequent interactive communication is controlled by the design and acquisition system, and the downlink message is sent to obtain the response message of the gas meter. The message interaction process is shown in Figure 5.

2.2.3Packet and unpack

According to the business scenario, certain arithmetic rules can be selected for the message processing between the gas meter and the acquisition system to meet the security requirements, as shown in Table 1.

Table 1. security strategy.

Function	Functional details	Security Strategy
Normal interactio n	Collect the non-sensitive information of the general negotiation between the system and the IoT meter, such as the overall picture information of the meter, etc.	plaintext+MAC or Ciphertext+MAC
Query collection	Remotely query and collect the running status of the IoT table, detailed records of daily gas consumption, daily summary records and monthly summary records	Ciphertext+MAC
Real-time reporting	Report the events and processing results in the meter of the IoT gas meter	Ciphertext+MAC
Parameter settings	Remotely set the operating parameters of the IoT meter, such as overdraft volume, meter reading interval, billing method, and tiered pricing	Ciphertext + Signature
Remote control	Remote control functions for IoT meters, such as remote valve opening and closing at the meter end, etc.	Ciphertext + Signature
Key update	Remotely update a specified key information	Ciphertext + Signature

2.2.4 Security chip driver implementation

The driver of the security chip mainly completes the realization of functions such as private instructions, package function API, file operation read and write API, etc.

3. Conclusion and Prospect

The IoT gas meter obtained the test certificate from Beijing Kunlun Gas Company. However, in the current scheme, the security chip is placed at the back of MCU, so the interaction efficiency is relatively low. The security chip can be considered to directly connect the communication module, which is placed in the front of the MCU, which can play the role of the Internet of Things firewall and resist the enemy on the front end.

Acknowledgments
Sponsor: School Scientific Research Project of Xi'an Eurasia University: Research on Fault Diagnosis Technology Based on Big Data (2021XJZK02).

References

[1] Geng,Y.Y.(2021) Application of gas Internet of things security chip in Information Security Center. China Petroleum and chemical industry standards and quality, 21:143-144.
[2] Quan,Y.Q. (2021) Intelligent gas meter information security management function standardization recommendations. City gas, 2021(S1):264-267.
[3] Su,L.F. (2020) Information security of IoT smart gas meter system. Gas and Heat, 40(11):24-28+45-46.
[4] Chen,G. (2021) The architecture and application of IoT data security system in smart gas industry based on cryptography. City Gas, 2021(S1):100-103.
[5] Gao,Z.R. (2021) A research and design of Internet of Things gas meter based on NB_IoT. Information Technology and Network Security, 40(07):90-94+107.
[6] Weng,Q.W. (2022) Safety protection system based on narrowband Internet of things digital gas meter. Industrial control computer,35(02):47-49+51.
[7] Shao,Z.H. (2022) Safety-cut-off smart gas meter with self-learning function.Gas and heat,42(05):39-40+46.

Intelligent algorithms for six degrees of freedom robot trajectory planning considering vibration suppression

fazhe LI, Jiayong CAO, shihao WU, peihua WU, ziwei ZHU

School of Mechanical Engineering, Shanghai Institute of Technology, 201418, China

E-mail: caojy@sit.edu.cn

Abstract. The residual vibration of the flexible joint manipulator during and after the movement will seriously affects the positioning accuracy of the manipulator. This paper discusses the trajectory planning method for suppressing the joint vibration. A trajectory optimal model which contains the minimization of both time and jerk was proposed. An optimization method based on comprehensive learning particle swarm optimization (CLPSO) algorithm was utilized for solving the optimal problem. Compared with other optimal solution method, CLPLO has the advantages of avoiding premature convergence. A numerical simulation case shows that the optimal trajectory obtained in this paper generated less joint vibration. The effectiveness of the proposed method was verified to some extent.

1. Introduction

Industrial robot is a kind of multidisciplinary and multidisciplinary comprehensive product of scientific progress and technological development [1].

To meet complex engineering tasks, joint systems are typically multiple coupling structures. Therefore, joint flexibility is unavoidable [2]. For flexible joint manipulators with flexible components, previous researchers have used lots of experiments to prove that ignoring the flexibility of flexible components will seriously affect the performance of the system [3].

Therefore, vibration suppression is essential to ensure the good operation of flexible robots. There are many methods to suppress residual vibration. Among them, trajectory planning method is a new research direction in the field of robot vibration suppression. Compared with other methods, it has simple and direct advantages. Sarkar [4] et al. studied the minimum residual vibration and the shortest joint trajectory planning. The method is divided into two steps. Firstly, only the time of the trajectory is minimized, and then the residual energy is reduced on this basis. Mohri [5] et al. also studied the suppression of residual vibration along a given trajectory, which is suitable for flexible robots that complete repetitive work in a fixed period of time.

The time-jerk optimization algorithm based on comprehensive learning particle swarm optimization (CLPSO) is used to balance the optimal time and the optimal vibration trajectory planning to suppress the vibration of the flexible manipulator. In theory, the jerk can be constrained to avoid the resonance of the manipulator, so that the vibration of the manipulator is reduced. Therefore, this study has a great role in studying the vibration problem of industrial robots in the process of operation, making industrial robots develop towards high precision, high speed and high quality.

2. Robot trajectory optimization model

Industrial robots should not only ensure the work efficiency, but also consider the influence of vibration factors, so the robot trajectory optimization model should consider both the working time and the residual vibration as far as possible. Therefore, the previous models are improved with the following equation:(1) The optimization goal is a comprehensive function after weighting the execution time and vibration effects, and the variables are the speed, acceleration, jerk, torque, etc. of the joint.

$$\begin{cases} min\, f(h) = K_T N \sum_{i=1}^{n-1} h_i + \alpha K_J \sum_{i=1}^{N} \sum_{i-1}^{n-1} \left[\frac{(\ddot{\theta}_{j,i+1} - \ddot{\theta}_{j,i})^2}{h_i} \right] \\ s.t. \\ max\{|\dot{\theta}_{j,i}(t_i)|, |\dot{\theta}_{j,i}(t_i *)|, |\dot{\theta}_{j,i}(t_{i+1})|\} - V_{jm} \le 0 \\ max\{|\ddot{\theta}_{j,1}(t_1)|, |\ddot{\theta}_{j,2}(t_2)|, \cdots, |\ddot{\theta}_{j,n}(t_n)|\} - A_{jm} \le 0 \;\; j = 1, \ldots, N \forall i = 1, \ldots, n-1 \\ max \left| \frac{\ddot{\theta}_{j,i}(t_{i+1}) - \ddot{\theta}_{j,i}(t_i)}{h_i} \right| - J_{jm} \le 0 \\ max|\tau_{j,i}(t_{i+1}) - \tau_{j,i}(t_i)| - \tau_{jm} \le 0 \end{cases} \tag{1}$$

where the values of K_T and K_J are adjusted by $K_T + K_J = 1$ to optimize the total execution time and vibration. Because the total execution time is proportional to the integration of the square of the jerk, which can vary greatly in the level of quantities, elasticity coefficient α is introduced to balance their effects. The significance of the remaining parameters is shown in Table 1 below.

Table 1. Meaning of symbols

Sign	Meaning	Sign	Meaning
N	Number of robotic arm joints	$\dot{\theta}_{j,i}(t_i)$	Velocity of the jth joint
n	Number of time periods of the joints	$\ddot{\theta}_{j,i}(t_i)$	Acceleration of the jth joint
K_T	Time-weighted coefficient	$\dddot{\theta}_{j,i}(t_i)$	Jerk of the jth joint
K_J	Vibrating weighted coefficient	V_{jm}	Speed limit of the jth joint
T_i	Time interval	A_{jm}	Acceleration limit of the jth joint
τ_{jm}	Torque limit of the jth joint	J_{jm}	Jerk limit of the jth joint

3. Trajectory optimization method of particle swarm optimization algorithm

3.1. Selection of Optimization Algorithms

In CLPSO, the p-best of all particles is used to update the particle velocity. Each dimension of a particle is learned from the p-best of different particles, rather than from the same sample of all dimensions, thus increasing the diversity of particles. The speed of the i particle is updated with the following equation :(2)

$$V_i^d = wV_i^d + c \times ran\, d_i^d \times (\text{p-best}_{fi(d)}^{\,d} - X_i^d) \tag{2}$$

where, $f_i(d) = [f_i(1), f_i(2), \ldots, f_i(D)]$ indicates whether the i particle follows the pbest_i^d of itself or other particles on each dimension d. The example for each dimension is determined by the value of learning probability Pc, and global particles and local development particles are specified according to these learning probability values. The Pc of each particle is calculated by the following formula:

$$Pc_i = a + b \times \frac{(exp((10(i-1))/(ps-1))-1)}{(exp(10)-1)} \tag{3}$$

where ps represents the particle size, $a = 0.05, b = 0.45$

The superiority of using CLPSO is verified by function test. The diversity of particle swarm determines whether the population has the ability of global optimization and local development, so single-peak function and multi-peak function on 30 dimensions are prepared to test its performance. The measurement of group diversity is based on the sum of equations.

$$\text{Diversity } (S(t)) = \frac{1}{N}\sum_{i=1}^{N}\sqrt{\sum_{d=1}^{D}\left(X_i^d(t) - \overline{X^d(t)}\right)^2} \tag{4}$$

$$\overline{X^d(t)} = \frac{\sum_{i=1}^{N} X_i^d(t)}{N} \tag{5}$$

In the test process, the standard particle swarm optimization (PSO)[6], chaotic-particle swarm optimization (CPSO)[7], heterogeneous comprehensive learning particle swarm optimization (HCLPSO)[8] and comprehensive learning particle swarm optimization (CLPSO)[9] are compared.

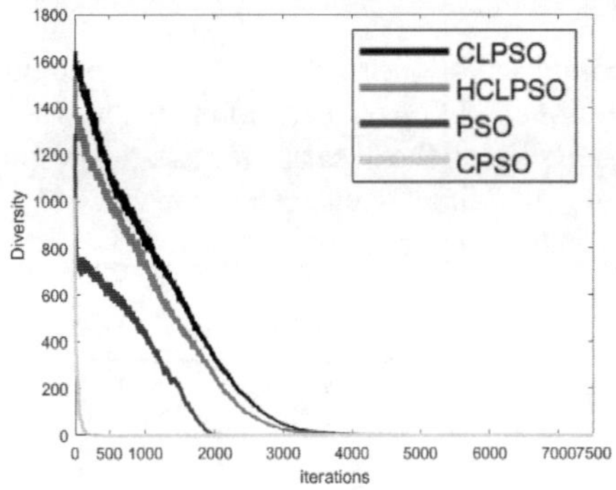

Figure 1. Comparison of single-peak test functions F_1

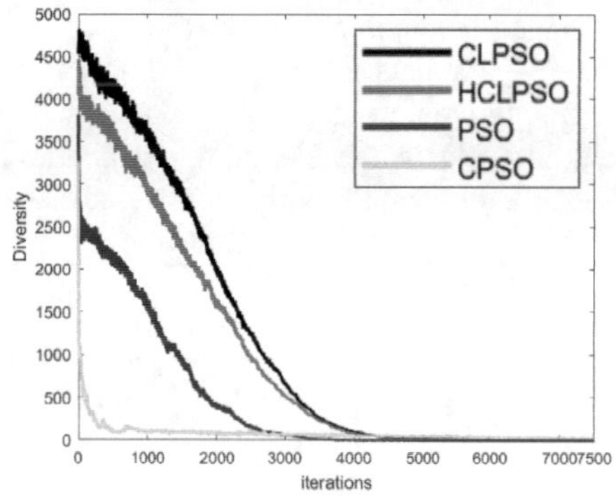

Figure 2. Comparison of Multi - peak Test Functions F_4

It can be seen from the test results that compared with the other three algorithms, the search accuracy of CLPSO algorithm in the test function is high, the convergence speed is faster, the diversity test effect is better, and the probability of local optimal value is small. CLPSO not only performs well in function sum, but also produces the best solution in function with HCLPSO. In summary, CLPSO has no worst performance in various test functions and has outstanding performance in multi-peak areas.

3.2. Time-jerk optimization algorithm based on CLPSO

This algorithm takes into account the advantages of the diversity of CLPSO 's searching ability in multi-peak and single-peak and solves the optimal time-jerk optimal path trajectory planning.

This algorithm randomly generates the initial value, so it is easy to find that the final result is not the optimal value. Therefore, the solution method proposed in this paper is, firstly, based on the description of the constraint problem, the problem is transformed into a logical term into a matrix form of 3-5-3 polynomials. Secondly, the randomly generated position particles retain the obtained optimal value. If they have reached the requirement, it means that they can directly withdraw from the cycle. If they do not meet this requirement, they must enter the cycle again. By using the method of learning value, p-best and g-best are updated to ensure that the optimal value is obtained and avoid falling into the cycle of local extremum. Finally, the speed and position of the iterative position particles are updated again to see whether the maximum number of iterations is reached, and the output of the loop is terminated immediately after the requirements are met; the obtained value is the optimal solution of the problem.

4. Optimization method and result analysis

4.1. Selection of interpolation methods

The 3-5-3 piecewise polynomial interpolation algorithm is used to solve the polynomial order optimization problem, reduce the mechanical resonance and prolong the service life. For the 3-5-3 polynomial of the ith joint, the general formula is:

$$\begin{aligned} h_{i1}(t) &= a_{i13}t_1^3 + a_{i12}t_1^2 + a_{i11}t_1 + a_{i10} \\ h_{i2}(t) &= a_{i25}t_2^5 + a_{i24}t_2^4 + a_{i23}t_2^3 + a_{i22}t_2^2 + a_{i21}t_2 + a_{i20} \\ h_{i3}(t) &= a_{i33}t_3^3 + a_{i32}t_3^2 + a_{i31}t_3 + a_{i30} \end{aligned} \tag{6}$$

where, $h_{i1}(t), h_{i2}(t), h_{i3}(t)$ represent the trajectories of three polynomials, respectively.

a_{i1j} ,a_{i2j} ,a_{i3j} are the coefficients of each segment number of the jth coefficient of the ith joint.

4.2. Main parameters of the robot

The pose of four points selected in the known space is composed of one starting point, the end point and two interpolation points, as shown in Table 2. The constraint conditions of each joint are shown in Table 4.5, including velocity, acceleration, jerk and torque. When solving the dynamic torque, the torque curve can be obtained by inputting the joint velocity, joint angle and joint angle acceleration.

Table 2. Robotic arm trajectory planning input data

joint	$\theta_{i0}(rad)$	$\theta_{i1}(rad)$	$\theta_{i2}(rad)$	$\theta_{i3}(rad)$
1	0	-6	-10	2
2	0	15	0	-18
3	0	12	4	-8
4	0	5	9	-5
5	0	6	12	-8
6	0	10	4	12

Table 3. Restrictions

joint	$V_{max}(rad/s)$	$A_{max}(rad/s^2)$	$J_{max}(rad/s^3)$	$N_{max}(N.m)$
1	38	86	80	600
2	24	50	90	480
3	35	67	98	300
4	60	65	138	160
5	80	85	160	100
6	100	110	180	80

4.3. Analysis of optimization results

The elastic coefficients α is 0.1, $K_T + K_J = 1$, K_T and K_J are adjusted as follows: K_J is obtained when K_T is 1, 0.8, 0.5, 0.2, 0. Through the values of these five groups of weighting coefficients, the trajectory planning curves corresponding to the optimization results are obtained as shown in Figure 3 and Figure 4.

Journal of Physics: Conference Series **2365** (2022) 012042 doi:10.1088/1742-6596/2365/1/012042

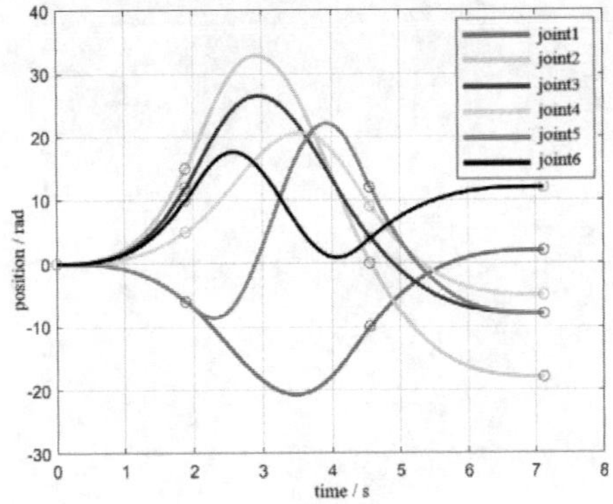

(a) position

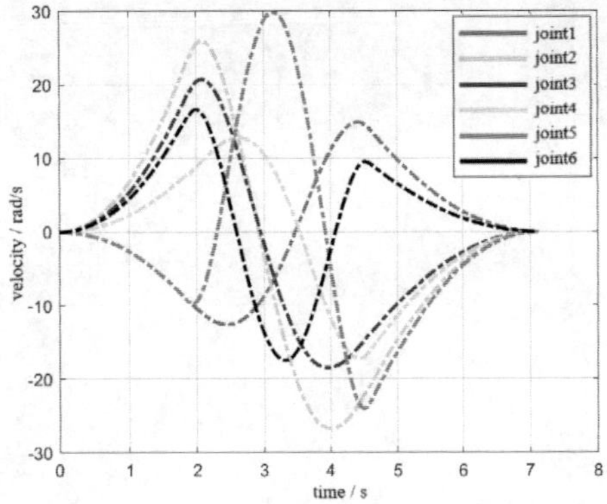

(b) velocity

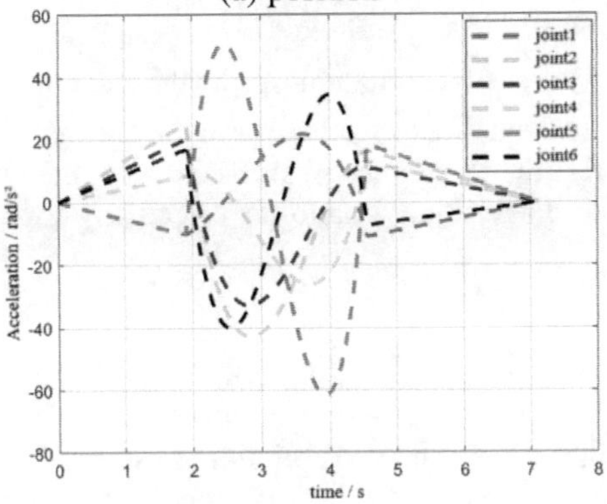

(c) acceleration

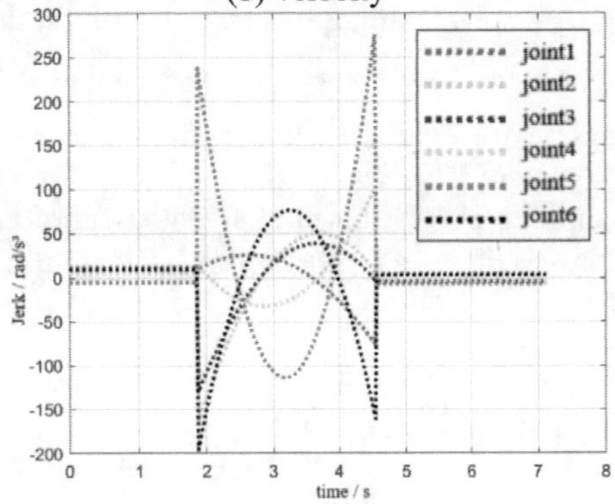

(d) jerk

Figure 3. Optimal joint trajectories when $K_T = 1, K_J = 0$

In Figure 3, the required movement time of each joint is obviously less than other results, but the fluctuation of the robot in this state is large and the value is also large. Therefore, the manipulator is prone to small vibration, keeping this working state for a long time will reduce the operation accuracy of the manipulator, affect the service life, and even cause accidents.

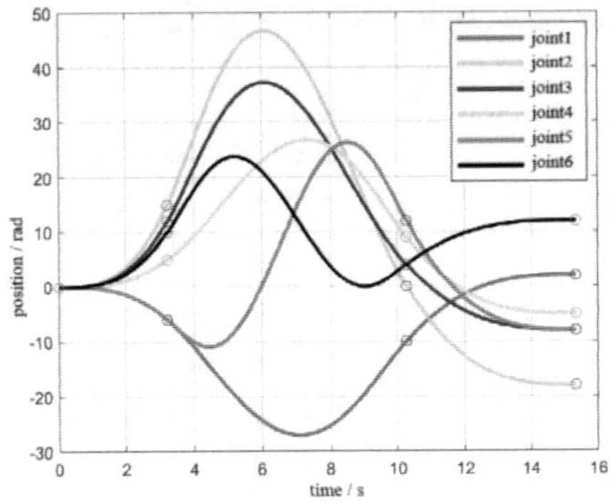

(a) position

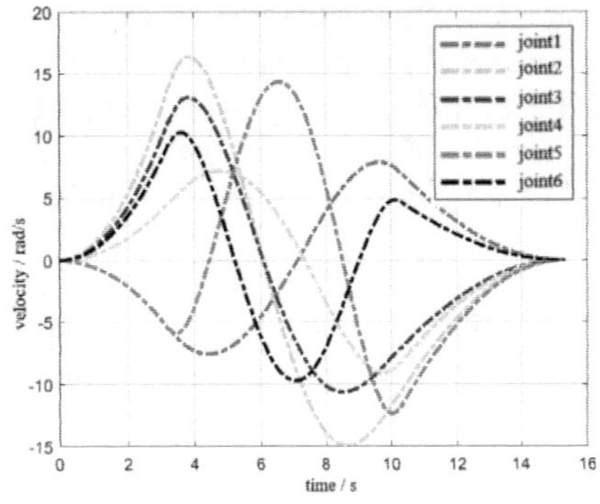

(b) velocity

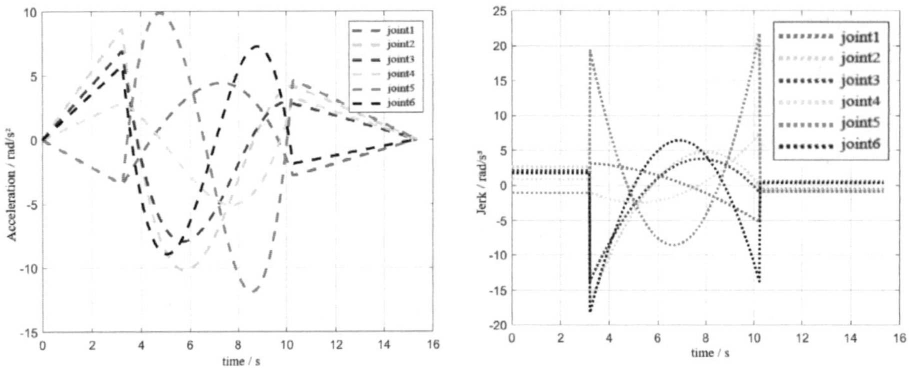

(c) acceleration (d) jerk

Figure 4. Optimal joint trajectories when $K_T = 0.5$, $K_J = 0.5$

When $K_T = 0.5$,planning time has increased, but the speed, acceleration, jerk curve fluctuations are gentle, compared with the previous optimization results have been greatly improved. By analyzing the jerk images in Figure 3 and Figure 4, it can be seen that the jerk mutation still exists, but the mutation amplitude is greatly reduced after optimization. Accordingly, the vibration is effectively suppressed, which proves the effectiveness of the algorithm.

5. Conclusions

In this paper, based on a large number of previous experiments show that the residual vibration of the flexible joint manipulator will seriously affect the performance of the manipulator in the process of movement and after the end of the movement. Aiming at the demand of vibration suppression of the flexible manipulator, a time-jerk optimization algorithm based on CLPSO is proposed to balance the optimal time and the optimal pulse trajectory planning to suppress the vibration of the manipulator. In order to fall into local optimum, the algorithm uses the method of learning value to update p-best and g-best to obtain the optimal value, avoiding the cycle of falling into local extremum.

Further, the part of the theory in this paper is verified by simulation experiments, which verifies that the size of the jerk directly affects the vibration amplitude of the manipulator.

The results show that the results obtained by the method in this paper are accurate and reliable, and the method is simple and convenient to operate. It can well achieve the purpose of suppressing the residual vibration of the robot, which provides a theoretical reference and practical basis for the study of vibration suppression by trajectory planning method. In the future, the experiment can be designed by combining the input shaping and trajectory optimization method to suppress the vibration of the robot in the process of motion, so that the optimization algorithm in this paper is more consistent with the actual, and increase the practical significance of the algorithm.

References

[1] Li B, Hou B, Yu W, et al., 2017.Applications of artificial intelligence in intelligent manufacturing. Frontiers of Information Technology & Electronic Engineering, 18(1):86-96.

[2] Bruno Siciliano, Oussama Khatib, 2008.Springer Handbook of Robotics.Springer, Berlin.

[3] Sweet L M, Good M C, 1984.Re-definition of the robot motion control problem: Effects of plant dynamics, drive system constraints, and user requirements. In: Proc.23rd IEEE Conf. on Decision and Control. Las Vegas.724-731.

[4] Sarkar P K, Yamamoto M, Mohri A, 1997.Sub-optimal trajectory planning of flexible manipulator along specified path, Proceedings of the 1997 IEEE/RSJ International Conference on Intelligent Robot and Systems. Innovative Robotics for Real-World Applications. IROS '97, 1997, pp. 1540-1545 vol.3, doi: 10.1109/IROS.1997.656563.

[5] Mohri A, Sarkar P K, Yamamoto M, 1998. An Efficient Motion Planning of Flexible Manipulator Along Specified Path. In: Robotics and Automation Proceedings. 1998 IEEE International Conference on. 1998:1 104-1109 vol.2.
[6] Kennedy J, Eberhart R., 1995.Particle swarm optimization. In: Proceedings of ICNN'95-international conference on neural networks. Perth.1942-1948.
[7] Shi Y, Eberhart R., 1998.A modified particle swarm optimizer. In: Proceedings of the 1998 IEEE International Conference on Evolutionary Computation Proceedings.69-73.
[8] Lynn N, Suganthan P N., 2015.Heterogeneous comprehensive learning particle swarm optimization with enhanced exploration and exploitation. Swarm and Evolutionary Computation,24: 11-24.
[9] Liang J J, Qin A K, Suganthan P N, et al., 2006.Comprehensive learning particle swarm optimizer for global optimization of multimodal functions. IEEE transactions on evolutionary computation,10(3): 281-295.

IOTAIMA-2022 IOP Publishing
Journal of Physics: Conference Series **2365** (2022) 012043 doi:10.1088/1742-6596/2365/1/012043

Simulation of three cylinders synchronous circuit based on RBF neural network optimized PID controller

Liang Tan[1], Jinfeng Huang[1], Xinghua Kou[1], Xingzhong Wang[1], Wanli Wang [2]

[1] China Ship Research and Design Center

[2] China University of Petroleum (East China)

E-mail: wwli0919@163.com

Abstract: RamRig hosting system is a common system for offshore and onshore drilling rigs, which uses ram cylinders to lift and lower the top drive on the drilling rig. In order to ensure the accuracy, stability and robustness of synchronous control of RamRig hosting system, RBF neural network is used to adjust PID parameters online, and a three cylinders synchronous loop simulation model based on RBF optimized PID controller is built in AMESim and Simulink. The simulation results show that the three cylinders synchronous hydraulic circuit based on RBF neural network optimized PID controller has higher accuracy and stability, and is fully capable of realizing the three cylinders synchronous output action of RamRig hosting system.

1. Introduction

With the increasing demand of national energy, the development of marine resources has become one of the key R & D projects. As an important equipment on the offshore drilling and production platform, RamRig hosting system has a complex and changeable working environment, so its accuracy and stability must be ensured. At present, domestic and foreign scholars have made mature research on the control of hydraulic cylinders, and conducted simulation research on the synchronous control of multiple hydraulic cylinders [1-2]. Liu Shaolong uses servo control mode to establish a simulation model of four cylinders synchronous hydraulic circuit based on AMESim software and MATLAB for research. The results show that the hydraulic cylinder has achieved good synchronous control effect [3]. Li honglong established the synchronous control model of double hydraulic cylinders, and compared and analyzed the control effects of fractional PID and conventional PID control methods on the control model. The results show that the fractional PID control model has stronger robustness [4]. Most scholars mainly study the application of PID control method to hydraulic control circuit, while the control method based on neural network algorithm optimization is rarely used in hydraulic synchronous control circuit.

According to the functional requirements of RamRig hosting system, a controller based on RBF neural network to optimize PID parameters is designed. The three cylinders synchronous hydraulic circuit model and controller model are established in AMESim and Simulink for joint simulation to verify the working performance of the designed controller.

2. RamRig hosting system modeling

2.1. RamRig hosting system structure

RamRig hosting system is a comprehensive hosting and compensation system, which can not only lift the load, but also compensate the heave movement. RamRig hosting system is mainly composed of ram cylinder, accumulator, hosting valve group, control valve and other elements. Ram cylinder is guided by ram guide and connected to RamRig HPU through cylinder manifold, and connected to boost and compensation accumulator through accumulator manifold.

2.2. Hydraulic control principle of RamRig hosting system

The hydraulic system of RamRig hosting system adopts the synchronous control principle. The controller judges and processes the real-time displacement signal of the hydraulic cylinder of the hydraulic system and the specified input signal to generate the displacement difference. The controller quickly responds and outputs the control signal to control the control valve and reduce the displacement deviation between the hydraulic cylinder and the specified position. The structural block diagram of hydraulic system of RamRig hosting system is shown in the figure.

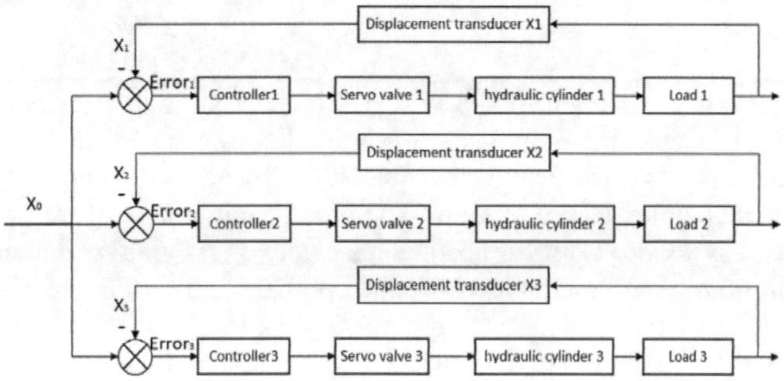

Figure 1 Structural block diagram of hydraulic system of RamRig hosting system

2.3. Mathematical model of three cylinders synchronous circuit

The transfer function of the displacement of the hydraulic cylinder piston to the input flow is:

$$\frac{X(s)}{Q(s)}=\frac{1}{\frac{mV_t}{AB_e}s^3+\left(\frac{K_{ce}}{A}+\frac{V_tB_p}{AB_e}\right)s^2+\left(A+\frac{K_{ce}B_e}{A}\right)s} \quad (1)$$

Where, X is the displacement of piston rod; Q is the input flow; A is the effective area of the piston; B_e is the bulk elastic modulus of oil; B_p is the viscous damping coefficient of oil; K_{ce} is the leakage coefficient; V_t is the area of high pressure oil chamber of hydraulic cylinder;

2.4. AMESim hydraulic system modeling

According to the structural block diagram of the three cylinders synchronous control hydraulic system of RamRig hosting system, a simulation model of the hydraulic system is established in AMESim [6], as shown in Figure 2; The parameter settings of the simulation model are shown in Table 1.

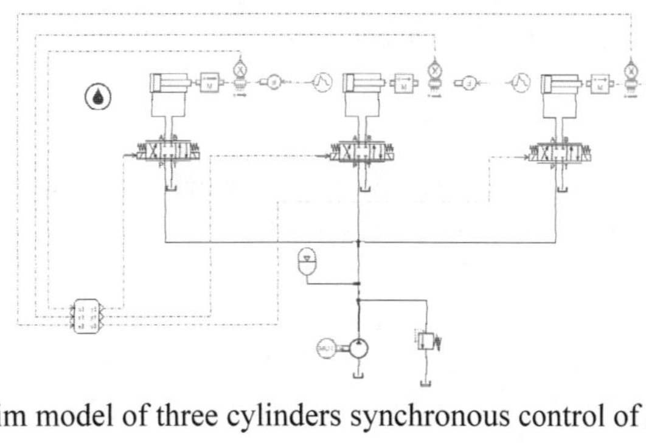

Figure 2 AMESim model of three cylinders synchronous control of RamRig hosting system

Table 1 Simulation parameters of three cylinders synchronous circuit

Element	Parameter
Motor speed/ r•min	1440
Relief valve set pressure /MPa	30
Quality of block /kg	18800
Sensor gain	10
Servo directional valve current /mA	40
Cylinder diameter /mm	500
Stem diameter /mm	360
Hydraulic cylinder stroke /m	18
Load1/N	10000
Load2/N	15000
Load3/N	20000

Set the parameters of the simulation model in AMESim according to the simulation parameters of the three cylinders synchronous circuit, and set the simulation time as 10s and the time step as 0.01s. Because AMESim software is insufficient in calculation speed and control system design, it is impossible to establish a complex control system. Therefore, Simulink software is used to establish the hydraulic circuit control system of RamRig hosting system, and it is jointly simulated with AMESim's three cylinders synchronous circuit model to verify the optimal control of the hydraulic system.

3. PID controller based on RBF neural network

The conventional PID controller has simple structure and good reliability, and is widely used in many fields. In order to meet the functional characteristics of high accuracy and high reliability of RamRig hosting system, the artificial neural network is combined with the conventional PID controller to realize the on-line tuning of PID parameters and improve the control performance of the hydraulic system.

Radial basis function neural network (RBF) algorithm is one of the artificial intelligence algorithms. It is a single-layer feedforward network model, which can approach the optimal solution without restriction. The main implementation process of RBF neural network algorithm is as follows: firstly, the research object is decomposed into several local objects; Secondly, in the hidden layer, each target is optimized according to the radial basis function; Finally, in the output layer, the cumulative sum of each target is calculated according to the connection weight[7-8]. The core of RBF neural network is radial basis function. The radial basis function selected in this paper is Gauss function, which has better smoothness and symmetry. Its formula is:

$$h_j = \exp\left(-\frac{\left\|x-c_j\right\|^2}{2b_j^2}\right), \quad j=1,2,\cdots,m \tag{2}$$

Where: $\left\|x-c_j\right\|$ is Euclidean norm; $c_j=[c_{j1},c_{j2},c_{j3}]^T$ is the center vector of the jth hidden layer neuron node; b_j is the width of the jth hidden layer neuron node; $\omega_j=[\omega_1,\omega_2,\cdots,\omega_m]^T$ is the output layer weight vector.

On the basis of retaining the characteristics of the conventional PID controller, the RBF neural network optimized PID controller can conduct self-learning and online tuning of PID parameters according to the actual operation of the hydraulic cylinder of RamRig hosting system, which improves the response characteristics of the whole system. The control structure is shown in Figure 3.

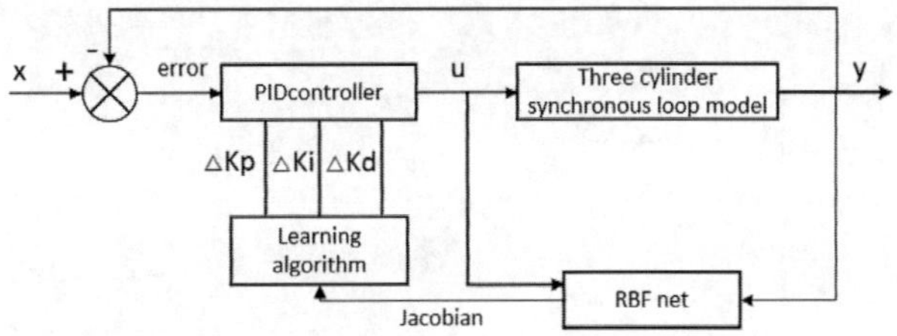

Figure 3 Optimization PID control structure block diagram based on RBF neural network

4. Research on joint simulation of AMESim and Simulink

4.1 Establishment of AMESim and Simulink joint simulation model

In AMESim, an interface for joint simulation with Simulink is established, which is set to three inputs (x1, X2, x3) and three outputs (Y1, Y2, Y3). Run the three cylinders synchronous circuit model in AMESim with CO Simulink, the system will automatically jump to the Matlab interface and display the AMESim model file compiled by VC. In Simulink, the interface file of AMESim model can be called through S-function. In this paper, the three cylinders synchronous loop control model of RamRig hosting system is established through the displacement feedback signal. The controller calculates the output control signal according to the difference between the displacement feedback signal and the input target signal, and adjusts the servo valve to make the operation of the hydraulic cylinder consistent with the expected target [9-10]. The control model of the whole system in Simulink is shown in Figure 4.

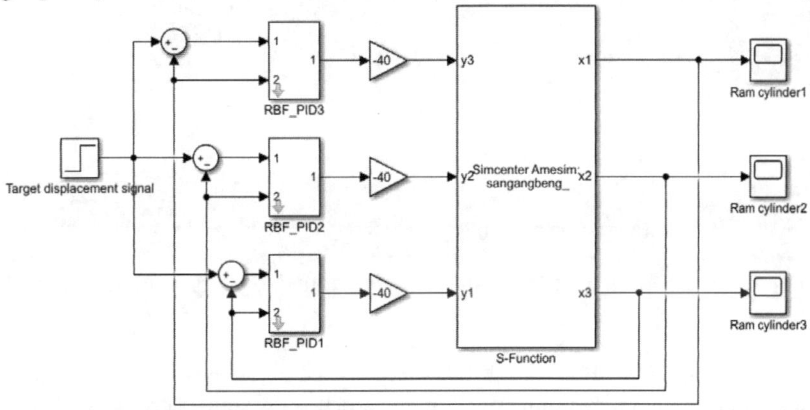

Figure 4. Simulink control model of three-cylinder synchronous loop based on RBF-PID controller

4.2 Joint simulation results of AMESim and Simulink

Adjust the model in AMESim to the simulation mode, run Simulink control model for joint simulation calculation, and the calculated three cylinders synchronous output displacement of the hydraulic system is shown in Figure 5. It can be seen from Figure 5 that there is a small deviation in the synchronous displacement of three cylinders of RamRig hosting system based on RBF neural network optimized PID controller, and the displacement of three hydraulic cylinders is almost completely synchronized.

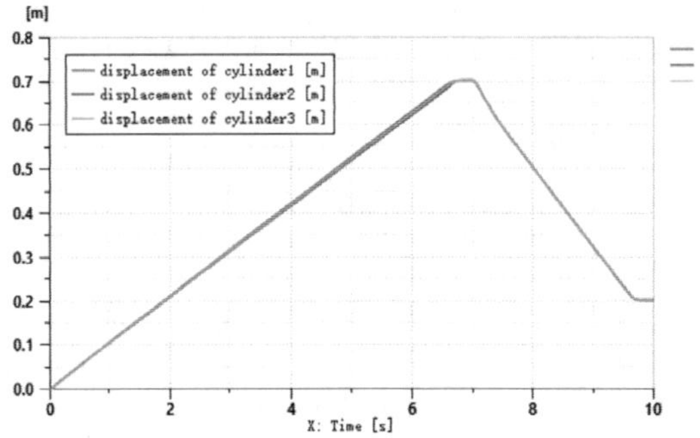

Figure 5 Three cylinders synchronous output displacement under the action of RBF-PID controller

In order to more clearly show the advantages of optimizing the PID controller based on RBF neural network, the RBF-PID controller and the conventional PID controller are respectively placed in the Simulink control model for calculation. The PID initial parameter settings are consistent. The calculation results of the two operations are shown in Figure 6. It can be seen from Figure 6 that the output displacement of hydraulic cylinder under the action of RBF-PID controller is obviously closer to the input target signal, with better accuracy and dynamic response characteristics.

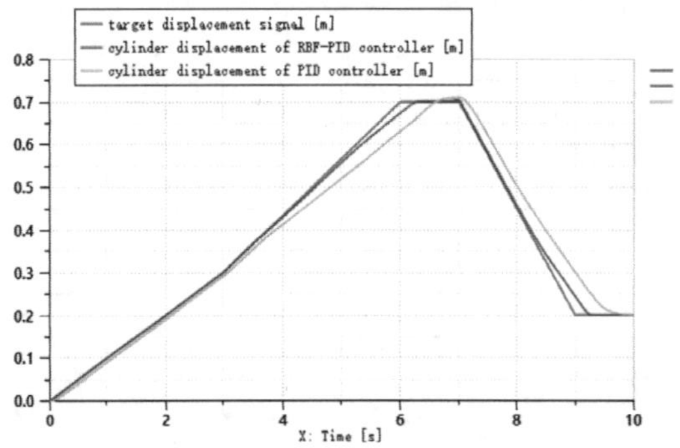

Figure 6 Output displacement of hydraulic cylinder under the action of two controllers

5. Conclusion

In this paper, AMESim and Matlab/Simulink software are used to jointly simulate and calculate the three cylinders synchronous hydraulic circuit of RamRig hosting system. Through the verification and comparison of output displacement signals, it is proved that the accuracy and stability of PID controller based on RBF neural network optimization are fully capable of realizing the synchronous output action of hydraulic cylinders of RamRig hosting system.

Reference

[1] Ma YC. Identification method of Hydraulic Cylinder position Servo System Based on MATLAB-AMESim[J]. Hydraulics pneumatics and seals, 2020,40(02):10-14+19.

[2] Zhou SX, Luo YL, Du L. Simulation design of Fuzzy PID synchronous Control for Driving Hydraulic Cylinder of paddy field Leveling Machine[J]. Hydraulics pneumatics and seals, 2021,41(11):29-32.

[3] Liu SL, Wei CM, Ding D. Simulation research of four-cylinder synchronous circuit based on AMESim/MATLAB[J]. Fluid power transmission and control, 2017(03):34-37.

[4] Li HL, Feng B. Research on Synchronous Control System of Two cylinders four col-umn Hydraulic Press[J]. Fluid power transmission and control, 2019,47(02):82-85+91.

[5] Zhao Y, Zhou QY, Shao NF, Lu RR, Hu XZ. Electro-hydraulic servo system based on RBF neural network fuzzy PID control[J]. Mechanical and electrical engineering, 2022,39(02):244-249.

[6] Liang Q, Su QY. AMESim Computer Simulation Guide for Hydraulic System[M]. China Machine Press: 201408.322.
[7] Ji P, Liu WT, Zhang Y, Wei HF, Zhou XW. Research on PID controller of ship power station based on RBF neural network[J]. Journal of Chongqing university of technology (natural science),2020,34(02):203-209.
[8] Tan ZP. PID parameter optimization based on adaptive weight neural network[J]. Computer CD software and application,2014,17(07):97-98.
[9] Tao L, He Q. Co-simulation of Multi-Functional Tunnel Trolley Hydraulic System Based on AMESim and MATLAB[J]. Machine tool & hydraulics,2015,43(22):108-110.
[10] Fu TT, Zhu YC, Gu Y. Fuzzy PID Control of Electro-hydraulic Servo System Based on MATLAB-AMESim [J]. Machine tool & hydraulics,2016,44(20):144-146+154.

Optimization prediction and experimental verification of cyclone performance based on BP neural network

Mochuan Sun[1], jiangang Wang[1*], fengqin He[2], zhiqiang Long[1]

[1]College of Mechanical Engineering, Shanghai Institute of Technology, Shanghai 201418, China

[2]the college of information, mechanical and electrical engineering, Shanghai normal university, Shanghai, 200234, China

*wjg@sit.edu.cn

Abstract: As a kind of centrifugal separation equipment, hydrocyclone has developed rapidly with the advantages of high economic benefits, good compactness, high processing efficiency, simple process and so on.Hydrocyclone has broad practical significance and application prospects in chemical, light industry, oil mining and refining, environmental protection, medicine, food processing, ship transportation and surface oil spill treatment.BP neural networks are mainly used in the following four aspects: function approximation, pattern recognition, classification, and data compression.In this paper, we introduce the principle of BP neural network, establish an analytical model, apply BP neural network to predict the performance of hydrocyclone, and conduct experimental verification to optimize the hydrocyclone under different working conditions according to the experimental results.

1. Introduction

Hydrocyclones are now widely used, and the model optimization design of hydrocyclones is of great significance for the study of hydrocyclones[1].BP network is the most widely used artificial neural network model, with self-learning ability, theoretically BP neural network can achieve arbitrary precision approximation continuous function[2].Many scholars at home and abroad have used artificial neural networks to predict the performance parameters such as solid-liquid separation efficiency and particle size of hydrocyclones, and summarize the methods and approaches of hydrocyclone prediction modeling[3][4][5].Neural network performance prediction is carried out by experimental research combined with neural network analysis, analyzing the influence of structural parameters, operating parameters and physical parameters on the separation performance of cyclones, and providing the field data required for neural network analysis.A neural network is applied to analyze and predict the performance of a cyclone and verify its prediction accuracy.

2. Selection and analysis of neural networks

The BP neural network has fast computation speed and high network accuracy, while the weights of each connection layer can be dynamically adjusted[6].Because every node is a neuron. There is Y=a*(W*X+b), where a is the activation function, w is the weight, and b is the offset.In the design of BP networks, the most classic design is that the transfer function from the input layer to the implicit layer selects the Sigmoid function, while the transfer function from the implicit layer to the output layer uses a linear function; the learning algorithm uses error backpropagation.Error backpropagation refers to the fact that although in the training of BP neural networks, the weights are modified in reverse along the gradient of the error performance function in BP network training, which is passed backwards along the input layer to the implicit layer and then to the output layer.As the BP neural network continues to learn to correct, the network becomes less and less error-prone, knowing that it will meet the error requirements or reach a set number of trainings[7].

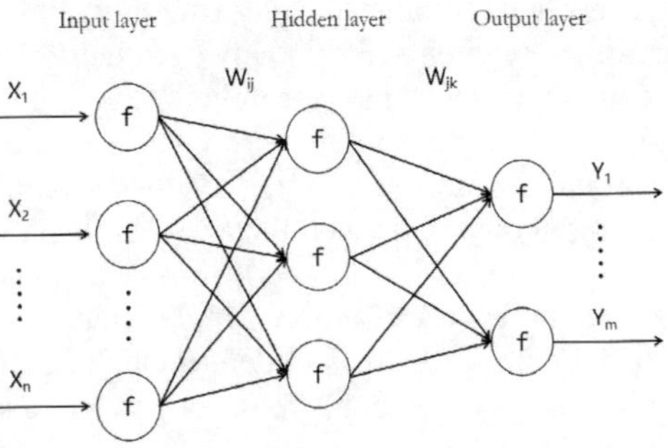

Figure 1. BP neural network structure diagram

The structure and performance of the BP neural network is affected by the number of layers of the implicit layer. The nonlinear mapping capability of a single implicit layer network is weak, and in order to achieve a predetermined mapping relationship, the implied layer nodes are more, so that the tunable parameters of the network are increased.We obtain three implied layers by constructing the method, and then verify the model prediction error one by one, selecting the one with the least error[8].

After determining the number of layers of the implicit layer, it is necessary to determine the number of neurons in the layer, if there are too many neurons, it will take a lot of training time, and it is easy to form an "overfit" phenomenon. if the number is too small, it cannot accurately and completely transmit information.

Empirical formula based on the number of hidden layer nodes:

$$h = \sqrt{m + n} + a \tag{1}$$

where h is the number of hidden layer nodes, m is the number of input layer nodes, n is the number of output layer nodes, and a is the adjustment constant of 1-10.

In the forward transfer process, the weight between node i and node j is set to w_{ij}, the threshold of node j is b_j, and the output value of each node is x_j, and the output value of each node is based on the output value of all nodes in the upper layer, the weight of the current node and all nodes in the upper layer, and the threshold of the current node, as well as the activation function. The specific calculation method is as follows:

$$S_j = \sum_{i=0}^{m-1} w_{ij}\, x_j + b_j \tag{2}$$

$$x_j = f\ (S_j) \tag{3}$$

Where f is the activation function, the S-type function or linear function is generally selected.

The reverse passing process, based on The Widrow-Hoff learning rules. Assuming that all the results of the output layer are d_j, the error function is as follows:

$$E(w, b) = \frac{1}{2}\sum_{j=0}^{n-1}(d_i - y_i)^2 \tag{4}$$

3. Cyclone performance test

After determining the structural parameters of the neural network, the cyclone performance experiment is performed to obtain the data required for neural network training. The structure of the swirl separator for oil-water separation is diverse, divided by the state of motion, mainly divided into two types: dynamic swirl separator and static swirl separator.The traditional oil-water separation cyclone is mainly a double cone cyclone separator and a single cone cyclone separator in a static cyclone separator. A double cone cyclone separator is used in this article, which is mainly divided into the following parts: overflow port, swirl chamber, inlet, large cone segment, small cone segment, and tailpipe.

Through a large number of comparative experimental studies, the double cone section in the swirl separator is more efficient than the oil-water separation efficiency of the single cone section.In recent years, a large number of researchers have conducted a large number of experimental studies on the angle and length of the large and small cone segments[9][10][11].However, there is less research on the connection method at the junction of the cone segment. Some researchers have used the connection part of the two cone segments of the curve transition and done a separation performance test, and found that compared with the ordinary cyclone, the separation accuracy of the cyclone separator is a little higher.Through a large number of literature data, it is shown that when the angle of the large cone section of the cyclone separator is 20°, the cyclone separation effect is good; the small cone section is between 1.5°and 6°Separation Efficiency is good. In this experiment, three angles were selected for experimental analysis, namely the small cone angle of 1.5°, 3°, 4.5°, so as to determine the best cone angle of the small cone section.Compared with the previous hydrocyclone, we use a new cone section and cone angle design method in order to obtain better cyclone performance effect.

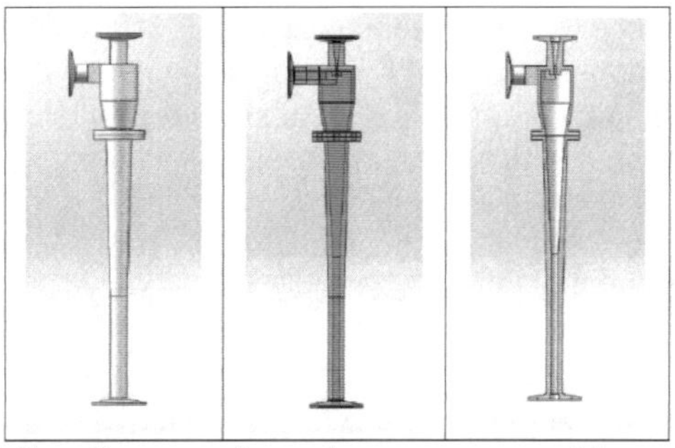

Figure 2. Modeling diagram of a cyclone separator with a cone angle of 3°

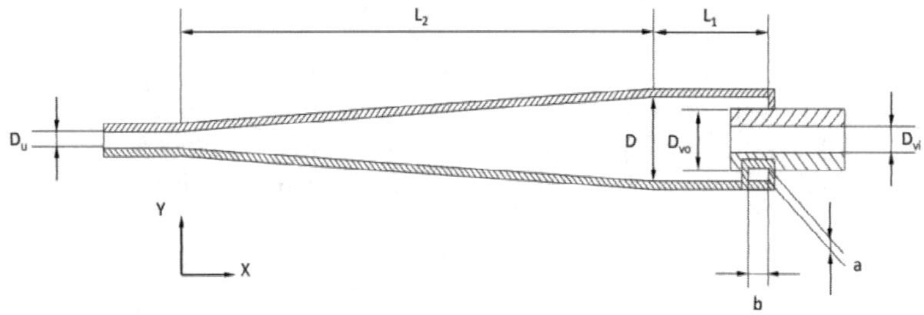

Figure 3. Dimensional drawing of the cyclone structure

We use our own hydrocyclones and 3D printing for performance tests, which are rare before and are both advanced and innovative.The nominal diameter of the experimental hydrocyclone is 25mm, and the remaining dimensions are determined according to the proportions in the following table. The separation performance of the degreasing cyclone is mainly the separation efficiency and the pressure drop of the fluid through the degreasing cyclone. Since the experiment uses a cyclone of the same specification in the same flow, the pressure drop of the sewage through the cyclone is basically the same can be ignored.Therefore, only the separation efficiency of the cyclone is expressed. The formula is calculated as follows:

$$\varepsilon = \frac{Q_{\text{yov}}}{Q_{yin}} = 1 - \frac{Q_{\text{yun}}}{Q_{yin}} = 1 - \frac{C_{\text{un}} Q_{un}}{C_{in} Q_{in}} \tag{5}$$

Table 1. Cyclone structure size ratio

D/mm	D / D_c	D_i / D_c	D_u / D_c	D_o / D_c	l_1 / D_c	l_2 / D_c
35	2	0.35	0.5	0.14	2	20

After the 3D modeling of the hydrocyclone, the 3D printing is carried out to generate the cyclone entity for performance testing. Using the method of controlling variables, the angle of the cone section of the cyclone is changed, and the relevant data of the cyclone performance are obtained and analyzed. Plot the operation parameters and performance parameters of the cyclone, and observe the variation characteristics of the cyclone performance parameters.

Table 2. Main performance parameters of cyclone

Design pressure/ MPa	Working pressure/ MPa	Operating temperature/ MPa	Pressure loss/ MPa
1.0	0.5	10~40	0.1~0.2

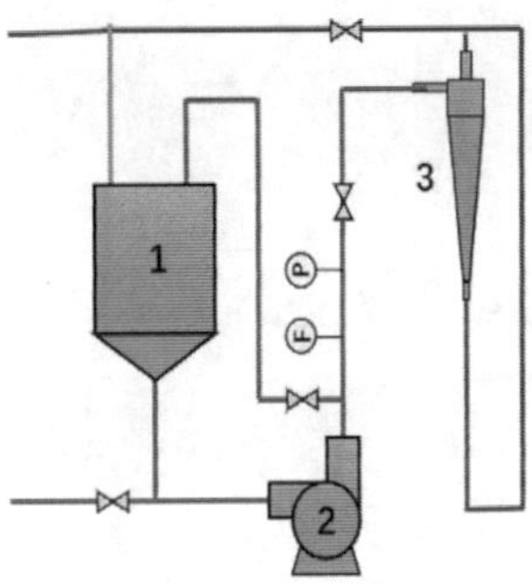

Figure 4. Cyclone performance test flowchart

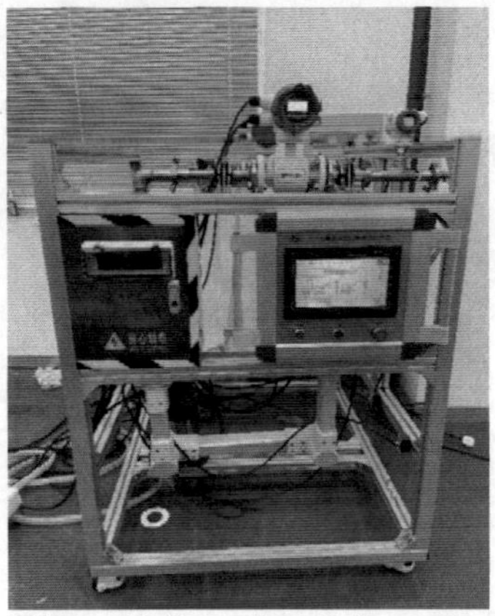

Figure 5. Experimental prototype site photo

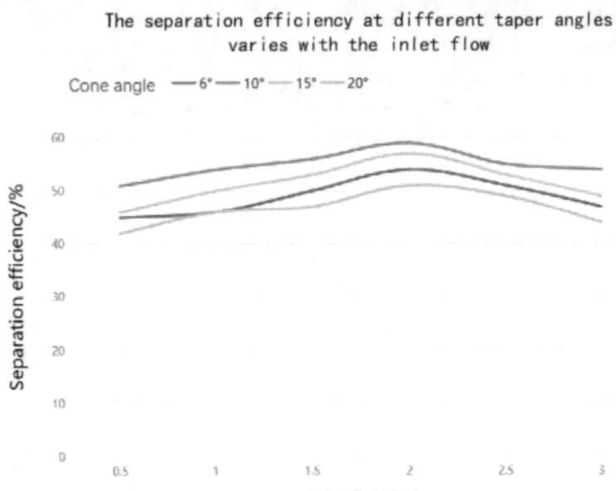

Figure 6. The separation efficiency at different taper angles varies with the inlet flow

4. Cyclone performance test prediction

According to the experimental results of the cyclone performance experiment, the performance parameters of the cyclone are designed under different working conditions[12][13]. In this experiment, we use all the structural and operating parameters of a cyclone as inputs to predict the separation efficiency and processing power under the same conditions. We use 50 sets of data under different working conditions as the training set, train the BP network model, and 10 sets of data as the test set to verify the accuracy of the model.Set the transfer function of the implicit layer to tansig, set the transfer function of the output layer to purelin, and choose trainscg for the training function of the network. At the same time, set the necessary network parameters, the number of training iteration steps epoch=5000, the learning rate lr=0.2, the learning target goal=0.001 After the neural network model is built, the training data is input to the network, and the training process is terminated when the stop condition is met.

Compare the simulation data of each cyclone model, establish a synergistic relationship model between structural parameters and separation performance, and select the best cyclone with the best comprehensive effect according to the optimal structural parameters under different needs (small cutting particle size, high separation efficiency, and low energy consumption). Repeat cyclone performance experiments to verify the simulation results.The cyclone with the best comprehensive effect optimized by simulation is processed into an experimental device, and the experimental study of cyclone separation performance is carried out, and the synergistic relationship model between the structural parameters and separation performance obtained by the simulation is verified, as well as the cyclone performance curve.

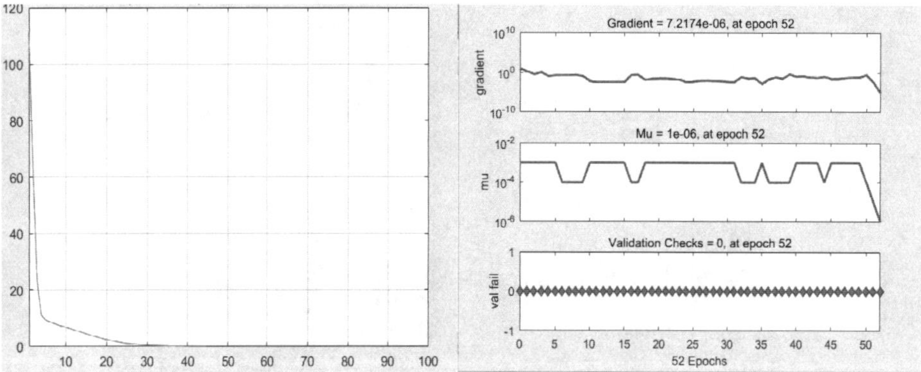

Figure 7. Prediction error analysis plot

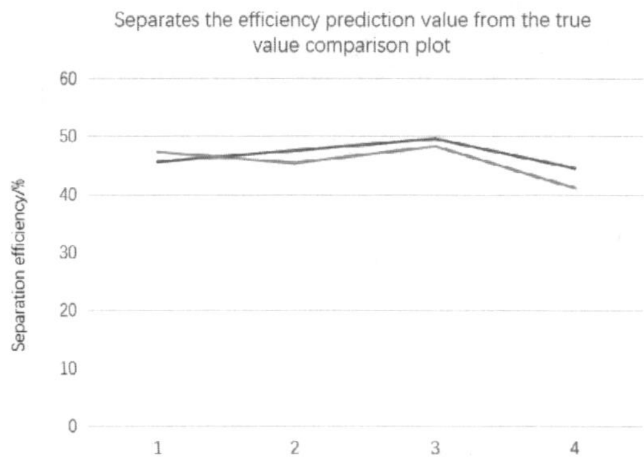

Figure 8. Separation efficiency prediction value compared to true value

5. conclusion

In this paper, through the 3D printing performance experiment of hydrocyclone, the results of the performance prediction of the BP neural network on the cyclone are verified, and the cyclone performance test device is constructed, focusing on the influence of cone angle on the separation efficiency of the hydrocyclone. The results show that the actual experimental data of the cyclone is not much different from the predicted value, which is in line with the experimental needs.

Acknowledgments

This work was sponsored by Natural Science Foundation of Shanghai (Grant No.19ZR1437400).

References

[1] Supachart P,Swasdisevi T,Wongsarivej P,Development of separation sharpness model for hydrocyclone[J].Chinese Journal of Chemical Engineering,2020,(8): 785-792.

[2] Kuang Y,System modeling and MATLAB implementation based on artificial neural networks[J]. Journal of Sichuan University of Science and Technology (Natural Science Edition). 2007(05)

[3] He F Q, Wang H L, Tan Y H.Predictive modeling of the performance of the hydrocyclone with different cone combination[C]Applied Mechanics and Materials. Trans Tech Publications Ltd, 2012, 190: 147-150.

[4] He F Q, Zhou P, Wang J G. Modeling on Hydrocyclone Separation Performance by Neural Network[C]Applied Mechanics and Materials. Trans Tech Publications Ltd, 2012, 105: 185-188.

[5] Zhang J, Xu W, Gai W, Research on Model Simulation of Hydrocyclone Based on BP Neural Network[J]. Coal Mine Machinery, 2008.

[6] Dong Y,Li X R.BP neural network is implemented in MATLAB[J] International Core Journal of Engineering Volume 5, Issue 8. 2019.

[7] Shuo D,Xiao H C, Qing H W.Approximation Performance of BP Neural Networks Improved by Heuristic Approach[[J].Applied Mechanics and MaterialsVolume 2700, Issue 411-414. 2013. PP 1952

[8] Liu M L. Oil-water pre-separation with a novel axial hydrocyclone[J]. Chinese Journal of Chemical Engineering.2018, 26(1): 60-66.

[9] Kharoua N.Hydrocyclones for De-oiling Applications-A Review[J].Petroleum Science and Technology,2010,28(7)738-755.

[10] F J Souza,L G M Vieira.Analysis of the influence of the filtering medium on the behaviour of the filtering hydrocyclone[J].Powder Technology,2000,107:259-267.

[11] Zhao L X,Li F,Wang Z C,Jiang M H.Study on the Flow Field Characteristics and Separation Characteristics of Liquid-Liquid Hydrocyclone (II)——The Effect of Cone Angle Change on Axial Velocity Field[J].Chemical Equipment Technology, 1999(05):1-3.

[12] Fung C C,Wong K W,Eren H.Developing a generalised neural-fuzzy hydrocyclone model for particle separation[C]. Instrumentation and Measurement Technology Conference.1998. IMTC/98.Conference Proceedings.IEEE.IEEE,1998,1:334-337.

[13] Eren H,Fung C C,Wong K W.Artificial neural networks in estimation of hydrocyclone parameter d50 c with unusual input variables[J].Instrumentation and Measurement,IEEE Transactions on,1997,46(4):908-912.

Journal of Physics: Conference Series **2365** (2022) 012045 doi:10.1088/1742-6596/2365/1/012045

Paper Defect Detection Algorithm Based on the Mathematical Morphology and Computer Vision

Zunhai Gao*, Yuzhan Qiu, Jicheng Zhou

School of Mathematics and Computer Science, Wuhan Polytechnic University, Wuhan, China

*Corresponding author: haigao007@whpu.edu.cn

Abstract. A method for detecting paper's surface defects is proposed, which is based on mathematical morphology algorithm and computer vision technique. Firstly, the mean filtering and wavelet denoising in image pre-processing are used to eliminate the image noise, and wavelet analysis can achieve image enhancement and increase the contrast between the target area and the background. And then, an improved threshold Canny operator is used to segment the image to extract the complete scratch line and other defect edges. Finally, the target region features are coloured by mathematical morphology processing. The method's stimulating result is verified in the Matlab software, from which we can conclude that the method's detecting effect is very effective.

1. Introduction

Due to the wear and tear of paper machine equipment, human operation errors and the quality problems of paper raw materials, some paper defects may occur in the production process, such as scratches, folds, holes, bright spots and so on. If these paper defects cannot be detected, then the paper products will bring different degrees of consequences in the subsequent use and processing production. The traditional human eye detection cannot adapt to the high-speed production line, and it is easy to miss detection, causing losses to enterprises. So the paper defect detection by machine vision technology provides a good scheme for paper quality detection.

A lot of researches have been done on the intelligent detection of paper defects. The Prewitt operator is used to detect holes, dirty spots and wrinkles in [1]. However, due to the diversity of paper defects, the recognition rate is insufficient. A method combining edge detection and wavelet fusion is proposed in [2]. When the contrast between the background and the target paper defects is low, it has accurate positioning and certain anti noise performance. However, the algorithm itself is too complex to meet the timeliness of detection. The geometric and gray features of paper defects are extracted and classified in [3], but only the differences of different features are described, and no specific classification method is given.

Defect detection of various objects has always been a research hotspot. Recently, researches based on machine learning and neural network methods have also begun to report [4-7]. However, there are not many articles on paper defect detection. Some methods based on different optimization algorithms [8-12] are proposed to detect paper surface defects, and there are other methods combined with mathematical morphology to deal with the problem of paper defects [13-15]. These methods have a certain effect on the detection of conventional paper defects, but the extraction and identification of subtle paper defects in paper still need to be improved.

This paper uses the combination of mathematical morphology algorithm and machine vision, and Matlab tools can also be used to realize image denoising, enhancement, segmentation and morphological processing in image processing algorithm, and analyzes and compares these algorithms after implementation. After selection and improvement, the detection method suitable for the detection of scratches, stains and other surface defects is designed.

2. Image pre-processing

The collected paper image mainly contains defects and Gaussian noise. In the Matlab platform, the colour image is transformed into gray image. After pre-processing such as filtering, wavelet transform is performed. Image denoising based on wavelet transform benefits from the following characteristics: low entropy, multi-resolution characteristics, decorrelation, and flexible choice of basis functions[16].

We apply wavelet threshold to image denoising. The basic principle is to generate a threshold through the noise intensity. Since the value of wavelet coefficient of signal must be greater than that of energy dispersive noise, we can use the threshold to remove some noise related components, so as to improve the signal-to-noise ratio and achieve the purpose of denoising. The realization method is to carry out wavelet transform on the noisy image, then select a threshold to choose or reject the wavelet coefficients, and finally reconstruct the original image through the inverse wavelet transform.

Similar to wavelet de-noising, the image can also be enhanced by wavelet transform, and then the enhanced signal can be transformed by inverse wavelet transform to obtain the enhanced image.

When Otsu selects the segmentation threshold, it is based on the image that the between-class variance is the largest or the intra-class variance is the smallest. Therefore, the selected threshold can increase the variance between the data of the two types of pixels, which indicates that the segmentation threshold can successfully distinguish the two types. Similarly, it is also hoped that the difference between the data in the class is as small as possible [17].

As for edge detection algorithm, we make use of Roberts operator, Sobel operator and Canny operator. These three algorithms are compared in experiments.

(a) Roberts operator

The calculation method of Roberts operator uses the principle of deviation points, also called the local difference method to find the edge [15] is given by the following formula

$$g(x,y) = \left[\sqrt{f(x,y)} - \sqrt{f(x+1,y)}\right]^2 + \left[\sqrt{f(x,y+1)} + \sqrt{f(x+1,y)}\right]^2. \quad (1)$$

(b) Sobel operator

Sobel edge operator is a kind of operator based on first-order differentiation, which is a processing method that firstly weights the average and then differentiates [18]. The Sobel operator is as follows

$$Sobel(y)=\begin{bmatrix}1 & 0 & -1\\ 2 & 0 & -2\\ 1 & 0 & -1\end{bmatrix},\ Sobel(x)=\begin{bmatrix}1 & 2 & 1\\ 0 & 0 & 0\\ -1 & -2 & -1\end{bmatrix}. \quad (2)$$

The definition of Gx and Gy is as follows

$$Gx=\begin{bmatrix}1 & 0 & -1\\ 2 & 0 & -2\\ 1 & 0 & -1\end{bmatrix}*A,\ Gy=\begin{bmatrix}1 & 2 & 1\\ 0 & 0 & 0\\ -1 & -2 & -1\end{bmatrix}*A. \quad (3)$$

Therefore, the pixel value in the original image can be obtained as

$$G=\sqrt{Gx^2+Gy^2}. \quad (4)$$

It can be abbreviated as

$$|G| = |Gx| + |Gy|. \quad (5)$$

After a new pixel value is obtained, a threshold value can be given, and the pixel can be divided into two parts and the pixel value smaller than the threshold value is converted into 0, and the pixel value larger than the threshold value is converted into 1.

(c) Canny operator

Step 1. Gaussian smoothing filter

$$H(x,y,\delta)=\frac{1}{2\pi\delta^2}e^{-\frac{x^2+y^2}{2\delta^2}}, \tag{6}$$

smoothing is achieved by Gaussian filter and image convolution, which can reduce the error rate and obtain

$$G(x,y)=H(x,y,\delta)*f(x,y). \tag{7}$$

Step 2. Reduce image noise.

Step 3. Perform non-maximum suppression on the gradient amplitude to eliminate edge false detection.

Step 4. Use the double threshold gate algorithm to achieve edge extraction.

Since the automatic selection of the threshold will produce some false edges, an algorithm for finding a suitable threshold can be designed. Process introduction: first use the sobel operator to find the partial derivatives in two directions, and then use

$$G=\sqrt{Gx^2+Gy^2} \tag{8}$$

to obtain the gradient amplitude, and set *T0* as the average gradient Amplitude, as the low threshold, then set *T1* as the high threshold, using the formula

$$T_H=\frac{E_H}{E_L}T_L \tag{9}$$

to obtain, whereE_H and $E_L$represent the mathematical expectation of high gradient amplitude and low gradient amplitude respectively [19].

Mathematical morphology algorithms include expansion, erosion, opening and closing. The two basic operations are expansion and erosion. They are different in binary and gray images and have their own characteristics. Morphological image processing is based on processing images through a set of structural elements, and the shape can be rectangle, circle, etc. The corrosion operation is equivalent to removing the connected pixels at the image boundary, and the expansion is just the reverse process of each other [20].

(1) Structure Element

The structure element is treated as a matrix, which acts as a filter for the image, and the processed image of the structure element is affected by its shape and size.

(2) Dilation operation

A uses B to expand and write $A\oplus B$, which is defined as

$$A\oplus B=\{x|(\hat{B})_x\cap A\neq\emptyset\}. \tag{10}$$

Where $\oplus$ is the expansion operator, and $(\hat{B})_x$ is the structural element.The expansion operation is equivalent to translating the structural element on the image set A and expanding the part of the structural element shape within the boundary of the intersection point. '*x*' represents the points included in the expanded set, and the expansion effect depends on the structure element.

(3) Erosion operation

A uses B to corrode, written as $A\ominus B$, which is defined as

$$A\ominus B=\{x|(B)_x\subseteq A\}. \tag{11}$$

The symbol "$\ominus$" means that set *A* is corroded by structural element *B* the erosion operation. Erosion operation is equivalent to taking *x* (*x* is the point in set *A*) as the centre of the structural element, removing those points whose structural elements are not completely contained by set A, and the remaining point x is the set composed of the operation results.

(4) Open Operation and Close Operation

A uses B to perform the turn-on operation, denoted as $A \circ B$, define as

$$A \circ B = (A \ominus B) \oplus B. \tag{12}$$

A is closed by B, denoted as A·B, which is defined as

$$A \cdot B = (A \oplus B) \ominus B. \tag{13}$$

Among of them, "∘ " is the open arithmetic operator which first erodes and then swells, " · " is a close operator which first swells and then erodes.

In a binary image, a set of pixels connected together according to a certain connection rule is regarded as a group, and each group contains a number of zero-value pixels and one-value pixels. These groups are the Unicom components. Common ones are 4 Unicom and 8 Unicom.

In matlab, the "bwlabel" function and "bwlabeln" function are used to identify binary images. The "bwlabel" function supports two-dimensional input, and the "bwlabeln" function supports input of any dimension. The output of the "bwlabel" function is a double type matrix, so the output matrix can be displayed with the index map "label2rgb" function. When displaying, add 1 to each element, and each pixel value is within the effective range of the index colour map. In this case, each target object presents a different colour, which is easy to distinguish.

3. Experiment comparison and analysis

The image collected in the experiment is composed of Gaussian noise and stains, and the line of the black signature pen. Figure 1 shows the original image and the gray image after conversion. Figure 2 shows the different mean filter with different values of window templates. In Figure 2 (a) and (b), it can be seen that the original image interfered by noise (here, Gaussian noise is added). After the mean filtering process, the improvement is obvious, and the larger the value of n, the better the denoising effect, but it is also clear that the shortcomings are the image becomes more and more blurred, and some details are also lost. Figure 3 compare the differences between the mean filter and the median filter in the same window template. It can be seen from the comparison chart in Figure 3 that under the pollution of Gaussian noise, compared with mean filtering, median filtering can enhance the noise, but cannot completely eliminate the noise.

Figure 4 shows the comparison between the mean filtering and threshold denoising of the n=3 template. The denoising effect of Figure 4(a) is good, but the details are missing. Figure 4(b) can highlight the depth and other details of the horizontal lines in the middle, but the denoising effect is not as good as the mean filtering. Therefore, the combination of mean filtering and wavelet denoising is used to image denoising. After image denoising, if the visual effect of the obtained image information is not obvious, image enhancement can be taken to improve the contrast between the target and the background, so that it conforms to the observation of human eyes.

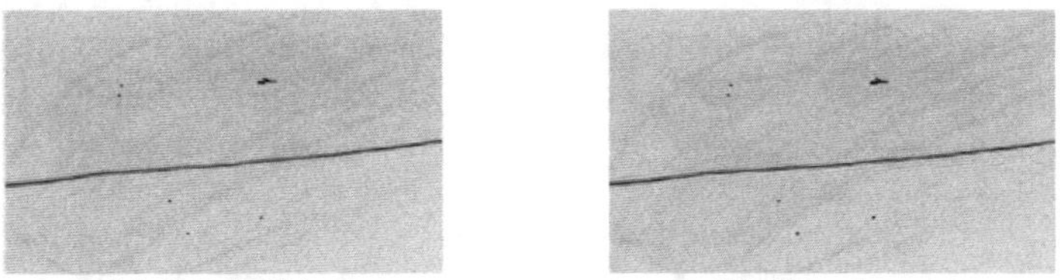

(a) Original image (b) Gray image

Figure 1. Original image and gray image

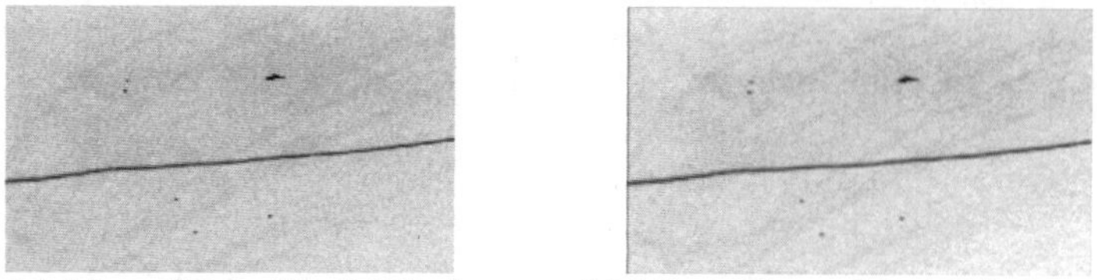

(a) Mean filter of 3×3 Window (b) Mean filter of 6×6 Window

Figure 2. Mean filter for different Windows

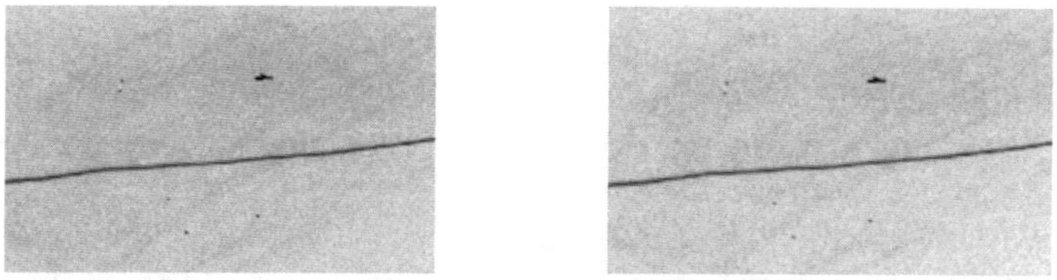

(a) Median filter of 3×3 Window (b) Median filter of 6×6 Window

Figure 3. Median filter for different Windows

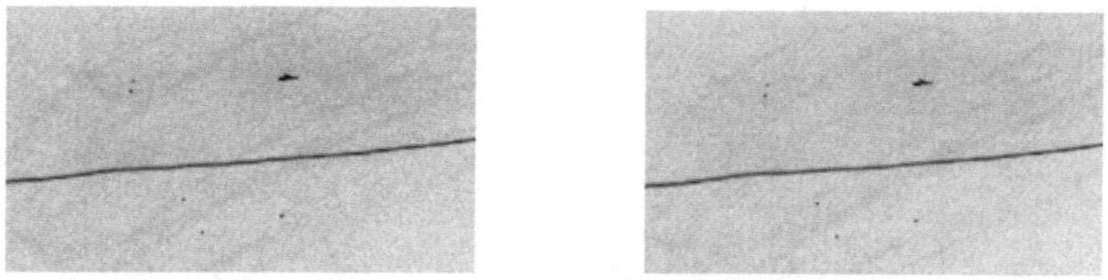

(a) Mean filter image (b) Threshold denoising image

Figure 4. Mean filter and threshold denoising

After the pre-processing, we need to segment the defective part of the defective image. According to the algorithm principle of threshold segmentation, the effect of Otsu segmentation is very different in the case of single background and complex background. According to the experiment, the two segmentation images in Figure 5 can be obtained. The segmentation effect in the case of multiple backgrounds is not as good as that in the case of single background. The wrong selection of the target region leads to poor segmentation results. Therefore, Otsu segmentation method is suitable for single background in this experiment. For image segmentation by iterative method, the experimental results are shown in Figure 6. Whether single background or multi background has little impact on the segmentation effect, provided that the ε select appropriate value.

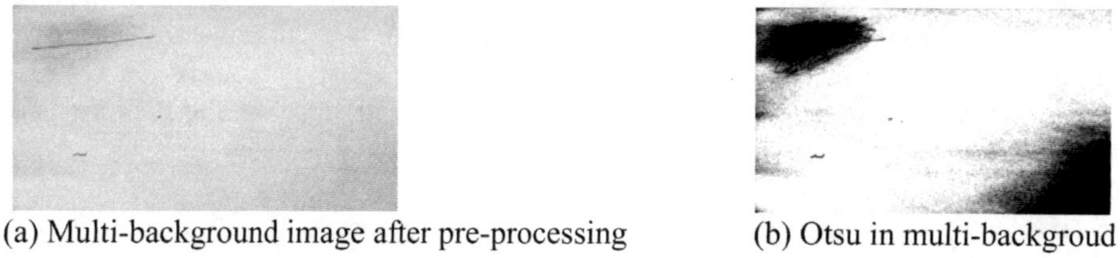

(a) Multi-background image after pre-processing (b) Otsu in multi-backgroud

(c) Single background image after pre-processing (d) Otsu in single background

Figure 5. Otsu in single background and multi-backgroud

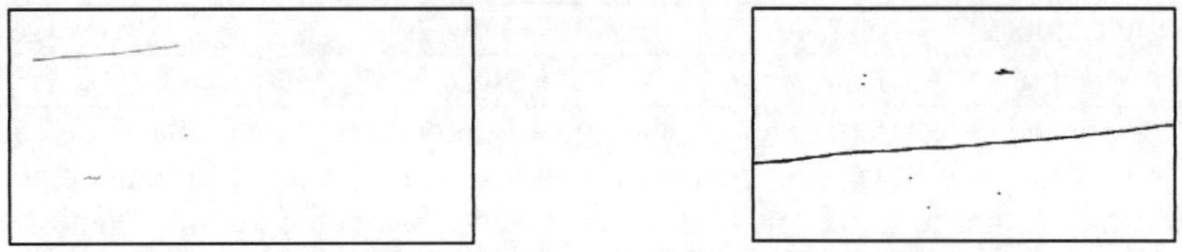

(a) Iterative method in **Multi-background** (b) Iterative method **in Single background**

Figure 6. Iterative method in single background and multi-background

Another method of image segmentation is edge detection. This paper selects three edge detection operators, namely Roberts operator, Sobel operator and Canny operator. In the case of automatic threshold selection, as shown in Figure 7, it can be seen that the continuity of edge detection of Roberts and Sobel operators is general, the accuracy of edge positioning is not high, and the segmentation effect is general. Canny operator is the operator with the best edge continuity and the most accurate location. Fig. 8 shows the results of edge extraction of Canny operators with different thresholds, in which the selection interval of automatic threshold is [0.0313, 0.0781]. Comparing (a), (b) and (c), it can be seen that the segmentation effect of automatic threshold is not as good as that of manually setting the threshold. It is obvious that false edges are generated. After the threshold is increased to a certain extent, most of the false edges disappear, and a good segmentation effect can be obtained. Therefore, the premise of using Canny operator for image segmentation is to use the algorithm to find the appropriate threshold. Figure 9 shows Canny edge detection on the selection of threshold. The result shows that the visual effect is better.

(a) Pre-processed image (b) Roberts operator (c) Sobel operator

Figure 7. Roberts operator and Sobel operator

(a) Canny operator 0.09 (b) Canny operator 0.15 (d) Canny operator automatic

Figure 8. Different thresholds' Canny operators

Figure 9. Improved threshold Canny operator detection in Single background

Table 1. Compared for edge detection algorithm with Roberts, Sobel and Canny operator.

	Continuity	Edge positioning accuracy	Segmentation effect
Roberts operator	Average	Low	Average

Sobel operator	Average	Low	Average
Canny operator	Better	High	Better

Finally, the extracted defect edges are morphologically processed. Firstly, the false edges generated by image pre-processing and edge segmentation are removed, and then the structure elements are used to expand the extracted image to make the edges continuous. Then the "infill" function is used to fill the holes in the closed area of the image. Finally, the image is corroded to select the appropriate structure elements to improve the vision of the defective area. Figure 10 is an image after morphological processing. Different connected domains are coloured and labelled with different colours. Therefore, it is necessary to output the matrix of the index map using the "label2rgb" function, and then add 1 to each matrix element. Each pixel value is within the effective range of the index colour map. In this way, the colours of each target object are different and can be easily distinguished. Figure 11 shows the paper defect image marked and coloured.

Figure 10. Morphological processing

Figure 11. Marked and coloured paper defect image

4. Conclusions

The digital image processing method is used to de-noise, enhance and segment the image of defective paper. And then, the mathematical morphology algorithm is used to realize the colour marking of the target region. And combined with the characteristics of the collected image surface, appropriate algorithm improvement is carried out. Finally, a paper surface defect detection algorithm suitable for the collected image is summarized. Through the experiment, the conclusion that common paper surface defects can be detected is obtained.

References

[1] Zhang X and Li J 2012 The Application of Edge Detection in Paper Defect Image. *Paper Science & Technology,* 31(6):121-125.
[2] Chen J and Wang Y 2013 Identification Algorithm of Low Contrast Paper Defects Based on Machine Vision. *Transactions of China Pulp and Paper,* 28(2):29-33.
[3] Yang B, Zhou Q and Zhang G 2011 Study on Algorithm of Paper Defect Detection Based on Geometric and Gray Feature. *China Pulp & Paper*, 30(9):50-53.
[4] Du W, Shen H and Fu J 2019 Approaches for improvement of the X-ray image defect detection of automobile casting aluminum parts based on deep learning. *NDT & E International*, 107,102144.
[5] Deng L, Li J and Han Z 2021 Online defect detection and automatic grading of carrots using computer vision combined with deep learning methods. *LWT*, 149,111832.
[6] Zhang T, Zhang C and Wang Y 2021 A vision-based fusion method for defect detection of milling cutter spiral cutting edge. *Measurement*, 177, 109248.
[7] Pan L, Rogulin R and Kondrashev S 2021 Artificial neural network for defect detection in CT images of wood. *Computers and Electronics in Agriculture*, 187, 106312.
[8] Gai S, Xu X and Xiong B 2020 Paper currency defect detection algorithm using quaternion uniform strength. *Neural Comput & Applic*, 32:12999–13016.
[9] Zhao X, He L and Yao B 2018 Algorithm of Paper Defect Detection Based on Connected-component Labeling. *Transactions of China Pulp and Paper*, 33(2):51-56.

[10] Wang Z, Zhou Q and Li M 2017 Optimal Control of Light Source in Paper Defect Detection System Based on Bacterial Colony Algorithm. *Transactions of China Pulp and Paper*, 32(4): 47-52.

[11] Wang X, Li Q and Li J 2011 An adaptive threshold segmentation method based on BP neural network for paper defect detection. *IEEE International Conference on Software Engineering & Service Science. IEEE.*

[12] Zhou Q, Chen Y and Shen T 2016 Review on the Development of Paper Defect Detection System Based on Machine Vision Technology. *China Pulp & Paper*, 35(05):72-79.

[13] Kang J and Gang Y 2010 The mathematical morphology of paper's defect detection method based on multi-scale and multi-structure elements morphology. *Proceedings of Spie the International Society for Optical Engineering*, 7820.

[14] Kang J, Pan S and Wang X 2017 Paper defect detection based on CB morphology and gray-code decomposition . *Computer Engineering and Applications*, 53(17):186-191.

[15] Zhang C and Tang W 2014 Paper Defect Detection Based on PSO and Mathematical Morphology. *Paper Science & Technology*, 33(02) :60-64.

[16] Alaa A H and Heba A E 2013 An Efficient Approach to Wavelet Image Denoising. *International Journal of Computer Science and Information Technology*, 5(3): 119-127.

[17] Yan H, Ma J and Feng Z 2021 Image Segmentation of Pitaya Disease Based on Genetic Algorithm and Otsu Algorithm. Journal of Physics: Conference Series, 1955(1).

[18] Han L, Tian Y and Qi Q 2020 Research on edge detection algorithm based on improved sobel operator. *MATEC Web of Conferences*, 309.

[19] Xu H, Xu X and Zuo Y 2019 Applying morphology to improve Canny operator's image segmentation method. *The Journal of Engineering*, 2019(23):8816-8819.

[20] Burgeth B, Kleefeld A and Naegel B 2019 Mathematical Morphology and Its Applications to Signal and Image Processing. *Springer Cham*.

Design of Intelligent Car Control Based on EEG Signal

Lili Tang*

College of Computer and Information Engineering, Zhixing College of Hubei University, Wuhan, 430011, P.R.China

*Corresponding author's e-mail: toney2001@126.com

Abstract. Based on the TGAM (ThinkGear Asic Module) EEG sensor, this paper mainly carried out the research and design of intelligent car control method. By collecting brainwave and analyzing TGAM output data, the required concentration parameters can be obtained. STM32F1 series MCU is used as the main controller to analyze data and control the car. The bluetooth module based on Bluetooth 4.0 technology is used to realize data communication. The power system includes L298N motor drive module, two independent DC motor and universal wheel. The software part mainly uses C language for development, the core part of the program is ThinkGear communication protocol analysis. Finally, the design realizes the control of the vehicle motion state through different values of concentration, which lays a foundation for further popularization and application.

1. Introduction

Brain wave is a representative type of bioelectric phenomenon, and the existence of brain wave is also the basic feature of life activities [1]. When EEG and electroencephalogram are used to measure brain waves, people can use different types of brain waves to measure brain signals. The direct communication channel formed between the brain and the external world is called "brain computer interface (BCI). In other words, it is to directly and specifically explain the small signals of human brain activities, understand human thoughts and infer human intentions, and then convert the purpose into corresponding sentence commands to control external devices [2]. EEG signal is the main brain computer interface acquisition signal at present. From the first EEG recorded by human beings in 1929 to now, the brain wave interaction technology has developed rapidly. In recent years, people have made a lot of progress through the research of EEG signals. Many laboratories in the world have realized the real brain computer interface. Many domestic universities have also set up research groups in related fields, and people's understanding of the brain is becoming more and more in-depth. At present, there are about three directions in the research of brain wave acquisition, including brain magnetism, EEG or directly inserting electrode chips into the brain. According to the wide application prospect, the non-invasive method must be the first choice. It is not an invasive brain computer interface. This scheme only needs a forehead stem electrode and an ear reference electrode to calculate the weak current signal on the scalp surface, and uses the stem electrode sensor to collect data. By closely contacting the forehead skin with the sensor electrode, the change signal of brain wave can be analysed and obtained from different neural activities, Easy to use, simple and convenient [3].

According to the understanding of EEG signal, the intelligent car system based on EEG control is realized. The NeuroSky TGAM brain wave sensor can monitor the brain wave of the forehead in real time and obtain the original data signal. The TGAT chip included in the TGAM module can process the collected original data technically. The collected data is wirelessly transmitted to the STM32 MCU processor through Bluetooth. The core controller STM32 MCU extracts and analyse the data packet, and finally realizes the effective control of the smart car. This design has certain entertainment and innovation. If it is further developed to improve the stability and accuracy of the system, it will have broad application prospects.

2. Overall design scheme

This design is an intelligent system integrating the collection of EEG signals, the analysis and processing of EEG signals and the control of EEG signals on the car. The control system is simple in structure and easy to operate. It can realize the left turn, right turn, forward and stop of the car through the change of human brain attention. The system uses the TGAM brain wave sensor launched by NeuroSky to collect and receive the brain wave signal data, analyse the collected data through the internal TGAT chip, and then transmit it to the core control board STM32 single chip microcomputer by Bluetooth. According to the comparison of performance parameters of each processor, stm32f1 Series MCU with good scalability and high cost performance is selected. Stm32f1 Series MCU is a microcontroller with flash memory, USB and can based on the 32-bit RISC core of high-performance arm Cortex-M3. Other main modules used include Bluetooth module, power module and motor drive module. The core processor STM32 single chip microcomputer processes the received data and converts it into the control signal of the single chip microcomputer to change the motion state of the car and realize the left turn, right turn, forward and stop of the car. The block diagram of intelligent car system based on brain wave control is shown in the figure below.

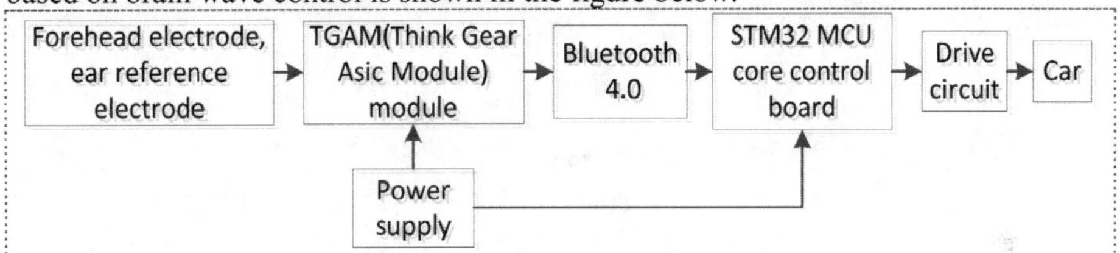

Figure 1. System block diagram.

3. Hardware design

3.1. Brain wave sensor module

TGAM brain wave module is an EEG acquisition module of NeuroSky company. It has simple structure, is not fixed, can move quickly, and is easy to carry and operate when detecting brain wave signals. The voltage stabilization circuit of the EEG sensor module is provided by two No. 7 dry batteries. When the load input, voltage, circuit parameters, ambient temperature and other factors change, the output voltage circuit of the module can still remain constant. With research level accuracy, this embedded EEG sensor provides a new solution for industries such as health, education and entertainment.

3.1.1. Principle of EEG technology. Scientific breakthroughs in recent years have shown that in daily life and work, the human brain will produce spontaneous electrophysiological activities. These spontaneous physiological electrical signals can be presented in the form of brain waves through professional instruments. According to the frequency of brain waves, the international brain wave Association and the neuroscience group divide brain waves into four main frequency bands, which is also the product of decades of research in the field of brain science α (alpha) wave β (beta) wave θ (theta) wave δ (delta) wave. In addition, another common brain wave, whose frequency is usually 30-100Hz and amplitude is different, is named γ (gamma) wave [4].

3.1.2. EEG signal acquisition. TGAM brain wave sensor collects the weak electrical signal generated by the brain through non-invasive dry electrode, puts the circular dry electrode piece on the skin on the forehead of the brain, clamps the two reference electrodes on the left ear and right ear respectively, clings to the skin at the earlobe or back of the ear, and measures the reference potential to minimize interference. In the process of collecting data, the brain wave sensor TGAM obtains the original brain wave data in the form of data packet [5]. About 513 data packets are collected every second. The output data stream of brain wave sensor TGAM is: 512 raw (small data packets) first, and then 513 large data packets. A small data packet is a piece of unprocessed original information, that is, an original data. The small data packet follows this format: aa aa 04 80 02 xxHigh xxLow xxCheckSum [6], the first to fifth bytes will not change, the last three bytes will change constantly, and the checksum bit xxCheckSum is the last bit. Because the original data is only composed of two bytes xxHigh and xxLow, a small data packet contains only one available raw rawdata. In the 513rd big data packet, there are 36 bytes, including signal value, 8 EEG power values, mediation relaxation value and attention concentration value.

3.1.3. Processing and output of EEG signals. Because the EEG signal is weak and easy to be disturbed, the EEG signal collected by the sensor cannot be directly used as effective data. The TGAM EEG module will preprocess the collected original signal data. First calculate the checksum from the collected data packet.

$$\mathrm{sum} = ((0x80 + 0x02 + xxHigh + xxLow) \wedge 0xFFFFFFFF)\ \&\ 0xff \tag{1}$$

If the checksum xxCheckSum and the calculated sum are the same, it indicates that the received packet is correct, and then calculate the original data [7].

$$\mathrm{rawdata} = (xxHigh\ << 8)\ |\ xxLow \tag{2}$$

$$\mathrm{if}(\mathrm{rawdata} > 32768)\quad \mathrm{rawdata} = 65536 \tag{3}$$

NeuroSky's patented algorithm eSense^{TM} is the core of the EEG sensor TGAT chip. This algorithm can express the mental state information of the human brain in numerical form. After the initial EEG signal can be processed, it can be transformed into a numerical value in the range of 1-100 to quantify the eSense parameters, which can be used to express people's concentration and relaxation. A Bluetooth 4.0 module is embedded in the TGAM EEG module. The TGAM EEG sensor will preprocess the collected EEG signal and send the signal in the form of data packet through Bluetooth communication

3.2. Core control module
The main control chip of the system is STM32F103C8T6 single chip microcomputer. It is a 32-bit STM32F1 series based on ARM architecture launched by Italian French semiconductor (ST), and a microcontroller with flash memory, USB and can (controller area network, can). STM32F103C8T6 microcontroller is mainly used to analyse the received EEG data in this design, and then control the external circuit to change the motion state of the car. The single chip microcomputer can fully meet the basic requirements and expansion requirements of this design, and can ensure the high cost performance of the system.

3.3. Bluetooth module
The system uses two Bluetooth DX-BT05 4.0 modules. The working mode is master-slave mode and follows V4.0 0 Bluetooth low energy (BLE) protocol specification. When used to send and receive information and data, the BLE 4.0 Bluetooth module has no byte limit, and the communication distance is greater than that of traditional Bluetooth. This module has the advantages of low power consumption and sensitive sending and receiving. It is very suitable for short-distance wireless data transmission.

3.4. Driver module

The selected motor drive module is L298N of ST company. It uses double H-bridge DC motor main drive chip, which has the characteristics of wide current and high voltage. The L298N drive module of the intelligent car system circuit mainly controls the rotation of the DC motor through its logic function, so as to realize the control of different motion states of the car, such as left turn, right turn, forward stop and so on.

3.5. Power module

Since the normal operation of the core control module of STM32F103C8T6 single chip microcomputer requires 3.3V stable voltage power supply, the step-down power supply module AMS1117-3.3 is selected. The step-down power supply module AMS1117-3.3 has the advantages of simple structure, convenient use, internal integration of current limiting and overheating protection circuits, good environmental protection and safety performance. It is an excellent choice for battery power supply.

4. Software design

4.1. System main program architecture

The system software design part is mainly divided into three steps. Firstly, the system initializes each module, then receives the EEG data through serial communication, analyses the EEG data, and finally controls the different motion states of the car through the converted values. As shown in Figure 2 below, it is the system program flow chart.

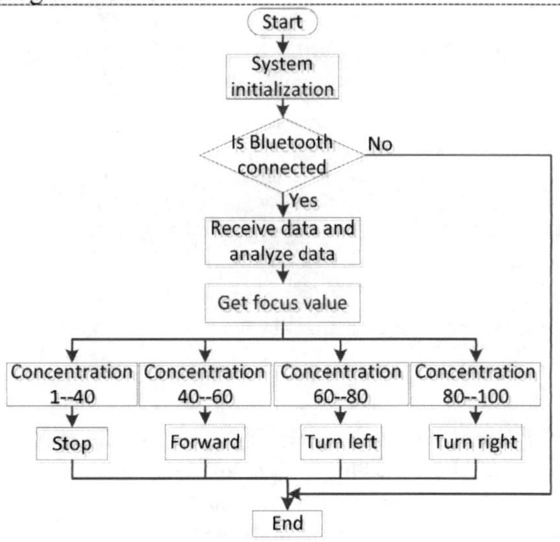

Figure 2. System program flow chart.

4.2. Data receiving and analysis

The receiving and analysis of Thinkgear data packet is mainly realized through the software program design of STM32F103C8T6 single chip microcomputer. Thinkgear data package mainly includes three parts: data package title, data package payload and data package payload checksum.

$$\frac{\text{[SYNC] [SYNC] [PLENGTH]}}{\text{^^^^^^^^^(Header)^^^^^^^^^}} \quad \frac{\text{[PAYLOAD...]}}{\text{^^(Payload)^^}} \quad \frac{\text{[CHKSUM]}}{\text{^(Checksum)^}} \tag{4}$$

In the EEG data packet, two [SYNC] bytes (0XAA) are used as the beginning of the data packet to prevent the received [sync] byte (0XAA) from being misunderstood as the header byte of the data packet. Although two consecutive [SYNC] bytes will still appear at a certain position in the middle of the data packet, the value of byte length [Plength] and checksum [Chksum] will ensure that they are read from the correct data packet header every time, Ensure that every time you receive a valid packet.

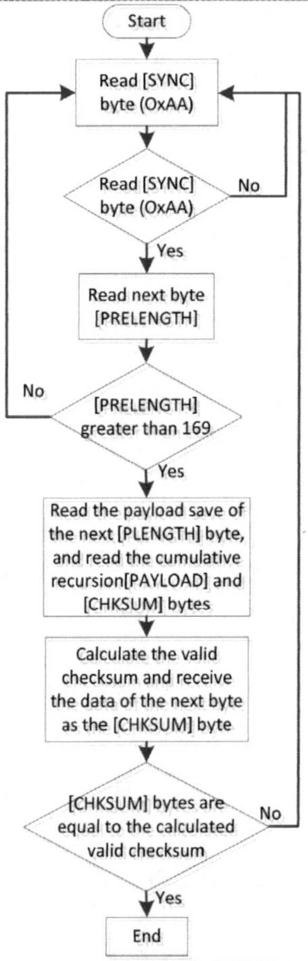

Figure 3. Flow chart of packet reception and analysis.

4.3. Car control program

When using TM32F103C8T6 single chip microcomputer to control the movement state of the car, it is necessary to quantify the concentration value range analysed after receiving into four different parameters and send them to the car. Through different parameter judgment, four different functions are configured to control the rotation of the motor, so as to realize the right rotation, left rotation, forward and stop of the car. The quantitative table of concentration value range is shown in Table 1 below.

Table 1. Quantification of concentration value range.

Focus value range	Parameters sent to car	Movement state of car
1~40	1	Stop
40~60	2	Forward
60~80	3	Turn left
80~100	4	Turn right

The "eSense Concentration Index" measures the strength of the user's psychological "attention" and "concentration". When the user is able to control his mental activities and achieve a highly focused mental world, the "eSense Concentration Index" increases, and the value ranges from 0 to 100. Agitation, anxiety and distraction are all mental states that lower the eSense Concentration Index.

5. Test

5.1. TGAM EEG module debugging

First test the TGAM brain wave sensor module. The module is equipped with a brain wave forehead electrode and two ear clips. When using, it must be in close contact with the skin. Next, ensure that the Bluetooth connection is normal, and the serial port will receive data, which will be presented in the form of data packet on the computer serial port/network data debugger software, as shown in Figure 4 below.

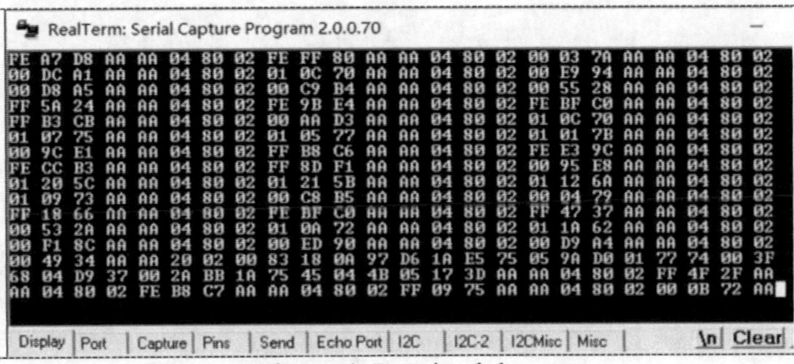

Figure 4. Received data.

0XAA of the first and second bits is the start bit. The third digit is the signal value, indicating the wearing status. A value of 0x00 indicates that the EEG sensor is well worn. A value greater than 0x00 indicates that the collected EEG contains clutter and the EEG sensor is not worn correctly. The fourth digit is concentration and the fifth digit is relaxation. Their maximum value is 0x64 and the minimum value is 0x00.

After analyzing the received data in the computer serial debugging software MindViewer, it will be presented in the form of waveform. Among them, the red integral graph is Noise, the green integral graph is Attention, and the blue integral graph is Meditation. Users can be trained to control their thoughts and improve their concentration.

In the process of brain wave acquisition, analysis and debugging, there will be different mental states to test the concentration signal value. The system is powered on, and the EEG sensor collects brainwave signals, and the waveform data after analysis are as follows: The state of highly concentrated attention is shown in Figure 5, the concentration fluctuates between 80 and 100, and the car will turn right. As shown in Figure 6, the concentration fluctuates between 60 and 80, and the car will be in a left turn state. The normal state is shown in Figure 7, the concentration will fluctuate between 40 and 60, and the car will move forward. The fatigue state is shown in Figure 8, when the concentration is between 1 and 40, the car will stop moving.

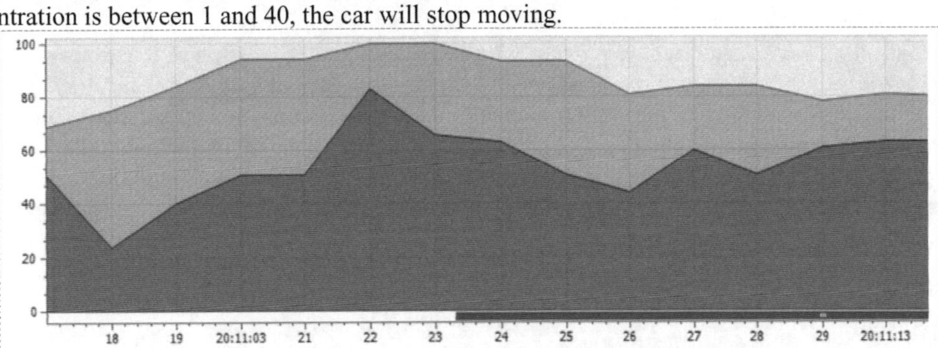

Figure 5. The state of highly concentrated attention.

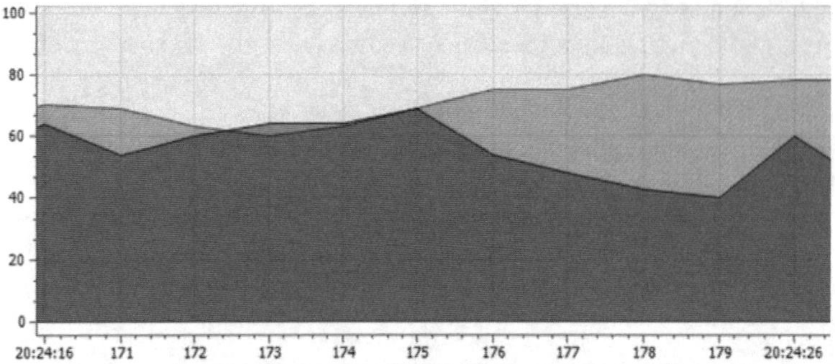

Figure 6. The state of concentrated attention.

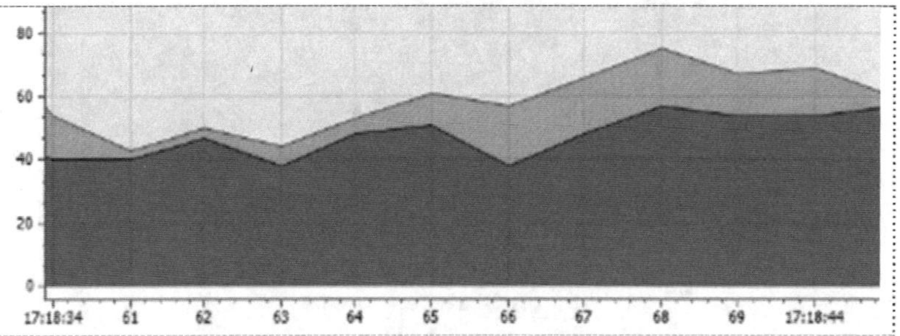

Figure 7. The normal state.

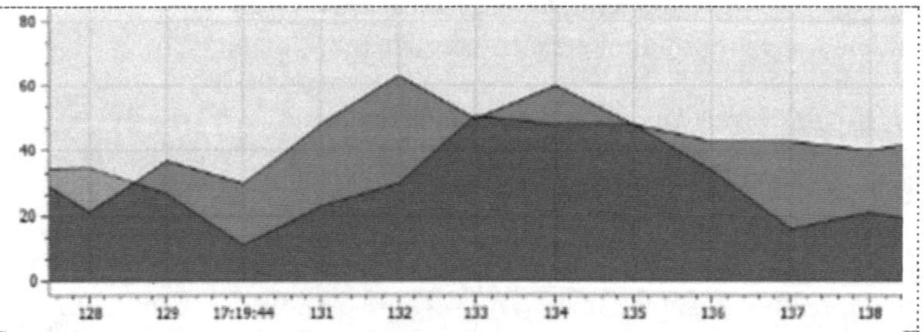

Figure 8. The fatigue state.

During the actual debugging, it is found that when wearing the EEG sensor, you must wipe your forehead and sensor electrode with a wet paper towel, otherwise, the skin sweat and oil will produce great interference, resulting in the inability of the EEG sensor to collect the EEG data.

5.2. Car movement state commissioning

When controlling the left and right rotation of the car, the differential rotation of the motors on the left and right sides is used. When the rotation speed of the right motor is faster than that of the left motor, the car turns left. When the rotation speed of the left motor is faster than that of the right motor, the car turns right. When the motor speed on the left and right sides of the car is the same, the car moves forward. Since it is a 4WD car, the rotation mode of the front and rear motors is the same. The single chip microcomputer outputs PWM pulse through I/O port to control the speed of car motor. It is necessary to debug the cycle and duty cycle of PWM wave, so as to control the car speed within an appropriate range.

6. Conclusion

This design introduces the intelligent car system based on brain wave control, studies the raw data rawdata and eSense concentration index read by TGAM brain wave sensor, and uses the mental concentration of human brain to control the basic motion state of intelligent car. In the past decades, brain wave technology has played an important role in intelligent development. Many countries are studying brain computer interface. Almost all technical methods that can be used to monitor brain activity information are used in the research of brain computer interface. At present, it still faces many challenges. In terms of basic research, there are too many things about brain science that have not been solved yet; In terms of engineering technology, there are also many problems in how to interpret signals; In terms of application and promotion, it also involves some complex ethical issues, which will be some difficulties and challenges in the process of research and development. Although this design has certain limitations on the control of intelligent car, this wireless control system has strong universality. If we study other kinds of brain wave information extracted by the chip, it is of great significance for other more complex mind control.

Acknowledgments
In this paper, the research was sponsored by the Science and Technology Research Program of 2021 Hubei Provincial Education Department (Project No: B2021410).

References
[1] Huang J.Q., Xu Q., Gao G., et al. (2014)Toy racing car based on brain wave control -- Italian world iCAN forum. Internet of things technology, 4 (03): 8-9.
[2] Gao S.K.. (2019)Current situation and future of brain computer interface. Robot industry, (5): 38-44.
[3] Wang F., Li Y.. (2020)Design of intelligent manipulator based on EEG control. Information and computer (theoretical Edition), 32 (04): 132-134.
[4] Chang Y., Yang F., Hao Q.. (2019)Intelligent wheelchair system based on EEG control. Micro special motor, 47(03): 82-86.
[5] Chen D.W., Wu F., Wang Z., et al. (2013)Design and implementation of intelligent car system based on brain computer interface. Information technology, 37(06): 80-82+87.
[6] Miao W.N., Chen X.J.. (2018)Car control system based on TGAM brain wave module. Automation technology application, 37(08): 112-115.
[7] Liu X., Leng J., Fu F.M.. (2020)Idea reading system based on TGAM. Digital technology and application, 38 (04): 150-151.

Structural optimization method of 6-DOF manipulator based on genetic algorithm

Chenyu Tu[a], long Zhang[b]

School of mechanical engineering, Nanjing University of Science and Technology, Nanjing, Jiangsu.

[a]1312418151@qq.com, [b]lzhang@njust.edu.cn.

Abstract. Most of the manipulator is designed under the constraint of motion reversibility, and the optimization of the manipulator structure is often carried out through the study of kinematics. In order to solve the structural optimization problem of a 6-Dof manipulator, this paper proposes a structural optimization method based on a 6-Dof manipulator with comprehensive indexes combining kinematics performance and dynamics performance. On the premise of satisfying the original workspace and taking dynamic performance as evaluation function, genetic algorithm was used to optimize the structural parameters of the manipulator, so that the optimized manipulator not only has the characteristics of the original workspace, but also consumes the least energy. The kinematic equation of the manipulator was derived, the dynamics model was established by Lagrange method, and then the fitness function and the structural optimization algorithm of the manipulator based on genetic algorithm were given.

1. Introduction

The research on the design criteria of manipulator has always been a hot spot in the field of robot research. Most of the early manipulator was designed under the constraints of kinematic reversibility. Gosselin et al. [1]took the condition number of the Jacobian matrix of the manipulator as the primary design criterion. The condition number of the Jacobian matrix as the design criterion met the requirements of the kinematic reversibility of the manipulator design, and also provided convenience for the subsequent path planning of the manipulator. On this basis, Gosselin et al. [2]proposed the combination of Jacobian condition number and workspace to obtain the global conditioning index (GCI), which is a performance index based on the distribution of Jacobian matrix condition number in the entire workspace of the manipulator. It measures the kinematic reversibility of the manipulator over its entire workspace. Hyeon-guk Kim et al. [3]proposed a method to determine the size of the link of the manipulator based on condition number, SLI and GCI indexes, and used this method to optimize and analyze the 7-DOF manipulator. High degree of freedom mechanical arm series manufactured by commonly used in aviation, space station tools very high demand for flexible and flexibility of task, the task requires not only meet the constraints of work space, mechanical arm in mechanical arm also requires to complete tasks at the same time shorten the execution time of tasks and to minimize the consumption of energy, especially the space station mission, There are high requirements for saving energy consumption. The design criteria of the manipulator studied in the above literature only meet the workspace and kinematic performance requirements of the manipulator, and do not make any requirements for the dynamic performance of the manipulator, especially the energy consumption. In reference on the basis of the above design principles, this paper puts forward a way to keep the original working space characteristics under the condition of optimization method based on energy consumption index, the method on the basis of the mechanical arm to keep the original structure, energy consumption of the mechanical arm to measure, and index of using genetic algorithm to optimize the structural parameters of mechanical arm. Based on a certain type of 6 dof serial manipulator as the research object, by GCI workspace performance reference index, though laser method from the Angle of energy, mechanical arm dynamics model is set up, put forward the measure of energy consumption, using genetic algorithm converts mechanical arm structure optimization problem with GCI index for energy consumption under the restriction of the minimum problem.

2. Workspace metrics and energy consumption metrics

In this paper, a 6-DOF manipulator of our research group was taken as the research object, and the GCI index proposed by Gosselin et al was taken as the constraint of the workspace index. Then, Lagrange method was used to establish the dynamics model of the manipulator and calculate the energy consumption index. Figure 1 shows the 6-DOF mechanical arm, Figure 2 shows the structure diagram of the mechanical arm, and Table 1 shows the D-H parameters of the mechanical arm.

Figure1 6-DOF mechanical arm

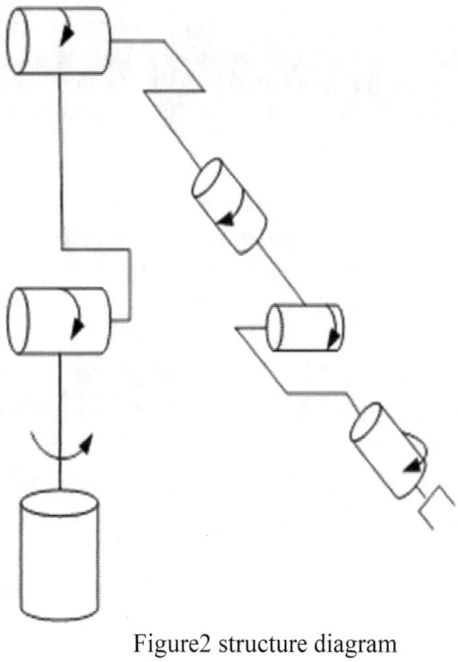

Figure2 structure diagram

Table 1 D-H parameters of manipulator

Joint	$\alpha_i/°$	a_i/mm	d_i/ mm	$q_i/°$
1	90	0	l_2	q_1
2	0	l_2	0	q_2
3	0	0	l_2	q_3
4	90	l_2	0	q_4
5	90	0	l_2	q_5
6	0	0	l_2	q_6

2.1 Workspace metrics

The workspace is premised on meeting the constraint of GCI index and is aimed at maintaining the volume characteristics of the original designed workspace. Formula (1) is GCI expression:

$$\eta = \int_W \left(\frac{1}{\kappa}\right) dV / V(n) \tag{1}$$

$V(n)$ is the volume of the workspace, κ is the condition number of the Jacobian matrix of the manipulator at the working point *W*. To simplify the calculation, the workspace calculation in the Cartesian coordinate system can be transformed into the calculation in the joint coordinate system, as shown in Formula (2) :

$$V(n) = \int_Q J(q) dq_6 ... dq_2 dq_1 \tag{2}$$

$\boldsymbol{q}$ is the joint intersection displacement vector, Q is the joint angular displacement constraint space, $\boldsymbol{q}=[q_1, q_2, q_3, q_4, q_5, q_6]^T \in Q$ .In order to meet the movement reversibility of the mechanical arm, the following conditions should be met:

$$0 < \eta < 1 . \tag{3}$$

Since the dynamic energy consumption of mechanical arm is mainly taken as the optimization objective in this paper, the inverse kinematics solution of mechanical arm and the solution of jacobian matrix condition number can be obtained from literature [4], which will not be detailed here. The subsequent optimization only needs to take GCI index as the constraint of energy consumption index function.

2.2 Energy consumption metrics

The energy consumption of the manipulator is mainly composed of the kinetic energy and potential energy of each link of the manipulator and the kinetic energy and potential energy of each joint motor. Therefore, the energy consumption index can be expressed by the ratio of total energy consumption after optimization to energy consumption before optimization:

$$f(n) = \frac{T(n) + P(n)}{T_0 + P_0} \tag{4}$$

n is the structural parameters of the connecting rod of the manipulator, which is also the optimization object, $n = [l_1, l_2, l_3, l_4, l_5, l_6]$.

In order to obtain the total energy of each connecting rod and motor of the manipulator, a simplified model of the manipulator is established, as shown in Figure 3.

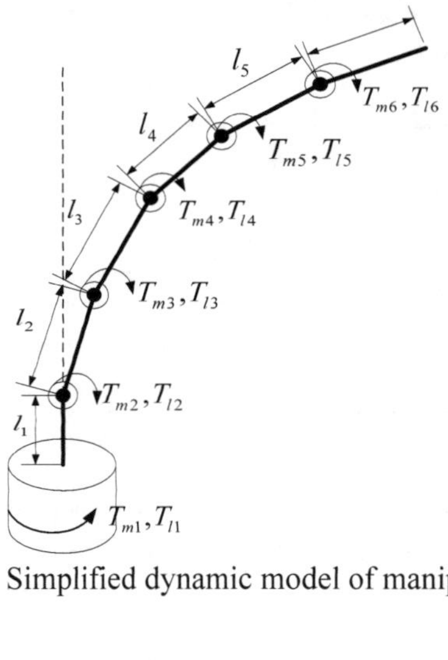

Figure3 Simplified dynamic model of manipulator

Lagrange method was used to derive the kinetic equation, and the kinetic energy and potential energy of each connecting rod and motor were obtained successively.[5]

$$T_{m1} = J_{m1} N^2 \dot{\theta}_1^2 / 2 \qquad T_{l1} = I_{l1} \dot{q}_1^2 / 2 \tag{5}$$

$$T_{m2} = J_{m2} N^2 \dot{\theta}_2^2 / 2 \qquad T_{l2} = \left\{ m_{l2} \left[\left(d_2 \dot{q}_2 \right)^2 + \left(d_2 s_2 \dot{q}_1 \right)^2 \right] + I_{l2} \dot{q}_2^2 \right\} / 2 \tag{6}$$

$$T_{m3} = \left\{ m_{r3} \left[\left(l_2 \dot{q}_2 \right)^2 + \left(l_2 s_2 \dot{q}_1 \right)^2 \right] + J_{m3} \left(N \dot{\theta}_3 + \dot{q}_2 \right)^2 \right\} / 2$$

$$T_{l3} = \left\{ m_{l3} \left[\left(l_2 s_2 + d_3 s_{23} \right)^2 \dot{q}_1^2 + \left(l_2 \dot{q}_2 + d_3 \dot{q}_{23} \right)^2 \right] + I_{l3} \dot{q}_{23}^2 \right\} / 2 \tag{7}$$

$$T_{m4} = \left\{ m_{r4} \left[\left(l_2 s_2 + l_3 s_{23} \right)^2 \dot{q}_1^2 + \left(l_2 \dot{q}_2 + l_3 \dot{q}_{23} \right)^2 \right] + J_{m4} \left(\dot{q}_{23} + N \dot{\theta}_4 \right)^2 \right\} / 2$$

$$T_{l4} = \left\{ m_{l4} \left[\left(l_2 s_2 + l_3 s_{23} + d_4 s_{24} \right)^2 \dot{q}_1^2 + \left(l_2 \dot{q}_2 + l_3 \dot{q}_{23} + d_4 \dot{q}_{24} \right)^2 \right] + I_{l4} \dot{q}_{24}^2 \right\} / 2 \tag{8}$$

$$T_{m5} = \left\{ m_{r5} \left[\left(l_2 s_2 + l_3 s_{23} + l_4 s_{24} \right)^2 \dot{q}_1^2 + \left(l_2 \dot{q}_2 + l_3 \dot{q}_{23} + l_4 \dot{q}_{24} \right)^2 \right] + J_{m5} \left(\dot{q}_{24} + N \dot{\theta}_5 \right)^2 \right\} / 2$$

$$T_{l5} = \left\{ m_{l5} \left[\left(l_2 s_2 + l_3 s_{23} + l_4 s_{24} + d_5 s_{25} \right)^2 \dot{q}_1^2 + \left(l_2 \dot{q}_2 + l_3 \dot{q}_{23} + l_4 \dot{q}_{24} + d_5 \dot{q}_{25} \right)^2 \right] + I_{l5} \dot{q}_{25}^2 \right\} / 2 \tag{9}$$

$$T_{m6} = \left\{ \begin{array}{l} m_{r6} \left[\left(l_2 s_2 + l_3 s_{23} + l_4 s_{24} + l_5 s_{25} \right)^2 \dot{q}_1^2 + \left(l_2 \dot{q}_2 + l_3 \dot{q}_{23} + l_4 \dot{q}_{24} + l_5 \dot{q}_{25} \right)^2 \right] + \\ J_{m6} \left(\dot{q}_{25} + N \dot{\theta}_6 \right)^2 \end{array} \right\} / 2$$

$$T_{l6} = \left\{ m_{l6} \left[\begin{array}{l} \left(l_2 s_2 + l_3 s_{23} + l_4 s_{24} + l_5 s_{25} + d_6 s_{26} \right)^2 \dot{q}_1^2 + \\ \left(l_2 \dot{q}_2 + l_3 \dot{q}_{23} + l_4 \dot{q}_{24} + l_5 \dot{q}_{25} + d_6 \dot{q}_{26} \right)^2 \end{array} \right] + I_{l6} \dot{q}_{26}^2 \right\} / 2 \tag{10}$$

Thus, the total kinetic energy of the manipulator is:

$$T(n) = \sum_{i=1}^{6} \left(T_{mi} + T_{li} \right) \tag{11}$$

The total gravitational potential energy of the manipulator is:

$$P(n)=g\begin{bmatrix} m_{l2}d_2c_2+m_{r3}l_2c_2+m_{l3}\left(l_2c_2+d_3c_{23}\right)+ \\ m_{r4}\left(l_2c_2+l_3c_{23}\right)+m_{l4}\left(l_2c_2+l_3c_{23}+d_4c_{24}\right)+ \\ m_{r5}\left(l_2c_2+l_3c_{23}+l_4c_{24}\right)+m_{l5}\left(l_2c_2+l_3c_{23}+l_4c_{24}+d_5c_{25}\right)+ \\ m_{r6}\left(l_2c_2+l_3c_{23}+l_4c_{24}+l_5c_{25}\right)+m_{l5}\left(l_2c_2+l_3c_{23}+l_4c_{24}+l_5c_{25}+d_6c_{26}\right) \end{bmatrix} \tag{12}$$

$$s_i=\sin\left(q_i\right),\quad s_{ij}=\sin\left(q_i+\ldots+q_j\right),\quad c_i=\cos\left(q_i\right),c_{ij}=\cos\left(q_i+\ldots+q_j\right),\dot{q}_{ij}=\dot{q}_i+\ldots+\dot{q}_j$$

T_{mi} is the kinetic energy of the motor i. T_{li} Is the kinetic energy of link i . J_{mi} is the moment of inertia of the i motor rotor. N is transmission ratio. I_i is the rotational inertia of the i connecting rod, m_{li} is the mass of the i connecting rod, m_{ri} is the mass of the i motor rotor, q_i is the i joint Angle, θ_i is the Angle of the i motor before deceleration, d_i、l_i and are the distance of the i connecting rod centroid and the length of the connecting rod respectively. The above symbols can be measured.

3. Structural parameter optimization method based on genetic algorithm

Using the space performance index and energy consumption index obtained above, the structural optimization problem of the manipulator is transformed into a constrained objective function optimization problem, as shown in the following equation:

$$\begin{cases} \min f(n)=\dfrac{T(n)+p(n)}{T_0+P_0} \\ s.t.V(n)\geq V_0, 0<\eta<1 \end{cases} \tag{13}$$

V_0、T_0、P_0 are respectively the working space volume, kinetic energy and potential energy of the manipulator before optimization. Genetic algorithm can well solve the above optimization problems. The optimization algorithm flow is shown in Figure 4, and the optimization algorithm is divided into the following steps:

(1) Randomly generate N working points in the selected workspace, plan the working path, and calculate the initial energy consumption.

(2) Initialize the population, set the distribution range of the structural parameters within the range of the original design parameters, calculate the Jacobian matrix, and screen the individuals of the population with the workspace index.

(3) Calculate the fitness function, and then select individuals for crossover and mutation operation. The fitness function is shown in Formula (14), $\delta=1\times10^{-6}$ $\alpha=1\times10^{6}$, the number of individuals in the initial population is set as 30, the evolution algebra is 100, crossover probability is 0.8, and mutation probability is 0.1.

$$J(n)=\frac{1}{f(n)+\alpha\sin(\eta)+\delta} \tag{14}$$

The optimization results are shown in Figure.5~6. Figure5 is the fitness function curve and Figure6 is the consumption index curve. It can be seen from Figure5 that when iteration reaches G=20, the optimal individual has the lowest energy consumption index, and the energy consumption is greatly reduced under the condition that its workspace and kinematic characteristics remain unchanged.

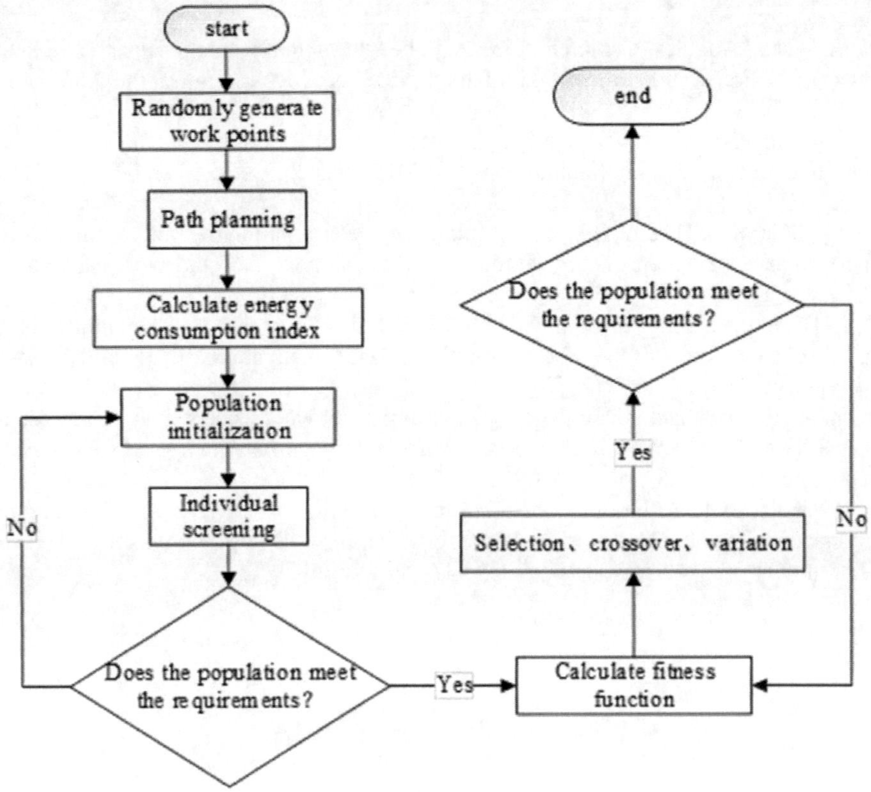

Figure 4 Flow chart of algorithm optimization

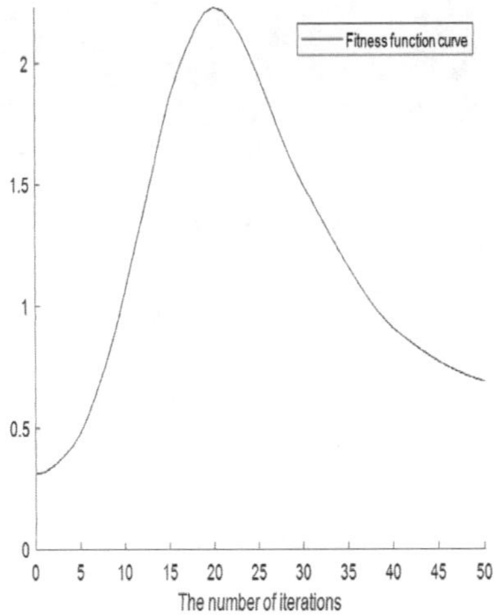

Figure 5 Fitness function curve

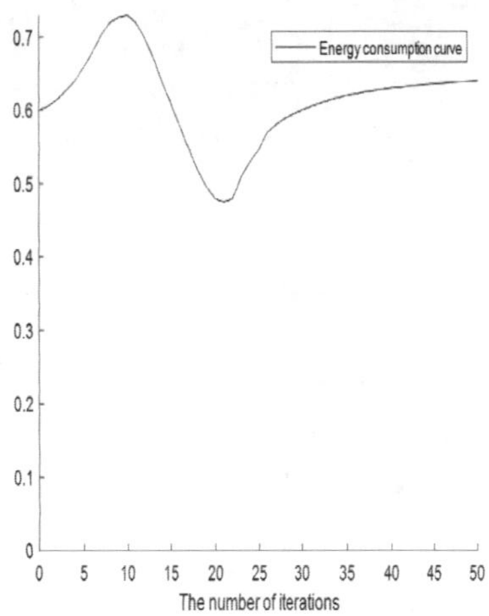

Figure 6 Energy consumption curve

4. conclusion

To sum up, this paper proposes a optimization of structural parameters optimization method, mechanical arm in arm of the original work under the limitation of space characteristics, through the optimization of parameters of connecting rod makes the mechanical arm at work of the energy consumption greatly reduced, thus verified the rationality and effectiveness of the optimization method, provides the structure optimization of mechanical arm more ideas.

References

[1] Gosselin, C., and Angeles, J. (March 1, 1988). "The Optimum Kinematic Design of a Planar Three-Degree-of-Freedom Parallel Manipulator." ASME. J. Mech., Trans., and Automation. March 1988; 110(1): 35–41.

[2] Gosselin, C., and Angeles, J. (September 1, 1991). "A Global Performance Index for the Kinematic Optimization of Robotic Manipulators." ASME. J. Mech. Des. September 1991; 113(3): 220–226.

[3] Kim, HG., Shin, KS., Hwang, SW. et al. Link length determination method for the reduction of the performance deviation of the manipulator: Extension of the valid workspace. Int. J. Precis. Eng. Manuf. 15, 1831–1838 (2014).

[4] Lim, H., Hwang, S., Shin, K. *et al.* Comparative study of optimization technique for the global performance indices of the robot manipulator based on an approximate model. *Int. J. Control Autom. Syst.* **10,** 374–382 (2012).

[5] F. Zacharias, C. Borst and G. Hirzinger, "Capturing robot workspace structure: representing robot capabilities," *2007 IEEE/RSJ International Conference on Intelligent Robots and Systems*, 2007, pp. 3229-3236, doi: 10.1109/IROS.2007.4399105.

[6] Yoon, J.W., Ryu, J. & Hwang, YK. Optimum design of 6-DOF parallel manipulator with translational/rotational workspaces for haptic device application. J Mech Sci Technol 24, 1151–1162 (2010).

Uncertainty measurement and research of Finite element analysis of Water Jet data based on Internet of Things

Changhu Wang[1a], Weiyun Meng[1b*], Jianxiang Wang[1c], Xiangmei Min[2d*]

[1]College of Mechanical and Electronic Engineering,Shandong Agriculture and Engineering University, Jinan, Shandong, 250100, China.

[2]Jining Industrial technician College, Jining,Shandong, 272100, China.

[a]z2013168@sdaeu.edu.cn, [b*]z2017019@sdaeu.edu.cn, [c]3381584317@qq.com, [d*]852754779@qq.com.

Abstract—Nowadays, the concept of energy saving and water saving is deeply rooted in people's minds, so it is necessary to promote efficient water-saving irrigation technology. Sprinkler irrigation is an efficient water-saving method, which is relatively uniform and does not affect the structure of the soil. It saves time and labor to irrigate fields, and has been used in more and more areas in China.The Internet of things technology is used to establish monitoring sites, collect and transmit water information, study and analyze crop fertilization and irrigation water volume, and achieve the purpose of saving water, fertilizer and labor. In test site traffic and sprinkler relationship between speed and pressure test, flow and in different pressure conditions were measured under different conditions of running speed, observe the fluid flows through the three links of pipe head loss and energy loss of the tee to the role of the internal pressure and axial force and other force, high stress concentration areas vulnerable to damage. The hydraulic characteristics in the process of water spraying are studied. The numerical simulation in this paper is completed by finite element software, and the hydraulic characteristics of the three-way pipe are analyzed by finite element to optimize the structure of the three-way pipe.

1. Introduction

The integration of water and fertilizer technology that fully stirred water and fertilizer, the mixture thus formed which goes through the pipeline to the crops, has not only improved the utilization efficiency of water and fertilizer, but also played a great role in increasing agricultural production and income.

1.1 Crop information collection

Researchers understand the growth of crops in a timely manner to take reasonable management means in the process of crop growth. Internet of things technology plays a significant part in collecting the information of crop aspect , monitoring the growth of crops closely, controlling of crops in soil and crop growth rule[1], taking reasonable throughput, watering adjustment according to the crop growth situation of crops needs moisture and nutrients[2].

1.2 Development trend of water and fertilizer integration in the Internet of Things

Large-scale agricultural planting enables the integration of water and fertilizer technology and web of things technology to be fully combined and utilized to be conscious of the intelligent development of agriculture[3]. The web of Things technology is used to build monitoring points for collecting and transmitting soil water content and other information according to irrigation water sources and irrigation methods in different regions. Through the accurate data, the fertilizer amount and irrigation water amount of crops are studied and analyzed to realize the integration of irrigation and fertilization and achieve the purpose of saving water and fertilizer and labor[4]. The web of things can be used to analyze and study fertilizer and water content to realize the integration of irrigation and fertilization in agricultural production, so that farmers can increase their income, save water, save energy and utilize water and fertilizer efficiently, and promote agricultural development [5]. To promote the use of water and fertilizer integration technology under the web of things, establish technology demonstration bases, better use in production, promote rural development and significantly improve the economic income of farmers.

1.3 Functions of various parts of the Internet of Things system

Agricultural lot users include farmers and technical professionals, and different people use different types of technology with different functions [6]. The application layer is the interface between the network and users to realize interconnected intelligent applications, including various intelligent information systems [7]. Terminal devices mainly refer to intelligent reading devices used by different users of the Web of Things, and the administration and management information system includes video monitoring and other functional modules [8]. The transmission layer mainly refers to the acquired data information of the sensing monitoring object [9], which is transmitted into the wireless transmission network by sensors and other components, and then into the monitoring management system through the network [10].

2. Finite element analysis of water jet data

Nowadays, the concept of energy saving and water saving is deeply rooted in people's minds, so it is necessary to promote efficient water-saving irrigation technology. Sprinkler irrigation is an efficient water-saving method, which is relatively uniform and does not affect the structure of the soil. It saves time and labor to irrigate fields, and has been used in more and more areas in China.

The pressure affects the water quantity of sprinkler head. The water quantity distribution model of pressure is established to study the water quantity of sprinkler and sprinkler head in a specific time in this paper. Soil permeability affects the intensity of sprinkler irrigation. When the intensity of sprinkler irrigation exceeds the infiltration rate of soil, water accumulation and runoff will be formed, resulting in waste of water, which is detrimental to the growth of crops.

2.1 Finite element analysis of tee pipe fittings

During sprinkler irrigation, the tee pipe is connected to different branch pipes. When fluid flows through the tee pipe, head loss and energy loss occur. Tee pipe to bear internal pressure and axial force and torque, complex structure of the force is more, the higher stress concentration of the part is vulnerable to damage.

In the study of hydraulic characteristics in the process of spraying water, the liquid flow in the pipe is a continuous single-phase flow. In this paper, the fluid in the pipe is liquid water. In this paper, the numerical simulation is completed by finite element software. Firstly, three-dimensional modeling is carried out in FIG. 1 to conduct finite element analysis on the fluid hydraulic characteristics of the tee pipe, so as to optimize the structure of the tee pipe.

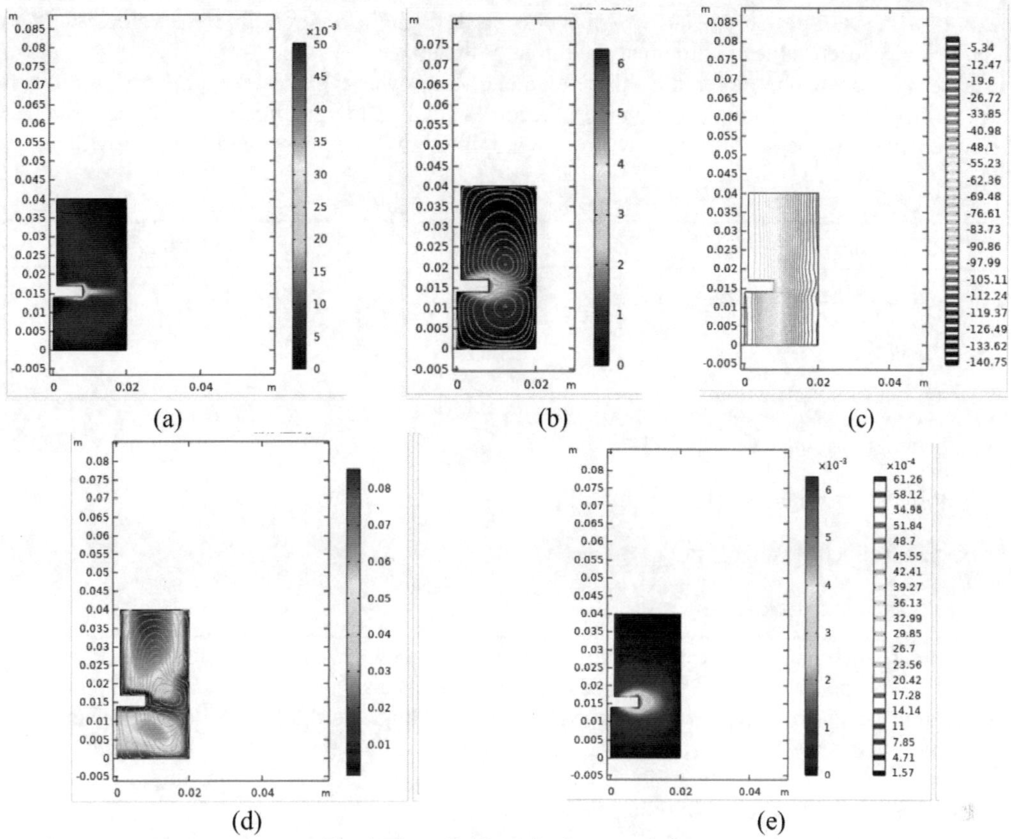

Fig.1 Flow field analysis at tee joint

In order to reduce the head loss at the water catchment, the structure of the water catchment of tee pipe fittings is optimized. In finite element analysis, when the flow rate is fixed, the cross section area of the pipeline is inversely proportional to the flow rate; when the cross section size of the pipeline increases, the quantity of flow decreases; when the cross section size decreases, the flow rate increases. The greater the flow rate, the greater the square of the flow rate, in direct proportion, the more water loss.

In the finite element analysis, it is planned to add a diversion plate at the water confluence of the tee fitting, and the diversion plate is made of the same material as the tee fitting. Through finite element flow field analysis, it can be seen that the area of backflow becomes smaller, indicating that the transverse velocity of water flow increases after the diversion plate is added.

After the diversion plate is added, the pressure at the junction of water flow will decrease, while the pressure required for the transverse part of pipe fittings will increase, and the maximum velocity will not change much, the velocity distribution will be more uniform, and the transverse velocity will increase faster.

2.2 Finite element analysis of nozzle flow field

The pressure and velocity cloud diagrams of transient flow in the nozzle. The pressure at both ends close to the cavity wall is relatively large, and different pressure distributions can be seen from the pressure diagrams. The pressure at the nozzle entrance is as high as about 0.81mpa, and the flow field distribution displayed in the downstream outlet of the nozzle is about 0.71mpa, and the lowest pressure is about 0.45mpa. Large pressure difference can be generated between the upstream and downstream of the nozzle, so that the water flows smoothly out of the nozzle.

The velocity distribution of the nozzle is similar to that of the pressure cloud diagram. The velocity in the area with small pressure is larger. After passing through the tapered tube, the velocity increases, and the velocity reaches the maximum at the throat of the nozzle.

In a certain range, the larger the throat diameter, the lower the pressure, the stronger the flow control ability. When the throat pipe diameter becomes smaller, the pipeline flow will be more stable, the flow change will be small, the flow growth amplitude will decrease, and the flow will be more stable.

Table. 1 Nozzle parameter

Index name	Technical specifications
Nozzle diameter（mm）	7.5
Inlet pressure（MPa）	0.59~0.96
Number of nozzles	14
Sprinkler spacing (m)	2.5
Flow rate of single nozzle（$m^3 \cdot h^{-1}$）	2~4
Nozzle effective range（m）	4~5
Tolerance of sprinkler range	≤±5%

Sprinkler head is very important in the water spraying device. Under the action of pressure, the water column is sprayed into the air, and under the action of air resistance, the water column will be dispersed into small droplets, evenly sprayed to the land. Improve the hydraulic performance of the nozzle,Make the nozzle low cost energy saving, efficient and durable, nozzle diameter can not be too large or too small. As shown in Table 1, when the nozzle diameter is 7.5mm, the jet nozzle works stably and has relatively good hydraulic performance.

3. Analysis of running speed and water collecting depth

In the test base, the relationship between the flow rate and the operation speed of the sprinkler and the pressure was tested. The inlet pressure of the sprinkler was 0.60 ~ 0.90MPa. The flow rate and operation speed of the sprinkler under different pressure conditions were measured, and the results were shown in Table 2.

Table.2 Operation speed of sprinkler under different pressure

Inlet pressure（MPa）	Inlet flow rate（$m^3 \cdot h^{-1}$）	Operating speed（$m \cdot h^{-1}$）
0.60	25.67	11.46
	26.75	13.27
	27.09	18.89
	27.43	21.69
0.70	28.46	12.31
	29.47	18.26
	29.89	26.53
	30.01	32.19
0.80	31.93	21.17
	32.76	28.67
	33.05	33.56
	33.87	39.86
0.90	34.65	28.17
	34.86	49.67

35.23	57.91
35.67	62.36

It can be seen from Table 2 that the operation speed of sprinkler under different pressures is different. The higher the pressure is, the faster the unit runs. At the same speed, the inlet flow increases with the increase of inlet pressure.

The sprinkler heads used in the test are refraction sprinkler heads, and the same arrangement of pipes, the working pressure at each sprinkler head is basically the same. When the flow and working pressure are constant, the faster the unit running rate is, the less the spray irrigation amount is, that is, the irrigation amount is inversely proportional to the operation rate.

Irrigation machine inlet pressure is 0.60, 0.70, 0.80, 0.90MPa, a total of four target pressure, analysis of irrigation machine operation state. Record the distance and time of water spray truck moving in different gears. Divide the rain gauge cylinders to collect water into four groups according to the sequence of the unit passing through, and arrange each group into the first row, second row, third row and fourth row respectively. Rain gauge cylinders are numbered outward from the middle position of the water spray truss trolley. Due to the influence of the wheel of the water spray trolley, one rain gauge cylinder is placed at 1m away from the center of the trolley, so the number is 1, 2... , no. 8. Analyze the rainfall collecting data of the rain gauge. When the working pressure is 0.60mpa, the collecting water depth of rain gauge cylinders at different gears is shown in Table 3.

Table. 3 The collecting depth of the rain bucket at different speeds

Speed ($m \cdot h^{-1}$)	Number of rain buckets	1	2	3	4	5	6	7	8
11.46	1	52.5	50.6	51.6	48.2	46.3	51.5	53.3	47.9
	2	52.7	51.3	52.2	46.3	46.4	48.2	55.2	51.2
	3	51.3	51.5	52.5	45.0	48.3	49.1	54.4	43.3
	4	49.5	52.7	51.2	43.3	45.5	44.4	54.7	46.6
13.87	1	43.8	41.8	43.4	40.5	40.2	43.5	45.2	40.6
	2	43.9	42.9	44.6	39.6	40.3	40.8	46.3	43.4
	3	44.2	41.6	44.0	38.3	41.4	41.2	47.5	36.5
	4	42.5	40.5	40.5	37.2	39.3	39.3	46.8	40.7
18.79	1	37.2	36.3	37.2	35.13	35.6	37.4	40.9	34.5
	2	38.7	37.4	38.5	3.4	35.2	35.6	41.3	37.3
	3	38.9	35.7	38.3	32.5	36.8	35.5	42.3	30.2
	4	36.6	35.6	35.7	32.3	33.7	34.3	40.7	34.8
21.59	1	30.8	30.3	31.5	30.6	30.1	31.9	34.2	26.9
	2	31.3	31.5	32.2	26.6	29.3	29.5	35.6	30.0
	3	30.5	30.6	31.6	26.2	30.6	28.2	36.7	24.5
	4	29.6	30.2	30.6	26.6	25.3	28.3	34.9	26.5

As can be seen from Table 3, the water depth of sprinkler irrigation is different with different speeds. Meanwhile, the water depth of sprinkler irrigation is higher near the central axis of sprinkler truck, while the water depth is lower at the position far from the central axis of sprinkler truck.Cucumber system based on Internet of things of agriculture by NongQing monitoring system, integrated water irrigation control system, implementation of farmland irrigation monitoring, farmland irrigation decision-making, production management, and other functions, farmland moisture as the core, realizing the comprehensive management of farmland and monitoring, to provide irrigation decision-making, planting operation management information platform.

4.Conclusion

Based on the results and discussions presented above, the conclusions are obtained as below:

(1) Finite element flow field analysis shows that when a diversion plate is added at the water confluence of tee pipe fittings, the area where reflux occurs becomes smaller, indicating that the transverse velocity of water flow increases after the diversion plate is added.

(2) Sprinkler water is different with different speeds. At the same time, the water depth near the central axis of sprinkler car is higher, and the water depth far from the central axis of sprinkler car is lower.

The integrated development of water and fertilizer still has high equipment cost and its applicability needs to be improved. With the development of society, the application will be gradually accelerated, thus driving the development of the integration of water and fertilizer.

Acknowledgments

This work was financially supported by the Science and Technology Poverty Alleviation Project of Shandong Agriculture and Engineering University (30220004) .

References

[1] TANG Yue, ZHU Xiangyuan, MEI Xingxin,et al. Per-formance test of water turbine of JP50 reel sprinkler[J]. China rural water and hydropower, 2014, (2):26—29.

[2] WANG Changwei, WANG Ruixin, ZHU Delan, et al. Development of automatic control device of translational sprinkling machine driven by solar energy[J]. Journal of agricultural mechanization research, 2014, 36(8):52—58.

[3] JIANG W, LIG, LIUPF, et al. Numerical investiga-tion of influence of the clocking effect on the unsteady pressure fluctuations and radial forces in the centrifugalpump with vaned diffuser[J]. International communica-tions in heat & mass transfer, 2016, 71:164—171.

[4] WANG Tao, WANG Chuan, KONG Fanyu, et al. Theoretical, experimental, and numerical study of special impeller used in turbine mode of centrifugal pump as turbine[J]. Energy, 2017, 130:473—485.

[5] ZAMIRI A, JINTC. Ability of URANS approach in prediction of unsteady turbulent flows in an unbaffled stirred tank [J]. International Journal of Mechanical Sciences, 2017, 133(11): 178—187.

[6] LI Zhantai, YANG Jungang, ZOU Guoyuan, et al. Current situation analysis of light and simplified production in greenhouse vegetable gardens in Beijing[J]. China Vegetables, 2019(8) : 68-75.

[7] Liu JINPING, Sun Feifei, Wang Xia, et al. Effects of fertilizer reduction on growth and quality of Tomato [J]. Northern Horticulture,2021(8):52-56.

[8] Han Xue, Qu Mei, Li Yinkun, et al. Effects of different fertilization levels on greenhouse tomato growth, nitrogen uptake and Soil and Fertilizer Science in China,2021(2):162-169.

[9] YUE Huanfang, LI Chao, LI Youli, et al. Studies on irrigation decision in substrate culture of high quality tomato[J]. China Vegetables, 2020(4) : 56-60.

[10] Pan Jing, GE Wen, Ye Yongwei. Integrated water Research on advantages, mechanism and significance of chemical technology [J]. New Agriculture, 2022 (3) : 16.

Journal of Physics: Conference Series **2365** (2022) 012049 doi:10.1088/1742-6596/2365/1/012049

Energy consumption analysis of intelligent temperature control steaming process of new concrete prestressed pipe piles

DongMin Zhang[1,2], YinCang Xu[1,2], PengLiang Huang[1,2]

[1]Shanghai Institute of Technology, College of Mechanical Engineering, Shanghai 201418, China

[2]School of Mechanical Engineering, Shanghai Engineering Research Center of Physical Vapor Deposition (PVD) Super Hard Coating and Equipment, Shanghai Institute of Technology, Shanghai 201418, China

1010778965@qq.com

Abstract: The traditional steam maintenance process of concrete prestressed pipe piles adopts steam maintenance, and the energy utilization rate is not high due to the inaccurate temperature automatic control system in the steaming pool and the serious loss of steam during the steaming process. In order to effectively ensure the strength and durability of concrete pre-products, shorten the production cycle at the same time, realize the automation of the steam pond temperature control system, the new maintenance process in the traditional maintenance process by manual temperature adjustment on the basis of intelligent control of temperature, thereby improving the accuracy of the steam pool temperature control system. In order to improve the energy utilization rate of the traditional steaming process of concrete prestressed pipe piles, the new concrete pipe pile maintenance process uses recycled water to maintain it, which has a certain energy-saving effect relative to the traditional steam maintenance, and calculates the energy consumption of the new steaming process and the energy consumption of the current steaming process of the enterprise to determine its energy saving.

1. Introduction

In recent years, China has vigorously advocated energy conservation and emission reduction. The revised GB/T 2589-2020 "General Principles of Comprehensive Energy Consumption Calculation" in 2020 has caused energy users across the country to continuously introduce new technologies and new processes to reduce production energy consumption [1]. Concrete prestressed pipe piles are widely used in construction, bridge, wharf and other industries. Steam curing is an essential part of its production process 70% of enterprises in China adopt steam curing [2]. Concrete pipe pile curing refers to the setting and hardening of concrete in normal or accelerated state to reach the design strength under a certain temperature and humidity environment [3].

Although the traditional steam curing method of concrete pipe pile effectively ensures the strength and durability of concrete pipe pile, its energy utilization rate is not high [4]. Therefore, how to reduce the energy consumption of steam curing process of concrete pipe pile has become a major topic in this industry [5]. In order to improve the energy utilization rate of concrete pipe pile steam curing process, this paper proposes a new water circulating concrete prestressed pipe pile steam curing process, calculates the energy consumption of the new steam curing process and compares it with that of the current steam curing process, and determines its energy-saving effect.

2. New steam curing process of concrete pipe pile

The new water circulation curing process uses hot water to cure the pipe pile, in which the hot water can be recycled. A steam curing cycle is also 7 hours, which is divided into four stages: resting, heating, constant temperature and cooling.

Design principle of new curing process of concrete pipe pile: effectively ensure the strength and durability of concrete prefabricated products, shorten the production cycle, and realize the automation of temperature control system of steam curing pool, that is, realize intelligent temperature control on the basis of manual temperature regulation of traditional steam curing process, so as to improve the accuracy of temperature control system of steam curing pool, which has a certain energy-saving effect compared with traditional steam curing.

3. Energy consumption calculation model of new water circulation steam curing process

3.1 Physical model

The steam curing tank is 16.1m long, 2.5m wide and 3.5m high. Each tank can steam curing 12 Φ600 pipe piles. The pool wall is made of 20cm thick concrete and the pool cover is made of 20cm thick rock wool, as shown in Figure 1.

Figure 1. Physical model of steam curing pond

3.2 Calculation model

Total energy consumption of the curing pool in a steam curing cycle = heat required for heating the pipe pile and pipe mold + heat loss of the pipe pile pool wall after opening the cover + heat loss of the air after opening the cover + heat loss of the heating pool cover + constant temperature heating during steam curing, as in equation (1).

$$Q_T = Q_P + Q_M + Q_A + Q_C + Q_W \tag{1}$$

Formula: Q_T is the total energy consumption of a steam curing cycle of traditional steam curing process, unit: kJ; Q_P refers to the heat required to heat concrete pipe pile and pipe formwork during steam curing, unit: kJ; Q_M is Heat required to maintain the constant temperature of the steam curing pool during the steam curing process, unit: kJ; Q_A is the heat consumption of the air in the free space of the steaming pool with the cover open for half an hour, unit: kJ; Q_C is the heat required for heating the tank cover in the process of steam curing, unit: kJ; Q_W is Heat loss from the tank wall in half an hour after the wall is opened, unit: kJ.

3.2.1 Heat required to maintain temperature during steam curing

The heat required to maintain constant temperature during steaming is calculated by equation, as in equation (2).

$$QM = 3.6AKt_1\omega(T_A - T_N) + 3.6AKt_2\omega(T_C - T_N) \tag{2}$$

Formula: Q_M is Heat required to maintain the constant temperature of the steam curing pool during the steam curing process, unit: kJ; A is the heat dissipation area, the surface area of the pool cover, unit: m^2; K is the heat transfer coefficient of the maintenance layer, in $w/(m^2 \cdot K)$; ω is the ventilation coefficient, taken as 1.3; T_A is the average temperature of the steaming tank, in °C; T_C is the constant temperature of concrete, in °C; t_1 is the time experienced in the heating stage, in S; t_2 is the time of constant temperature stage, in S; T_N is usually the normal temperature of the external environment, in °C.

$$K = \frac{1}{R} \tag{3}$$

$$R = \sum R_j \tag{4}$$

$$R_j = \frac{d_j}{\lambda_j} \tag{5}$$

Formula: R_j is thermal resistance of each layer in the flat wall, unit $(m^2 \cdot K)/W$; λ_j is the thermal conductivity of each layer in the flat wall, in $w/(m \cdot K)$; d_j is thickness of each layer in the flat wall, unit: m. as in equation(3-5).

3.2.2 Heat loss of tank wall after opening the cover for half an hour

Physical model: It can be seen from Figure 2 that the pipe piles are placed in four layers in the steam curing pool, three in each layer, and lifted out within 15min. Therefore, one layer is lifted out every 3.75min, and the heat dissipation depth of each layer is increased by 0.8m.

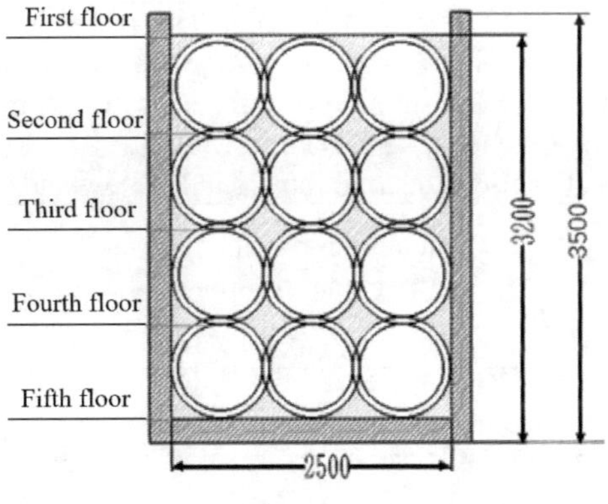

Figure 2. Side view of steam curing pond

Calculation model: The heat dissipation of the tank wall for half an hour after the cover is opened is mainly divided into two parts: One is thermal convection: the heat transfer phenomenon occurs when the ambient air outside the steam curing pool flows through the pool wall and the surface of the pool cover due to the temperature difference[6].The second is thermal radiation: the cover and wall of the steaming pool emit electromagnetic waves to the external environment of the steaming pool and other objects in the workshop, resulting in heat transfer[7].

Convective heat transfer between pool wall and external environment, as in equation (6-7).

$$Q_{TC1} = F_1 h (T_C - T_N) t \tag{6}$$

$$F_1 = 2(L_1 + L_2 + L_3) \tag{7}$$

Formula: Q_{TC1} refers to the convective heat dissipation between the pool wall and the external environment when the first layer of pipe piles are brought out after the cover is opened, in kJ; F_1 is heat dissipation area of the first heat dissipation layer, unit: m^2; t is heat dissipation time of the first heat dissipation layer, unit: S; h is convective heat transfer coefficient; L_3 is heat dissipation depth of the first heat dissipation layer, unit: m; L_1 is heat dissipation length of the first heat dissipation layer, unit: m;L_2 is heat dissipation width of the first heat dissipation layer, unit: m.

Similarly, the convective heat transfer between the second, third and fourth layer pool wall and the external environment can be calculated.

Radiation heat transfer between pool wall and external environment, as in equation (8-9).

$$Q_{TR1} = \varepsilon_W C F_1 t\left[(T_C)^4 - (T_N)^4\right] \tag{8}$$

$$F_1 = 2(L_1 + L_2)L_3 \tag{9}$$

Formula: Q_{TR1} refers to the radiant heat dissipation of the pool wall and the environment when the first layer of pipe piles are pulled out after the cover is opened, in kJ; ε_w is the blackness of concrete pool wall, generally 0.92; C is the blackbody radiation coefficient, taken as 5.67×10^{-8}, unit: w / ($m^2\cdot$ K).

Similarly, the radiant heat between the second, third and fourth layer pool wall and the external environment can be calculated.

4. Comparison of energy consumption between new and old steam curing processes

This energy consumption comparison is carried out under the conditions of ensuring the same floor area, steam curing time and output of the old and new steam curing processes.

Enterprise measurement method: using the conservation of mass, the quantity of boiler water burned in a steam curing cycle is the quality of steam flowing into the steam curing pool, and the temperature of steam flowing into the steam curing pool is 95 ℃.

In this paper, the energy consumption results of the new steam curing process are calculated and compared with the measured energy consumption of the current steam curing process. In a steam curing cycle, each tank saves 19.8% energy, as shown in Table 1.

Table 1. Comparison of energy consumption results of new and old curing processes

Curing process	Calculation results of new steam curing process	Actual measurement of traditional steam curing process enterprises
Total energy consumption / kJ	4950000	6170000
Equivalent to 95 ℃ steam / kg	2085	2600

5. Conclusion

In order to improve the energy utilization rate of the steam curing process of concrete prestressed pipe piles and save the energy consumption of enterprises, this paper proposes a new water circulation concrete prestressed pipe pile curing process. The energy consumption results of the new curing process are calculated and compared with the measured energy consumption of the current steam curing process. In a steam curing cycle, each tank saves 19.8% energy, which provides a reference and data basis for the optimization of the steam curing process of concrete prestressed pipe piles.

Reference

[1] Chen H H, Li P C, Wang Z H. Interpretation of GB/T 2589-2020 national standard of "General Principles of Comprehensive Energy Consumption Calculation"[J]. Standards Science, 2021 (12): 107-111.

[2] Chen L, Chen G X, Su F. Research status of mechanical properties of steam cured concrete at home and abroad [J]. Comprehensive utilization of fly ash, 2016 (05): 61-64.

[3] Zhang G Z, Ge J C, Zhang C X. Influence of curing system on the formation mechanism of concrete microstructure [J]. Material guide, 2021,35 (15): 15125-15133.
[4] Yi L, Li Y Y, Peng B. Analysis of energy consumption status and energy saving potential of ready mixed concrete [J]. Concrete, 2012 (11): 87-90+93.
[5] Liu M S. ZEK new steam curing system and energy saving [J]. Concrete, 1993 (04): 50-52
[6] Liu H, Jiang L F, Zhao Q. Calculation model of variable temperature heat dissipation of LED radiator [J]. Semiconductor optoelectronics, 2012,33 (04): 467-469+502 DOI:10.16818/j.issn1001-5868.2012.04.003.
[7] Zeng X. Calculation of heat dissipation loss of steam heating network and analysis of its influencing factors [J]. Energy and environmental protection, 2021,35 (04): 80-84.

Intelligent 3D Visual defect Detection for Industrial Applications

Rui Yang[1,2], Hualin Ke[1*], Lin Gan[1], Yuhong Li[1] and Ruihan Hu[1,3,4,5*]

[1] Guangdong Institute of Intelligent Manufacturing, Guangdong Key Laboratory of Modern Control Technology, Guangzhou 510070, China

[2]Cloud Computing Center Chinese Academy of Sciences, Dongguan 523808, china

[3]Guangdong Key Laboratory of Modern Control Technology, Guangdong Institute of Intelligent Manufacturing, Guangzhou, China

[4]Engineering Research Center of Intelligence Perception and Autonomous Control, Ministry of Education, Beijing, China

[5]Xinjiang Production and Construction Corps Key Laboratory of Modern Agricultural Machinery, Shihezi University, Shihezi, China

[*]Corresponding author's e-mail: hl.ke@giim.ac.cn, rh.hu@giim.ac.cn

Abstract. Currently, new energy industries such as electric vehicles and energy storage batteries are experiencing rapid growth throughout the world. As a recognized ideal energy storage element, lithium batteries have also attracted considerable interest. To improve image contrast, the industry employs image preprocessing algorithms. Due to the high requirement of image consistency in the template matching method, which is difficult to meet in practice, and the large apparent difference between defects, it is challenging to design features. Therefore, the framework for defect detection is based on a method of deep learning with strong feature expression capability. In light of the severe imbalance in the number of samples in defect detection, data augmentation and generation methods are used to simulate real samples in order to improve the training effect of deep neural networks and alleviate the burden of data annotation to some extent.

1. Introduction

Using deep convolution neural networks, an intelligent defect detection framework based on deep learning is proposed. The first step in image preprocessing is to enhance the contrast of the image in order to make it easier to distinguish between defect and non-defect areas in the image. Afterwards, a deep convolutional neural network is used to extract the features of the entire image. [1-3] A feature pyramid was constructed to combine the low-level features of fine resolution with the high-level features containing abstract semantic information to enhance the semantic information of features and improve their discrimination ability. In addition, three-dimensional depth information is introduced to identify defects like multiple materials, pits, convex surfaces, and scratches, which are difficult to identify with two-dimensional plane images. Due to the nonlinear mapping ability of depth neural networks, plane image information and depth information can be mapped to the same space, and adaptive fusion occurs at the feature level in order to take full advantage of the three-dimensional input information and extract defects' characteristics. Considering the relatively low defect rate in practice, defect detection faces serious sample shortages and imbalances. [4-5] In order to enhance the training effect of deep neural networks and prevent over-fitting, data augmentation and generative methods are employed in the training phase. Furthermore, the method of generating countermeasure networks is used to further resolve the problem of small sample sizes. [6-7] A generation model is used to generate new training data automatically and randomly from the existing training samples.

2. Lithium ion battery manufacturing defect detection

In an industrial environment, machine vision-based defect detection technology is one of the most important links in ensuring product quality. Despite these obstacles, intelligent defect detection technology continues to face numerous obstacles, such as low contrast between defect areas and non-defect areas, similarity between noise and subtle defects, slow detection speed, and low recognition accuracy. Due to the poor performance consistency of defects, the template matching method cannot be used, and traditional pattern recognition methods find it difficult to define defects based on the design characteristics. Since it simulates the visual processing mechanism of the human brain, the deep convolution neural network is capable of automatically learning sample characteristics. Therefore, it is appropriate for image processing applications.

2.1. Manufacturing segment - burr detection

As shown in the typical morphology of the electrode slice slitting section, the main particles of the coating peel off and break. However, the fluid collector has been cut and torn. Some particles of the electrode coating are cut off when the compaction density of the electrode coating increases and the binding force between the particles increases. Lithium batteries are greatly damaged by burrs, particularly metal burrs. A large metal burr penetrates the diaphragm directly, resulting in a short circuit between the positive and negative electrodes. During the manufacturing process of lithium ion batteries, the electrode slitting process is the primary source of burrs. (Fig. 1).

Figure 1. Burr

2.2. Visual inspection difficulties

This scenario involves the following conditions: coil linear speed: ≥ 1.5mm/s; Coil jitter ± 0.5mm; The difficulty lies in the process requirements: ≥ 0.005mm burrs; Polishing: for this burr, only the bottom light has the best imaging effect; camera: high resolution is required to achieve this accuracy, and the sampling frequency of high resolution is not sufficient; lens: high resolution with a small field of view can achieve accuracy. Small fields of view have a small depth of field, and the measurement result is affected by jitter; Operation: from data collection to data processing to result output, it should be completed in milliseconds. (Fig. 2).

Journal of Physics: Conference Series **2365** (2022) 012050 doi:10.1088/1742-6596/2365/1/012050

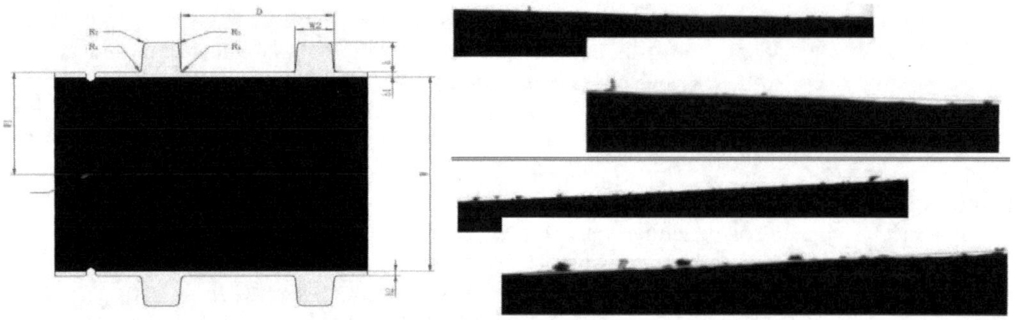

Figure 2. Burr image

3. Intelligent detection of laser 3D imaging

A 3D laser imaging system can be used for detecting hidden targets, mapping terrain, building virtual environments, urban modeling, and recognizing targets, among other applications. It is also practical to use real-time 3D sensing in regional imaging technology in addition to more conventional technologies such as stereo vision and structured lighting (Fig. 3-4).

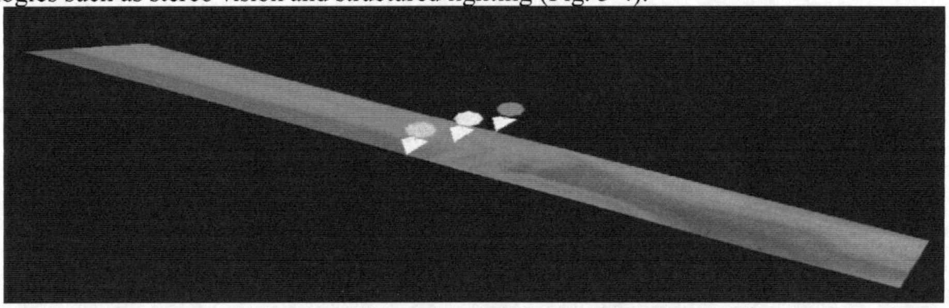

Figure 3. Horizontal measurement

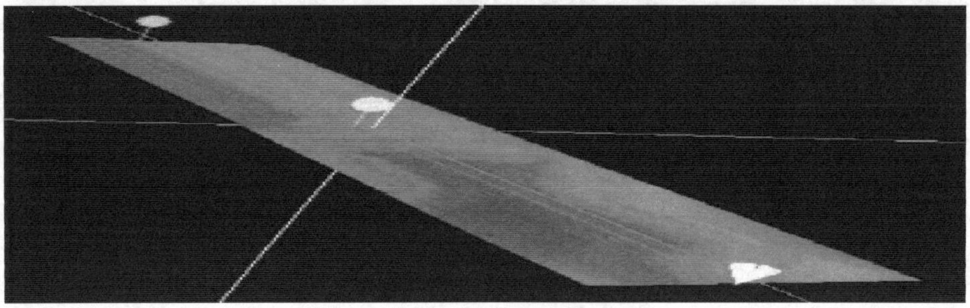

Figure 4. Vertical measurement

4. Conclusion

After systematic researches on the newest computer vision coordinate measuring technology and the image acquiring and processing technology, this paper brought forward a vision coordinate measuring technology based on the rectangular grid laser mark, and put it into use in a real vision coordinate measuring system. This technology filled up the blank of related fields in our country. It overcome the shortcomings of being unstable and low efficiency that other measuring means have.The sample, for instance, can be observed to contain a burr of 5um after intelligent measurement(see table 1) and observation of the coordinates on the right side (Figs. 5-7) .

Table 1. Demand accuracy.

Requirement item	**performance index**
Width of striped pole piece	±0.1mm

Polar ear height	±0.1mm
Shangdaolu / intaglio(1)	±0.1mm
Shangdaolu / intaglio(2)	±0.1mm
Polar ear spacing	±0.1mm
Polar ear width	±0.1mm

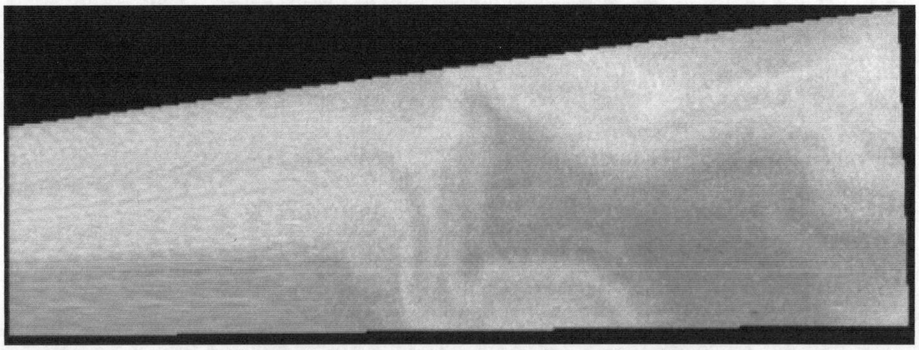

Figure 5. An example of an original drawing

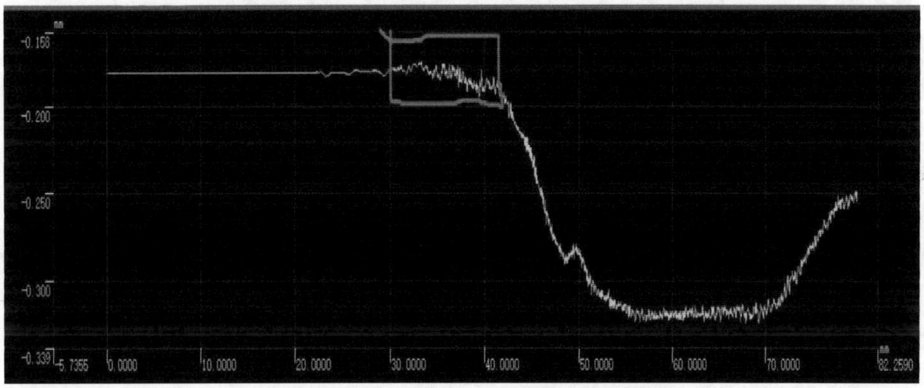

Figure 6. Intelligent detection results

Acknowledgments
This work was supported in part by the National Natural Science Foundation of China Youth Fund under Grant 6210023461, in part by the Natural Science Foundation of Guangdong province under Grant 2022A1515011749, in part by the Guangdong Academy of Sciences' (GDAS') Project of Science and Technology Development under Grant 2017GDASCX-0115 and Grant 2018GDASCX-0115, in part by the Guangdong Academy of Science for the Special Fund of Introducing Doctoral Talent under Grant 2021GDASYL-20210103087, in part by the Opening Foundation of Xinjiang Production and Construction Corps Key Laboratory of Modern Agricultural Machinery under Grant BTNJ2021003.

References

[1] Ivanova Bojidarka, Michael Spotteller. Stochastic dynamic quantitative and 3D structural matrix-assisted laser desorption/ionization mass spectrometric analyses of a mixture of nucleosides [J]. Journal of Molecular Structure, 2022, 1260.
[2] Wang Xin, Zhao Dong, Yao el, etc. 3D-printing of pink bioceramic scaffolds for bone tumor tissue therapy [J]. 2022,27 Applied Materials Today

[3] Hua Jianguan, Liang Shuyu, Chenel, etc. Femtosecond Laser 3D Sculpturing with FreeForm MicroOptics from Crystals [J]. Advanced Functional Materials,2022, 32 (26). Miller, A. T., and Allen, P. K., Graspit! A Versatile Simulator for Robotic Grasping [J]. IEEE Robotics and Automation Magazine, vol. 11, no. 4, pp. 110-122, 2004.
[4] Cheng Qisen,Qu Shuhui,Lee Janghwan. 72 ‐ 3: Deep Learning Based Visual Defect Detection in Noisy and Imbalanced Data[J]. SID Symposium Digest of Technical Papers,2022,53(1).
[5] Nandy Manali,Lahiri B.B.,Yadhukrishna C.H.,Philip John. Poly acrylic acid stabilized magnetic nanoemulsions for visual defect detection: Effect of pH on detection sensitivity and colloidal stability[J]. Journal of Molecular Liquids,2021,336.
[6] Hongyao Shen,Wangzhe Du,Weijun Sun,Yuetong Xu,Jianzhong Fu. Visual Detection of Surface Defects Based on Self-Feature Comparison in Robot 3-D Printing[J]. Applied Sciences,2019,10(1).
[7] Huangpeng Qizi,Zhang Hong,Zeng Xiangrong,Huang Wenwei. Automatic Visual Defect Detection Using Texture Prior and Low-Rank Representation[J]. IEEE Access,2018,6.

Journal of Physics: Conference Series **2365** (2022) 012051 doi:10.1088/1742-6596/2365/1/012051

Research on the path planning of transportation UAV in rasterization scene based on ant colony algorithm

Lihong Cheng, Yanping Xiao, Chensheng Ma

Systems Engineering Research Institute, Academy of Military Sciences

928756645@qq.com

Abstract: Aiming at the difficulty of material distribution on the island, this paper uses the ant colony algorithm to simulate the path of the transport UAV based on the two-dimensional gridded island scene. Since the ant colony algorithm has significant advantages of strong robustness and high scalability, this paper fully verifies the feasibility of the transportation UAV path planning in the rasterized island scene.

1. Introduction

China has the vast sea area and many islands, many of which are far away from the mainland coastline,which makes it difficult for garrison personnel to provide logistical support. Nowadays, the rapid development of the transport -UAV has become a new star in the logistics distribution resources. Due to its low cost, strong site adaptability and low requirements for space and time, the use of transport-UAV will greatly improve the material support ability of our personnel stationed on the island. Due to the late start of material delivery using transportation UAV, there are many problems in path planning, task allocation and other fields. Based on the ant colony algorithm, this paper verifies the feasibility of using transportation UAV to implement material transportation path planning in island scenarios.

Aiming at the problems of cross execution of general multi UAV collaborative planning algorithm simulation system or insufficient expansibility of the algorithm, G. Gao, Y. Zhang, J. Qian and others developed a multi UAV task collaborative planning simulation system based on Web technology. The system can bypass the specific independent operating system and allow the browser to directly access the server, which greatly improves the practicability of the algorithm system. The system framework is divided into four aspects: user management, scheme configuration, algorithm simulation and management algorithm. This method not only solves the problem of poor scalability of general algorithms, but also effectively solves the problem of conflict in the running environment of algorithms.[1]

J. Chen, F. ye, T. Jiang and others believe that ACO ant colony algorithm is a swarm intelligence algorithm in which ants and their related information elements cooperate. It has good stability and good algorithm expansibility. In this paper, the author successively increases the number and complexity of obstacles, and the ant colony algorithm can still optimize the path of UAV, which reflects the practicality of the algorithm.[2]

In order to compensate for the slow convergence speed of the ant colony algorithm and the relative ease of falling into the local optimal solution, L. Huan, Z. Ning and L. Qiang optimized and improved the basic ant colony algorithm from the aspects of node pheromone updating and guidance factor updating, that is, the reciprocal of the distance from the starting point to the target point was added to the route planning, and a new guidance factor was added to the original guidance factor, This can make the path planning give priority to the selection of the straight line between two points, and compared with the path planned by the basic ant colony algorithm, the path planned by the improved algorithm is smoother and closer to the optimal solution.[3]

2. Principle analysis of ant colony algorithm

Ants can subconsciously choose the optimal path on the way to forage. According to research, ants will release pheromone pheromones on the way of foraging. Over time, the pheromone in the path will increase or decrease. Ants choose The probability of a certain path is positively correlated with the number of pheromones on the path. When there are many pheromones on a certain path, the probability of ants choosing this path is high, and gradually the pheromone concentration on this path will become higher and higher. This will also allow more ants to choose the path, and it is precisely because of this positive feedback mechanism that the ant colony will find the optimal path. [4]

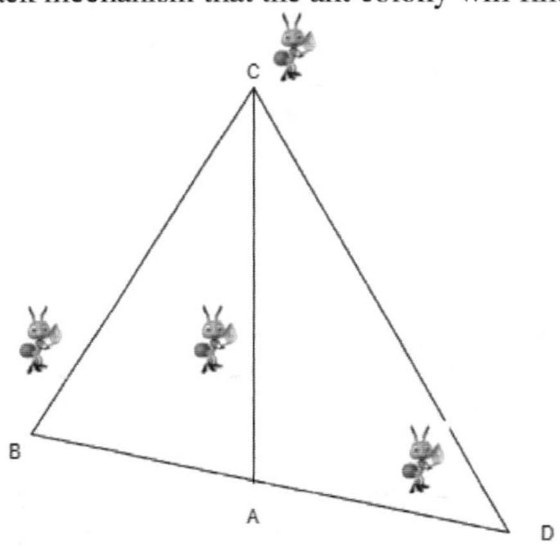

Figure 1 Schematic diagram of the principle of ant colony algorithm

A to point C. Initialize the process and assign an ant to each path. The length of each path is $D_{AC}=\frac{1}{2}D_{ABC}$, $D_{AC}=\frac{1}{3}D_{ADC}$. Assuming that each ant travels at the same speed, it takes two Unit time, it takes four units of time along A BC , and 6 units of time along A DC . When t = 4 , the AC segment ants have walked back and forth, and the A BC segment ants just walked from A to C, ADC The segment ants are right at the ADC segment $\frac{2}{3}$. When t = 24 , the ants in the A and C segments have traveled back and forth in the A and C segments 6 times, the A BC segment ants have shuttled back and forth in the A BC segment 3 times, and the A DC segment ants have shuttled back and forth in the ADC segment 2 times. At this time, the information on the three paths The element content ratio is: $D_{AC}:D_{ABC}:D_{ADC}=2:3:6$, according to the positive feedback mechanism of the ant colony, gradually the ants of other paths will finally choose the shortest path A C .

When modeling, it is necessary to add a taboo table to each ant to record the city nodes that the ant has walked through (the path cannot be repeated). The first position of the taboo table represents the landmark of the ant at the initial time. When the ants have walked all the city nodes, all the cities The nodes are also added to the taboo list, that is, the ant completes a one-week trip [5]. $\Gamma_{ij}(t)$ Represents the amount of pheromone between city nodes $i-j$. At the initial time, the amount of pheromone is constant.

According to the ant-cycle model , the moving probability of ant k can be known:

$$P_{ij}^{k}=\frac{(\Gamma_{ij}^{\alpha})(\eta_{ij}^{\beta})}{\sum_{z\in allowedx}(\Gamma_{ij}^{\alpha})(\eta_{ij}^{\beta})} \tag{1}$$

P_{ij}^{k} Represents k the probability that the ant chooses to follow $i-j$, α represents the concentration or density of pheromone pheromone , which is positively correlated with the probability of the ant choosing the relevant path. When the α value is larger, the ants are more inclined to take the relevant path. When α value is smaller, the ant colony is likely to be limited to local optimum; β it indicates the relative importance of the heuristic factor. The larger the value, the β greater the possibility that the ant chooses the local optimal path at a certain point; the heuristic factor $i-j$ is the expected degree of the ant 's follower, $\eta_{ij}=(d^{ij})^{-1}$, the $i-j$ is the reciprocal of the distance ,which means that the greater the distance between the two places is, the lower the expected value of the ants passing through the two places is;

When all ants visit each node, the pheromone concentration on the path will change. At this time, we should update the pheromone of the ant colony in real time:

$$\Gamma_{ij}(t+1)=(1-\rho)\Gamma_{ij}(t)+\Delta\Gamma_{ij} \tag{2}$$

Among them $\Delta\Gamma_{ij}=\sum_{i=1}^{m}\Delta\Gamma_{ij}^{k}$, it represents in the $i-j$ the sum of the pheromone released by the ants on the path ; according to the ant-cycle model $\Delta\Gamma_{ij}^{k}=\frac{Q}{L_k}$, it represents the pheromone released by the $i-j$ ants on the path k; $\Gamma_{ij}(t)$ represents the state of the t pheromone on the path at ρ time ; $i-j$ The volatilization coefficient over time, also representing the concentration of pheromone, ρ generally has a value range about $\rho<1$; $(1-\rho)$ representing the remaining amount of pheromone.

3. Experimental Analysis of Ant Colony Algorithm Parameters

When the ant colony algorithm is used to find the optimal solution, its parameters such as α、β、ρ、m、Q the definition and value have a great influence on the results. The following is a simulation experiment with matlab software, taking one parameter as a variable and other parameters unchanged to obtain the results to compare and analyze the optimal solution. value range.

First set the number of destinations $n=30$. According to literature analysis, the number of ants is m generally 1-1.5 times the number of destinations. In this experiment, the number of ants is constant at 45 , the number of iterations is constant at 150 , and the pheromone intensity is Q constant at 100 . [6]

(1) Experiment 1 the optimal solution is compared by changing the value of the heuristic factor α and the value of the expectation factor β

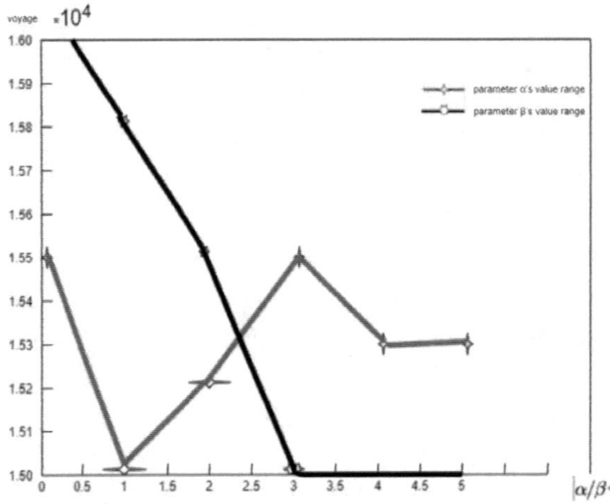

Figure 2 Relationship between heuristic factor α/expectation factor β and optimal path length

(2) Experiment 2 : The other parameters remain unchanged, and the optimal solution is compared by changing the value of the pheromone coefficient ρ:

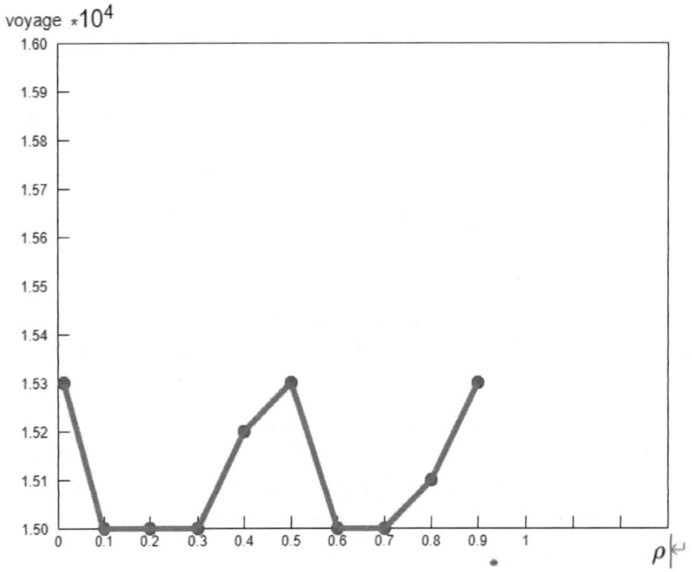

Figure 3 Relationship between pheromone coefficient ρ and optimal path length

Through simulation experiments, the relationship between the range of parameters and the corresponding optimal path length can be obtained. In experiment 1 , the path obtained by the ant colony algorithm was the shortest at that time, and this was the optimal solution; in experiment 2, $\alpha=1\ \beta\in[3,5]$ the path obtained by the ant colony algorithm was the shortest at that time, and this was the optimal solution; in the experiment In the third , the path obtained by the ant colony algorithm at that time is the shortest, and it is the optimal solution at this time. $\rho\in[0.1,0.3]\cup[0.6,0.7]$

Through the simulation and comparative analysis of the above series of experiments, the ideal value range of the parameters of the ant colony algorithm is obtained. Carry out a simulation experiment to find the optimal route. This experiment using ant colony algorithm to plan the route ignores the elevation in the actual environment, and conducts a simulation experiment in a two-dimensional environment, analyzes the advantages and disadvantages of the algorithm and the track, and proposes an improvement plan for the problem. In order to accurately compare the defects of the algorithm and improve the authenticity of the simulation results, this experiment was carried out 25 times.

3.1 The computing environment of ant colony algorithm under grid conditions

Operating system Windows 10 64 bit ; processor AMD Ryzen 5 2500U with Radeon Vega Mobile Gfx 2.00 GHz ; system type 64-bit operating system, x64-based processor ; processor model : AMD Ryzen 7 5700 , main frequency : 1.8GHz maximum Turbo frequency : 4.3GHz ; Matlab, R2021a ; memory 512 GB .

3.2 Establish a simulation environment

In this simulation experiment, it is assumed that the speed of the UAV is constant at 8m/s, and the constructed environment model is 20×20 a grid topographic map simulating an island. There are 400 grids to divide the terrain. The numbers are 1-400 from top to bottom . In the running code, the terrain matrix is divided into 0 and 1, where 1 represents the mountain or local threat area or meteorological threat area, 0 represents the flyable area, and each square represents 1 unit length : the grid terrain map is as follows:

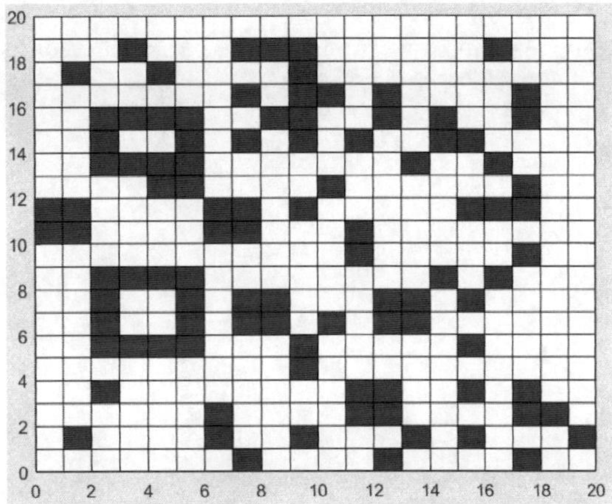

Figure 4 Grid simulation environment diagram

3.3 Set the ant colony algorithm parameter value as

The number of iterations K=100; the number of ants M=60; the parameter representing the importance of the pheromone Alpha= 1 ; the parameter representing the importance of the heuristic factor Beta = 4 ; the pheromone evaporation coefficient Rho = 0.6; the pheromone increase intensity coefficient Q = 1

1. Write and modify the grid terrain code of the threatened and set obstacles and the flyable area in the scene and the ant colony algorithm code to find the optimal path under the grid terrain.
2. Obtain the initial track through the basic ant colony algorithm.
3. Carry out 25 simulation experiments to output the operation results and analyze them.

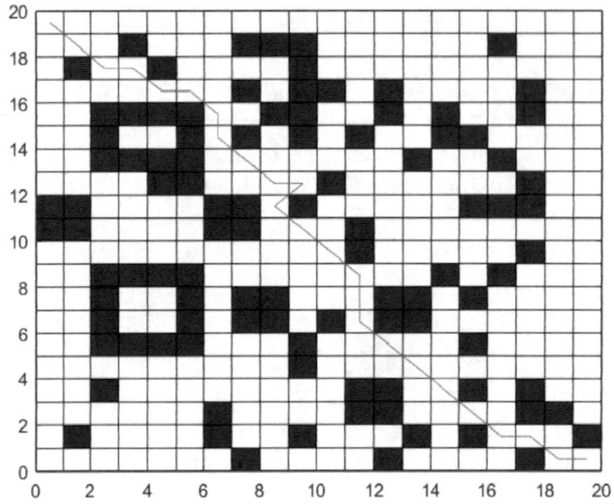

Figure 5 Route planning diagram under the basic ant colony algorithm

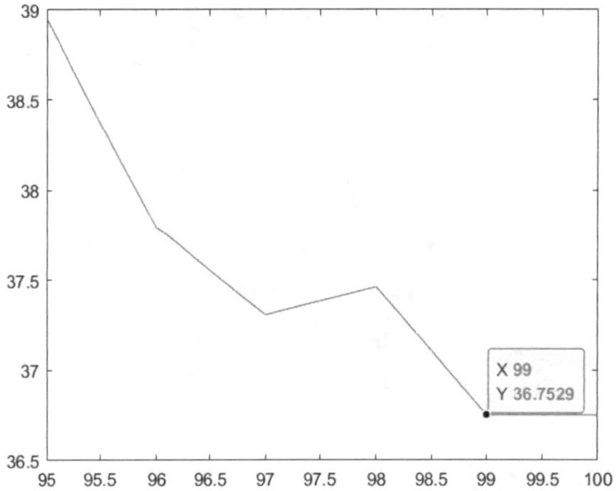

Figure 6 Fitness evolution curve/number of iterations under the basic ant colony algorithm

Through many experiments, we can see that although the basic ant colony algorithm can reasonably avoid obstacles to achieve a complete track search, the route fluency is low, and there are some invalid paths, which greatly increases the actual route length. The main reason is that the pheromone released by the ants when they randomly search the path on the past path has an impact on the subsequent ant search, making more ants mistakenly believe that it is the best route in the current path, It leads to the optimization of operation results under local conditions. In addition, the basic Ma group algorithm in the fitness curve finds the optimal solution after 99 pairs of iterations, which is slow and takes a long time. In order to further optimize the route of the transport UAV in the island scenario, we need to further improve and optimize the algorithm in the next step.

4. Conclusion

In order to help my country's strategy of strengthening China's maritime power and effectively promote the application of transportation drones in the logistics and material support of my country's border defense islands, this paper takes the simulated rasterized island as the scene, and through the improvement of the ant colony algorithm to verify that the transportation drone is in use. Feasibility of path planning in rasterized scenarios. Through comparison, it is found that the improved ant colony algorithm shortens the voyage and navigation time of the transport drone, which greatly improves the efficiency of transporting materials in this scenario. However, due to the continuous development of artificial intelligence technology, transportation The algorithm of the UAV will become more intelligent, which is the defect of the experiment in this paper, and it is also an important direction for the development of the transportation UAV system in the future.

References

[1] G. Gao, Y. Zhang, J. Qian and Y. Sun, "The Lightweight System for UAV Cooperative Mission Planning Simulation," 2019 International Conference on Cyber-Enabled Distributed Computing and Knowledge Discovery (CyberC), 2019, pp. 371-374, doi: 10.1109/CyberC.2019.00070.

[2] J. Chen, F. Ye and T. Jiang, "Path planning under obstacle-avoidance constraints based on ant colony optimization algorithm," 2017 IEEE 17th International Conference on Communication Technology (ICCT), 2017, pp. 1434-1438, doi: 10.1109/ICCT.2017.8359869.

[3] L. Huan, Z. Ning and L. Qiang, "UAV Path Planning Based on an Improved Ant Colony Algorithm," 2021 4th International Conference on Intelligent Autonomous Systems (ICoIAS), 2021, pp. 357-360, doi: 10.1109/ICoIAS53694.2021.00070.

[4] Longwang Yue,Hanning Chen. Unmanned vehicle path planning using a novel ant colony algorithm[J] . EURASIP Journal on Wireless Communications and Networking . 2019 (1)

[5] Qiang Luo,Haibao Wang,Yan Zheng,Jingchang He. Research on path planning of mobile robot based on improved ant colony algorithm[J] . Neural Computing and Applications . 2020 (6)

[6] Xiao Jing,Li Chaoqun,Zhou Jie.Minimization of Energy Consumption for Routing in High-Density Wireless Sensor Networks Based on Adaptive Elite Ant Colony Optimization[J]. ,Rivero-Angeles Mario E.. Journal of Sensors . 2021

Using an Improved Artificial Hummingbird Algorithm for Vision-Guided Optimization And Grasping of Multi-Objective Robots

Rui Yang[1], Qian Zhou[1,5*], Qinglong Mo[1], LinGan[1] and Ruihan Hu[1,2,3,4*]

[1] Guangdong Institute of Intelligent Manufacturing, Guangdong Key Laboratory of Modern Control Technology, Guangzhou 510070, China

[2]College of Electronic Countermeasures, National University of Defense Technology, Hefei 230037, China

[3]State Key Laboratory of Pulsed Power Laser, Hefei 230037, China

[4]Key Laboratory of Electronic Restriction of Anhui Province, Hefei 230037, China

[5]Chaozhou Zhongchao intelligent manufacturing innovation center, Chaozhou, China

[*]Corresponding author's e-mail: q.zhou@giim.ac.cn, rh_hu13534262@163.com

Abstract. AHA (artificial hummingbird algorithm) is a meta-heuristic optimization algorithm that simulates the unique flight and feeding patterns of hummingbirds. By combining the excellent strategy of the artificial hummingbird algorithm with the framework of the multi-objective optimization algorithm, the multi-objective artificial hummingbird algorithm is created, improved, and further enhanced. Therefore, the research on vision-guided optimization and grasping control of multi-objective robots based on an enhanced artificial hummingbird algorithm has significant academic significance and vast potential for application.

1. Introduction

The manufacturing industry is a pillar industry in China, accounting for approximately 35% of China's GDP. The 19th of May, 2015 marked the official release of made in China 2025 by the State Council, which aims to further improve the manufacturing industry's overall development level and international competitiveness. The nation is in a crucial stage of transition from a large manufacturing nation to a powerful manufacturing nation. With the gradual decline of China's labor advantage, intelligent robots will play an increasingly vital role in discrete manufacturing. Engelberger and DeVol created the first industrial robot in the 1950s. Since then, the robot has garnered significant interest from scholars around the world. After more than 60 years of rapid development, robots are now widely used in a variety of industries, including industrial production, family services, and danger rescue. According to survey data from the International Federation of Robotics (IFR)[1-3], the global robot market reached $29.41 billion in 2019, with industrial robots accounting for $15.92 billion and service robots for $9.46 billion. From 2014 to 2019, the average annual growth rate is 12.3%, and the robot market will continue to expand at a rapid rate for a long time to come. Even though the market for robots is substantial and expanding rapidly, the intelligence of the majority of robots is still quite low, and they can only perform relatively simple functions in a structured environment. For instance, in industrial production, robot grasping is accomplished primarily through preprogramming, which lacks autonomous initiative and adaptive ability in complex environments and makes it challenging to meet the requirements of modern industrial intelligent manufacturing and flexible manufacturing. As a fundamental ability of robots, grasping is crucial to their intelligent development. If robots are used to replace humans in completing various tasks, it is inevitable that robots will be required to independently perform more complex grasping tasks. Therefore, research on intelligent grasping technology for robots in unstructured environments has become a significant topic. The robot grasping methods can be divided into two categories, analysis-based and data-driven, based on the calculation form of the grasping pose of the target object. The analysis-based robot grasping method[4-6] must first obtain an accurate geometric model of the object and a dynamic model of the mechanical claw, and then establish a complex objective optimization function through a variety of constraints. The optimal grasping position of the target object can now be determined. In actual robot grasping applications, however, these conditions are often difficult to meet, limiting the scope of this grasping method's application. Vision is known as the "eye of wisdom" of robots, and the objective of the data-driven grasping algorithm is to determine the grasping pose of a target object using a visual recognition method.

Meta heuristic algorithm features a simple concept, straightforward implementation, absence of gradient information, and multi-disciplinary applicability. It has been widely employed to solve practical engineering issues. The natural heuristic-based meta heuristic algorithm is proposed by simulating biological or physical phenomena. First, the initial population is generated at random, and then each generation's members are modified according to certain iterative rules. Population update techniques include the evolution-based method, the physics-based method, and the population-based method. These methods achieve the objective of optimization by continuously selecting the optimal individuals to populate the following generation. On the basis of species evolution theory, evolution based heuristic algorithms are primarily proposed, such as genetic algorithm (GA), probability based incremental learning (PBIL), evolution strategy (ES), genetic programming (GP), and biogeography based optimizer (BBO)[7-9]. ro:ray optimization algorithm, sa:simulated annealing algorithm, acroa:artificial chemical reaction optimization algorithm, bh:black hole algorithm, small world optimization algorithm, Gbsa:galaxy -based search algorithm, and cso:curved space optimization are examples of physics-based heuristic algorithms that simulate the physical rules of the universe[11-18]. Primarily, the group-based method mimics the social behavior of animal groups. Particle swarm optimization (PSO), developed by Kennedy and Eberhart[19] is the most popular method. The PSO algorithm obtains the optimal solution, namely the optimal position, based on the flight of multiple particles throughout the search space. In each iteration, the particles are accelerated toward the optimal position. In addition, ant colony optimization (ACO) simulates the social intelligence behavior of ants in ant colonies in order to find the path closest to the nest and food source[20-26]. The population-based algorithm has more advantages than the evolution-based and physics-based algorithms, retains search information during the iterative process, and has fewer operational variables, making it easier to implement. Modern heuristic algorithms have distinct mechanisms for optimization, but similar optimization processes. They all begin with a set of initial solutions, establish key parameters, generate multiple solutions based on certain rules, update the current state based on acceptance criteria, and conduct multiple iterative searches until the optimal results are found. Widespread use of meta heuristic algorithms to solve practical engineering problems. In 2013, the Hummingbird algorithm was first proposed. It optimizes and affects the Google search engine. Its function is to perform a comprehensive comparison between the contents of the Google index database and the keywords or queries entered by users of the Google search engine.

2. Detecting and capturing 3D targets with an intelligent robot in an unstructured environment around discrete manufacturing

Analyzing and defining the advantages and disadvantages of the current multi-objective artificial hummingbird algorithm, this paper explores an improved algorithm and its basic function guided optimization effectiveness verification method. Based on improved multi-objective artificial hummingbird algorithm, optimal guidance control method of visual robot is investigated for multi-objective workpiece recognition and grasping in 3D unstructured environment. Taking into consideration the control based on grasping contact force and position, a gripper of a flexible automatic manipulator is designed based on the control based on grasping contact force and position.

2.1. Improvements to the multi-objective artificial hummingbird algorithm and its basic function optimization guidance

Weiguo Zhao et al. proposed the artificial hummingbird algorithm (AHA) in 2021. Figure 4 illustrates the unique flying skills and intelligent foraging strategies of hummingbirds in nature. The artificial hummingbird algorithm mimics the guiding foraging strategy of the hummingbird, combines the excellent strategy of the artificial hummingbird algorithm with the multi-objective optimization algorithm framework to form the multi-objective artificial hummingbird algorithm, and improves the multi-objective artificial hummingbird algorithm and its verification of its basic function optimization guidance effectiveness (Fig. 1).

In the search process of artificial bee colony algorithm, initialization is required first, including determining the population number, the maximum number of iterations MCN, the control parameter limit and determining the search space, that is, the range of solutions. The initial solutions Xi (i=1,2,3,..., Sn) are randomly generated in the search space. SN is the number of food sources. Each solution Xi is a d-dimensional vector, and D is the dimension of the problem. After initialization, the whole population will repeat the search process of leading bee, following bee and reconnaissance bee until the maximum number of iterations MCN or the allowable error value is reached ε.

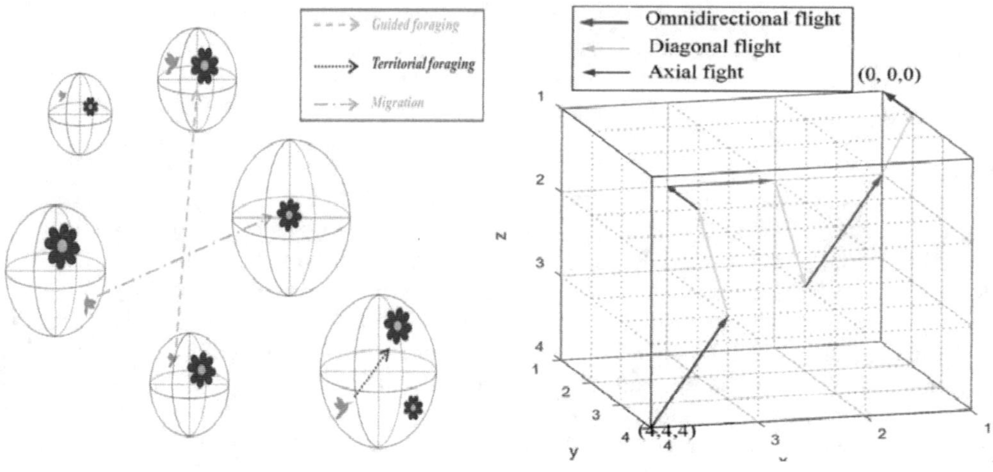

Figure 1. Schematic diagram of strategy principle and guiding foraging diagram

2.2. Design of a gripper for a flexible automatic manipulator that controls grasping force and position

Based on grasping force/position control, the flexible automatic manipulator develops a feedback sensing machine gripper. By organically integrating off-line training results with in-depth reinforcement learning, it facilitates the in-depth reinforcement learning process, complements each other, and improves the efficiency and real-time performance of in-depth reinforcement learning to complete high-precision and personalized disordered sorting of industrial production (Fig. 2).

Journal of Physics: Conference Series **2365** (2022) 012052 doi:10.1088/1742-6596/2365/1/012052

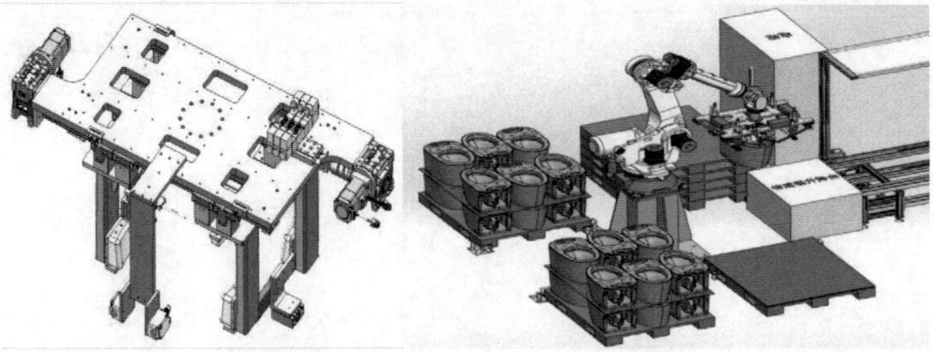

Figure 2. Design and control of the gripper and target grasping of a flexible automatic manipulator based on grasping contact force and position control

3. Visual grab toilet

The food source quality of artificial bee colony algorithm is generally the greater the better, that is, the greater the fitness value, the better. Corresponding to the problem to be optimized, it needs to be considered in two cases: the minimum value problem and the maximum value problem. Let fi be the objective function of the optimization problem, so if the minimum value problem is optimized, the fitness function is the deformation of fi, which is generally expressed by formula (1); If the maximum value problem is optimized, the fitness function is the objective function.(see Figure 4)

$$fiti=\{1+abs(fi)\ fi>=01/1+fi\ fi>0 \tag{1}$$

This can result in a reduction in the labor intensity on site as well as a reduction in the number of operators. It is necessary for two production workers to carry the toilet from the production line to the bottom bracket in turn before the robot intelligent toilet grab is launched. The weight of one toilet is approximately 100 kilograms. The task of carrying 400-500 toilets per day is extremely labor-intensive. With the launch of the robot intelligent toilet grab, the enterprise will be able to save two production workers. A variety of toilets can be grasped by changing the polyurethane rubber block quickly. (See Figure 3).

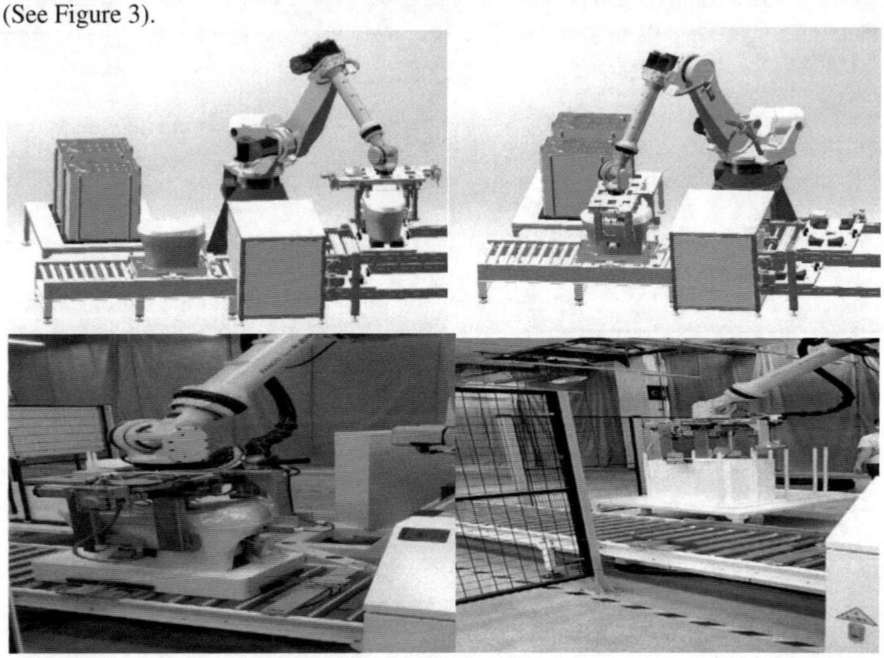

Figure 3. Grabbing and lowering

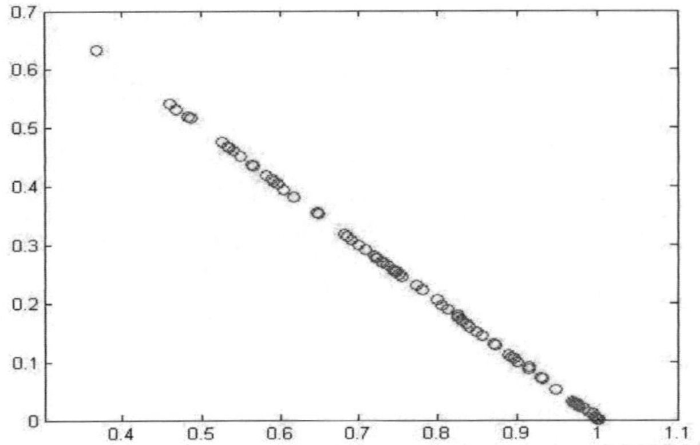

Figure 4. Operation results

4. Conclusion

An improved multi-objective artificial hummingbird algorithm is used to guide and optimize deep reinforcement learning based on the results of off-line learning. Using differential dynamic programming, guidance samples are generated by off-line programming so that deep reinforcement learning does not fall into local optimization when multiple parameters are involved. Based on the target pose deviation, the off-line depth learning controller obtains the initial torque of the manipulator, which is adjusted by the on-line depth reinforcement learning controller, and together with the correction of the feedback torque of the torque observer, the actual torque of the final manipulator is obtained to guide the grasping process.

Acknowledgments

This project is supported by the National Natural Science Foundation of China Youth Fund under Grant 6210023461, in part by the Natural Science Foundation of Guangdong province under Grant 2022A1515011749, in part by the Guangdong Academy of Sciences' (GDAS') Project of Science and Technology Development under Grant 2017GDASCX-0115 and Grant 2018GDASCX-0115, in part by the Guangdong Academy of Science for the Special Fund of Introducing Doctoral Talent under Grant 2021GDASYL-20210103087.

References

[1] https://ifr.org/free-downloads/.
[2] Bohg J, Morales A, Asfour T, et al. Data-driven Grasp Synthesis-A Survey[J].IEEE Transactions on Robotics, 2014, 30(2) : 289-309.
[3] Murray R M, Li Z, Sastry S S, et al. A Mathematical Introduction to Robotic Manipulation [M]. CRC press, 1994.
[4] Mason M T, Salisbury Jr J K. Robot Hands and the Mechanics of Manipu- lation[M]. 1985.
[5] Miller A T, Allen P K. Graspit! A Versatile Simulator for Robotic Grasping[J]. IEEE Robotics and Automation Magazine, 2004, 11(4): 110-122.
[6] Gen M, Cheng R. Genetic Algorithms and Engineering Design[M]. 1997.
[7] Rechenberg I. Evolutions strategien[M]. Springer Berlin Heidelberg. 1978.
[8] Dasgupta D, Michalewicz Z. Evolutionary Algorithms in Engineering Applications[M]. DBLP, 1997.
[9] Koza J R, Poli R.Genetic Programming[J]. 2005.

Path Planning of Mobile Robot Based on Improved Genetic Algorithm

Yongdong Wei[1], Jihe Feng[1], Yuming Huang[1], Kaiwei Liu[1] and Bin Ren[2*]

[1] School of Computer Science and Technology, Dongguan University of Technology, Dongguan, 523000, China

[2] International School of Microelectronics, Dongguan University of Technology, Dongguan, 523000, China

*Corresponding author's e-mail: renbin@dgut.edu.cn

Abstract. With the continuous development of science and technology, mobile robots are gradually applied to people's routine life. As one of the key technologies of mobile robots, the performance of path planning algorithms is a significant factor affecting its efficiency. On the basis of using a traditional Genetic Algorithm for path planning, combined with specific application scenarios, this paper proposes a new path planning algorithm for mobile robots. The path smoothness and the number of path corners are added to the fitness function to evaluate the path, so as to select a more superior path. Simulation experiments in MATLAB show the validity and feasibility of the modified method. By comparison to the traditional GA and other classical path planning algorithms, the optimal path selected has fewer turning angles under the same path length, which has a better application effect and greatly improves the working efficiency and stability.

1. Introduction

With the advent of the era of Industry 4.0 [1], it has become a general trend to use intelligent robots to replace manpower for logistics transportation and warehousing management. Among them, mobile robots are widely used due to their advantages of high-cost performance and strong environmental applicability. As a research hotspot of mobile robots, path planning determines the operation efficiency and application effect of mobile robots and is the key technology for mobile robots to achieve autonomous navigation [2]. Its task goal is to find a collision-free path from the start point to the endpoint in an environment with barriers [3], according to certain evaluation standards such as time, length and consumption of resources.

In order to detect the optimal path, many solutions have been proposed. Common path planning methods include the Dijkstra algorithm, A star algorithm and D star algorithm [4][5][6][7]. Dijkstra algorithm solves the shortest path by searching the global space, but its running process is long and it is difficult to meet the requirement of rapidity. The A star algorithm adds a heuristic function on its basis, which reduces the search space and the D star algorithm can use the information calculated by the A star to quickly update the trajectory after the map obstacle or the position of the robot changes. However, because the three search methods are based on the grid eight-neighborhood search method, the obtained paths have many turning points and poor smoothness, which is not suitable for robotic applications. Planning algorithms based on swarm intelligence optimization include Particle Swarm algorithm, Simulated Annealing algorithm, Genetic Algorithm, etc. ZHANG et al. [8] made use of Particle Swarm

Journal of Physics: Conference Series **2365** (2022) 012053 doi:10.1088/1742-6596/2365/1/012053

Optimization (PSO) to improve motion efficiency. This algorithm can effectively optimize the path length, path smoothness and security, but the algorithm is prone to premature convergence. Simulated Annealing (SA) algorithm proposed by GONG et al. [9] can effectively avoid some path exposure problems in node layout. Genetic Algorithm [10][11] is the most widely used optimal solution processing algorithm, however, the fitness function used in the traditional Genetic Algorithm only considers the length of the path. When specifically applied to mobile robots, there are problems such as not considering the actual constraints of the robot and unreasonable planning paths.

This paper proposes an improved Genetic Algorithm for mobile robots to address the problem. Firstly, by improving the fitness function used in the traditional Genetic Algorithm to make it more suitable for the actual task requirements, then establishing a grid model to simulate the specific application environment and then carrying out the selection, crossover and mutation operation of the Genetic Algorithm and iterate continuously, and calculate the path with short path length and few path corners. MATLAB software is used to verify the simulation experiment and make a comparison of its effect with the traditional Genetic Algorithm and other classical algorithms.

2. Path planning model

Before path planning, a specific map model needs to be established to represent the working environment of the mobile robot. In this paper, the grid method is an important method for modeling the environment. MATLAB software is used to model the environment map, and the specific model is shown in Figure 1, where the black part is the obstacle grid and the white part is the free grid.

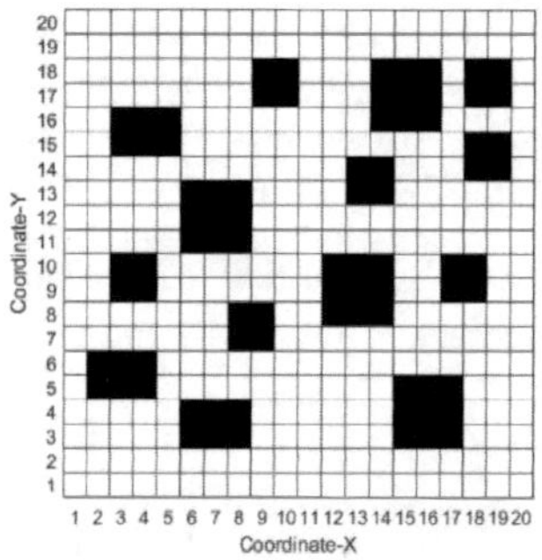

Figure 1. Environment modeling diagram

3. Improved Genetic Algorithm

Genetic Algorithm (GA) was proposed in the 20th century and has been used to solve lots of practical problems. The content of traditional Genetic Algorithm includes population initialization, fitness function and genetic operation.

3.1 Population initialization

Population initialization [12][13] is the random generation of multiple initial individuals that can reach the goal. In this paper, initializing the population means randomly generating multiple feasible paths. The idea is that in initialization, one free grid is randomly selected in each row in order, and then all the selected free grids are connected into a continuous path as an initial individual.

3.2 Fitness function

In the process of biological genetic evolution, individuals with higher fitness are more likely to be inherited by the next generation [14]. The population probability of the next generation is determined by the fitness function of the parent generation. In the traditional path planning problem, the total length of the taken path is a very important indicator [15]. In this paper the path length is used as a component of the fitness function, which is calculated as:

$$d=\sum_{i=1}^{n-1}\left[(x_{i+1}-x_i)^2+(y_{i+1}-y_i)^2\right]^{1/2} \tag{1}$$

where n is the number of grids in the individual path, d is the length of the individual path. In general, the shorter the length means that the path is better. Therefore, the inverse proportional function is chosen as the fitness function, and its formula is:

$$F_1=1/d \tag{2}$$

In practical applications, the time required for a robot to perform corner actions and the required operations are more complex than straight lines. This paper firstly judges whether all the adjacent 3 points in the path form an angle, and imposes a penalty if there is an angle. The basic logic is to find the sum of the distance between the first two points and the distance between the last two points as D1; the distance between the first point and the last point is D2, if D1= D2, it means that the three points do not form an angle, otherwise it proves that an angle is formed, and the *sum* of the corners is increased by 1. Its specific formula is:

$$D1= \left[(x_1-x_2)^2+(y_1-y_2)^2\right]^{1/2}+\left[(x_3-x_2)^2+(y_3-y_2)^2\right]^{1/2} \tag{3}$$

$$D2= \left[(x_1-x_3)^2+(y_1-y_3)^2\right]^{1/2} \tag{4}$$

Among them, x_1, x_2, x_3, y_1, y_2 and y_3 represent the horizontal and vertical coordinates of the three points. A certain penalty can be imposed on the number of corners to plan a better path. The number of corners obtained by the path is *sum*. In this paper, a penalty of 100 is imposed on each additional angle in the path, and the reciprocal of the sum of all penalties is used as a component of the fitness function to optimize the final path. The calculation formula is:

$$F_2=1/(sum*100) \tag{5}$$

In actual operation, the angle of robot turns should not be too large to ensure speed and attitude stability. The smoother the path, the larger the angle formed by 3 adjacent points in the path, and the larger the angle the larger the distance between 3 adjacent points, so the distance of all adjacent 3 points in the path is added to the fitness function. Its specific formula is:

$$d'=\sum_{i=1}^{n-2}\left[(x_{i+2}-x_i)^2+(y_{i+2}-y_i)^2\right]^{1/2} \tag{6}$$

In robot dynamics, the best smoothness is the straight angle, followed by the obtuse angle and right angle, while the case of acute angle does not satisfy the constraints of robot dynamics. We choose to penalize the remaining three cases except for the straight angle and add different penalty coefficients to each case to optimize the individuals that do not satisfy the constraint and obtain a better path. The penalty coefficients selected in this paper are 1, 10 and 1000. In this paper, the reciprocal of the sum of all penalties in the path is used as a component of the fitness function, which is calculated as follows:

$$F_3=1/d' \tag{7}$$

In the improved Genetic Algorithm proposed in this paper, the final fitness function consists of the path length, the number of corners and the smoothness of the path, with the following formula:

$$F=F_1+F_2+F_3 \tag{8}$$

3.3 Genetic manipulation

3.3.1 Selection function

Selection function is used to apply the fitness function in the current population for evaluation to determine the probability of being selected for retention. In this paper, the roulette wheel [16][17] approach is chosen as the selection function. When the population size is *n*, the probability of individual *i* being selected for an inheritance to the next generation is:

$$P_i=F_i(\sum_{i=1}^{n} F_i)^{-1} \quad (9)$$

3.3.2 Crossover function and mutation function

The crossover function and mutation function [18][19] are the way used to generate new individuals. The single-point crossover is used in this paper [20]. Firstly, we generate a random number between 0 and 1 and determine the crossover probability *pc*, then make a comparison of them. If the random number is smaller, it means that the crossover operation is performed and if the crossover operation is required, 2 individuals with common crossover points are selected as parents; then the grid paths after exchanging the crossover points of these 2 individuals form new individuals to join the population. Same as above, first the mutation probability *pm* is determined and a random number from 0 to 1 is generated, perform the mutation operation if the random number is smaller. The steps are: randomly select a feasible path, randomly select a grid on the path as the mutation grid according to the mutation probability, then mutate the mutation grid into any feasible non-repeating free grid in its surrounding upper, lower, left and right, and form a new feasible path as a new individual to join the population.

4. MATLAB simulation experiment

The initial starting point of the robot in the simulation experiment is set as (0, 0) and the endpoint is (20, 20). The number of iterations of the algorithm is taken as 200, the number of populations n=300, the crossover probability pc=0.8 and the mutation probability pm=0.2. Simulation is performed by MATLAB, and the path of the traditional Genetic Algorithm is shown in Figure 2, and the iterative process of the algorithm is shown in Figure 3. The picture shows that the whole algorithm converges after about 45 iterations, and the optimal path length is 29.2132, and there are 14 corners in total. In particular, there is an acute angle in the path, which is to be avoided by the robot motion.

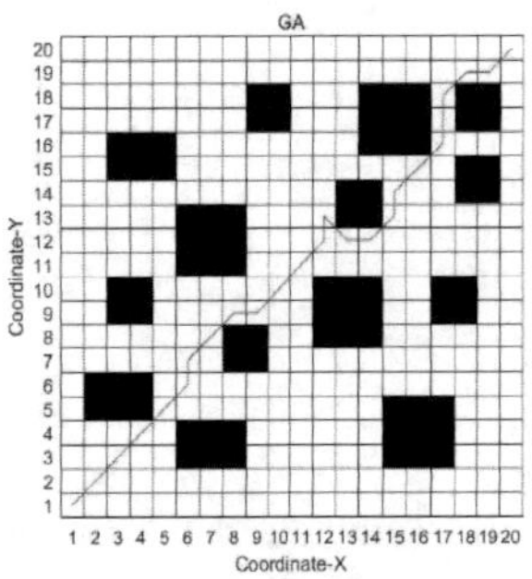

Figure 2. Genetic Algorithm path

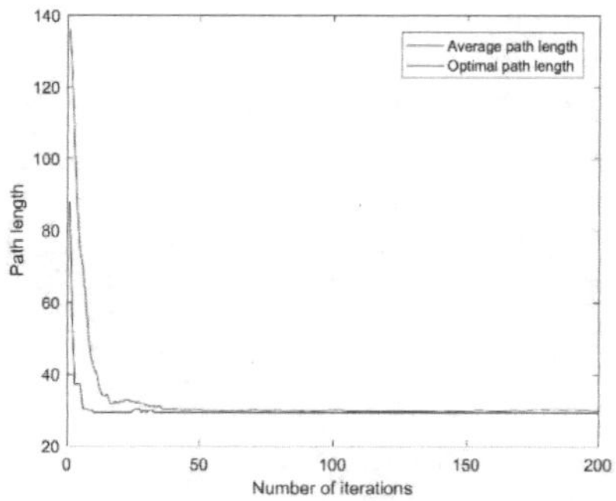

Figure 3. Genetic Algorithm iteration

In the simulation of this paper, the smoothness of the path and the number of path corners are added as an important evaluation index of the improved fitness function. The results of the improved Genetic Algorithm are shown in Figure 4, and the algorithm iterations are shown in Figure 5. We can see clearly from the figure that the algorithm has converged after about 20 iterations and the optimal path length is 29.2132, while there are only 8 corners in the whole path. It is obvious to see that the number of path corners is less and no small-angle turns in the path if we make a comparison of the effect of the two Genetic algorithms.

Journal of Physics: Conference Series **2365** (2022) 012053 doi:10.1088/1742-6596/2365/1/012053

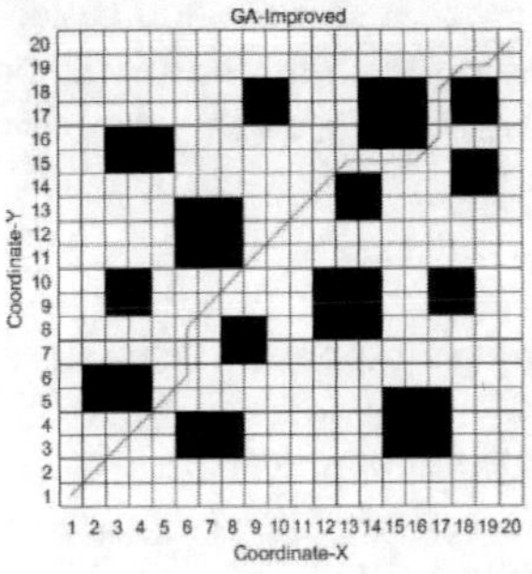

Figure 4. Improved Genetic Algorithm path

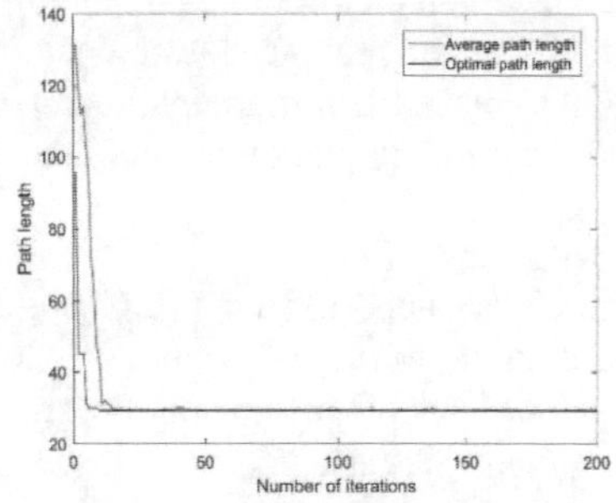

Figure 5. Improved Genetic Algorithm iteration

To further verify the superiority of the improved algorithm, simulation experiments of the problem are performed on the MATLAB platform using the classical A star algorithm and D star algorithm. The effect of the A star algorithm simulation is shown in Figure 6 and the effect of the D star algorithm simulation is shown in Figure 7.

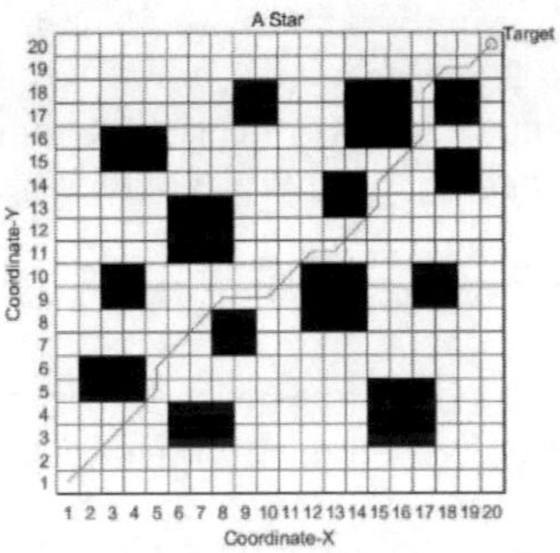

Figure 6. A star algorithm path

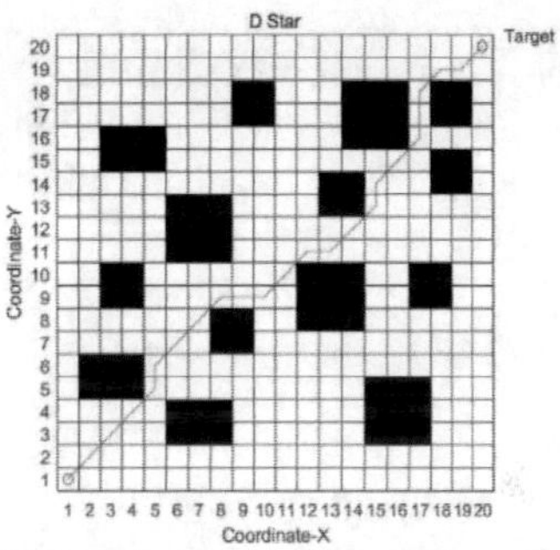

Figure 7. D star algorithm path

The effects of the traditional Genetic Algorithm, the improved GA, the A star and D star algorithm are compared on the basis of optimal path length and the average number of corners respectively, and the comparison of the algorithm effects are shown in Table 1. It is obvious to see that the algorithm we improved has better performance.

Table 1: Comparison of algorithm effects

Algorithm	Optimal path length	Average number of corners
GA	29.2132	14
GA-Improved	29.2132	**8**
A star	29.2132	12
D star	29.2132	12

5. Conclusion

In this paper, we propose a path planning algorithm for mobile robots. On the basis of the traditional Genetic Algorithm, the path smoothness and the number of path corners are added to the fitness function as other evaluation indicators of the fitness function, achieve the goal of selecting a smoother path with fewer corners, and improve the use efficiency of mobile robots in the application process. In MATLAB, the simulation experiment effect of the improved algorithm and other classical path planning algorithms are compared. The comparison conclusion shows that the improved Genetic Algorithm proposed in this paper generated a path with fewer corners and has a better effect under the condition of the same shortest length.

In the future mobile robot path planning algorithm design, it will be a trend to add better algorithms for optimization or advanced technical means such as deep learning and reinforcement learning technology to make them more intelligent and efficient on the current basis, and this paper will further expand the research in this direction in the future work.

Acknowledgments
The research is supported by the Guangdong Provincial Basic and Applied Basic Research Fund Regional Joint Fund under Grant No. 2020B1515120095, Key Project of Dongguan Science and Technology of Social Development Program under Grant No.20211800904692 and No.2021180090449 2， Dongguan Sci-tech Commissioner Program under Grant No.20211800500012, Guangdong Provincial Enterprise Key Laboratory Project under Grant No. 2020B121202001.

References
[1] Zezulka, F. , Marcon, P. , Vesely, I. , & Sajdl, O. . (2016). Industry 4.0 – an introduction in the phenomenon. IFAC-PapersOnLine, 49(25), 8-12.
[2] Mac, T. T. , Copot, C. , Tran, D. T. , & Keyser, R. D. . (2016). Heuristic approaches in robot path planning: a survey. Robotics and Autonomous Systems, 86.
[3] Zhao, X., Wang, Z., Huang, C., & Zhao, Y. . (2018). Mobile robot path planning based on an improved a* algorithm. Robot.
[4] Wang, J., Zhang, X., Chen, B., Beijing, Wuhan, & China. (2007). A heuristic optimization path-finding algorithm based on Dijkstra algorithm. Journal of University of Science & Technology Beijing.
[5] Skiena, S. . (1990). Dijkstra's algorithm. encyclopedia of operations research & management science.
[6] Saian, P. , Suyoto, & Pranowo. (2017). Optimized A-Star algorithm in hexagon-based environment using parallel bidirectional search. International Conference on Information Technology & Electrical Engineering. IEEE.
[7] Kadry, S. , Alferov, G. , & Fedorov, V. . (2020). D-star algorithm modification. International Journal of Online and Biomedical Engineering (iJOE), 16(8), 108.
[8] Zhang, J. H., Yong, Z., & Yong, Z. . (2018). Path planning of mobile robot based on hybrid multi-objective bare-bones particle swarm optimization with differential evolution. IEEE Access, 6, 1-1.
[9] Gong, D. W. Zeng, X. F., & Zhang, Y. . (2013). Global path planning method of robot based on modified simulated annealing algorithm. Journal of System Simulation.
[10] Holland, J. . (1975). Adaptation in natural and artificial systems : an introductory analysis with application to biology. Control & Artificial Intelligence.
[11] Al-Zaydi, Z. , Al-Jamali, N. , & Ismail, K. S. . (2012). A modified genetic algorithm path planning for intelligent autonomous mobile robot. inventi rapid algorithm.
[12] Msu, A. Mm, B., & Ss, C. . (2021). A hybrid genetic algorithm and tabu search for minimizing makespan in flow shop scheduling problem. Journal of King Saud University - Computer and Information Sciences.
[13] Kazimipour, B. , Li, X. , & Qin, K. . (2014). A review of population initialization techniques for evolutionary algorithms. IEEE. IEEE.
[14] Bottaci, L. . (2001). A Genetic Algorithm Fitness Function for Mutation Testing.
[15] Bailey, J. P. , Nash, A. , Tovey, C. A. , & Koenig, S. . (2021). Path-length analysis for grid-based path planning. Artificial Intelligence, 301(4), 103560.
[16] Kumar, R. , & Jyotishree. (2012). Blending roulette wheel selection & rank selection in genetic algorithms.
[17] Huang, K. . (2009). Analysis and improvement on roulette wheel method of genetic algorithm. Computer Applications in Engineering Education, 45(28), 60-63.

[18] Guo, G. , & Wang, Y. . (2015). Interactive genetic algorithm containing crossover and mutation operation. Computer Engineering.

[19] Damia, A. , Esnaashari, M. , & Parvizimosaed, M. . (2021). Adaptive Genetic Algorithm Based on Mutation and Crossover and Selection Probabilities. 2021 7th International Conference on Web Research (ICWR).

[20] Joseph, N. P. , & Ramadoss, B. . (2013). A genetic algorithm applying single point crossover and uniform mutation to minimize uncertainty in production cost. World Applied Sciences Journal, 23(8), 1013-1017.

Event-Triggered Regulation for Enhanced Synchronization of Recurrent Neural Networks using Time-Varying Delay

Lei Yu[1,a], Guici Chen[1,2,b]

[1]College of Science, Wuhan University of Science and Technology, Wuhan, 430065, People's Republic of China

[2]Hubei Province Key Laboratory of System Science in the Metallurgical Process, Wuhan University of Science and Technology, Wuhan, 430065, People's Republic of China

[a] yulei@wust.edu.cn, [b]chenguici@wust.edu.cn

Abstract. This research is primarily concerned having the synchronization issue of time-varying delay recurrent neural networks (RNNs) with event-triggered regulation (ETR). ETR may legitimately reduce the number of control activities performed and the amount of energy expended. Via the L-K functional, Schur complement lemma, linear matrix inequalities (LMIs), and other analytic techniques, a sufficient condition for the synchronization of the drive-response system is achieved. Consequently, a more sensible and effective controller is implemented. Using a Matlab-made example and graph, we can show that our inferred results are correct and we can ensure the rationality of the set controller.

1. Introduction

In recent decades, RNNs have continued to evolve and have been effectively used in numerous fields, including artificial intelligence, management, and research [1]. To ensure the reliability of the RNN system, people have conducted extensive research on system control and stability and developed many processing approaches, such as dissipative control [2], fuzzy control [3], etc. These controlling approaches are offered to a great degree to resolve system oscillation and asynchrony issues and technological advancements. Networking and intelligence have become hot topics pursued by contemporary society, and synchronization, as a common method in networked control, has also become the object favored by researchers. On the basis of previous research, we composed this essay by consulting the data, synthesizing, conducting mathematical derivation and matlab drawing. In this paper, the control of state feedback is used to look into the problem of synchronization in RNN systems.

Only the magnitude of the time delay phenomena in a regulated system varies in real applications. The impact of time delay complicates the investigation of the system. Existing investigations and computational techniques for time-delay systems have reached a highly developed state in published works. Because of the variety of techniques, several time-delay programs have been designed, having the memristive neural network time method [4] and random time-delay mechanisms [5] among the most prevalent.

In the majority of neural network control systems, the controller delivers information based on a timer. Thus, every individual is detected and computed at a certain period, which will increase network stress. This article suggests an ETR technique to address this issue [6]. Through the supplied event-triggered mechanism, it is determined whether every event has occurred, and only if the trigger criteria are not fulfilled will the information be delivered. Using this approach substantially reduces the quantity of data transfer and network transmission strain [7].

The contribution of this paper is in two aspects: on the one hand, it is the processing of the time delay, and the time delay is scaled by the upper and lower bounds of the time delay; on the other hand, the nonlinear terms in the obtained matrix are calculated by some theorems. Transform, convert to linear terms.

In light of the preceding research, this paper focuses primarily on the synchronization issue of RNNs with time delay under event-triggered control. Using the L-K function and LMIs, suitable conditions for system synchronization are produced while accounting for system time delay. The remainder of the article is as follows: In Section 2, you will create mathematical models and explain the developed models. Section 3 demonstrates the stability of the system employing mathematical approaches. In Section 4, instances and correlations are provided. Conclusions are drawn in Section 5.

2. Mathematical notation and indications

For the purpose of the article's elegance, we include the next explanations: The transpose and inverse of a matrix are denoted by "T" as well as "-1" correspondingly; R^n represent the n-dimensional Euclidean space; the identity matrix is denoted by "I"; and "*" is often utilized to designate a symmetric term.

Take the next RNNs drive system with time-varying latency (TVL), which is outlined below:

$$\dot{x}_i(\beta) = -c_i x_i(\beta) + \sum_{j=1}^{n} a_{ij} f_j\left(x_j(\beta)\right) + \sum_{j=1}^{n} b_{ij} g_j\left(x_j\left(\beta - d(\beta)\right)\right) + I_i, \tag{1}$$

In this Equation, $i, j \in \mathrm{N} = \{1, 2, \cdots, \mathrm{n}\}$. $x_i(\beta) \in R^n$ represents the voltage of the capacitor C_i; $\sum_{j=1}^{n} a_{ij}$ and $\sum_{j=1}^{n} b_{ij}$ represent self-feedback coefficients. $f_j(\cdot)$ and $g_j(\cdot)$ denote bounded activation function;. $d(\beta)$ represents the time-varying delay in signal transmission satisfying $0 \le d(\beta) \le d_{\mathrm{M}}$ (d_{M} represents the maximum possible delay constant boundaries). I_i represents the ith neuron's external constant input; and $x_i(\beta) = \psi_i(\beta)$ represents the initial condition of the drive system

The structure of the feedback mechanism matching to (1) is as shown in:

$$\dot{y}_i(\beta) = -c_i y_i(\beta) + \sum_{j=1}^{n} a_{ij} f_j\left(y_j(\beta)\right) + \sum_{j=1}^{n} b_{ij} g_j\left(y_j\left(\beta - d(\beta)\right)\right) + I_i + u_i(\beta), \tag{2}$$

In this Equation, $y_i(\beta)$ represents the state variable linked to the ith neuron. $y_i(\beta) = \varphi_i(\beta)$ represents the starting state of the drive system, and $u_i(\beta)$ represents the appropriate outer control input with bounds.

We specify

$$x(\beta) = \left(x_1(\beta), x_2(\beta), \cdots, x_n(\beta)\right)^T, \; y(\beta) = \left(y_1(\beta), y_2(\beta), \cdots, y_n(\beta)\right)^T, \tag{3}$$

The error signal is written as follows: $e(\beta) = y(\beta) - x(\beta)$. (4)

In other words, we can conclude:

$$\dot{e}(\beta) = -\tilde{C}e(\beta) + \tilde{A}\hat{\theta}\left(e(\beta)\right) + \tilde{B}\varpi\left(e(\beta - d(\beta))\right) + u(\beta), \tag{5}$$

In this equation,

$$f(x(\beta))=\left(f_1(x_1(\beta)),\cdots,f_n(x_n(\beta))\right)^T, g(x(\beta-d(\beta)))=\left(g_1(x_1(\beta-d(\beta))),\cdots,g_n(x_1(\beta-d(\beta)))\right)^T,$$
$$f(y(\beta))=\left(f_1(y_1(\beta)),\cdots,f_n(y_n(\beta))\right)^T, g(y(\beta-d(\beta)))=\left(g_1(y_1(\beta-d(\beta))),\cdots,g_n(y_1(\beta-d(\beta)))\right)^T,$$
$$\hat{\theta}(e(\beta))=f(y(\beta))-f(x(\beta)),\varpi(e(\beta-d(\beta)))=g(y(\beta-d(\beta)))-g(x(\beta-d(\beta))),$$
$$u(\beta)=(u_1(\beta),u_2(\beta),\cdots,u_n(\beta))^T, \tilde{C}=diag\{c_1,c_2,\cdots,c_n\}, \tilde{A}=(a_{ij})_{n\times n}, \tilde{B}=(b_{ij})_{n\times n}, \tag{6}$$

The state error and transmission error decide whether the sample state $e(\beta_\alpha)=y(\beta_\alpha)-x(\beta_\alpha)$ can be conveyed. The error between the present system state and the previous sampled state is referred to as $e(\beta)-e(\beta_\alpha)$. State error represents the difference between the present state of the system and the equilibrium point, which is indicated as $e(\beta_\alpha)$. When $e(\beta_\alpha)$ is transmitting, in order to determine the sampling state at the next moment, $e(\beta_\alpha)$ also needs to be kept in the controller. The data will not be updated until the next sampling state does not meet the trigger circumstance. Assume that for each integer $\alpha>0$, there is $\beta_{\alpha+1}-\beta_\alpha\le h$, In this Equation, h represents the sampling interval's upper limit. The ETR mechanism described in this research is expressed as the next. (7)

$$\beta_{\alpha+1} = \beta_\alpha + min\left\{\beta | \left(e(\beta)-e(\beta_\alpha)\right)^T \Omega\left(e(\beta)-e(\beta_\alpha)\right) \ge \delta e^T(\beta_\alpha)\Omega e(\beta_\alpha)\right\}, \tag{7}$$

In this Equation, $\delta\in[0,1]$ represents the parameter of the ETR mechanism and $\Omega>0$ represents a positive, definite, weighted symmetric matrix

The controller is designed like this: $u(\beta) = \hat{K}e(\beta_\alpha)$, In this Equation, $\hat{K}$ represents the control gain.

Some of the subsequent assumptions and lemmas will be utilized in the third subsection's proving procedure. For the purpose of the paper's neatness, we provide it beforehand.

Assumption 1: For all i, $i,j\in \mathrm{N}=\{1,2,\cdots,\mathrm{n}\}$. the neuron activation function $\hbar_j(\cdot)$ and $\lambda\!\!\!^-_j(\cdot)$ represents a continuous bounded function, and for every satisfied $\ell_1\neq\ell_2\in R$, the corresponding inequalities apply:

$$\sigma_j^- \le \frac{\hbar_j(\ell_1)-\hbar_j(\ell_2)}{\ell_1-\ell_2} \le \sigma_j^+, \rho_j^- \le \frac{\lambda\!\!\!^-_j(\ell_1)-\lambda\!\!\!^-_j(\ell_2)}{\ell_1-\ell_2} \le \rho_j^+. \tag{8}$$

In this Equation, $\sigma_j^-,\sigma_j^+,\rho_j^-,\rho_j^+$ represent known constant.

Lemma 1[8]: In the case of determined matrices $\Pi=\Pi^T$, J and G with appropriate dimensions, $\Pi+J\bar{F}(\dagger)G+G^T\bar{F}(\dagger)^T J^T<0$, is valid for all $\bar{F}(\dagger)^T\bar{F}(\dagger)\le I$ when and ony when there is $\epsilon>0$ and we can conclude that $\Pi+\epsilon^{-1}JJ^T+\epsilon G^T G<0$.

Lemma 2[9]: For every constant positive matrix $U\in R^{n\times n}$, scalar $0\le d(\beta)\le d_M$, and vectors function $\dot{\chi}(\beta):[-d_M,0]\to R^n$, and we can conclude that the next integration can be written as:

$$-d_M\int_{\beta-d_M}^{\beta}\dot{\chi}^T(v)R_1\dot{\chi}(v)dv \le \begin{bmatrix}\chi(\beta)\\ \chi(\beta-d(\beta))\\ \chi(\beta-d_M)\end{bmatrix}^T \begin{bmatrix}-U_1 & U_1 & 0\\ * & -2U_1 & U_1\\ * & * & -U_1\end{bmatrix} \begin{bmatrix}\chi(\beta)\\ \chi(\beta-d(\beta))\\ \chi(\beta-d_M)\end{bmatrix}. \tag{9}$$

Lemma 3[10]: Assuming a constant matrix M is positive and symmetric, if $\varsigma>0$ and function $\mathbb{S}(.)$ exist, we can obtain $\left(\int_0^\varsigma \mathbb{S}(\vartheta)d\vartheta\right)^T M\left(\int_0^\varsigma \mathbb{S}(\vartheta)d\vartheta\right) \le \varsigma\int_0^\varsigma \mathbb{S}^T(\vartheta)M\mathbb{S}(\vartheta)d\vartheta$.

Lemma 4[11]: In the case of matrices $\wp_1, \wp_2, \wp_3$, In this Equation, $\wp_1, \wp_2, \wp_3$ represent constant, $\wp_1 = \wp_1^T > 0$ & $\wp_2 = \wp_2^T > 0$. Therefore, we can obtain $\wp_1 + \wp_3^T \wp_2^{-1} \wp_3 < 0$ when and only when $\begin{bmatrix} \wp_1 & \wp_3^T \\ \wp_3 & -\wp_2 \end{bmatrix} < 0.$

3. Findings

In this part, the criteria for the stability of the error system are shown using mathematical techniques, and the nonlinear variables in the conditions are transformed. Assume the error system below::

$$\dot{e}(\beta) = -\tilde{C}e(\beta) + \tilde{A}\hat{\theta}\left(e(\beta)\right) + \tilde{B}\varpi\left(e(\beta - d(\beta))\right) + \hat{K}e_\alpha(\beta). \tag{10}$$

Theorem 1: In the case of determined parameters $d_M, \mu,$ the error system (10) is asymptotically stable. If matrix $P > 0$, $Q_i > 0(i = 1,2)$ exists, $R_i > 0(i = 1,2,3)$, positive scalars ε_1, ε_2, and we can conclude that the following LMIs are valid.

$$\hat{\Gamma}_1 = \begin{pmatrix} \hat{\Theta}_{11} & R_1 & 0 & 0 & -P+Z & PA & PB \\ * & -(1-\mu)Q_1 + \varepsilon_2 L_2^T L_2 - 2R_1 & R_1 & 0 & 0 & 0 & 0 \\ * & * & -Q_2 - R_1 & 0 & 0 & 0 & 0 \\ * & * & * & -R_3 + d_M^2 R_1 + W & 0 & 0 & 0 \\ * & * & * & * & -R_2 & 0 & 0 \\ * & * & * & * & * & -\varepsilon_1 I & 0 \\ * & * & * & * & * & * & -\varepsilon_2 I \end{pmatrix}, \tag{11}$$

In this Equation, $\hat{\Theta}_{11} = -2PC + \varepsilon_1 L_1^T L_1 + Q_1 + Q_2 - R_1 - R_2$

Proof: The next six L-K functional is selected for (10) as evidence:

$$V_1(\beta) = e^T(\beta)Pe(\beta), \tag{12}$$

$$V_2(\beta) = \int_{\beta-d(\beta)}^{\beta} e^T(l)Q_1 e(l)dl, \tag{13}$$

$$V_3(\beta) = \int_{\beta-d_M}^{\beta} e^T(l)Q_2 e(l)dl, \tag{14}$$

$$V_4(\beta) = d_M \int_{-d_M}^{0} \int_{\beta+\eta}^{\beta} \dot{e}^T(l)R_1\dot{e}(l)dld\eta, \tag{15}$$

$$V_5(\beta) = h\int_{-h}^{0} \int_{\beta+\eta}^{\beta} \dot{e}^T(l)R_2\dot{e}(l)dld\eta, \tag{16}$$

$$V_6(\beta) = \int_{\beta}^{\beta_{\alpha+1}} \dot{e}^T(\beta)R_3\dot{e}(\beta)dl, \tag{17}$$

Therefore, we can obtain

$$\mathcal{L}V_1 = 2e^T(\beta)P\dot{e}(\beta) = 2e^T(\beta)P\left(-Ce(\beta) + A\hat{\theta}\left(e(\beta)\right) + B\varpi\left(e(\beta - d(\beta))\right) + \hat{K}e_\alpha(\beta)\right), \tag{18}$$

$$\mathcal{L}V_2 \leqslant e^T(\beta)Q_1 e(\beta) - (1-\mu)e^T\left(\beta - d(\beta)\right)Q_1 e\left(\beta - d(\beta)\right), \tag{19}$$

$$\mathcal{L}V_3 \leqslant e^T(\beta)Q_2 e(\beta) - e^T\left(\beta - d_M\right)Q_2 e\left(\beta - d_M\right), \tag{20}$$

$$\mathcal{L}V_4 \leqslant d_M^2 \dot{e}^T(\beta)R_1\dot{e}(\beta) - d_M \int_{\beta-d_M}^{\beta} \dot{e}^T(l)R_1\dot{e}(l)dl, \tag{21}$$

$$\mathcal{L}V_5 \leqslant h^2 \dot{e}^T(\beta)R_2\dot{e}(\beta) - h\int_{\beta-h}^{\beta} \dot{e}^T(l)R_2\dot{e}(l)dl, \tag{22}$$

$$\mathcal{L}V_6 \leqslant -\dot{e}^T(\beta)R_3\dot{e}(\beta), \tag{23}$$

We can obtain the following Equations according to Assumption 1 & Lemma 1.

$$2e^{\mathrm{T}}(\beta)PA\hat{\theta}(e(\beta)) \le \varepsilon_1^{-1}e^{\mathrm{T}}(\beta)PAA^T Pe(\beta)+\varepsilon_1 e^T(\beta)L_1^T L_1 e(\beta), \tag{24}$$

$$2e^{\mathrm{T}}(\beta)PB\varpi(e(\beta-d(\beta))) \le \varepsilon_2^{-1}e^{\mathrm{T}}(\beta)PBB^T Pe(\beta)+\varepsilon_2 e^T(\beta-d(\beta))L_2^T L_2 e(\beta-d(\beta)), \tag{25}$$

According to Lemma 2, we can obtain

$$-d_M\int_{\beta-d_M}^{\beta}\dot{e}^T(l)R_1\dot{e}(l)dl \le \begin{bmatrix} e(\beta) \\ e(\beta-d(\beta)) \\ e(\beta-d_M) \end{bmatrix}^T \begin{bmatrix} -R_1 & R_1 & 0 \\ * & -2R_1 & R_1 \\ * & * & -R_1 \end{bmatrix} \begin{bmatrix} e(\beta) \\ e(\beta-d(\beta)) \\ e(\beta-d_M) \end{bmatrix}, \tag{26}$$

Based on Lemma 3, we can obtain:

$$-h\int_{\beta-h}^{\beta}\dot{e}^T(l)R_2\dot{e}(l)dl \le -(\beta-\beta_\alpha)\int_{\beta_\alpha}^{\beta}\dot{e}^T(l)R_2\dot{e}(l)dl \le -[e(\beta)-e(\beta_\alpha)]^T R_2[e(\beta)-e(\beta_\alpha)], \tag{27}$$

Define $\xi(\beta) = \left[e^T(\beta) \quad e^T(\beta-d(\beta)) \quad e^T(\beta-d_M) \quad \dot{e}^T(\beta) \quad e^T(\beta_\alpha)\right]^T$, (28)

By adding (12) to (28), we can obtain $\mathcal{L}\mathbf{V} \le \xi^T(\beta)\Gamma_1\xi(\beta)$, (29)

In the following Equation,

$$\Gamma_1 = \begin{pmatrix} \Theta_{11} & R_1 & 0 & 0 & R_2+P\hat{K} \\ * & -(1-\mu)Q_1+\varepsilon_2 L_2^T L_2-2R_1 & R_1 & 0 & 0 \\ * & * & -Q_2-R_1 & 0 & 0 \\ * & * & * & -R_3+d_M^2R_1+h^2R_2 & 0 \\ * & * & * & * & -R_2 \end{pmatrix}, \tag{30}$$

Using Schur's complement lemma plus congruential transformation, it is possible to demonstrate that $\hat{\Gamma}_1 < 0$.

This concludes the demonstration of Theorem 1.

4. Statistical illustration using simulations

The following case shows that the inferences made in the last section were correct and that the set controller is effective.

Assume the RNNs to be the propulsion system (1) with n = 2.

$$\begin{aligned}\dot{x}_1(k) &= -c_1x_1(k)+a_{11}f_1(x_1(k))+a_{12}f_2(x_2(k))+b_{11}g_1(x_1(k-d(k)))+b_{12}g_2(x_2(k-d(k)))+I_1 \\ \dot{x}_2(k) &= -c_2x_2(k)+a_{21}f_1(x_1(k))+a_{22}f_2(x_2(k))+b_{21}g_1(x_1(k-d(k)))+b_{22}g_2(x_2(k-d(k)))+I_2\end{aligned} \tag{31}$$

In this Equation, $c_1 = 0.9$, $c_2 = 1.1$, $a_{11} = 1.05$, $a_{12} = -0.5$, $a_{21} = 2.9$, $a_{22} = -2$, $b_{11} = -1.7$, $b_{12} = 1.8$, $b_{21} = -0.3$, $b_{22} = -2.5$ .The function of activation $f_j(x_j) = g_j(x_j) = \tanh(x), j = 1,2$. If $d_M = 1, I_1 = I_2 = 0$ .The response RNNs (2) are written as

$$\begin{aligned}\dot{y}_1(k) &= -c_1y_1(k)+a_{11}f_1(y_1(k))+a_{12}f_2(y_2(k))+b_{11}g_1(y_1(k-d(k)))+b_{12}g_2(y_2(k-d(k)))+I_1+u_1(k) \\ \dot{y}_2(k) &= -c_2y_2(k)+a_{21}f_1(y_1(k))+a_{22}f_2(y_2(k))++b_{21}g_1(y_1(k-d(k)))+b_{22}g_2(y_2(k-d(k)))+I_2+u_2(k)\end{aligned} \tag{32}$$

Based on known conditions (7), we assume $\delta = 0.08$, and $\Omega = \begin{bmatrix} 20.3537 & 7.2546 \\ 7.2546 & 17.4190 \end{bmatrix}$. According to (11), using the LMIs, we can come up having the next solution $\epsilon_1 = 0.2777, \epsilon_2 = 0.2112$,

$$P=\begin{bmatrix}0.0487 & 0.0008\\ 0.0008 & 0.0232\end{bmatrix}, Q_1=\begin{bmatrix}0.2781 & 0.0003\\ 0.0003 & 0.2781\end{bmatrix}, Q_2=\begin{bmatrix}0.2633 & 0.0002\\ 0.0002 & 0.2629\end{bmatrix},$$

$$R_1=\begin{bmatrix}0.2392 & -0.0001\\ -0.0001 & 0.2381\end{bmatrix}, R_2=\begin{bmatrix}0.8650 & -0.0005\\ -0.0005 & 0.8800\end{bmatrix}, R_3=\begin{bmatrix}0.8859 & -0.0001\\ -0.0001 & 0.8851\end{bmatrix}, \hat{K}=\begin{bmatrix}-17.7755 & 0.6245\\ 0.6256 & -37.8747\end{bmatrix}$$

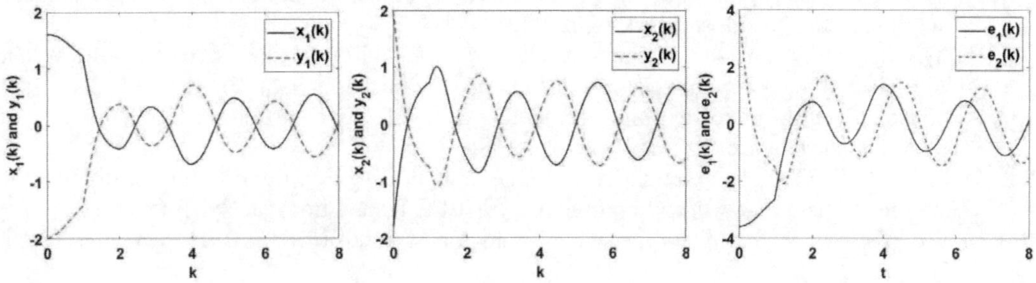

(a) (b) (c)

Fig.1. (a) Graphs of variation for RNNs (31) and (32) with no controller.(b) Variation graphs for $x_2(t)$ and $y_2(t)$ for RNNs (31) and (32) without a controller. (c) Variation graphs of synchronization error $e(t)$ for RNNs (31) and (32) lacking a controller.

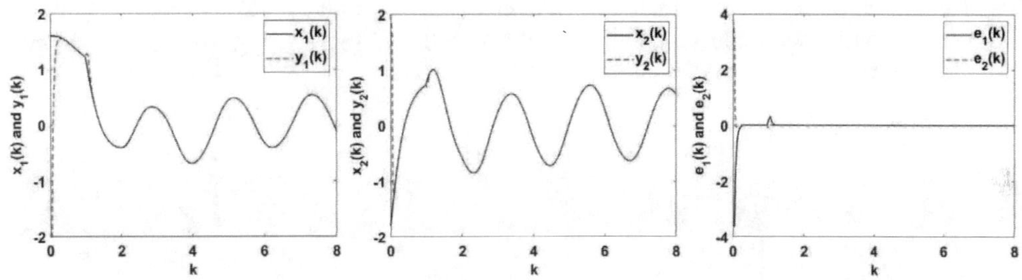

(a) (b) (c)

Fig.2. (a) RNN variation graphs for $x_1(t)$ and $y_1(t)$ for RNNs (31) and (32) with controller; (b)Variation graphs for $x_2(t)$ and $y_2(t)$ for RNNs (31) and (32) having the controller. (c) RNN synchronization error $e(t)$ for RNNs (31) and (32) with controller

5. Conclusions

This essay investigates the synchronization issue of RNNs with TVL using ETR. By making a good L-K functional and a good ETR, a flexible condition is made for the synchronization problem of RNNs with TVL, and a graph is used to show how reliable the results are and how flexible the ETR is.

Acknowledgments

This research was sponsored by grants 61473213 and 61473212 from the National Natural Science Foundation of China.

References

[1] M S Ali, M Marudai. Stochastic stability of discrete-time uncertain recurrent neural networks with Markovian jumping and time-varying delays. Mathematical and computer modelling, 2011, 54(9-10): 1979-1988.

[2] S Yu, Y Pang, G Chen, et al. Dissipativity Analysis for a Class of Discrete-Time Neutral Stochastic Nonlinear Systems with Time Delay. Discrete Dynamics in Nature and Society, 2021, 2021.
[3] A Chandrasekar, R Rakkiyappan. Impulsive controller design for exponential synchronization of delayed stochastic memristor-based recurrent neural networks. Neurocomputing, 2016, 173: 1348-1355.
[4] F Wei, G Chen, W Wang. Finite-time synchronization of memristor neural networks via interval matrix method. Neural Networks, 2020, 127: 7-18.
[5] Y ZHANG, H Yong, W Min. Delay-dependent robust stability for uncertain stochastic systems with interval time-varying delay. Acta Automatica Sinica, 2009, 35(5): 577-582.
[6] P Tabuada. Event-triggered real-time scheduling of stabilizing control tasks. IEEE Transactions on Automatic Control, 2007, 52(9): 1680-1685.
[7] W J Lin, Y He, C K Zhang, et al. Event-triggered fault detection filter design for discrete-time memristive neural networks with time delays. IEEE Transactions on Cybernetics, 2020.
[8] M Wu, Y He, J H She. Stability analysis and robust control of time-delay systems. Berlin: Springer, 2010.
[9] C Peng, Y C Tian . Delay-dependent robust stability criteria for uncertain systems with interval time-varying delay. Journal of Computational and Applied Mathematics, 2008, 214(2): 480-494.
[10] K Gu. An integral inequality in the stability problem of time-delay systems//Proceedings of the 39th IEEE conference on decision and control (Cat. No. 00CH37187). IEEE, 2000, 3: 2805-2810.
[11] G Nagamani, S Ramasamy. Stochastic dissipativity and passivity analysis for discrete-time neural networks with probabilistic time-varying delays in the leakage term. Applied Mathematics and Computation, 2016, 289: 237-257.

Discrete spatiotemporal network synchronization based on adaptive control

HaiFeng Wang, BiGang XU

Shanghai Institute of Technology, College of Mechanical Engineering, Shanghai 201418, China

836311803@qq.com

Abstract: This paper first introduces the basic concept of network synchronization and several common synchronization types. Secondly, aiming at the synchronization control problem of a class of discrete spatiotemporal networks, a standard synchronization control strategy and a synchronization controller are proposed based on Lyapunov stability theory. In order to further verify the effectiveness of the synchronization theory, the spatiotemporal network model is selected, the coupling matrix and the identification rate of unknown parameters are designed for numerical simulation. Finally, the simulation results show the feasibility of the theory.

1. Introduction

At present, network synchronization is an important subject in the research of network dynamics. Generally speaking, synchronization refers to the process that two or more networks adjust their dynamic properties to achieve the same properties under the external drive or mutual coupling. People realize that the influence of topology on the dynamic characteristics of the network is equally important. Under certain conditions, each node in the network can achieve synchronization by adjusting the coupling to an appropriate strength[1]. At present, the research of complex networks has entered a relatively mature period. However, whether it is the synchronization of nodes in the network or the synchronization between multiple networks, it is found that single-layer networks can not represent all the complex systems. Because there will be some correlation between real networks. At this time, people put forward multi-layer complex networks based on of single-layer networks, and the research of multi-layer network synchronization is in a rapid development period[2].

From the point of view of node state relations, the achieved synchronization can be roughly divided into four types: If the motion trajectories are completely consistent, it is full synchronization[3]; In particular, when the mapping relationship is a proportional function, the corresponding synchronization is projection synchronization[4,5]; Delay synchronization is also called lag synchronization, because a node reaches a certain state, and the corresponding point will reach the state after a constant time delay[6,7]. And with the researchers' in-depth study of synchronization between complex networks, many effective control methods have been found.

There are two main directions to study the synchronization in complex networks[8]: One is to study the synchronization ability brought by the network's own properties, change its topology or design adaptive coupling strength ,so that the network has better synchronization ability[9]; Another way is to add mutual coupling or control between systems to make two or more systems achieve dynamic synchronization. When the coupling between networks can not achieve synchronization, it is necessary to design appropriate controllers[10].

On the basis of previous studies on synchronization of complex networks and Lyapunov stability principle, this paper designs an adaptive controller. The controller uses a one-dimensional acousto-optic bistable coupled map lattice system model as a response and drive network composed of nodes for synchronization control, and puts forward some basic conditions for synchronization. Finally, the feasibility of this method is verified by numerical simulation.

2.Theoretical analysis of synchronization

Consider a discrete spatiotemporal network composed of N nodes. The state equation of the ith node is as follows:

$$x_i(m,n+1) = F(x_i(m,n)) + \varepsilon_i \sum_{j=1}^{N} a_{ij} x_j(m,n) \tag{1}$$

Equation (1) represents the drive network and constructs the corresponding response network:

$$y_i(m,n+1) = F(y_i(m,n)) + \varepsilon_i \sum_{j=1}^{N} a_{ij} y_j(m,n) + u_i(m,n) \tag{2}$$

Where $x_i(m,n) \in R^n, y_i(m,n) \in R^n$ are the state variables. m and n represent the spatial and temporal variables of the system respectively. $F : R^n \to R^n$ is a nonlinear function mapping. ε_i is the coupling strength between network nodes. $A = (a_{ij})_{N \times N}$ is the coupling matrix, indicates the connection mode and coupling strength between different nodes in the network. When there is a connecting edge between nodes i and $j(i \neq j)$, $a_{ij} \neq 0$; conversely, $a_{ij} = 0$. The diagonal elements of A are defined a $a_{ii} = -\sum_{j=1, j \neq i}^{N} a_{ij}$. $u_i(m,n)$ is a network synchronization controller.

Define the error equations of the drive network and the response network:

$$e_i(m,n) = y_i(m,n) - x_i(m,n) \tag{3}$$

Bring equations (1) and (2) into the above equation to obtain:

$$\begin{aligned} e_i(m,n+1) &= y_i(m,n+1) - x_i(m,n+1) \\ &= F(y_i(m,n)) - F(x_i(m,n)) + \varepsilon_i \sum_{j=1}^{N} a_{ij} e_j(m,n) + u_i(m,n) \end{aligned} \tag{4}$$

In order to achieve the purpose of synchronous control, the following definition needs to be made: If the condition is met $\lim_{n\to\infty} \| e_i(m,n) \| = \lim_{n\to\infty} \| y_i(m,n) - x_i(m,n) \| = 0$, synchronization between the drive network and the response network is achieved.

Theorem 1: when the identification rate of the synchronization controller and the important parameters in the network satisfies the following equation, the synchronization between the spatiotemporal network and the target signal can be realized, the design controller $u_i(m,n)$ is:

$$\begin{aligned} u_i(m,n) &= -F(y_i(m,n)) + F(x_i(m,n)) - \varepsilon_i \sum_{j=1}^{N} a_{ij} e_j(m,n) \\ &+ (L - \hat{\alpha}_i(n) - 1) Q_i(m,n) - \alpha_i e_i(m,n) \end{aligned} \tag{5}$$

$$\hat{\alpha}_i(n+1) = \alpha_i + \varphi_i(m,n)(\hat{\alpha}_i(n) - \alpha_i) \tag{6}$$

In the equation, $Q_i(m,n)$ and $\varphi_i(m,n)$ are pending synchronization function. L is the adjust parameter. α_i is the adaptive adjustment parameters. $\hat{\alpha}_i(n)$ is the parameter recognition rate.

Theorem 2: if we can determine a Lyapunov function $V(n)$, then the function variation ΔV can be obtained, and it's negative definite, all points are satisfied with $\Delta V < 0$. Then the equation is asymptotically stable.

Proof The Lyapunov function is constructed as follows:

$$V(n) = \sum_{i=1}^{N} | e_i(m,n) + Q_i(m,n) | + \sum_{i=1}^{N} \frac{1}{\xi} | \hat{\alpha}_i(n) - \alpha_i | \tag{7}$$

$$Q_i(m,n+1) = Q_i(m,n) + Le_i(m,n) \tag{8}$$

Combine equation (7) and (8):

$$\begin{aligned} \Delta V &= V(n+1) - V(n) \\ &= \sum_{i=1}^{N} | e_i(m,n+1) + Q_i(m,n+1) | + \sum_{i=1}^{N} \frac{1}{\xi} | \hat{\alpha}_i(n+1) - \alpha_i | \\ &\quad - \sum_{i=1}^{N} | e_i(m,n) + Q_i(m,n) | - \sum_{i=1}^{N} \frac{1}{\xi} | \hat{\alpha}_i(n) - \alpha_i | \end{aligned} \tag{9}$$

It can be deduced from the equation (4) (5), (6), (8) and (9):

$$\begin{aligned} \Delta V &= \sum_{i=1}^{N} | F(y_i(m,n)) - F(x_i(m,n)) \\ &\quad + \varepsilon_i \sum_{j=1}^{N} a_{ij} e_j(m,n) + u_i(m,n) + Q_i(m,n) + Le_i(m,n) | \\ &\quad - \sum_{i=1}^{N} | e_i(m,n) + Q_i(m,n) | + \sum_{i=1}^{N} \frac{1}{\xi} | \hat{\alpha}_i(n+1) - \alpha_i | - \sum_{i=1}^{N} \frac{1}{\xi} | \hat{\alpha}_i(n) - \alpha_i | \\ &= \sum_{i=1}^{N} | (L - \hat{\alpha}_i(n))(Q_i(m,n) + e_i(m,n)) + (\hat{\alpha}_i(n) - \alpha_i) e_i(m,n) | \\ &\quad - \sum_{i=1}^{N} | e_i(m,n) + Q_i(m,n) | + \sum_{i=1}^{N} \frac{1}{\xi} (| \varphi_i(m,n) | - 1) | \hat{\alpha}_i(n) - \alpha_i | \end{aligned} \tag{10}$$

$$\begin{aligned}\Delta V &\le \sum_{i=1}^{N} | L-\hat{\alpha}_i(n) || Q_i(m,n)+e_i(m,n) | + \sum_{i=1}^{N} | \hat{\alpha}_i(n)-\alpha_i || e_i(m,n) | \\ &-\sum_{i=1}^{N} | e_i(m,n)+Q_i(m,n) | + \sum_{i=1}^{N} \frac{1}{\xi}(| \varphi_i(m,n) |-1) | \hat{\alpha}_i(n)-\alpha_i | \\ &\le \sum_{i=1}^{N} [| L-\hat{\alpha}_i(n) |-1] | Q_i(m,n)+e_i(m,n) | \\ &+\sum_{i=1}^{N} \frac{1}{\xi}(| \varphi_i(m,n) |-1+\xi | e_i(m,n) |) | \hat{\alpha}_i(n)-\alpha_i | \end{aligned} \tag{11}$$

It is observed that in equation (10) and (11), $| Q_i(m,n)+e_i(m,n) | \ge 0$, $| \hat{\alpha}_i(n)-\alpha_i | \ge 0$, When the adaptive adjustment parameters and the pending synchronization function satisfy the following conditions:

$$| L-\hat{\alpha}_i(n) |-1 \le 0 \tag{12}$$

$$| \varphi_i(m,n) |-1+\xi | e_i(m,n) | = 0 \tag{13}$$

Finally get:

$$\Delta V \le 0 \tag{14}$$

According to Lyapunov stability theory, that is, theorem 2. When the adjustment parameters satisfy $| L-\hat{\alpha}_i(n) |-1 \le 0$, $| \varphi_i(m,n) | = 1-\xi | e_i(m,n) |$. The unknown parameters in the network are identified, and the spatiotemporal network realizes the synchronization between networks under the action of Theorem 1. The proof is completed.

3. Numerical simulation

In order to further verify the correctness of the previous theoretical analysis, the one-dimensional coupled map lattice is now used as a model, and the equation of state can be expressed as:

$$x(m,n+1) = (1-\varepsilon) f(x(m,n)) + \frac{\varepsilon}{2}(f(x(m+1,n)) + f(x(m-1,n))) \tag{15}$$

In the equation (15), *m* is the space coordinate, and *n* is the discrete time coordinate. $m = 1,2,...,L$, L is the system scale and $L = 100$. ε is a system parameter. $f(x(m,n))$ is a nonlinear dynamic function, $x(m,n)$ is a state variable. The boundary condition is $x(1,n) = x(L,n) = 0$. The iteration relationship is as follows:

$$f(x(m,n)) = A\sin^2(x(m,n)-x_b) \tag{16}$$

According to equations (15) and (16), the following system equations are constructed:

$$\begin{aligned} x(m,n+1) &= (1-\varepsilon)A\sin^2(x(m,n)-x_b) + \frac{\varepsilon}{2}[A\sin^2(x(m+1,n)-x_b) \\ &+A\sin^2(x(m-1,n)-x_b)] \end{aligned} \tag{17}$$

Where parameters are $A = 3$, $x_b = 0.85\pi$, $\varepsilon = 0.13$. The spatiotemporal evolution diagram with spatiotemporal changes is shown in Figure 1.

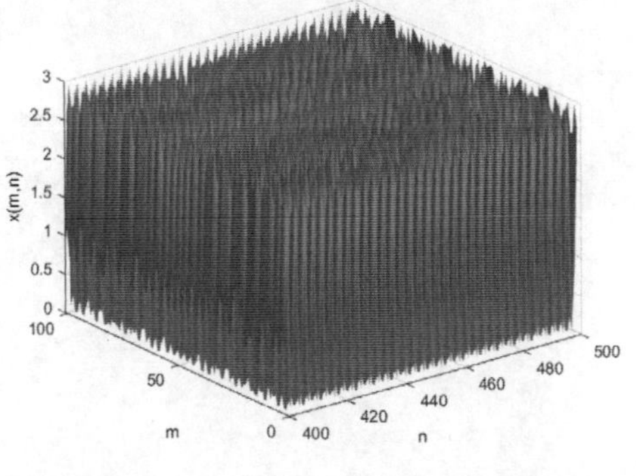

Figure 1. Spatiotemporal evolution of system state variables

In the process of simulation, the complex spatiotemporal network model is constructed by taking equation (17) as the network node equation, The number of nodes is taken as $N=5$. So the drive network is:

$$x_i(m,n+1)=(1-\varepsilon)A\sin^2(x_i(m,n)-x_b)+\frac{\varepsilon}{2}[A\sin^2(x_i(m+1,n)-x_b)$$
$$+A\sin^2(x_i(m-1,n)-x_b)]+\varepsilon_i\sum_{j=1}^{N}a_{ij}x_j(m,n) \tag{18}$$

The response network is:

$$y_i(m,n+1)=(1-\varepsilon)A\sin^2(y_i(m,n)-x_b)+\frac{\varepsilon}{2}[A\sin^2(y_i(m+1,n)-x_b)$$
$$+A\sin^2(y_i(m-1,n)-x_b)]+\varepsilon_i\sum_{j=1}^{N}a_{ij}y_j(m,n)+u_i(m,n) \tag{19}$$

Coupling matrix $A=(a_{ij})_{5\times5}$ is:

$$A=\begin{bmatrix}-1 & 1 & 0 & 0 & 0\\ 0 & -1 & 1 & 0 & 0\\ 0 & 0 & -1 & 1 & 0\\ 0 & 0 & 0 & -1 & 1\\ 0 & 0 & 0 & 0 & -1\end{bmatrix} \tag{20}$$

According to equation (5) of the controller, adjust L to satisfy $|L-\hat{\alpha}_i(n)|-1\le 0$. In actual simulation, $L=1$, $\xi=2$ and the initial value of the unknown parameter is set to $\alpha_i(n)=0.5$. Then network (18) and (19) can be synchronized. Figures 2-6 represents the numerical simulation results of inter network errors. The error fluctuates obviously before $n=25$ due to different initial values. But in a short time, the error tends to zero and remains stable.

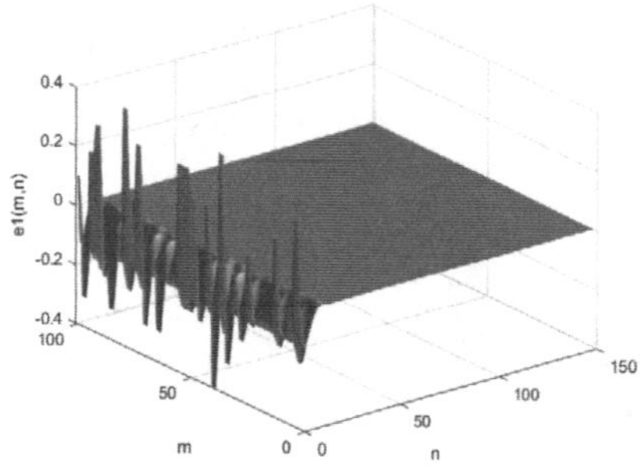

Figure 2. Error graphs of Node 1

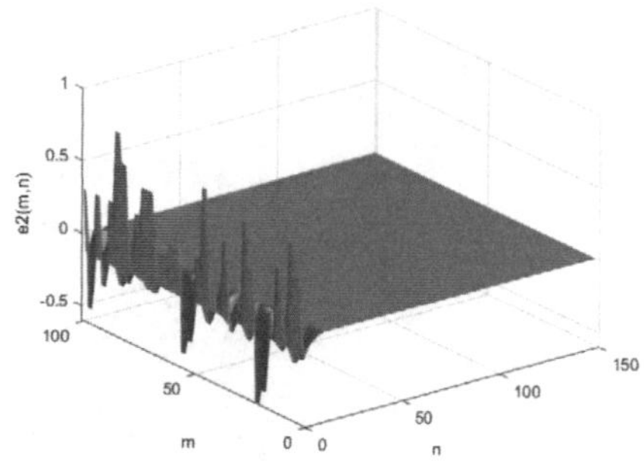

Figure 3. Error graphs of Node 2

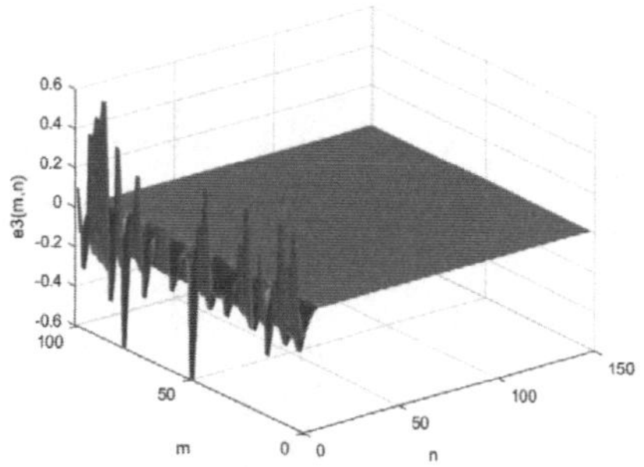

Figure 4. Error graphs of Node 3

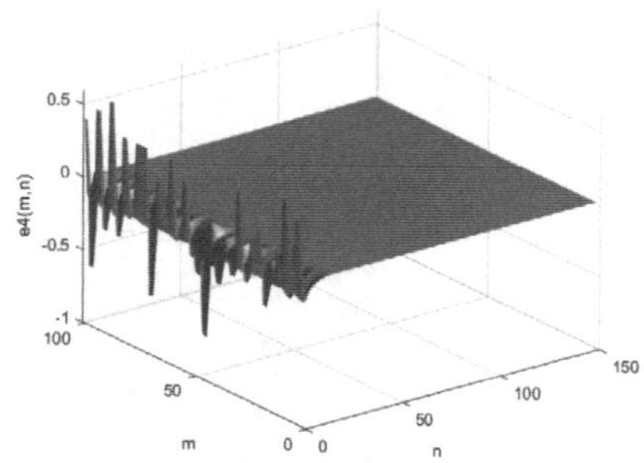

Figure 5. Error graphs of Node 4

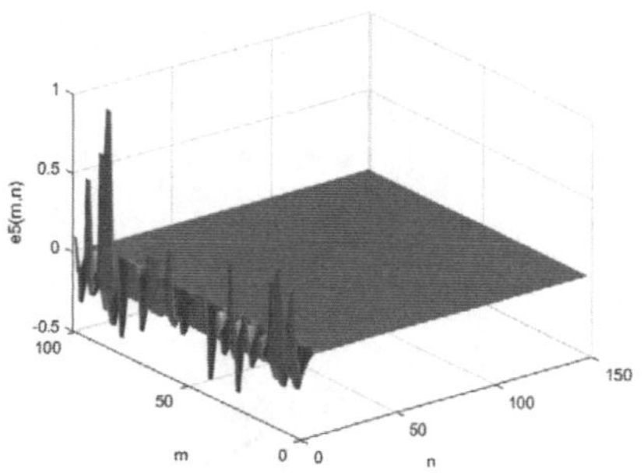

Figure 6. Error graphs of Node 5

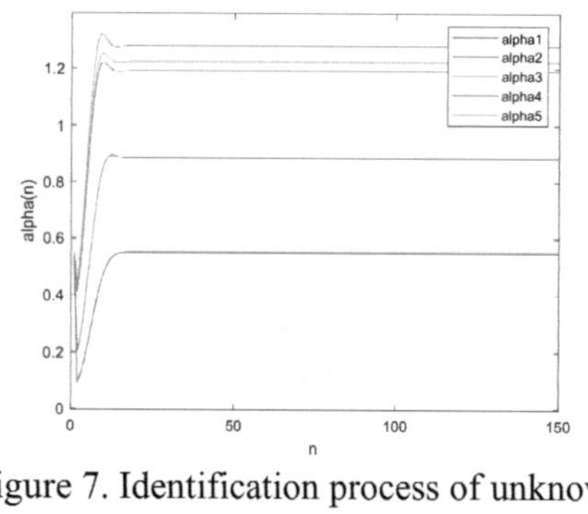

Figure 7. Identification process of unknown parameters

Figure 7 shows the identification process of unknown parameters in the controller. It can be seen from the figure that although the initial values of the unknown parameters are different, they tend to be stable in a short time, and it in the network synchronization controller are accurately identified.

4. Conclusion

This paper mainly discusses the synchronization between two spatiotemporal networks with the same coupling structure. The spatiotemporal chaotic system is used as the network node equation to construct a complex network. According to the Lyapunov stability principle, a network synchronization controller with unknown parameters is designed, and the recognition rate of the unknown parameters is given. In order to verify the feasibility of the network synchronization, the one-dimensional acousto-optic bistable coupled map lattice model is selected as the network node for numerical simulation, and the appropriate value of the adjustment parameter L is selected to make the corresponding node error tend to be stable after a short fluctuation, which verifies the effectiveness of the control scheme in this paper.

References

[1] Wu C W, Chua L O. Synchronization in an array of linearly coupled dynamical systems[J]. IEEE Transactions on Circuits and Systems I: Fundamental Theory and Applications, 2002, 42(8): 430-447.

[2] Liu J, Wang J Z, Liu B H. Community detection of multi-layer attributed networks via penalized alternating factorization[J]. Mathematics, 2020, 8(2): 239.

[3] Rui L, Chu T. Complete Synchronization of Boolean Networks[J]. IEEE Transactions on Neural Networks & Learning Systems, 2012, 23(5):840-846.

[4] Quan X, Xu X, Zhuang S, et al. New complex projective synchronization strategies for drive-response networks with fractional complex-variable dynamics[J]. Applied Mathematics and Computation, 2018, 338:552-566.

[5] Wang Z, Wu H. Projective synchronization in fixed time for complex dynamical networks with nonidentical nodes via second-order sliding mode control strategy[J]. Journal of the Franklin Institute, 2018, 355(15):7306-7334.

[6] Yang W G, Yu W W, et al. Global exponential stability and lag synchronization for delayed memristive fuzzy Cohen-Grossberg BAM neural networks with impulses[J]. Neural Networks the Official Journal of the International Neural Network Society, 2018, 98:122-153.

[7] Abdurahman A, Jiang H, Teng Z. Lag synchronization for Cohen-Grossberg neural networks with mixed time-delays via periodically intermittent control[J]. International Journal of Computer Mathematics, 2017, 94(1-4):275-295.

[8] Qian F, Gao H J, et al. Synchronization in complex networks and its application - A survey of recent advances and challenges[J]. Annual Review in Control, 2014.

[9] Albert R, Barabasi A L. Statistical mechanics of complex networks[J]. Review of Modern Physics, 2001, 74(1):xii.

[10] Vesna B. Complexity and dynamics of topological and community structure in complex networks[J]. The European Physical Journal Special Topics, 2017, 226(10):2205–2218.

AUTHOR INDEX